AMERICAN MATHEMATICAL SOCIETY
COLLOQUIUM PUBLICATIONS

VOLUME 41

A FORMALIZATION OF SET THEORY WITHOUT VARIABLES

BY
ALFRED TARSKI
and
STEVEN GIVANT

AMERICAN MATHEMATICAL SOCIETY
PROVIDENCE, RHODE ISLAND

1985 *Mathematics Subject Classification.* Primary 03B; Secondary 03B30, 03C05, 03E30, 03G15.

Library of Congress Cataloging-in-Publication Data

Tarski, Alfred.
A formalization of set theory without variables.

(Colloquium publications, ISSN 0065-9258; v. 41)
Bibliography: p.
Includes indexes.
1. Set theory. 2. Logic, Symbolic and mathematical. I. Givant, Steven R. II. Title. III. Series: Colloquium publications (American Mathematical Society); v. 41.
QA248.T37 1987 511.3'22 86-22168
ISBN 0-8218-1041-3 (alk. paper)

The paper used in this book is acid-free and falls within the guidelines
established to ensure permanence and durability. ♾

Contents

Explanation of section interdependence diagrams

The diagrams on the next two pages indicate the essential interdependencies of the various sections of this book. In general, the dependence of a section on earlier sections is determined by following upwards the lines leading to the section's box. For example, Section 4.8 depends on Sections 3.8–3.9 (and possibly on sections above them, such as 3.7, 3.1–3.3, 2.1–2.3, and 1.2–1.4), as well as on Sections 4.1–4.4 (and possibly on sections above them). A small part of it also depends on part of Section 3.10; this more limited dependence is indicated by a dotted line. For a second example, Section 4.7 depends on 3.6 (which in turn depends on earlier sections), as well as on 4.6 (and possibly on some of the sections above it, such as 4.1–4.5). As a final example, Section 6.2 depends on 6.1 and 4.4. (A line flows to 6.2 from the box labeled 4.1–4.5, but we have indicated parenthetically that only 4.4 is really important.) A small part of 6.2 also uses some results from 3.8–3.9.

Diagrams of Section Interdependence

Equipollence Results

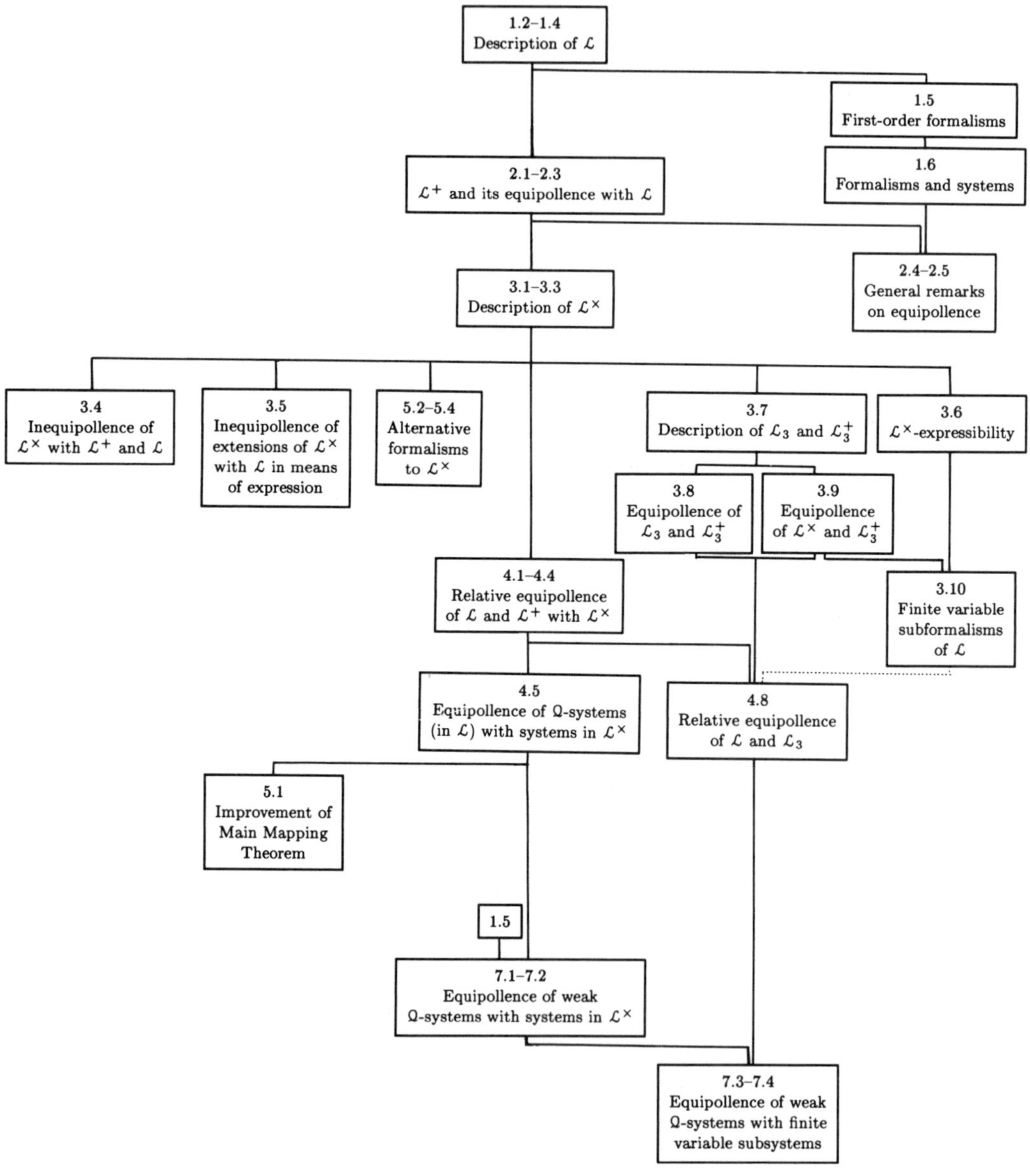

Applications

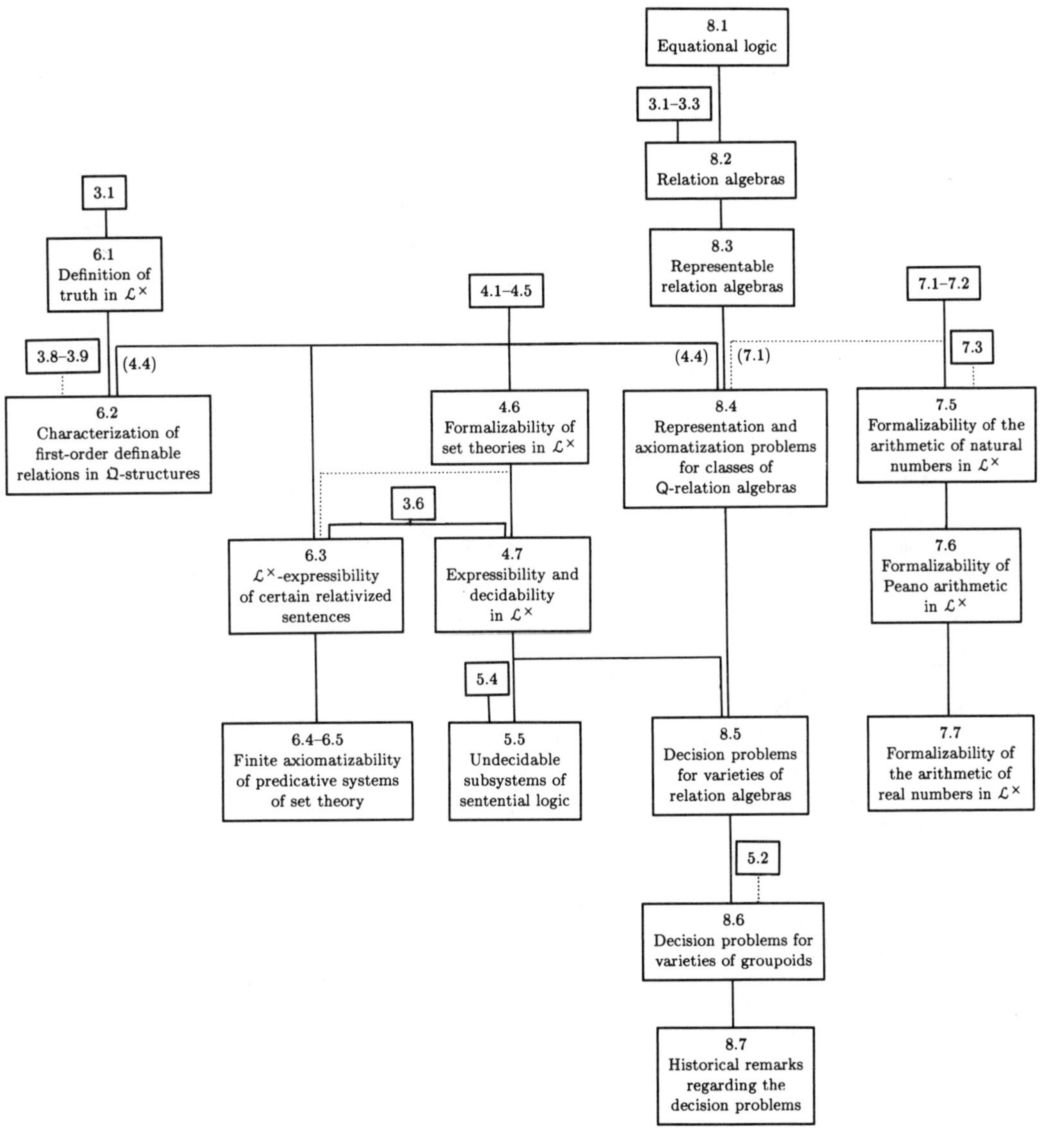

Preface

In this work we shall show that set theory and number theory can be developed within the framework of a new, different, and very simple formalism, $\mathcal{L}^{\times}$. $\mathcal{L}^{\times}$ is closely related to the equational theory of abstract relation algebras essentially given in Chin–Tarski [**1951**]. Its language contains no variables, quantifiers, or sentential connectives. There are two basic symbols, $\mathring{\mathbf{1}}$ and $\boldsymbol{E}$, intended to denote the identity and the (set-theoretic) membership relations. Compound expressions are constructed from the basic symbols by means of four operation symbols, $\odot$, $\smile$, $+$, and $^{-}$, that denote the well-known operations (on and to binary relations) of relative product, conversion, Boolean addition, and complementation. All mathematical statements in $\mathcal{L}^{\times}$ are formulated as (variable-free) equations between such expressions. The deductive apparatus of $\mathcal{L}^{\times}$ is based upon ten logical axiom schemata that are the analogues of the equational postulates for abstract relation algebras essentially given in Chin–Tarski [**1951**], p. 344. There is just one rule of inference, namely, the rule familiar from high school algebra of replacing equals by equals. A (deductive) system in $\mathcal{L}^{\times}$ is given by a set of nonlogical axioms, i.e., equations of $\mathcal{L}^{\times}$, and can be identified with the theory in $\mathcal{L}^{\times}$ generated by these axioms, i.e., with the set of all equations derivable from these axioms, the logical axioms of $\mathcal{L}^{\times}$, and equations of the form $A = A$, by means of the single rule of replacing equals by equals.

$\mathcal{L}^{\times}$ appears to be quite weak in its powers of expression and proof. Even the simple statement that there exist at least four elements cannot be equivalently expressed in $\mathcal{L}^{\times}$, as follows at once from a result of Korselt given in Löwenheim [**1915**], p. 448 (see below). Furthermore, $\mathcal{L}^{\times}$ is *semantically incomplete* in the sense that there are semantically valid equations in $\mathcal{L}^{\times}$ which are not derivable. This follows readily from a result of McKenzie [**1970**] (sharpening an earlier result of Lyndon [**1950**]) by which the postulates for abstract relation algebras given in Chin–Tarski [**1951**] do not even yield all of the equations with just one variable that are true in every concrete algebra of relations. We shall show, in fact, that $\mathcal{L}^{\times}$ is equipollent (in a natural sense) to a certain fragment, $\mathcal{L}_3$, of first-order logic having one binary predicate and containing *just three variables.* ($\mathcal{L}_3$ is also semantically incomplete.) It is therefore quite surprising that $\mathcal{L}^\times$ proves adequate for the formalization of practically all known systems of set theory, and hence for the development of all of classical mathematics.

As a language suitable for the formalization of most set-theoretical systems, we take the first-order logic $\mathcal{L}$ with equality and one nonlogical binary predicate $\boldsymbol{E}$. (For technical reasons we use "$\mathring{\mathbf{1}}$" instead of "$=$" as the name of the symbol denoting the relation of equality between individuals.) A system in $\mathcal{L}$ is given by a set of nonlogical axioms, and, as before, can be identified with the theory in $\mathcal{L}$ generated by these axioms.

It proves convenient to consider also an auxiliary formalism $\mathcal{L}^+$ that is a kind of definitional extension of $\mathcal{L}$. In addition to the basic predicates $\mathring{\mathbf{1}}$ and $\boldsymbol{E}$ of $\mathcal{L}$, the vocabulary of $\mathcal{L}^+$ contains as logical constants (of a new kind) the symbols $\odot$, $^\smile$, $+$, and $^-$ from $\mathcal{L}^\times$, by means of which (compound) *predicates* are constructed from $\mathring{\mathbf{1}}$ and $\boldsymbol{E}$; specifically, if A and B are predicates, then so are $A \odot B$, $A^\smile$, $A + B$, and A^-. The vocabulary of $\mathcal{L}^+$ also contains the second equality symbol, $=$, from $\mathcal{L}^\times$, intended to denote the relation of equality between binary relations. Atomic formulas of $\mathcal{L}^+$ are expressions of the form xAy and $A = B$, where x, y are individual variables and A, B are predicates. Arbitrary formulas are constructed from atomic ones in the usual way. In addition to a set of logical axioms similar in character to those of $\mathcal{L}$, $\mathcal{L}^+$ has five axiom schemata that can be regarded as possible definitions of the constants $\odot$, $^\smile$, $+$, $^-$, and $=$. For example, the schemata for $\odot$ and $=$ are respectively

$$\forall_{xy}(xA \odot By \leftrightarrow \exists_z(xAz \land zBy)),$$

and

$$A = B \leftrightarrow \forall_{xy}(xAy \leftrightarrow xBy),$$

where A, B are arbitrary predicates of $\mathcal{L}^+$. The rules of inference for $\mathcal{L}^+$ and the notion of a system in $\mathcal{L}^+$ are taken just as in $\mathcal{L}$. A "definitional extension" of $\mathcal{L}$ which essentially includes $\mathcal{L}^+$ is discussed in Quine [**1969**], pp. 15–27, under the name of "the virtual theory of classes".

With the help of $\mathcal{L}^+$ we shall compare the powers of expression and proof of $\mathcal{L}$ and $\mathcal{L}^\times$, and also of systems developed in these formalisms. Since $\mathcal{L}^\times$ has no variables, we must replace familiar notions like "definitionally equivalent" by suitable analogues. In each of the formalisms $\mathcal{L}$, $\mathcal{L}^+$, and $\mathcal{L}^\times$ (and, more generally, in every system developed in these formalisms) there is the notion of *sentence* and the notion of *derivability*, i.e., of a *sentence being derivable* (in the formalism or system) *from a set of sentences.* Suppose $\mathcal{S}_1$ and $\mathcal{S}_2$ are formalisms (or systems) that have these two notions. We say that $\mathcal{S}_2$ is an *extension* of $\mathcal{S}_1$ if every sentence of $\mathcal{S}_1$ is a sentence of $\mathcal{S}_2$ and derivability in $\mathcal{S}_1$ implies derivability in $\mathcal{S}_2$. Such an extension $\mathcal{S}_2$ is called an *equipollent extension* of $\mathcal{S}_1$ if also the following two conditions hold: (1) (equipollence in means of expression) for every sentence X of $\mathcal{S}_2$ there is a sentence Y of $\mathcal{S}_1$ that is equivalent to X in $\mathcal{S}_2$, i.e., Y is derivable from X, and X from Y, in $\mathcal{S}_2$; (2) (equipollence in means of proof) for every sentence X and set of sentences Ψ of $\mathcal{S}_1$, if X

is derivable from Ψ in $\mathcal{S}_2$, then it is so derivable in $\mathcal{S}_1$. (We avoid the term "definitional extension" because it usually involves a version of (1) applying to arbitrary formulas, and not just to sentences.) Finally, $\mathcal{S}_1$ and $\mathcal{S}_2$ are said to be *equipollent* if they have a common equipollent extension. When we wish to emphasize the role of a particular common equipollent extension $\mathcal{S}_3$, we shall say that $\mathcal{S}_1$ and $\mathcal{S}_2$ are *equipollent relative to* $\mathcal{S}_3$. It is not difficult to show that when $\mathcal{S}_1$ and $\mathcal{S}_2$ are equipollent, there is a natural one-one correspondence between the theories (i.e., the deductively closed sets of sentences) in $\mathcal{S}_1$ and $\mathcal{S}_2$ that preserves various important properties of theories such as consistency and completeness (cf. Theorem 2.4(viii) below); a theory is *consistent* if it does not coincide with the set of all sentences, and it is *complete* if it is a maximal consistent set of sentences.

It is readily seen (and follows from what is in Quine [**1969**]) that $\mathcal{L}^+$ is an equipollent extension of $\mathcal{L}$ (and even more; cf. §2.3). It is also easy to show (using, e.g., the semantic completeness of $\mathcal{L}^+$) that $\mathcal{L}^+$ is an extension of $\mathcal{L}^\times$. However, it is not an equipollent extension, and in fact both (1) and (2) fail. The failure of (1) is a direct consequence of Korselt's result, cited above (see Theorem 3.4(iv)), while the failure of (2) is due to the aforementioned semantic incompleteness of $\mathcal{L}^\times$ (see Theorem 3.4(vi)). Regarding (1), we shall actually prove (in Theorem 3.5(viii)) the following much stronger result.

(i) *Even if we enrich $\mathcal{L}^\times$ (and $\mathcal{L}^+$) by adjoining any finite number of new constants, all of which are intended to denote operations on and to binary relations, or else relations between binary relations (over the universe of any realization of $\mathcal{L}^\times$), and which are "logical" in the sense that the denoted operations and relations are preserved under all permutations of the universe, there will still be sentences of $\mathcal{L}$ that are not equivalent (in the enriched $\mathcal{L}^+$) with any sentence of the enriched $\mathcal{L}^\times$.*

Thus, the inadequacy of the expressive powers of $\mathcal{L}^\times$ is not due to a faulty choice of the set of fundamental notions; there is no way of extending this set in a finite and "logical" way so as to achieve equipollence with $\mathcal{L}$ in means of expression. We shall also prove (in §§3.8 and 3.9) the theorem, referred to before, that:

(ii) *$\mathcal{L}^\times$ is equipollent to a certain three-variable fragment, $\mathcal{L}_3$, of $\mathcal{L}$ (relative to a similar fragment, $\mathcal{L}_3^+$, of $\mathcal{L}^+$).*

As we have seen, $\mathcal{L}^\times$ is weaker than $\mathcal{L}$ both in means of expression and proof. Nevertheless, as stated above, we are going to establish the surprising result that $\mathcal{L}^\times$ is adequate for the development of classical mathematics. This will follow from two theorems, (iii) and (iv), which we now describe.

In (iii) we shall show that for certain special systems in $\mathcal{L}$ called $\mathcal{Q}$-systems we can construct equipollent systems in $\mathcal{L}^\times$. A system $\mathcal{S}$ in $\mathcal{L}$ is called a $\mathcal{Q}$-system if there are formulas D and E (of $\mathcal{L}$) containing at most three distinct variables, and just two free variables, such that in every model of $\mathcal{S}$, the two binary relations defined by D and E form a pair of conjugated quasiprojections, i.e., are functions with the following additional property: for every pair of elements x, y (in the universe of the model) there is a z which is mapped to x by the first function and to y by the second; the element z should be thought of as representing the ordered pair $\langle x, y \rangle$. Our main equipollence theorem (which is established in §§4.4 and 4.5, and upon which most of the later results in the book are based) is as follows.

(iii) *Every $\mathfrak{Q}$-system $\mathcal{S}$ in $\mathcal{L}$ is equipollent with a system in $\mathcal{L}^{\times}$ (relative to a system in $\mathcal{L}^{+}$); moreover, the system in $\mathcal{L}^{\times}$ will be, e.g., finitely axiomatizable or decidable iff $\mathcal{S}$ is.*

In (iv) this theorem is generalized to *weak* $\mathfrak{Q}$-systems developed in *arbitrary* first-order formalisms with finitely many nonlogical constants. The definition of a weak $\mathfrak{Q}$-system is obtained from that of a $\mathfrak{Q}$-system by dropping the restriction on the number of distinct variables occurring in formulas D and E. We prove, in fact (in Theorem 7.2(iv)), that:

(iv) *Every weak $\mathfrak{Q}$-system $\mathcal{U}$ developed in a first-order formalism with finitely many nonlogical constants is equipollent with a system in $\mathcal{L}^{\times}$; again, this latter system is, e.g., finitely axiomatizable or decidable iff $\mathcal{U}$ is.*

Both (iii) and (iv) seem very specialized. However, we shall show (in §4.6) that the hypothesis of (iii) holds for almost every known system of set theory, and (in §§7.5–7.7) that the hypothesis of (iv) applies, e.g., to the (full) elementary theory of natural numbers, to its well-known, recursively axiomatized subtheory, first-order Peano arithmetic, and to the elementary theory of the real numbers (with the set of natural numbers as a distinguished subset). Thus each of these systems is equipollent to a system in $\mathcal{L}^{\times}$.

With the help of the equipollence theorems (ii)–(iv) we shall also investigate a variety of other problems, quite apart from the one of formalizing mathematical systems in $\mathcal{L}^{\times}$. These concern, for example, the construction of undecidable subsystems of sentential logic (in §5.5), the relatively simple definition of truth for the formalism $\mathcal{L}^{\times}$ (in 6.1(i),(ii)), a characterization of the first-order definable binary relations in models of set theory and arithmetic (in 6.2(ix) and 7.4), the finite axiomatizability of predicative versions of systems of set theory (in 6.4(vi) and 6.5(iv)), the adequacy of first-order formalisms with only finitely many variables for the development of various mathematical disciplines (in 4.8(xi),(xii), 7.3(ii),(iii), and 7.5(vi)), the first-order definitional equivalence of number theory and the theory of hereditarily finite sets (in 7.5(v) and 7.6(ii)), the representation problem for relation algebras with a pair of quasiprojective elements, and the nonfinite axiomatizability of the equational theory of these algebras (in 8.4(iii),(vii)), the undecidability of the equational theory of several important classes of relation algebras (in 8.5(xii)), and what seems to be the first construction of a finitely based, essentially undecidable equational theory (and in fact a theory of groupoids—see 8.5(xi) and 8.6(x)).

We now make some historical remarks regarding the above theorems and their relation to results in the literature. The mathematics of the present work is rooted in the calculus of relations (or the calculus of relatives, as it is sometimes called) that originated in the work of A. De Morgan, C. S. Peirce, and E. Schröder during the second half of the nineteenth century. The universe of discourse of this calculus is the collection of all binary relations on an arbitrary but fixed set U, i.e., the set of all subsets of $U \times U$. There are six fundamental operations on and to (binary) relations, and four distinguished relations. Specifically, there are the four binary operations of forming, for any two relations R and S, their absolute sum, which is simply the union $R \cup S$, their absolute product, which is the intersection $R \cap S$, their relative sum, $R \dagger S$, consisting of all pairs $\langle x, y\rangle$ such that for every z either xRz or zSy (i.e., either $\langle x, z\rangle$ is in R or $\langle z, y\rangle$ is in S), and their relative product, $R|S$, consisting of all pairs $\langle x, y\rangle$ such that for some z, both xRz and zSy. Further, there are two unary operations of forming, for every relation R, its complement, $\sim R$, with respect to $U \times U$, and its converse, R^{-1}, consisting of all pairs $\langle x, y\rangle$ such that yRx. Finally, the distinguished relations are the absolute zero, which is the empty relation $\varnothing$, the absolute unit, which is the universal relation $U \times U$, the relative zero, which is the diversity relation Di on U consisting of all pairs $\langle x, y\rangle$ such that $x \neq y$, and the relative unit, which is the identity relation Id on U consisting of all pairs $\langle x, y\rangle$ such that $x = y$. This is the framework of the calculus as finally presented in Peirce [**1882**] after several earlier versions.

Both Peirce and, later, Schröder, who extended Peirce's work in a very thorough and systematic way in Schröder [**1895**], were interested in the expressive powers of the calculus of relations and the great diversity of laws that could be proved. They were aware that many elementary statements about (binary) relations can be expressed as equations in this calculus. (By an "elementary statement" about relations $R, S, \ldots$ (over U) we mean a first-order statement about the structure $\langle U,\ R, S, \ldots\rangle$.) To give an example, the (elementary) statement that a relation R is transitive,

$$\textit{for every} \quad x, y, z, \quad \textit{if} \quad xRy \quad \textit{and} \quad yRz, \quad \textit{then} \quad xRz,$$

can be expressed by the equation

$$(R|R) \cup R = R.$$

Similarly, the more complicated statement that R is a one-one function mapping U onto itself is rendered by the equation

$$[(R|R^{-1}) \cap Di] \cup [(R^{-1}|R) \cap Di] \cup [\sim R \dagger \varnothing] \cup [\varnothing \dagger \sim R] = \varnothing.$$

Schröder seems to have been the first to consider the question whether all elementary statements about relations are expressible as equations in the calculus of relations, and in Schröder [**1895**], p. 551, he proposed a positive solution to the problem. A critique of Schröder's proposed solution appeared in Löwenheim [**1915**], p. 450, along with a negative solution due to Korselt that was referred to above. A far-reaching extension of Korselt's result, formulated in (i) above for $\mathcal{L}^{\times}$ (but also true in the more general setting of the calculus of relations), was first announced in Tarski [**1941**].

In the same paper Tarski posed the problem of proving that there is no algorithm for deciding in every particular case whether an elementary statement about relations is expressible in the calculus (as an equation). Michael Kwatinetz finally settled the problem around 1971, with the help of (ii) above (restated for the calculus of relations), by showing that the set of elementary statements which can equivalently be formulated using just three variables is not recursive (see Kwatinetz [**1981**] for the proof).

Despite the weak expressive powers of the calculus of relations, Tarski was able to establish a kind of relative equipollence in means of expression between it and the elementary theory of relations. Namely, if we assume we have a pair of conjugated quasiprojections, then for any elementary statement X we can effectively construct an equation X^* in the calculus that is equivalent to X. This is a preliminary and much weaker form of (iii) above; it does not concern itself with the problem of equipollence in means of proof. With its help, Tarski proved that any decision procedure for the set of true equations in the calculus of relations would bring with it a decision procedure for the elementary theory of relations, in contradiction to a result of Church [**1936**] and Kalmár [**1936**]; hence the set of true equations in the calculus of relations is not recursive (see 8.5(xii) below). This theorem was announced in Tarski [**1941**], p. 88. Lemmas I–III of the abstract Tarski [**1953**] give a rough outline of Tarski's original proof.

Tarski [**1941**] presented an interesting formalization of the calculus of relations as a deductive discipline. The language contained (binary) relation variables, but no individual variables or quantifiers, and although sentential connectives were present, it was pointed out on p. 87 of *op. cit.* that an equivalent formalization involving only equations, i.e., without sentential connectives, could be given. (Such a formalization was essentially carried out in Chin–Tarski [**1951**].) Tarski proposed a finite set of axioms for the calculus (essentially equivalent to the set of axioms given in Chin–Tarski [**1951**], as noted in *ibid.*, p. 352, footnote 10), indicated that he could derive all the hundreds of laws occurring in Schröder [**1895**] on the basis of these axioms, and asked whether every true law (true for all domains of individuals) in the calculus was so derivable. As mentioned above, this problem was subsequently answered negatively by Lyndon [**1950**], and, in fact, Monk [**1964**] showed that the set of true equations of the calculus is not finitely axiomatizable at all.

Nevertheless, just as in the case of expressibility, Tarski was able to establish a kind of relative equipollence in means of proof between his axiomatization of the calculus and the elementary theory of relations. This is essentially the result stated above in (iii) (when reformulated for the calculus of relations). From this Tarski concluded that the set of equations derivable from his set of axioms for the calculus of relations is not recursive (see 8.5(xii)). Further, since his theorem reduced every problem concerning the derivability of a mathematical statement from a set of axioms to the problem of whether an equation is derivable from a set of equations in the calculus of relations, in principle the whole of mathematical research could be carried out within the framework of this calculus. These theorems were obtained by Tarski during the period 1943–1944, and presented for the first time in his seminar on relation algebras at the University of California, Berkeley, during the year 1945. References to these theorems, as well as to the Berkeley seminar, can be found in Chin–Tarski [**1951**], pp. 341–343; see also Chin [**1948**], pp. 2–3. The abstracts Tarski [**1953**], [**1953a**], [**1953b**], [**1954**], [**1954a**] contain announcements of these theorems and several of the other results (also dating from the 1943–1944 period) that were referred to in the first part of this foreword.

Roughly speaking, the formalism $\mathcal{L}^{\times}$ that is the central focus of this work is obtained from Tarski's equational formalization of the calculus of relations by introducing the constant $\boldsymbol{E}$ and deleting all variables.

Tarski's formalization of set theory in $\mathcal{L}^{\times}$ was certainly not the first attempt to eliminate the use of variables in formalizing mathematics. Probably the earliest results in this direction appeared in Schönfinkel [**1924**]. There, a kind of calculus of unary functions was developed. Basically, Schönfinkel considered three distinguished unary functions (later called combinators by Curry), C, S, and U, and one binary operation on and to unary functions: that of applying a unary function f to an argument x, the result being represented by juxtaposing the two, as in "fx". The definitions of C, S, and U are not simple. Each of them has the property that, when applied to a unary function, it yields another unary function; thus each of them takes on unary functions as both arguments and values. For example, C is the function that, when applied to any unary function f, yields a constant unary function, and in fact the function constantly equal to f, i.e., Cf is the unary function which, when applied to any argument x, yields f, in symbols $(Cf)x = f$. The definitions of S and U are still more involved; the reader is referred to Schönfinkel's paper. By means of the binary operation of functional application we can construct further unary functions (called compound combinators) from the basic three, for example, CC, CS, $(CS)C$, etc. Schönfinkel indicated how not only bound individual variables, but also bound variables of higher orders can be eliminated from mathematical statements with the help of combinators. Thus the expressive power of his calculus reaches far beyond the domain of first-order logic.

Schönfinkel made no attempt to set up a deductive apparatus for his calculus. This task was taken up by Curry and his collaborators, starting in the late 1920s, and proved to be quite involved. We shall not attempt to describe their many achievements—the reader is referred to books on combinatory logic (as this domain is now called), and in particular to Curry–Feys [**1958**] and Curry–Hindley–Seldin [**1972**], which contain extensive bibliographies. Rather, we shall briefly contrast the character of the results presented in this book with those that have been obtained in the domain of combinatory logic; moreover, we contrast them only as regards the specific problem of developing parts of mathematics within a variable-free formalism.

First of all, in contrast to the expressive powers of combinatory logic, those of $\mathcal{L}^{\times}$ do not overreach first-order logic, and (as was pointed out above) actually comprise but a weak part of it, namely the first-order logic of three variables. Only under certain additional assumptions (satisfied by most systems of set theory and various systems of arithmetic) does $\mathcal{L}^{\times}$ become equipollent with first-order logic in means of expression. Similar remarks apply to the deductive powers of $\mathcal{L}^{\times}$. Secondly, the method presented in this book for formalizing a given first-order system within $\mathcal{L}^{\times}$ is quite general; it can be applied almost mechanically to many different mathematical theories. In contrast to this, the various attempts to develop different parts of mathematics within combinatory logic have been quite specific in character, and the approaches used have depended on the particular theory to be formalized. Finally, each of our correlated systems in $\mathcal{L}^{\times}$ is shown to be equipollent with the original first-order system in a strong and precisely defined sense that entails, e.g., the equiconsistency, equicompleteness, and equidecidability of the two systems. It is not at all clear to what extent various first-order systems and their combinatory analogues are equipollent. Indeed the very problem of the consistency, or relative consistency, of systems formalized in combinatory logic has traditionally posed difficulties; in several cases the answer proved to be negative and some of these problems are still open.

In the late 1940s and early 1950s there began some work which has a bearing on the problem of formalizing mathematics without variables. Various algebraic theories were developed that are analogues of first-order logic, much as Boolean algebra is an algebraic analogue of the sentential calculus. The creation of the theories of relation algebras by Tarski, and of projective algebras by Everett-Ulam [**1946**] may be viewed as preliminary steps in this direction. The theory of cylindric algebras, perhaps the most extensively developed of such theories, was created by Tarski in collaboration with his former students Louise Chin (Lim) and Frederick Thompson during the period 1948–52, and further developed by Tarski, Henkin, and Monk. A detailed presentation of various portions of this theory can be found in Henkin–Monk–Tarski [**1971**], [**1985**]. The closely related theory of polyadic algebras was created by Halmos in the mid 1950s; the relevant papers can be found in Halmos [**1962**]. Other noteworthy theories of this type can be found in Bernays [**1959**] and Craig [**1974**]. The specific problem of using such algebraic theories to construct formalisms which contain no (bound) variables, quantifiers, or sentential connectives, and which are equipollent in some sense to first-order logic, is addressed in these last two works and in Quine [**1960**], [**1971**]. An excellent discussion of various approaches to this problem can be found in Quine [**1971**].

With the exception of some algebraic notions and results in Chapter 8, this work is intended to be largely self-contained, and accessible not only to mathematicians and logicians, but also to computer scientists, philosophers, and others who may be interested in foundational research.

In §§1.2–1.5, 2.1–2.3, and 3.1–3.4 we respectively describe the formalisms $\mathcal{L}$, $\mathcal{L}^+$, and $\mathcal{L}^\times$, and their interrelationship. After reading these portions of the book, it is possible to proceed directly to the main equipollence results presented in §§4.1–4.5 and 7.1–7.2, omitting the intervening text. §§4.6 and 7.5–7.7, concerning the formalizability of various systems of set theory and arithmetic in $\mathcal{L}^\times$, essentially depend only on §§4.1–4.5 and 7.1–7.2, respectively. A more detailed picture of the interdependence of different sections of the book is presented in the diagrams following the table of contents.

Alfred Tarski
Steven Givant
Berkeley, California
October, 1983

Postscript

Alfred Tarski died on October 27, 1983, shortly after the manuscript for this work was completed. With his passing, the world has lost a great logician and an inspiring teacher, and I have lost a loyal friend. In the period since his death it has become apparent that certain small additions should be made to the text. For example, in the last few years a number of interesting results have been obtained that have a direct bearing on some of the open problems stated here. In particular, after receiving preliminary copies of the manuscript, Roger Maddux, Hajnal Andréka, and István Németi solved several of these open problems. In addition, various relevant results in the literature that appeared in the late 1970s and early 1980s, and were overlooked by us, have been called to my attention by Andréka, Maddux, and Németi as well as by George McNulty. Rather than amending the text itself at this point, I have decided to include some additional footnotes, indicated by an asterisk (*), to discuss these results.

Steven Givant
Berkeley, California
January, 1986

Acknowledgments

It is a great pleasure to acknowledge our indebtedness and to express our gratitude to those who helped us at various stages during the preparation of this book. Both Professor George McNulty, for a period of several months, and Professor Roger Maddux, for almost a year, assisted Tarski in his work on certain portions of the manuscript. In addition, they read over the final draft and made many valuable suggestions. Professor John Corcoran also read through the final draft in a very careful and thorough way. He pointed out several technical inaccuracies and made innumerable detailed suggestions as to how the style and exposition might be improved. Drs. Hajnal Andréka and István Németi carefully checked many proofs in the final draft. Professor Robert Vaught reviewed preliminary versions of the introduction and helped us very much to improve its presentation. The author, subject, and symbol indices were meticulously prepared by Mr. Peter Baker. Professor Leon Henkin helped us to settle many practical problems that arose in connection with our work on this book.

We also wish to acknowledge the support we received from the National Science Foundation through grants to the University of California at Berkeley (Grant Nos. GP-27920, GP-35844X, MCS74-223878, and MCS77-22913).

CHAPTER 1

The Formalism $\mathcal{L}$ of Predicate Logic

Axiomatic systems of set theory developed in the formal language $\mathcal{L}$ of the (first-order) predicate logic with identity and in some other languages with different formal structures are the central topic of the present monograph. In the main results of our work these systems and languages will be compared with respect to their powers in means of expression and proof. This first chapter contains a sketchy description of the formalism $\mathcal{L}$. In the last two sections we make some general observations concerning the formalisms discussed in the book.

1.1. Preliminaries

The discussion is conducted throughout the book within an appropriate *metasystem.* In the *metalanguage*, i.e., the language of the metasystem, we have at our disposal various logical, set-theoretical, and metalogical symbols and notions. In general we adhere to the notation adopted in Henkin–Monk–Tarski [**1971**] (see in particular pp. 25–46), with some deviations motivated by the specific needs of the present discussion. To facilitate the understanding of our discussion, we begin with a clarification of a few of the set-theoretical notions that will be employed.

The metasystem and its language are not assumed to be formalized. The set-theoretical notions occurring in the metasystem are sometimes employed in a way which is usually described by the phrase "in the sense of naive set theory". (A suitable formalization of the metasystem would present no essential difficulties. In that case any one of several variants, and actually appropriate extensions, of the well-known Zermelo–Fraenkel system could be used to form a set-theoretical basis for the metasystem. In particular, we could use for this purpose the system in Morse [**1965**], or a more conventional version of it that is outlined in Kelley [**1955**].)

Among metalogical notions of the metasystem we find, in particular, symbolic designations of all expressions occurring in formal languages to which the discussion refers. No symbols, i.e., expressions appearing in our metalogical discussion, should be interpreted as belonging to formal languages themselves.

In principle, the shape of symbols and expressions occurring in formal languages is irrelevant for our purposes, and no conclusions concerning this shape can be drawn from our discussion. In practice, we shall mostly employ boldface symbols as metalogical designations of corresponding lightface symbols commonly used in formal languages; for instance the symbol " $\boldsymbol{\rightarrow}$ " will serve as the metalogical designation of the common implication symbol "$\rightarrow$". This device will facilitate the understanding of our metalogical notations and will give the reader some idea of the intended shape of symbols and expressions used in constructing the formal languages discussed.

We now review some of the set-theoretical notions, notations, and terminology that will be used in this book. For a more detailed discussion see Henkin–Monk–Tarski [**1971**], pp. 25–39.

Formulas such as "$x, y, \ldots \in A$" and "$x, y, \ldots \notin A$" respectively express the facts that $x, y, \ldots$ are members of, or are not members of, the class A. Given a formula X with a free variable x, we use the symbolic expression $\{x\colon X\}$ to denote the class of all elements x which *satisfy* the formula X. The formula X may have, besides x, other free variables which are regarded as parameters. Using this notation we define, in particular,

$$\begin{aligned} \varnothing &= \{x\colon x \neq x\} && \text{(the \textit{empty set})},\\ \{x\} &= \{y\colon y = x\} && \text{(the \textit{singleton} of } x),\\ \{x, y\} &= \{z\colon z = x \text{ or } z = y\} && \text{(the \textit{unordered pair} } x, y), \end{aligned}$$

etc. As usual, several variants of this notation are also quite convenient. For example, if t is a symbolic expression which contains x as a free variable and which represents an object for each value of x satisfying X, then we let

$$\{t\colon X\} = \{y\colon y = t \text{ for some element } x \text{ satisfying } X\}.$$

Inclusion and *proper inclusion* are represented by $\subseteq$ and $\subset$ respectively; the symbols $A \subseteq B$ and $B \supseteq A$ will be used interchangeably, and similarly for $A \subset B$ and $B \supset A$. SbA denotes the class of all subsets of A. $A \cup B$ and $A \cap B$ denote the *union* and *intersection* of A and B. More generally $\bigcup C$ and $\bigcap C$ denote the union and intersection of all members of the class C. So, in particular, $\bigcap\varnothing$ is the *universal class*, i.e., the class of all elements; if, however, in a given context we are dealing exclusively with subsets of a fixed class U, then the notation $\bigcap\varnothing$ will be used to denote U. Finally, $A \sim B = \{x\colon x \in A \text{ and } x \notin B\}$ is the set-theoretical *difference* of A and B; if the class A is fixed throughout a certain portion of our discussion and $B \subseteq A$, then we sometimes suppress "A" in our notation and write simply "$\sim B$", in words, the *complement* of B (with respect to A).

The *ordered pair* with *first term* x and *second term* y is the set $\langle x, y\rangle = \{\{x\}, \{x, y\}\}$; the *ordered triple* with *first term* x, *second term* y, and *third term* z is defined to be $\langle x, y, z\rangle = \langle\langle x, y\rangle, z\rangle$, etc. (In some portions of our work we shall find it convenient to construct sequences in which one or more of the terms are classes that are not sets. Such constructions are well known and present no essential difficulties; the interested reader can find an outline of such a construction in §4.6 of the present work.)

By a (binary) *relation* we understand a class of ordered pairs. For example, $Id = \{\langle x, y\rangle : x = y\}$ is the *identity relation*, and $Di = \{\langle x, y\rangle : x \neq y\}$ is the *diversity relation.* If R is a relation we shall sometimes write "xRy" instead of "$\langle x, y\rangle \in R$". For any relations R and S we let

$R|S = \{\langle x, y\rangle : \text{for some } z,\ xRz \text{ and } zSy\}$ (the *relative product* of R and S);
$R^{-1} = \{\langle x, y\rangle : yRx\}$ (the *converse* of R);
$DoR = \{x : \text{for some } y,\ xRy\}$ (the *domain* of R);
$RnR = \{y : \text{for some } x,\ xRy\}$ (the *range* of R).

For any relation R and class A we let

$A{\upharpoonleft}R = \{\langle x, y\rangle : x \in A \text{ and } xRy\}$ (the *restriction*, or *domain restriction*, of R to A);

$R^*A = \{y : \text{for some } x,\ x \in A \text{ and } xRy\}$ (the *R-image* of A).

For any classes A, B we let

$$A \times B = \{\langle x, y\rangle : x \in A \text{ and } y \in B\}.$$

$A \times B$ is called the (*Cartesian*) *product* of A and B; when employing geometric terminology we refer to A as the *first axis*, and to B as the *second axis*, of the (*Cartesian*) *space* $A \times B$.

A *function* is a relation F such that for every $x \in DoF$ there is exactly one y with $\langle x, y\rangle \in F$. This unique y is referred to as the xth *value* of F and is usually denoted by Fx; alternative notations are $F(x)$, F_x, F^x, $F^{(x)}$, etc. A *system indexed by I* is just a function F with domain I; when employing this terminology we shall often use the notation $\langle F_i : i \in I\rangle$ to refer to F, and we shall speak of F_i as the ith *term* of the system. In case F and G are functions we write $F \circ G$ for the function obtained by composing F with G; thus $F \circ G = G|F$, and $(F \circ G)x = F(Gx) = FGx$ for every $x \in Do(F \circ G)$. We say that a function F *maps A into* (or *onto*) B if $DoF = A$ and $RnF \subseteq B$ (or $RnF = B$); alternately we say F is a *mapping from A into* (or *onto*) B. If both F and F^{-1} are functions we say that F is *one-one* and we refer to F^{-1} as the *inverse* of F. In case F is a one-one function mapping A onto B we sometimes refer to F as a one-one correspondence between A and B. Finally, we denote by ${}^AB$ the set of all functions mapping A into B. ${}^AB$ is referred to as the Ath (*Cartesian*) *power* of B.

Ordinal numbers, or *ordinals* for short, are denoted by lower case Greek letters. We assume that the notion of an ordinal has been defined in such a way that each ordinal α is a set whose members are precisely the ordinals smaller than α. Thus on ordinals the membership relation $\in$ coincides with the natural well ordering of ordinals. The least ordinals are $0 = \varnothing$, $1 = \{0\} = \{\varnothing\}$, $2 = \{0, 1\} = \{\varnothing, \{\varnothing\}\}$, etc. The least ordinal different from 0 and for which no immediately smaller ordinal exists is ω.

The *cardinality* or *power* of a set A, in symbols $|A|$, is the smallest ordinal α that can be mapped in a one-one way onto A. An ordinal α is called a *cardinal number*, or simply a *cardinal*, if $|\alpha| = \alpha$. A set A is *finite* or *infinite* according as $|A| < \omega$ or $|A| \geq \omega$; it is *denumerable* if $|A| = \omega$ and *countable* if $|A| \leq \omega$. Finite ordinals coincide with finite cardinals and we identify them with *natural numbers*. Thus ω is the set of all natural numbers.

By an *α-termed sequence* or a *sequence of length α* we understand any function F with $DoF = \alpha$. We usually write "F_ξ" instead of "$F\xi$" for $\xi \in \alpha$, and we call F_ξ the ξth *term* of the sequence F; thus a sequence of length α is treated as a system indexed by α. If a sequence (as a function) is one-one, then we refer to it as a sequence *without repeating terms.* For any α-termed sequence x we have

$$x = \langle x_\xi \colon \xi < \alpha \rangle = \langle x_\xi \rangle_{\xi < \alpha} = \langle x_0, \ldots, x_\xi, \ldots \rangle_{\xi < \alpha}.$$

We also use the notation $\langle x_0, \ldots, x_{\alpha-1} \rangle$ in case $0 < \alpha < \omega$. Analogously, the range of a sequence x is expressed by $\{x_\xi \colon \xi < \alpha\}$, $\{x_0, \ldots, x_\xi, \ldots\}_{\xi < \alpha}$, etc., and by $\{x_0, \ldots, x_{\alpha-1}\}$ in case $0 < \alpha < \omega$.

In many situations we shall identify a 1-termed sequence $\langle x_\xi \colon \xi < 1 \rangle$ with the element x_0, a 2-termed sequence $\langle x_\xi \colon \xi < 2 \rangle$ with the ordered pair $\langle x_0, x_1 \rangle$, whose first and second terms are x_0 and x_1 respectively, etc. The set of all α-termed sequences x with $Rnx \subseteq A$ coincides with ${}^\alpha A$.

Any class R of α-termed sequences is referred to as an *α-ary relation* and α is referred to as the *rank* of R, in symbols $\rho R = \alpha$. (See Henkin–Monk–Tarski [**1971**], p. 34, for some comments regarding the rank of the empty relation.) Since $\emptyset$ is the only 0-termed sequence, the only 0-ary relations are $0 = \emptyset$ and $1 = \{\emptyset\}$. An *α-ary relation* on a set A is just a subset of ${}^\alpha A$. An *α-ary operation* on a set A is any function O from ${}^\alpha A$ into A, i.e., any member of ${}^{{}^\alpha A}A$; α is called the *rank* of O, in symbols $\rho O = \alpha$. In case $0 < \alpha < \omega$ and x is an α-termed sequence, we frequently write "$O(x_0, \ldots, x_{\alpha-1})$" instead of "$Ox$". In general we shall mostly be concerned with *finitary* operations, i.e., operations of rank smaller than ω. Operations of rank 0, i.e., members of ${}^{\{0\}}A$, are identified with members of A.

In §1.5 we shall introduce the notion of an *algebraic structure.* Only in the final chapter, Chapter 8, shall we find it necessary to make use of more involved algebraic notions such as "homomorphism", "subdirect product", "free algebra", etc. The interested reader is referred to Henkin–Monk–Tarski [**1971**], Chapter 0, for the appropriate definitions.

1.2. Symbols and expressions of $\mathcal{L}$

The reader is certainly well acquainted with the formalism $\mathcal{L}$ of predicate logic, and many details in the immediately following description and discussion of the formalism may seem redundant. Our main aim in giving these details is to facilitate the discussion of other formalisms with which we shall concern ourselves in later chapters.

The vocabulary of $\mathcal{L}$ is assumed, as usual, to contain symbols of three different kinds: *variables*, *logical constants*, and *nonlogical constants.* We disregard the question whether $\mathcal{L}$ contains also some technical symbols, such as parentheses and commas.

The set Υ of all variables of $\mathcal{L}$ is denumerable. The elements of Υ are arranged in an infinite sequence $\langle \boldsymbol{v}_0, \boldsymbol{v}_1, \ldots, \boldsymbol{v}_k, \ldots\rangle$ (without repeating terms). As abbreviations we introduce metalogical constants "$\boldsymbol{x}$", "$\boldsymbol{y}$", "$\boldsymbol{z}$", "$\boldsymbol{u}$", "$\boldsymbol{w}$", "$\boldsymbol{p}$", "$\boldsymbol{q}$", "$\boldsymbol{r}$", "$\boldsymbol{s}$", and "$\boldsymbol{t}$" by setting

$$\boldsymbol{x} = \boldsymbol{v}_0,\ \boldsymbol{y} = \boldsymbol{v}_1,\ \boldsymbol{z} = \boldsymbol{v}_2,\ \boldsymbol{u} = \boldsymbol{v}_3,\ \boldsymbol{w} = \boldsymbol{v}_4,$$
$$\boldsymbol{p} = \boldsymbol{v}_5,\ \boldsymbol{q} = \boldsymbol{v}_6,\ \boldsymbol{r} = \boldsymbol{v}_7,\ \boldsymbol{s} = \boldsymbol{v}_8,\ \boldsymbol{t} = \boldsymbol{v}_9.$$

Also, we use letters "a", "b", … ,"x", "y", … as metalogical variables ranging over arbitrary variables of $\mathcal{L}$. (In those parts of the discussion, primarily in Chapters 7 and 8, where we shall use language, and not metalanguage, the italic lower case letters are used as variables ranging over various objects of the discussion, especially when these objects are not assumed to be sets themselves, but are treated as elements of certain sets involved in the discussion. As variables ranging over the sets involved, we shall frequently use capital italics.) Given an $x \in \Upsilon$, the *index* of x, $\mathit{in}\, x$, is the unique natural number n such that $x = \boldsymbol{v}_n$. This one-one correspondence between variables and their indices induces in an obvious way a natural well ordering on the set Υ of all variables. We refer implicitly to this ordering, e.g., when speaking of the variable immediately succeeding or preceding a given variable, when speaking of the first variable having a given property, or when calling a sequence of variables (strictly) increasing.

There are four logical constants in $\mathcal{L}$: two sentential connectives, the *implication symbol* $\rightarrow$ and the *negation symbol* $\neg$, the *universal quantifier* $\forall$, and the *identity symbol.* As the metalogical designation of the latter symbol it would be natural to use "$=$". This, however, proves inconvenient for our purposes as soon as we pass in the next chapter from $\mathcal{L}$ to a richer formalism $\mathcal{L}^+$. We shall use "$\mathring{1}$" instead, reserving "$=$" as a designation for a related symbol that occurs in $\mathcal{L}^+$ but not in $\mathcal{L}$.

In the prevailing part of our discussion we assume that $\mathcal{L}$ contains only one nonlogical constant, the *membership symbol* $\boldsymbol{E}$. The symbols $\mathring{1}$ and $\boldsymbol{E}$, because of their role in constructing formulas and their semantical interpretation, are referred to as *predicates* or *relation symbols* and, more specifically, as binary predicates (predicates of rank 2).

Letters "A", "B", …,"X", "Y",… will be used as metalogical variables ranging over arbitrary expressions (i.e., symbols and concatenations of symbols) of our formal language.

The most important among compound expressions are *formulas.*

(i) *The expressions $x\mathring{1}y$ and $x\boldsymbol{E}y$ (where $x, y \in \Upsilon$) are called atomic formulas of $\mathcal{L}$. An expression X is said to be a formula of $\mathcal{L}$ if it belongs to every set Ω of expressions such that*

(α) *all atomic formulas are in* Ω;
(β) $\neg Y$, $Y \to Z \in \Omega$ *whenever* $Y, Z \in \Omega$;
(γ) $\forall_x Y \in \Omega$ *whenever* $Y \in \Omega$ *and* $x \in \Upsilon$.

The set of all formulas of $\mathcal{L}$ is denoted by "$\mathbf{\Phi}[\mathcal{L}]$". Usually, however, we shall omit an explicit reference to $\mathcal{L}$ in symbolic notation; thus, in the present case we shall simply write "$\mathbf{\Phi}$" instead of "$\mathbf{\Phi}[\mathcal{L}]$".

From the definition of formulas it follows that $Y \to Z$, $\neg Y$, $\forall_x Y \in \mathbf{\Phi}$ whenever $Y, Z \in \mathbf{\Phi}$ and $x \in \Upsilon$. Thus, $\to$, $\neg$, and $\forall_x$ can be and will be treated here as (metalogical) operations on and to formulas, in fact, $\to$ as a binary operation (formation of conditionals), and $\neg$ and $\forall_x$ as unary operations (formation of negations and universal quantification with respect to the variable x).

In terms of $\to$, $\neg$, $\forall_x$, further operations on formulas are defined in a familiar way, for instance, the operations $\vee$, $\wedge$, $\leftrightarrow$ (formation of disjunctions, conjunctions, biconditionals), and $\exists_x$ (existential quantification with respect to x).

Whenever parentheses indicating the order of operations in symbolic designations of formulas are lacking, it will be understood that the usual conventions apply. For example, each of the unary operations $\neg$, $\forall_x$, and $\exists_x$ has priority over the binary operations (i.e., each unary operation applies to as small a formula as possible); also, each of the binary operations $\wedge$ and $\vee$ has priority over $\to$ and $\leftrightarrow$.

The binary operations $\vee$, $\wedge$ are extended recursively to operations on arbitrary nonempty finite sequences of formulas $\langle Y_0, \dots, Y_{n-1}\rangle$; the results of the extended operations are denoted by "$Y_0 \vee \cdots \vee Y_{n-1}$" and "$Y_0 \wedge \cdots \wedge Y_{n-1}$". We do not extend these operations to the case of an empty sequence ($n = 0$); we stipulate, however, that in this case

$$Y_0 \vee \cdots \vee Y_{n-1} \vee Z \;=\; Z \;=\; Y_0 \wedge \cdots \wedge Y_{n-1} \wedge Z$$

for every $Z \in \mathbf{\Phi}$.

Given any finite sequence $\langle x_0, \dots, x_{n-1}\rangle$ of variables, $\forall_{x_0 \cdots x_{n-1}}$ and $\exists_{x_0 \cdots x_{n-1}}$ are compositions of the corresponding sequences of unary operations, $\langle \forall_{x_0}, \dots, \forall_{x_{n-1}}\rangle$ and $\langle \exists_{x_0}, \dots, \exists_{x_{n-1}}\rangle$; in case $n = 0$, $\forall_{x_0 \cdots x_{n-1}}$ and $\exists_{x_0 \cdots x_{n-1}}$ coincide with the identity operation.

It is known under what conditions a variable x occurring at some place in a formula Y *occurs* at this place *bound* or *free*. $\Upsilon\phi Y$ is the set of all variables occurring free at some places in Y. If Y contains no quantifiers, and hence no variable occurs bound in it, Y is called *quantifier-free*.

Every formula Y determines uniquely an increasing sequence $\langle x_0, \dots, x_{n-1}\rangle$ of variables such that $\{x_0, \dots, x_{n-1}\} = \Upsilon\phi Y$. $\langle x_0, \dots, x_{n-1}\rangle$ is referred to as the *canonical sequence* of Y. By the *closure* of the formula Y, symbolically $[Y]$, we understand the formula $\forall_{x_0 \cdots x_{n-1}} Y$ where $\langle x_0, \dots, x_{n-1}\rangle$ is the canonical sequence of Y. Obviously, $[Y]$ is a *sentence*, i.e., a formula without any free variable, and we have $[Y] = Y$ iff Y is a sentence. (Throughout the book "iff" is used as an abbreviation for "if and only if".) The set of all sentences of $\mathcal{L}$ is denoted by "$\Sigma[\mathcal{L}]$" or simply by "Σ".

Using the notion of a free variable we define a new operation on formulas which essentially consists in replacing in any given formula X some distinct variables $x_0, \ldots, x_{n-1}$ by other variables $u_0, \ldots, u_{n-1}$ at all those places (if any) in X at which $x_0, \ldots, x_{n-1}$ occur free. This operation is referred to as (*simultaneous*) *substitution of* $u_0, \ldots, u_{n-1}$ *for* $x_0, \ldots, x_{n-1}$, and the result of performing this operation on X is denoted by "$X[x_0/u_0, \ldots, x_{n-1}/u_{n-1}]$". The operation plays an important role in metamathematical discussions, if only because of its close connection with the fundamental notion of derivability which will be introduced below in 1.3(ii). In fact, the notion of substitution is defined in such a way that, for every formula X, the expression $X[x_0/u_0, \ldots, x_{n-1}/u_{n-1}]$ is a formula whose closure is derivable from the closure of X. Throughout the present work, however, the notion of substitution will not be extensively used, but will prove very helpful in some fragments of our discussion.

The substitution $X[x_0/u_0, \ldots, x_{n-1}/u_{n-1}]$ is essentially the formula Y obtained from X by simultaneously replacing all x_k's $(0 \leq k < n)$ with the corresponding u_k's at all places where the x_k's occur free in X. It may happen, however, that as a consequence of this replacement some u_k occurs bound in the resulting formula Y at a place where x_k occurs free in X; this is a situation which we want to avoid for rather obvious reasons, in particular, because in this case the connection between substitution and derivability pointed out above may easily fail. If such an undesirable situation occurs, the construction becomes somewhat more complicated. We first select new variables $w_0, \ldots, w_{n-1}$ not occurring in X and such that $w_k = w_l$ iff $u_k = u_l$, for $0 \leq k < l < n$; to make the selection unambiguous we assume that $w_0, \ldots, w_{n-1}$ are, in order, the first variables with these properties. To obtain $X[x_0/u_0, \ldots, x_{n-1}/u_{n-1}]$ we now replace all free occurrences of the x_k's by the corresponding u_k's and, at the same time, all bound occurrences of the u_k's by the corresponding w_k's.

In case $x_0 = \boldsymbol{v}_0, \ldots, x_{n-1} = \boldsymbol{v}_{n-1}$, we use an abbreviated notation, setting

$$X[u_0, \ldots, u_{n-1}] = X[\boldsymbol{v}_0/u_0, \ldots, \boldsymbol{v}_{n-1}/u_{n-1}].$$

Thus, in particular,

$$X[u] = X[\boldsymbol{x}/u], \ X[u,v] = X[\boldsymbol{x}/u, \boldsymbol{y}/v], \ X[u,v,w] = X[\boldsymbol{x}/u, \boldsymbol{y}/v, \boldsymbol{z}/w],$$

etc. Obviously

$$X = X[\boldsymbol{x}] = X[\boldsymbol{x}, \boldsymbol{y}] = \cdots.$$

1.3. Derivability in $\mathcal{L}$

An important step in the formalization of predicate logic is the selection of logical axioms. The set of logical axioms is necessarily infinite, but it is usually described as the set of all particular instances of a few simple axiom schemata. Several axiomatizations of predicate logic which are adequate for our purposes can be found in the literature. To fix the ideas, we select as the set of logical axioms for $\mathcal{L}$ the set Σ_1 described (in a somewhat different symbolism) in Tarski [**1965**], pp. 67–68; the set will be denoted here by "$\mathbf{\Lambda}[\mathcal{L}]$".

A precise definition of $\mathbf{\Lambda}[\mathcal{L}]$ follows.

(i) $\mathbf{\Lambda}[\mathcal{L}]$, *or simply* $\mathbf{\Lambda}$, *is the set of sentences* S *in* Σ *such that, for some* $X, Y, Z \in \mathbf{\Phi}$ *and some* $x, y \in \mathbf{\Upsilon}$, S *coincides with one of the following sentences:*

(AI) $[(X \rightarrow Y) \rightarrow ((Y \rightarrow Z) \rightarrow (X \rightarrow Z))]$;

(AII) $[(\neg X \rightarrow X) \rightarrow X]$;

(AIII) $[X \rightarrow (\neg X \rightarrow Y)]$;

(AIV) $[\forall_x \forall_y X \rightarrow \forall_y \forall_x X]$;

(AV) $[\forall_x (X \rightarrow Y) \rightarrow (\forall_x X \rightarrow \forall_x Y)]$;

(AVI) $[\forall_x X \rightarrow X]$;

(AVII) $[X \rightarrow \forall_x X]$, *where* $x \notin \Upsilon\phi X$;

(AVIII) $[\neg \forall_x (\neg x \mathring{\mathbf{1}} y)]$, *where* $x \neq y$;

(AIX) $[x \mathring{\mathbf{1}} y \rightarrow (X \rightarrow Y)]$, *where* X *is any atomic formula in which* x *occurs, and* Y *is obtained from* X *by replacing a single occurrence of* x *with* y.

It is known from *op. cit.*, pp. 71–74, that if the restriction $x \neq y$ in (AVIII) is omitted, then (AVI) can be deleted altogether, since in this case all particular instances of it become derivable from the remaining logical axioms. However, it doesn't seem to be possible to delete (AVI) in logical formalisms provided with only a finite number of variables, which we shall consider in a part of the subsequent discussion, primarily in Chapter 3.

This axiomatization of predicate logic has two virtues: (1) just as in an earlier axiomatization due to Quine [**1951**], we can use *modus ponens* as the only direct, or primitive, rule of inference to derive all logically valid sentences from the axioms; (2) the description of the axioms does not involve the general notion of substitution. These features of the set of logical axioms simplify to some extent the most involved argument which is carried through in this work, that is, the proof of Theorem 4.4(xxxiv), including the proofs of the auxiliary statements in §4.4, upon which 4.4(xxxiv) is based.

In terms of logical axioms we define a fundamental metalogical notion, that of derivability.

(ii) *A sentence* $X \in \Sigma$ *is said to be derivable from a set* $\Psi \subseteq \Sigma$, *in symbols* $\Psi \vdash X$ (*or* $\Psi \vdash X\ [\mathcal{L}]$), *if* X *belongs to every set* $\Omega \subseteq \Sigma$ *such that*

(α) $\mathbf{\Lambda} \subseteq \Omega$,

(β) $\Psi \subseteq \Omega$,

(γ) $Z \in \Omega$ *whenever there is a* Y *such that* Y, $Y \rightarrow Z \in \Omega$.

Sets $\Omega \subseteq \Sigma$ that satisfy condition (ii)(γ) above are said to be closed under *modus ponens* (the *operation of detachment*). Instead of "$\{Y\} \vdash X$" we write "$Y \vdash X$" (and we proceed similarly in other analogous situations).

The definition of derivability just formulated suggests a general method of showing that every sentence derivable from a given set Ψ of sentences possesses a certain property: we take Ω to be the set of sentences possessing this property, and show successively that Ω satisfies conditions (α)–(γ) in (ii). Hence we conclude that Ω contains all sentences derivable from Ψ. We shall refer to this method of proof as *induction on derivable sentences.* Similarly, the definition of a formula in $\mathcal{L}$ given in §1.2 suggests an analogous method of showing that every formula possesses a certain property; this method may be analogously referred to as *induction on formulas.*

We say that a set Ω is *derivable* from a set Ψ, in symbols $\Psi \vdash \Omega$, if $\Psi \vdash X$ for every $X \in \Omega$. We write "$\vdash X$" (or "$\vdash \Omega$") instead of "$\varnothing \vdash X$" (or "$\varnothing \vdash \Omega$") and we say in this case that X is a *logically provable* sentence (or Ω is a set of logically provable sentences).

Given a $\Psi \subseteq \Sigma$, we set

$$\mathbf{\Theta\eta}\Psi \text{ (or } \mathbf{\Theta\eta}\Psi[\mathcal{L}]) = \{X : X \in \Sigma \quad \text{and} \quad \Psi \vdash X\}.$$

$\mathbf{\Theta\eta}\Psi$ is referred to as the *theory generated by,* or *based upon* Ψ. A set Ω is called a *theory* if $\Omega = \mathbf{\Theta\eta}\Psi$ for some $\Psi \subseteq \Sigma$ or, equivalently, if $\Omega = \mathbf{\Theta\eta}\Omega$. Every set $\Psi \subseteq \Sigma$ such that $\Omega = \mathbf{\Theta\eta}\Psi$ is said to be a *base,* or a (*possible*) *axiom set* of Ω; Ω is called *finitely based* or *finitely axiomatizable* if $\Omega = \mathbf{\Theta\eta}\Psi$ for some finite set Ψ.

From the above definition it is seen that $\mathbf{\Theta\eta}\varnothing$ coincides with the set of all logically provable sentences and is the least theory in $\mathcal{L}$. It can be called the *logic of* $\mathcal{L}$ or the (*predicate*) *logic of one binary relation.*

A set $\Psi \subseteq \Sigma$ is called *consistent* if $\mathbf{\Theta\eta}\Psi \neq \Sigma$ and hence $\mathbf{\Theta\eta}\Psi \subset \Sigma$. A set $\Phi \subseteq \Sigma$, or a sentence $X \in \Sigma$, is said to be *compatible* with Ψ if the set $\Psi \cup \Phi$, or $\Psi \cup \{X\}$, is consistent. Ψ is called *complete* if Ψ is consistent and if every sentence which is compatible with Ψ is derivable from Ψ.

Given three sets Ψ, Ω, and Φ, we say that Ω is *derivable from* Ψ *on the basis of* (or *relative to*) Φ, symbolically $\Psi \vdash_\Phi \Omega$, if $\Phi \cup \Psi \vdash \Omega$. Ψ and Ω are said to be *equivalent on the basis of* (or *relative to,* or *under*) Φ, in symbols $\Psi \equiv_\Phi \Omega$, if $\Psi \vdash_\Phi \Omega$ and $\Omega \vdash_\Phi \Psi$. In case $\Phi = \varnothing$, we call Ψ and Ω *logically equivalent,* in symbols $\Psi \equiv \Omega$.

We recall here a simple but important result concerning the notion of derivability.

(iii) *For any* $X, Y \in \Sigma$ *and* $\Psi \subseteq \Sigma$, *if* $\Psi \cup \{X\} \vdash Y$, *then* $\Psi \vdash (X \rightarrow Y)$.

The proof, by induction on sentences derivable from $\Psi \cup \{X\}$, is straightforward.

This is the so-called *deduction theorem.* It could also be called the *threefold-implication theorem,* since it involves three different (though related) notions of implication expressed respectively by the symbol "$\vdash$", the symbol "$\rightarrow$", and the phrase "*if..., then ...*" (which, of course, could be replaced by a symbol, e.g., "$\Rightarrow$", and probably would be so replaced, were the metasystem to be formalized). The mutual relationship of these three notions has been frequently a source of confusion.

The converse of (iii) is a trivial consequence of the definition of derivability. We thus arrive at a stronger form of the deduction theorem:

(iv) *For any* $X, Y \in \Sigma$ *and* $\Psi \subseteq \Sigma$, *we have* $\Psi \cup \{X\} \vdash Y$ *iff* $\Psi \vdash (X \rightarrow Y)$.

An immediate corollary of (iv) is:

(v) *For any* $X, Y \in \Sigma$ *and* $\Psi \subseteq \Sigma$, *we have* $X \equiv_\Psi Y$ *iff* $\Psi \vdash (X \leftrightarrow Y)$.

Corollary (v) suggests a natural way of extending the relation of equivalence under a set Ψ—and, in particular, that of logical equivalence—from sentences to arbitrary formulas. In fact, we can stipulate that, for any formulas $X, Y \in \mathbf{\Phi}$ and any set of sentences $\Psi \subseteq \Sigma$,

$$X \equiv_\Psi Y \quad \text{iff} \quad \Psi \vdash [X \leftrightarrow Y],$$

and, in particular,

$$X \equiv Y \quad \text{iff} \quad \vdash [X \leftrightarrow Y].$$

We do not extend the relations $\equiv_\Psi$ and $\equiv$ to arbitrary sets of formulas.

We assume that the notion of recursiveness and related notions have been appropriately extended to sets of expressions of $\mathcal{L}$, and to relations between and operations on these expressions. As a consequence, such sets as $\mathbf{\Upsilon}$, $\mathbf{\Phi}$, Σ prove to be recursive. On the other hand, the notions of recursiveness and related notions cannot be applied to the relation $\vdash$ since its domain, consisting of all subsets of Σ, is nondenumerable. However, these notions can be applied to various denumerable subsets of $\vdash$, for instance to the *finitary part of derivability*, i.e., to the relation

$$\{\langle \Xi, Y\rangle : \Xi \subseteq \Sigma,\ |\Xi| < \omega,\ Y \in \Sigma, \ \text{ and } \ \Xi \vdash Y\},$$

and to the *singleton part of derivability*, i.e., to the relation

$$\{\langle X, Y\rangle : X, Y \in \Sigma \ \text{ and } \ X \vdash Y\},$$

and both of these relations are well known to be recursively enumerable, though not recursive. This applies also to the logic of $\mathcal{L}$, i.e., to the theory

$$\mathbf{\Theta\eta}\varnothing = \{Y : Y \in \Sigma \ \text{ and } \ \varnothing \vdash Y\}$$

(see Church [**1936**] and Kalmár [**1936**]; cf. also Tarski–Mostowski–Robinson [**1953**], p. 62).

In general, for every theory Θ the following two conditions are equivalent: (1) Θ is recursively enumerable, and (2) Θ has a recursively enumerable base; by Craig [**1953**] each of them is also equivalent to the condition: (3) Θ has a recursive base.

In application to theories, the terms "recursive" and "decidable" are used interchangeably, and so are the terms "not recursive" and "undecidable". Recall that a theory Θ is defined to be *essentially undecidable* if Θ and every consistent theory (in $\mathcal{L}$) extending Θ is undecidable; Θ is said to be *hereditarily undecidable* if it, and each of its subtheories is undecidable. The following important consequences of the deduction theorem (iii) are well known.

(vi) *If a theory Ψ is decidable and a set Ω of sentences is finite, then the theory $\boldsymbol{\Theta\eta}(\Psi \cup \Omega)$ is also decidable.*

(vii) *Every finitely based theory which is undecidable is hereditarily undecidable.*

(viii) *If a theory Ψ is consistent and decidable, then there is a complete and decidable theory Θ such that $\Psi \subseteq \Theta$.*

Cf. Tarski–Mostowski–Robinson [**1953**], pp. 15–17, where (vi) and (viii) are stated and proved; (vii) is an obvious corollary of (vi).

If we now wish to describe a system $\mathcal{S}$ of set theory (or any other mathematical system) which is formalized in $\mathcal{L}$, the only thing that remains to be done is to specify a base Φ (i.e., a specific set of nonlogical axioms) for $\mathcal{S}$. Using the set Φ, we relativize to $\mathcal{S}$ various notions defined above. Thus we call a sentence $X \in \Sigma$ *provable in* $\mathcal{S}$ if $\Phi \vdash X$. We say that X is *derivable in* $\mathcal{S}$ *from* a set $\Psi \subseteq \Sigma$ if $\Psi \vdash_{\Phi} X$. Two sentences $X, Y \in \Sigma$ are said to be *equivalent in* $\mathcal{S}$ if $X \equiv_{\Phi} Y$. $\boldsymbol{\Theta\eta}\Phi$ is referred to as the *theory of* $\mathcal{S}$; more generally, we can call Θ a *theory in* $\mathcal{S}$ if Θ is a theory in $\mathcal{L}$ and $\Phi \subseteq \Theta$.

It is important to notice that, as we use the term "system", Φ uniquely determines $\mathcal{S}$, but is not uniquely determined by $\mathcal{S}$. Without changing $\mathcal{S}$ we can replace Φ by any other set $\Psi \subseteq \Sigma$ which is logically equivalent to Φ. On the other hand, it will sometimes be convenient to fix Φ for some portion of our discussion. We refer then to Φ as the *axiom set* of $\mathcal{S}$, in symbols $\mathbf{A}\boldsymbol{\xi}[\mathcal{S}]$ or simply $\mathbf{A}\boldsymbol{\xi}$. Actually, our discussion until the middle of Chapter 4 does not require us to fix the set Φ, and thus to specify $\mathbf{A}\boldsymbol{\xi}[\mathcal{S}]$. The discussion has a general metalogical character: connections between some formal languages are studied, and the study is not influenced by the fact that various systems of set theory can be formalized in these languages. On the other hand, in later parts of the book we shall need some specific information about systems $\mathcal{S}$ of set theory to which the results of the earlier discussion can be applied. But even then no detailed description of the axiom sets $\mathbf{A}\boldsymbol{\xi}[\mathcal{S}]$ will be required; the desired information will be provided in the form of some general and rather weak assumptions imposed on these sets.

1.4. Semantical notions of $\mathcal{L}$

In this section we concern ourselves with semantical notions related to the language $\mathcal{L}$.

Roughly speaking, semantical notions are introduced to provide an interpretation for a formal language, to ascribe meanings to its symbols and expressions. In the case of the language $\mathcal{L}$ the only kind of compound expressions to which meanings are ascribed are formulas. Just as other formal languages, $\mathcal{L}$ admits many interpretations. Technically, as a base for each interpretation we take an arbitrary relational structure $\mathfrak{U} = \langle U, E\rangle$, where U, the *universe* of $\mathfrak{U}$, is any nonempty set, and E, the *fundamental relation* of $\mathfrak{U}$, is any binary relation on U, i.e., between elements of U, in symbols $E \subseteq U \times U$. Every such structure is referred to as a *(possible) realization* of $\mathcal{L}$. By choosing $\mathfrak{U}$ as a base for interpreting $\mathcal{L}$ we implicitly assume that U is the common range of all variables of $\mathcal{L}$ and that E is the relation denoted by the unique nonlogical constant $\boldsymbol{E}$ of $\mathcal{L}$. The meanings ascribed to the logical constants of $\mathcal{L}$ are essentially independent of $\mathfrak{U}$; we interpret these constants implicitly by treating them as synonymous with some expressions of common language. For instance, the logical predicate $\mathring{\mathbf{i}}$ is treated as a synonym of the phrase "is identical with"; we can also say that $\mathring{\mathbf{i}}$, just as $\boldsymbol{E}$, denotes a binary relation, in fact the identity relation Id restricted to elements of U, in symbols $U \rceil Id$.

Under these assumptions, the basic semantical notion for the language $\mathcal{L}$, that of *satisfaction*, acquires a clear intuitive meaning and can be precisely defined. Satisfaction is a relation between a formula X and certain elements of U arranged in a sequence x; it is formally expressed by the phrase "X is satisfied (in $\mathfrak{U}$) by the sequence x". Intuitively, it is natural to assume that a sequence x satisfying X is always a finite sequence with the same length n as the canonical sequence $\langle \boldsymbol{v}_{k_0}, \ldots, \boldsymbol{v}_{k_{n-1}} \rangle$ of X. For technical reasons, however, it proves more convenient first to define what it means for a simple infinite sequence $y = \langle y_0, \ldots, y_n, \ldots \rangle$ to satisfy X (although only finitely many terms of this sequence, in fact $y_{k_0}, \ldots, y_{k_{n-1}}$, determine whether y actually satisfies X). The definition proceeds by recursion on formulas; details are well known from the literature (see, e.g., Tarski [**1956**], p. 193). We can then define satisfaction for finite sequences by stipulating that $x = \langle x_0, \ldots, x_{n-1} \rangle$ satisfies X just in case there is an infinite sequence y satisfying X such that $x_i = y_{k_i}$ for $i = 0, \ldots, n-1$. (In the treatment of satisfaction for finite sequences we deviate slightly from the terminology of Henkin–Monk–Tarski [**1971**], p. 44.)

In terms of satisfaction other semantical notions are defined. The most important among them are the closely related notions of truth and model. A sentence $Y \in \Sigma$ is said to be *true* of $\mathfrak{U}$ or to *hold* in $\mathfrak{U}$ if every sequence $x = \langle x_0, \ldots, x_n, \ldots \rangle$ of elements of U satisfies Y. Under the same condition $\mathfrak{U}$ is called a *model* of Y. More generally, $\mathfrak{U}$ is a *model of* a set $\Psi \subseteq \Sigma$ iff it is a model of every sentence X in Ψ.

From the definition of truth it easily follows that the set Θ of all sentences which are true of a given realization $\mathfrak{U}$ of $\mathcal{L}$ is a theory in the sense of §1.3; it is called the *(first-order) theory of* $\mathfrak{U}$, in symbols $\mathbf{\Theta\rho}\mathfrak{U}$. More generally, given a class K of such structures $\mathfrak{U}$, the *theory of* K, $\mathbf{\Theta\rho}\mathsf{K}$, is the intersection of all $\mathbf{\Theta\rho}\mathfrak{U}$ for $\mathfrak{U} \in \mathsf{K}$.

A sentence $X \in \Sigma$ is called a *consequence* of a set $\Psi \subseteq \Sigma$, and Ψ is said to *imply* X, in symbols $\Psi \models X$, if every model of Ψ is a model of X.

Using the notion of consequence we introduce a series of notions entirely analogous to those defined in §1.3 in terms of derivability. In opposition to the new, semantical notions, those previously defined are referred to as syntactical. To obtain a technical term for a semantical notion, we frequently use the term introduced for the corresponding syntactical notion, qualifying it with "semantical(ly)"; however, to avoid clumsy expressions, we sometimes omit a

part of the original term, or else we introduce a new term, without following the general rule.

To give a few examples, the relation $\models$ between a set $\Psi \subseteq \Sigma$ and a sentence $X \in \Sigma$ is extended to the cases when Ψ is replaced by a sentence $Y \in \Sigma$ or X is replaced by a set $\Phi \subseteq \Sigma$. In case $\Psi = \varnothing$, we write "$\models X$" instead of "$\varnothing \models X$"; the sentence X is then called *logically valid* or *logically true.*

Two sets $\Psi, \Omega \subseteq \Sigma$ are said to be *semantically equivalent relative to* a set $\Phi \subseteq \Sigma$ if $\Phi \cup \Psi \models \Omega$ and $\Phi \cup \Omega \models \Psi$; they are called simply *semantically equivalent* (and not *semantically logically equivalent*) in case $\Phi = \varnothing$. For semantical equivalence we shall sometimes use the same symbol "$\equiv$" as for syntactical equivalence; we shall then state explicitly, however, that the symbol is being used in the semantical sense.

Given a system $\mathcal{S}$ formalized in $\mathcal{L}$, a sentence X is said to be *true in* $\mathcal{S}$, or to *hold in* $\mathcal{S}$, if $\mathbf{A\xi}[\mathcal{S}] \models X$; more generally, a sentence X is a *consequence in* $\mathcal{S}$ *of* a set Ψ of sentences if $\Psi \cup \mathbf{A\xi}[\mathcal{S}] \models X$. A model of $\mathbf{A\xi}[\mathcal{S}]$ is simply called a *model of* $\mathcal{S}$.

The most important result in metalogic is the well-known completeness theorem of Gödel [**1930**]:

(i) *The relations* $\vdash [\mathcal{L}]$ *and* $\models [\mathcal{L}]$ *coincide.*

This result splits naturally into two parts.

(ii) *For every* $\Psi \subseteq \Sigma$ *and* $X \in \Sigma$, *if* $\Psi \vdash X$, *then* $\Psi \models X$.

(iii) *For every* $\Psi \subseteq \Sigma$ *and* $X \in \Sigma$, *if* $\Psi \models X$, *then* $\Psi \vdash X$.

The implication (ii) is quite elementary; its converse (iii) has a much deeper character.

Obviously, (i) implies that every semantical notion defined in terms of $\models$ coincides with the corresponding syntactical notion. This yields, in particular, the following conclusions.

(iv) *A sentence* $X \in \Sigma$ *is logically provable iff it is logically true, i.e., true of every realization of* $\mathcal{L}$.

(v) *Two sets* $\Psi, \Omega \subseteq \Sigma$ *are equivalent relative to a set* $\Phi \subseteq \Sigma$ *iff they are semantically equivalent relative to* Φ, *i.e., every model of* Φ *which is a model of one of the sets* Ψ, Ω, *is also a model of the other. In particular, the sets* Ψ, Ω *are logically equivalent iff they are semantically equivalent, i.e., have all models in common.*

(vi) *A set* $\Psi \subseteq \Sigma$ *is consistent iff it is semantically consistent, i.e., it has some model.*

(vii) *A set* $\Psi \subseteq \Sigma$ *is complete iff it is semantically complete, or equivalently, iff there is a realization* $\mathfrak{U}$ *of* $\mathcal{L}$ *such that* $\mathbf{\Theta\eta}\Psi = \mathbf{\Theta\rho}\mathfrak{U}$.[1]

[1] *The reader will notice that the term "semantically complete" is used in two rather different senses in this book. When applied to a set of sentences Ψ it means that every

The completeness theorem and its consequences such as (iv)–(vii) lead to considerable simplifications of various metalogical arguments concerning $\mathcal{L}$.

Since we are particularly interested in $\mathcal{L}$ as a framework for the development of formal systems $\mathcal{S}$ of set theory, we pay special attention to *set-theoretical realizations* of $\mathcal{L}$, i.e., to structures $\mathfrak{U} = \langle U, E\rangle$ in which E is the membership relation $\in$ restricted to members of U; if such a $\mathfrak{U}$ is a model of a set $\Phi \subseteq \Sigma$ or of a system $\mathcal{S}$, we refer to it as a *set-theoretical model* of Φ, or of $\mathcal{S}$. In case we concentrate our attention on a particular system $\mathcal{S}$, we may be inclined to single out a set-theoretical model $\mathfrak{U}' = \langle U', E'\rangle$ and call it the *standard model* of $\mathcal{S}$ and also the *standard realization* of $\mathcal{L}$; U' is then sometimes called the *universe of discourse* of $\mathcal{L}$ and $\mathcal{S}$. By selecting $\mathfrak{U}'$ as the standard realization we emphasize that the interpretation of $\mathcal{L}$ based upon $\mathfrak{U}'$ is for us, intuitively, the proper interpretation under which all symbols, formulas, and sentences of $\mathcal{L}$ acquire their intended meanings. An explicit relativization to the standard model $\mathfrak{U}'$ is usually omitted; e.g., sentences true of $\mathfrak{U}'$ are simply referred to as true sentences.

Actually, we are not planning to single out a standard realization of $\mathcal{L}$ in this work. (Apart from other difficulties, this would certainly require a much more detailed and precise description of the metasystem and its set-theoretical basis.) Nevertheless, in our informal remarks we shall sometimes use semantical terms without explicitly mentioning a realization $\mathfrak{U}$ to which these terms should be relativized. It should then be understood that the remarks refer (not to the standard realization, but) to any given realization fixed in advance.

Nothing that has been said above in this section implies that all the axiomatic systems of set theory which will be involved in our discussion have actual set-theoretical models in our metasystem.

1.5. First-order formalisms

The formalism $\mathcal{L}$ is a member of the important class of formalisms called *formalisms of first-order predicate logic*, or simply *first-order formalisms.* The prevailing part of contemporary metamathematics is concerned with these kinds of formalisms.

This class can be conceived of in many different ways. For our purposes it is convenient to adopt a formally rather restricted conception. Thus we assume that every first-order formalism $\mathcal{P}$ has the same variables and logical constants as $\mathcal{L}$. However, in opposition to $\mathcal{L}$, the vocabulary of $\mathcal{P}$ contains, not necessarily one, but finitely many (and sometimes even infinitely many) nonlogical constants. These constants are (first-order) relation or operation symbols of any finite rank. They are assumed to be arranged in a system $\boldsymbol{C} = \langle C_i : i \in I\rangle$, indexed by the elements of an arbitrary set I, without repeating terms, i.e., $C_i \neq C_j$ whenever

sentence semantically compatible with Ψ is a consequence of Ψ (or, equivalently, that $\boldsymbol{\Theta\eta}\Psi$ is the theory of some model). When applied to a formalism or a system $\mathcal{S}$ it means that the completeness theorem holds for $\mathcal{S}$, i.e., that the notions of derivability and consequence in $\mathcal{S}$, $\vdash [\mathcal{S}]$ and $\models [\mathcal{S}]$, coincide.

$i \neq j$. Without any essential loss of generality we could assume that the index set I is always an ordinal number α (and hence, as usual in modern set theory, that it coincides with the set of all ordinals smaller than α); we shall make use of this assumption whenever it proves convenient.

With each $i \in I$ we associate a natural number ρi called the *rank* of $\mathbf{C}_i$. Realizations of $\mathcal{P}$ are arbitrary (first-order) *algebraic structures* of the form $\mathfrak{U} = \langle U, Q \rangle$, where U, the universe of $\mathfrak{U}$, is any nonempty set, $Q = \langle Q_i : i \in I \rangle$, and for each $i \in I$, Q_i is either an n-ary relation on U (i.e., a subset of ${}^nU$) or an n-ary operation on U (i.e., a function from ${}^nU$ to U) according as $\mathbf{C}_i$ is a relation or operation symbol of rank $\rho i = n$; the symbol $\mathbf{C}_i$ is used to denote in $\mathfrak{U}$ the relation or operation Q_i. The case $\rho i = 0$ is not excluded. However, predicates of rank 0, which are also referred to as *sentential constants*, are rarely used and will be disregarded in the subsequent discussion. On the other hand, operation symbols of rank 0 can be identified with *individual constants*, i.e., symbols denoting particular elements of U. Instead of "$\langle U, Q \rangle$" we shall sometimes write "$\langle U, Q_i \rangle_{i \in I}$", or just "$\langle U, Q_{i_0}, \dots, Q_{i_n} \rangle$" when $i_0, \dots, i_n$ are the distinct elements of I.

If $\mathbf{C}_i$ is a predicate of rank $\rho i > 0$ and $x_0, \dots, x_{\rho i - 1}$ are arbitrary variables, then the expression formed by the string of symbols $\mathbf{C}_i, x_0, \dots, x_{\rho i-1}$ and represented here by $\mathbf{C}_i(x_0, \dots, x_{\rho i-1})$ is an atomic formula of $\mathcal{P}$. In case $\rho i = 2$ we can use, as we have done already, $x_0 \mathbf{C}_i x_1$ instead of $\mathbf{C}_i(x_0, x_1)$; a similar remark applies to the logical binary predicate $\mathring{\mathbf{1}}$. In case some of the symbols $\mathbf{C}_j$ are operation symbols, the definition of atomic formula undergoes some complications. We first define by recursion the notion of a term: variables and operation symbols of rank 0 are the atomic terms; if $\mathbf{C}_j$ is an operation symbol of rank $\rho j > 0$ and $t_0, \dots, t_{\rho j-1}$ are terms, then the expression $\mathbf{C}_j(t_0, \dots, t_{\rho j-1})$ is a term. Notice that, although terms are, in general, compound expressions of a language, we use in this work lower case italics, preferably "p", "q", "r", "s", and "t", as metamathematical variables ranging over terms. We thus deviate here from our general convention by which upper case italics are used as variables ranging over compound expressions (cf. §1.2). Atomic formulas are now defined as expressions of the forms $\mathbf{C}_k(t_0, \dots, t_{\rho k-1})$ and $s_0 \mathring{\mathbf{1}} s_1$, where $\mathbf{C}_k$ is a predicate of rank $\rho k > 0$ and $t_0, \dots, t_{\rho k-1}, s_0, s_1$ are arbitrary terms. Formulas in $\mathbf{\Phi}[\mathcal{P}]$ are formed in the familiar way from atomic formulas.

The description of the logical axioms and the definition of derivability, given in §1.3 for $\mathcal{L}$, remain in $\mathcal{P}$ virtually unchanged. (For a detailed formalization of predicate logic with operation symbols, following the lines of §1.3, see Kalish–Montague [**1965**].) The notions, notations, and results of §§1.2–1.4 can now be carried over from $\mathcal{L}$ to $\mathcal{P}$ with only minor modifications. Just as $\mathcal{L}$, the formalism $\mathcal{P}$ is semantically complete.

For the discussion in our work, the assumption that $\mathcal{L}$ has just one nonlogical constant is not essential. With very minor and obvious changes the results extend to first-order formalisms whose set of nonlogical constants consists of finitely many, and often even infinitely many, binary predicates, and to mathematical systems developed in such formalisms. The assumption that all nonlogical constants are binary predicates is essential. Nevertheless, we shall see in Chapter 7 that with certain modifications, some of the fundamental results in this work can be extended to a much wider class of formalisms, namely the class of all first-order formalisms with just finitely many nonlogical constants.

1.6. Formalisms and systems

As was mentioned at the outset we shall be concerned in the subsequent chapters of this book not only with $\mathcal{L}$, but also with some other formalisms. Those closest to $\mathcal{L}$ are the arbitrary formalisms of first-order predicate logic discussed in §1.5. In Chapters 2 and 3 we shall construct some further formalisms which are not formalisms of predicate logic, and which in some cases, such as $\mathcal{L}^{\times}$ in Chapter 3, differ considerably in their structure from the latter. We wish to emphasize that we are not interested here in developing a general theory of formalisms. It is hard to imagine a precise definition of formalism that would fit all discussions in which this term in involved.

When speaking informally of arbitrary formalisms, we have in mind primarily all the formalisms dealt with in this book (possibly including others with closely related formal structures). When using the term "formalism" in formal parts of our discussion we shall in each case determine unambiguously the class of formalisms involved.

Still, we shall try to give the reader some idea of the logical status of the notion of "formalism". Just as in the case of $\mathcal{L}$, we assume that, for any formalism $\mathfrak{F}$ with which we shall concern ourselves, four metalogical objects have been determined. Two of these objects are syntactical: the set Σ (or more precisely, $\Sigma[\mathfrak{F}]$) of sentences in $\mathfrak{F}$, and the relation $\vdash$ (or $\vdash [\mathfrak{F}]$) of derivability in $\mathfrak{F}$, between subsets of Σ and members of Σ. The other two are semantical: the first is the class RE (or $\mathsf{RE}[\mathfrak{F}]$) of realizations of $\mathfrak{F}$, which consists of algebraic structures; the second is the function MO (or $\mathsf{MO}[\mathfrak{F}]$) which associates with each sentence X in Σ a subclass of RE, referred to as the class of all *models* of X. The reader can easily check that the observations just made agree with the way in which these notions have actually been introduced for $\mathcal{L}$ in §§1.3 and 1.4. For formalisms with which we shall be concerned in this book, RE will be the class of *all* first-order algebraic structures of a certain similarity type (referred to as relational structures in Henkin–Monk–Tarski [**1971**], pp. 36–37). This applies even to those formalisms such as $\mathcal{L}^{\times}$ which are not first-order. Moreover, the class RE will coincide with the union of all classes $\mathsf{MO}X$ for $X \in \Sigma$, so that the notion of a realization could be dispensed with as one of the fundamental components of formalisms.

For various discussions it is adequate to assume that, conversely, these four objects uniquely determine the formalism involved. Thus a formalism $\mathfrak{F}$ can be construed as an ordered quadruple,

$$\mathfrak{F} = \langle \Sigma, \vdash, \mathsf{RE}, \mathsf{MO} \rangle.$$

The objects Σ, $\vdash$, RE, MO may be called *fundamental components* of $\mathfrak{F}$.

We can now define the notion of consequence in terms of MO just as in §1.4, and can extend to $\mathfrak{F}$, without changes, the definitions of other syntactical and semantical notions and notations which have been formulated in terms of derivability and consequence in §§1.3 and 1.4.

A formalism $\mathfrak{F}$ is said to be *semantically sound* if it satisfies the condition 1.4(ii) (with Σ, $\vdash$, and $\models$ referring to $\mathfrak{F}$ and not to $\mathcal{L}$). It is called *semantically adequate* if it satisfies 1.4(iii), and *semantically complete* if it satisfies 1.4(i), i.e., if it is both semantically sound and semantically adequate. It should be emphasized that the notion of a semantically complete formalism just defined is not closely related to that of a semantically complete set of sentences introduced implicitly in §1.4.

All the formalisms with which we shall be concerned are assumed to be semantically sound, and are usually easily seen to be so. Not all of them, however, are semantically adequate. Hence, in the formalisms which will be discussed here the semantical notions introduced in §1.4 do not coincide in general with the corresponding syntactical notions. Instead, a kind of implication or inclusion holds as a rule between any two corresponding notions. For instance, if two sentences $X, Y \in \Sigma[\mathfrak{F}]$ are syntactically equivalent, then they are also semantically equivalent, i.e., they have all models in common. However, the implication in the opposite direction may fail.

In the present work, syntactical notions play a much more essential role than the semantical ones. Indeed a large part of our development could be carried out without the use of any semantical notions whatsoever. It therefore seems natural to investigate the syntactical properties common to all formalisms in which we shall be interested. It turns out that in every such formalism $\mathfrak{F}$ the following five postulates hold (for any $X \in \Sigma$ and any $\Phi, \Psi \subseteq \Sigma$):

(FI) *If* $X \in \Phi$, *then* $\Phi \vdash X$;
(FII) *If* $\Phi \vdash Y$ *for each* $Y \in \Psi$, *and if* $\Psi \vdash X$, *then* $\Phi \vdash X$;
(FIII) *If* $\Phi \vdash X$, *then* $\Delta \vdash X$ *for some finite* $\Delta \subseteq \Phi$;
(FIV) Σ *is countable*;
(FV) *There is a finite set* $\Delta \subseteq \Sigma$ *such that* $\Delta \vdash Y$ *for each* $Y \in \Sigma$.

Various properties of notions defined in terms of $\vdash$ that will be implicitly involved and applied in our discussion can be established with the exclusive help of (FI)–(FV). Since the semantical components play no role in this development, we can refer the results obtained, not just to formalisms in the original sense, but also to formalisms $\mathfrak{G}$ provided exclusively with syntactical components, and thus construed as ordered pairs

$$\mathfrak{G} = \langle \Sigma, \vdash \rangle.$$

Such formalisms are often referred to as *syntactical* or *uninterpreted formalisms* (as opposed to the original, *semantical* or *interpreted formalisms*). We can thus arrive at an abstract theory of syntactical formalisms, i.e., the theory of all ordered pairs $\langle \Sigma, \vdash \rangle$ (where $\vdash$ is a binary relation between subsets of Σ and elements of Σ) satisfying (FI)–(FV); see a very closely related development in Tarski [**1956**], Article V, pp. 60ff. We have never considered an analogous, mathematically interesting, abstract theory of semantical formalisms. Suppose $\mathfrak{F}$ and $\mathfrak{G}$ are respectively a semantical and syntactical formalism; in case $\Sigma[\mathfrak{F}] = \Sigma[\mathfrak{G}]$ and $\vdash[\mathfrak{F}] = \vdash[\mathfrak{G}]$, we can refer to $\mathfrak{G}$ as the *syntactical part* of $\mathfrak{F}$, and to $\mathfrak{F}$ as a *semantical expansion* of $\mathfrak{G}$.

Given two (syntactical or semantical) formalisms $\mathfrak{F}$ and $\mathfrak{G}$, we say that $\mathfrak{G}$ is a *subformalism* of $\mathfrak{F}$, and $\mathfrak{F}$ an *extension* of $\mathfrak{G}$, provided

$$\Sigma[\mathfrak{G}] \subseteq \Sigma[\mathfrak{F}]$$

and

$$\Psi \vdash X \ [\mathfrak{G}] \quad \text{implies} \quad \Psi \vdash X \ [\mathfrak{F}] \quad \text{whenever} \quad \Psi \subseteq \Sigma[\mathfrak{G}] \quad \text{and} \quad X \in \Sigma[\mathfrak{G}].$$

In case $\mathfrak{F}$ and $\mathfrak{G}$ are both semantical formalisms one could think of supplementing the definition of subformalism with further conditions, for example, that $\mathsf{RE}[\mathfrak{G}] \supseteq \mathsf{RE}[\mathfrak{F}]$ and $\mathsf{MO}X[\mathfrak{G}] \supseteq \mathsf{MO}X[\mathfrak{F}]$ for every $X \in \Sigma[\mathfrak{G}]$. However, these supplementary conditions are not satisfied in all situations within the present work where the term "subformalism" seems appropriate. Moreover, no simple semantical conditions that would be adequate for all of our purposes seem to suggest themselves.

For many purposes the conception of formalisms discussed here may prove to be too simple. The list of fundamental components may have to be supplemented by other objects. As additional syntactical notions we may have to consider the set of formulas or the set of logical axioms; as an additional semantical notion we may need to introduce the relation of satisfaction.

In case the definition of formalism is supplemented with further syntactical components, for instance, the set $\mathbf{\Phi}$ of all formulas, it may be desirable to supplement the definition of subformalism with corresponding conditions, for instance, $\mathbf{\Phi}[\mathfrak{G}] \subseteq \mathbf{\Phi}[\mathfrak{F}]$.

In this work we use the terms "formalism" and "formal language" (or "language" for short) interchangeably. In other contexts it may be useful to differentiate between the meanings of these two terms. Formal languages would then be construed as structures with a different list of fundamental components; the list would include some notions referring to the intrinsic structure of sentences, such as the vocabulary of a language.

There is another notion of a general character, closely related to the notion of a formalism, that will frequently be used in this work, namely the notion of a system. Just as in the case of formalisms, we are not planning to construct a general theory of systems. Actually, for the purposes of this work, we restrict ourselves to those systems which are developed in a formalism or, what amounts to the same thing, to systems obtained by relativizing a formalism to a certain set of sentences. Thus we shall speak of the system of Zermelo set theory as a system *developed* (or *formalized*) in $\mathcal{L}$, or else as a system obtained by relativizing

$\mathcal{L}$ to a well-known set of Zermelo's axioms. Similarly, Peano arithmetic can be referred to as a system developed in a first-order formalism (with appropriate nonlogical constants), or else as a system obtained by relativizing this formalism to the set of Peano's axioms.

(i) *Let $\mathcal{F}$ be the formalism*

$$\mathcal{F} = \langle \Sigma, \vdash, \mathsf{RE}, \mathsf{MO} \rangle,$$

and let Φ be a subset of Σ. By the system $\mathcal{S}$, with base Φ, developed (or formalized) in $\mathcal{F}$ we understand the quadruple

$$\mathcal{S} = \langle \Sigma', \vdash', \mathsf{RE}', \mathsf{MO}' \rangle,$$

where

$$\Sigma' = \Sigma,$$

for every $\Psi \subseteq \Sigma'$ and $X \in \Sigma'$ we have $\Psi \vdash' X$ iff $\Psi \cup \Phi \vdash X$ (i.e., $\vdash' = \vdash_\Phi$),

$$\mathsf{RE}' = \bigcap \{\mathsf{MO}X : X \in \Phi\},$$

and

$$\mathsf{MO}'X = \bigcap \{\mathsf{MO}Y : Y \in \Phi \quad or \quad Y = X\} \quad whenever \quad X \in \Sigma'.$$

We shall also refer to $\mathcal{S}$ as the relativization of $\mathcal{F}$ to Φ.

(ii) *We shall say that $\mathcal{S}$ is a system developed in a formalism $\mathcal{F}$, or that $\mathcal{S}$ is a relativization of $\mathcal{F}$, if there is a set Φ of sentences of $\mathcal{F}$ such that $\mathcal{S}$ is the system in $\mathcal{F}$ with base Φ.*

(iii) *Finally, we shall simply say that $\mathcal{S}$ is a system if there is a formalism $\mathcal{F}$ such that $\mathcal{S}$ is a system developed in $\mathcal{F}$.*

As a direct consequence of (i)—(iii) we obtain

(iv) *Every formalism $\mathcal{F}$ is a system, and in fact the system in $\mathcal{F}$ with base $\varnothing$.*

(Since, by definition, $\mathsf{MO}X[\mathcal{F}]$ is a subclass of $\mathsf{RE}[\mathcal{F}]$, we are assuming that

$$\bigcap \{\mathsf{MO}X[\mathcal{F}] : X \in \varnothing\} = \mathsf{RE}[\mathcal{F}];$$

cf. the remarks in §1.1.) By (iv) every result established for all systems holds *a fortiori* for all formalisms.

Various notions and notations introduced for the formalism $\mathcal{L}$, and carried over in the first part of this section to arbitrary formalisms $\mathcal{F}$, can be extended in an obvious way to arbitrary systems $\mathcal{S}$. Thus we shall write, e.g., $\Sigma[\mathcal{S}]$ and $\vdash [\mathcal{S}]$ respectively for the set of all sentences and the derivability relation in $\mathcal{S}$. Similarly, we extend the notion of subformalism.

(v) *A system* $\mathcal{S}$ *is a subsystem of a system* $\mathcal{T}$, *and* $\mathcal{T}$ *an extension of* $\mathcal{S}$, *if*

$$\Sigma[\mathcal{S}] \subseteq \Sigma[\mathcal{T}]$$

and

$$\Psi \vdash X\ [\mathcal{S}] \quad \textit{implies} \quad \Psi \vdash X\ [\mathcal{T}] \quad \textit{whenever} \quad \Psi \subseteq \Sigma[\mathcal{S}] \quad \textit{and} \quad X \in \Sigma[\mathcal{S}].$$

Notice that the notion of system is used here in the sense of (iii); thus the systems $\mathcal{S}$ and $\mathcal{T}$ in (v) need not be developed in the same formalism. It may also be pointed out that, for two formalisms, $\mathcal{F}$ and $\mathcal{G}$, $\mathcal{F}$ is a subformalism of $\mathcal{G}$ just in case it is a subsystem of $\mathcal{G}$.

Let $\mathcal{F}$ be any formalism and Φ any subset of $\Sigma[\mathcal{F}]$. As in the case of systems in the formalism $\mathcal{L}$, the set Φ uniquely determines the corresponding system $\mathcal{S}$ in $\mathcal{F}$, but is not uniquely determined by it. In fact, any other set $\Psi \subseteq \Sigma[\mathcal{F}]$ determines the same system as Φ in $\mathcal{F}$ iff $\Phi \equiv \Psi[\mathcal{F}]$; here we are using our implicit assumption that all formalisms $\mathcal{F}$ under discussion are sound and satisfy (FI)–(FV).

We obviously have

$$\mathbf{\Theta\eta}\varnothing[\mathcal{S}] = \mathbf{\Theta\eta}\Phi[\mathcal{F}] = \mathbf{\Theta\eta}\Phi[\mathcal{S}],$$

i.e., the set $\mathbf{\Theta\eta}\Phi$ coincides with the theory generated in $\mathcal{S}$ by $\varnothing$, and is therefore the least theory in $\mathcal{S}$. Hence, following the terminology introduced in §1.3 for $\mathcal{L}$, we should refer to $\mathbf{\Theta\eta}\Phi$ as the logic of $\mathcal{S}$. We prefer however to use instead the term "the theory of $\mathcal{S}$". Consider the function T from systems in $\mathcal{F}$ to theories in $\mathcal{F}$ and defined for every system $\mathcal{S}$ in $\mathcal{F}$ by the formula

$$T\mathcal{S} = \mathbf{\Theta\eta}\varnothing[\mathcal{S}].$$

Obviously this function maps the set of all systems in $\mathcal{F}$ in a one-one way onto the set of all theories in $\mathcal{F}$. Thus we could entirely eliminate systems from our discussion, replacing everywhere a system $\mathcal{S}$ by its theory $T\mathcal{S}$. However we find it convenient to have the present notion of system available for our discussion.[2]

In certain situations it proves desirable to single out from among all possible bases for a given system $\mathcal{S}$ a definite base Φ which we fix for the entire discussion of $\mathcal{S}$. In such a case we refer to Φ as the *axiom set* of $\mathcal{S}$, and denote it by "$\mathbf{A\xi}[\mathcal{S}]$", or simply "$\mathbf{A\xi}$". In most cases we shall consider, $\mathbf{A\xi}$ will be a recursive set, and we will refer then to $\mathcal{S}$ as an *axiomatic system.* As is well known, the theory of an axiomatic system developed, say, in a formalism of first-order logic, is not necessarily recursive, but is at any rate recursively enumerable.

[2]In earlier papers of Tarski (see, e.g., Tarski [**1956**], Articles V and XII) the notion of a system as we understand it here did not play an important role. Rather, the discussions centered around our present notion of a theory. However, this may not be quite clear to a casual reader since in those papers the term "system" did not have its present meaning, but was rather used as a name for what we are now calling "theory". (In addition to "system", the terms "closed system" and "deductively closed system" were used in the same sense.) On the other hand, the term "theory" was used in a rather loose and informal way, with a meaning rather close to that of "system" in the present work.

However, in this book we shall also concern ourselves with systems in formalisms of predicate logic whose theories are not recursively enumerable. To obtain one such example consider the algebra

$$\mathfrak{A} = \langle A, +, \cdot \rangle$$

where A is the set of natural numbers, and $+$ and $\cdot$ are respectively the familiar operations of addition and multiplication on A. An appropriate apparatus for studying $\mathfrak{A}$ is the formalism $\mathcal{F}$ of predicate logic with two nonlogical constants, the binary operation symbols $\mathbf{+}$ and $\boldsymbol{\cdot}$. We are interested in constructing a system $\mathcal{A}$ in $\mathcal{F}$ in which the set of all logically provable sentences consists just of those sentences that are true of $\mathfrak{A}$. The only natural choice for a base of $\mathcal{A}$ seems to be the set of all sentences in $\mathcal{F}$ that are true of $\mathfrak{A}$, i.e., the whole theory $\mathbf{\Theta\rho}\mathfrak{A}$. Thus we take $\mathbf{A\xi}[\mathcal{A}] = \mathbf{\Theta\rho}\mathfrak{A}$. It is known that $\mathbf{\Theta\rho}\mathfrak{A}$ is not recursively enumerable, and consequently that there is no recursive or even recursively enumerable base for $\mathcal{A}$. Instead of $\mathfrak{A}$ we could of course take many other algebraic structures or classes of similar algebraic structures.

One could also think of introducing the notion of a system developed in an uninterpreted formalism $\mathcal{F}$, i.e., a relativization of $\mathcal{F}$ to some set of sentences. Such relativizations turn out again to be uninterpreted formalisms, contrary to the situation for relativizations of interpreted formalisms. Thus, consideration of such systems does not lead us to anything new. In consequence, we shall only use the term "system" in application to interpreted formalisms.

CHAPTER 2

The Formalism $\mathcal{L}^+$, a Definitional Extension of $\mathcal{L}$

We shall now discuss a formalism $\mathcal{L}^+$ which is obtained from the formalism $\mathcal{L}$ by slightly enriching the logical base of the latter. $\mathcal{L}^+$ is not a formalism of first-order logic in the strict sense in which we introduced this term in §1.5, but it may still be regarded as a variant of a first-order formalism.

2.1. Symbols and expressions of $\mathcal{L}^+$

We obtain the vocabulary of $\mathcal{L}^+$ by including in that of $\mathcal{L}$ five new logical contants: the *absolute sum*, or *absolute addition*, *symbol* $+$, the *complement symbol* $^-$, the *relative product*, or *relative multiplication*, *symbol* $\odot$, the *converse symbol* $^\smile$, and the *second identity symbol* $=$. The first four symbols are (*second-order*) operation symbols; for simplicity we shall almost always call them *operators*; $+$, $\odot$ are binary operators (operators of rank 2), and $^-$, $^\smile$ are unary ones (operators of rank 1). While in $\mathcal{L}$ just two binary predicates, $\boldsymbol{E}$ and $\mathring{\mathbf{1}}$, are available, in $\mathcal{L}^+$, using $\boldsymbol{E}$, $\mathring{\mathbf{1}}$, and the operators, we construct infinitely many symbolic expressions which are also regarded as binary predicates (i.e., as designations of binary relations). The new identity symbol, $=$, is not to be confused with the old one, $\mathring{\mathbf{1}}$; when combined with any two predicates, $=$ yields a formula, and hence it can be regarded as a predicate of higher order. The intuitive meanings of the new constants will be clarified in §2.2.

The variables in $\mathcal{L}^+$ are the same as in $\mathcal{L}$. The membership symbol $\boldsymbol{E}$ continues to be the only nonlogical constant. However, $\mathcal{L}^+$ contains new compound expressions that do not occur in $\mathcal{L}$, namely (compound) predicates.

(i) *$\boldsymbol{E}$ and $\mathring{\mathbf{1}}$ are called atomic predicates (or relation symbols) of $\mathcal{L}^+$. An expression is called a predicate (or relation term) of $\mathcal{L}^+$ if it belongs to every set Γ of expressions of $\mathcal{L}^+$ such that*

(α) $\boldsymbol{E}, \mathring{\mathbf{1}} \in \Gamma$,

(β) $A+B$, A^-, $A\odot B$, $A^\smile \in \Gamma$ *whenever* $A, B \in \Gamma$.

The set of all predicates is denoted by "$\mathbf{\Pi}[\mathcal{L}^+]$" or simply by "$\mathbf{\Pi}$".

In analogy with the definitions of a formula and of derivability, the above definition of a predicate carries with it a method of showing that every predicate in $\mathcal{L}^+$ possesses a certain property; this method of proof may be referred to as *proof by induction on predicates* (cf. remarks after 1.3(ii)).

By the definition of predicates just stated, we have $A \mathbin{\mathbf{+}} B$, A^-, $A \odot B$, $A^\smile \in \Pi$ whenever $A, B \in \Pi$. Thus $\mathbf{+}$, $\odot$ can be treated as binary operations, and $^-$, $^\smile$ as unary operations, on and to predicates. In terms of these we define further operations of the same kind, and also, using $\mathring{1}$, we single out some particular predicates by setting:

(ii) $A \cdot B = (A^- \mathbin{\mathbf{+}} B^-)^-$ *and* $A \oplus B = (A^- \odot B^-)^-$ *for any* $A, B \in \Pi$;[1]

(iii) $\mathring{0} = \mathring{1}^-$, $1 = \mathring{1} \mathbin{\mathbf{+}} \mathring{1}^-$, *and* $0 = (\mathring{1} \mathbin{\mathbf{+}} \mathring{1}^-)^-$.

The operations $\cdot$ and $\oplus$ are referred to as *absolute multiplication* and *relative addition.*

Regarding the use of parentheses in symbolic designations of predicates, we shall adhere to the following rules. Whenever parentheses indicating the order of operations are totally or partially lacking, it will be understood that: (1) each of the unary operations, $^-$ and $^\smile$, has priority over each of the binary operations; (2) each of the two multiplications, $\cdot$ and $\odot$, has priority over each of the two additions, $\mathbf{+}$ and $\oplus$; (3) in case neither (1) nor (2) is applicable, the operations should be performed successively from left to right.

Each of the binary operations mentioned above extends by the usual recursion to an operation on finite sequences of predicates. Thus, for every sequence $\langle A_0, \ldots, A_{n-1} \rangle$ of predicates we set

$$
\begin{aligned}
A_0 \mathbin{\mathbf{+}} \cdots \mathbin{\mathbf{+}} A_{n-1} &= \mathbf{0} && \text{if } n = 0, \\
&= A_0 && \text{if } n = 1, \\
&= (A_0 \mathbin{\mathbf{+}} \cdots \mathbin{\mathbf{+}} A_{n-2}) \mathbin{\mathbf{+}} A_{n-1} && \text{if } n = 2, 3, \ldots .
\end{aligned}
$$

By changing, respectively, in these formulas "$\mathbf{+}$" to "$\cdot$", "$\oplus$", "$\odot$", and "$\mathbf{0}$" to "$\mathbf{1}$", "$\mathring{0}$", "$\mathring{1}$", we obtain recursive extensions of the operations $\cdot$, $\oplus$, $\odot$. Exclusively in the case of $\odot$ we shall use exponential notation; i.e., we set

$$B^n = A_0 \odot \cdots \odot A_{n-1}$$

in case each of the predicates $A_0, \ldots, A_{n-1}$ coincides with B.

The expressions xAy and $A = B$, where $x, y \in \Upsilon$ and $A, B \in \Pi$, are the atomic formulas of $\mathcal{L}^+$; arbitrary formulas are constructed from atomic ones just as in the case of $\mathcal{L}$. The notions of free and bound variables, canonical sequence, and closure of a formula extend automatically from $\mathcal{L}$ to $\mathcal{L}^+$. Notice that some sentences of $\mathcal{L}^+$, as opposed to sentences of $\mathcal{L}$, are quantifier-free. For instance, the atomic formulas $A = B$, which are sometimes called *predicate*

[1] *Thus, in opposition to its use in Henkin–Monk–Tarski [**1971**], here the symbol "$\oplus$" does not denote the operation of symmetric difference from Boolean algebra.

equations or simply *equations*, are quantifier-free sentences; in general, quantifier-free sentences are just those formulas which can be obtained from equations by a repeated application of the operations $\rightarrow$ and $\neg$. Inclusions are treated as special kinds of equations: $A \leq B$ is defined by the formula

(iv) $A \leq B \;=\; (A + B = B)$ *for any* $A, B \in \mathbf{\Pi}$.

$\mathbf{\Phi}[\mathcal{L}^+]$ and $\Sigma[\mathcal{L}^+]$ are, respectively, the set of all formulas and that of all sentences of $\mathcal{L}^+$. Usually, however, occurrences of "$\mathcal{L}^+$" in metalogical symbolic expressions will be omitted, and the expressions will be provided instead with a conveniently placed superscript "+"; e.g., we shall write "$\mathbf{\Phi}^+$" instead of "$\mathbf{\Phi}[\mathcal{L}^+]$". (In the case of "$\mathbf{\Pi}[\mathcal{L}^+]$" we have agreed to use simply the abbreviation "$\mathbf{\Pi}$", since no corresponding symbol has been introduced for the language $\mathcal{L}$.)

2.2. Derivability and semantical notions of $\mathcal{L}^+$

The set $\mathbf{\Lambda}^+$ (or $\mathbf{\Lambda}[\mathcal{L}^+]$) of logical axioms of $\mathcal{L}^+$ is constructed in the following way. First, we include in this set all the sentences described in the same way as the logical axioms of $\mathcal{L}$ (cf. §1.3)—with the provision, however, that every reference to formulas, or atomic formulas, occurring in this description is interpreted as a reference to formulas, or atomic formulas, of $\mathcal{L}^+$. Secondly, we supplement the set thus obtained with all the sentences of the following five kinds, where A, B are arbitrary predicates:

(DI) $\forall_{xy}[xA + By \leftrightarrow (xAy \vee xBy)]$,

(DII) $\forall_{xy}(xA^{-}y \leftrightarrow \neg xAy)$,

(DIII) $\forall_{xy}[xA \odot By \leftrightarrow \exists_z(xAz \wedge zBy)]$,

(DIV) $\forall_{xy}(xA^{\smile}y \leftrightarrow yAx)$,

(DV) $A = B \leftrightarrow \forall_{xy}(xAy \leftrightarrow xBy)$.

Thus, (DI)–(DV) are new axiom schemata which have been adjoined to those for $\mathcal{L}$ to form the full set of logical axiom schemata for $\mathcal{L}^+$. Because of their form and content, (DI)–(DV) can be referred to as *possible definitions* or, more precisely, *possible definition schemata* in $\mathcal{L}$ of those logical constants which belong to the vocabulary of $\mathcal{L}^+$ but not to the vocabulary of $\mathcal{L}$. They obviously clarify the intuitive meanings of those constants. In fact, we see from (DI)–(DIV) that the operators $+$, $\odot$, $^{-}$, and $^{\smile}$ are intended to denote, respectively, certain familiar operations on and to binary relations: the formation of unions (absolute sums) $R \cup S$ and relative products $R|S$ of pairs of relations R, S, as well as the formation of complements $\sim R$ and converses R^{-1} of single relations R. (The complement $\sim R$ is usually taken with respect to some fixed unit relation $U \times U$.) Similarly, by (DV) the new identity symbol $=$ is intended to denote the relation of (extensional) identity between binary relations.

As a consequence of the above observations the intuitive meanings of the operations on predicates and of the special predicates defined in 2.1(ii),(iii) also become clear. We see, e.g., that $A \cdot B$ denotes the intersection (absolute product)

$R \cap S$ of the relations R and S denoted respectively by A and B, and that $\mathring{\mathbf{0}}$ denotes the diversity relation Di, usually restricted to elements of a fixed set U, i.e., the relation $(U \times U) \cap Di$.

The relation of derivability between a set $\Psi \subseteq \Sigma^+$ and a sentence $X \in \Sigma^+$ is expressed symbolically by "$\Psi \vdash^+ X$", and is defined for $\mathcal{L}^+$ in terms of logical axioms just as this was done for $\mathcal{L}$ in 1.3(ii). The notions and results discussed in §1.3, including the deduction theorem, extend automatically from $\mathcal{L}$ to $\mathcal{L}^+$. The theory $\boldsymbol{\Theta\eta}^+\varnothing$, the logic of $\mathcal{L}^+$, may be called the *extended predicate logic of one binary relation.*

We shall state a few interesting logical equivalences which hold between quantifier-free sentences of $\mathcal{L}^+$ (where A, B are assumed to be arbitrary predicates in $\mathbf{\Pi}$).

(i) $(A = B) \equiv^+ (A \cdot B + A^- \cdot B^- = \mathbf{1})$.

(ii) $(A = \mathbf{1} \rightarrow B = \mathbf{1}) \equiv^+ (\mathbf{1} \odot A^- \odot \mathbf{1} + B = \mathbf{1})$.

(iii) $(\neg A = \mathbf{1}) \equiv^+ (\mathbf{1} \odot A^- \odot \mathbf{1} = \mathbf{1})$.

(iv) $(A = \mathbf{1} \vee B = \mathbf{1}) \equiv^+ (A \oplus \mathbf{0} \oplus B = \mathbf{1})$.

(v) $(A = \mathbf{1} \wedge B = \mathbf{1}) \equiv^+ (A \cdot B = \mathbf{1})$.

The proof of (i)–(v) is quite elementary. From (i)–(iii) we easily derive by induction on formulas the following important theorem.

(vi) *For every quantifier-free* $X \in \Sigma^+$ *there is a* $C \in \mathbf{\Pi}$ *such that* $X \equiv^+ (C = \mathbf{1})$.

The results (ii)–(vi) originate with Schröder [**1895**], pp. 150–153.

We now deal briefly with the problem of introducing semantical notions for the language $\mathcal{L}^+$.

As opposed to $\mathcal{L}$, $\mathcal{L}^+$ contains two kinds of compound expressions which are intended to be meaningful: formulas and predicates. As a consequence, in developing the semantics of $\mathcal{L}^+$ we need two basic notions: the notion of *satisfaction* applying to formulas and that of *denotation* applying to predicates. Actually we first introduce the latter notion, and with its help we define the former.

Since the binary predicate $\boldsymbol{E}$ occurs in both $\mathcal{L}$ and $\mathcal{L}^+$ as the only nonlogical constant, the same structures $\langle U, E\rangle$ with $U \neq \varnothing$ and $E \subseteq U \times U$ are used as realizations for both languages. Consider any such structure $\mathfrak{U} = \langle U, E\rangle$. Keeping 2.1(i) in mind, we first define by a recursion on predicates what is meant by saying that a given relation $R \subseteq U \times U$ is denoted (in $\mathfrak{U}$) by a given $A \in \mathbf{\Pi}$. In case A is an atomic predicate, we stipulate that the relation denoted by A is E if $A = \boldsymbol{E}$, and is $U{\upharpoonright}Id$ if $A = \mathring{\mathbf{1}}$. Then, in full agreement with the content of Axioms (DI)–(DIV), we make four recursive stipulations, one for each of the four operators occurring in $\mathcal{L}^+$. Thus, for example, we stipulate that, in case $A = B + C$ $(B, C \in \mathbf{\Pi})$, the relation denoted by A is the union $R \cup S$ of the relations R and S respectively denoted by B and C.

Just as in the case of $\mathcal{L}$, the notion of satisfaction is defined for $\mathcal{L}^+$ by a recursion on formulas. In defining this notion for atomic formulas we make essential use of the notion of denotation. In fact, we stipulate that an infinite sequence $x = \langle x_0, \ldots, x_n, \ldots \rangle$ of elements of U satisfies (in a given realization $\mathfrak{U} = \langle U, E \rangle$) the formula $\boldsymbol{v}_k A \boldsymbol{v}_l$ (with $k, l = 0, 1, 2, \ldots$ and $A \in \Pi$) iff the relation R denoted by A holds between x_k and x_l, that is, in set-theoretical notation, iff $\langle x_k, x_l \rangle \in R$. Similarly, x satisfies $A \boldsymbol{=} B$ with $A, B \in \Pi$ iff A and B denote the same relation. The remaining parts of the definition of satisfaction, referring to nonatomic formulas, are routine; cf. Tarski [**1956**], p. 193.

Other semantical notions, such as truth and model, are defined in terms of satisfaction just as in the case of $\mathcal{L}$. The completeness theorem and its consequences stated in §1.4, as well as the subsequent remarks in that section, easily extend to $\mathcal{L}^+$.

2.3. The equipollence of $\mathcal{L}^+$ and $\mathcal{L}$

We have thus completed the description of the formalism $\mathcal{L}^+$. As we have seen, $\mathcal{L}^+$ has been constructed by extending the original formalism $\mathcal{L}$ in two directions: five new logical constants, say C_1–C_5, have been included in the vocabulary of $\mathcal{L}$, and an equal number of new axiom schemata, S_1–S_5 (actually (DI)–(DV)), have been adjoined to the set of logical axiom schemata for $\mathcal{L}$. Because of its form, each schema S_k, $k = 1, \ldots, 5$, can be regarded as a possible definition (schema) of the corresponding constant C_k in $\mathcal{L}$; more specifically, as a possible definition of C_k in terms of logical constants of $\mathcal{L}$. We can sum up the situation by saying that, in a certain sense, $\mathcal{L}^+$ is a definitional extension of $\mathcal{L}$. (For the notion of definitional extension in different but related contexts compare 2.4(xiii) below and also Tarski [**1968**], p. 277.) It is a matter of common belief that when passing from a formalism to its definitional extensions we do not enrich the means of expression and proof of the formalism involved. In the case of semantically complete formalisms this belief appears to be fully justified, and the method by which it can be established for $\mathcal{L}$ and $\mathcal{L}^+$ is routine. (In §3.7 we shall come across some semantically incomplete formalisms for which the belief is, in all likelihood, not justified.[2]) Nevertheless, we shall state here the relevant results in a formal and detailed way, since this discussion will serve as a model and a source of reference for analogous but more involved discussions in later chapters, particularly in Chapter 4.

We begin with some obvious statements to the effect that $\mathcal{L}$ is a *subformalism* of $\mathcal{L}^+$ (in the sense of §1.6), and hence $\mathcal{L}$ is at most as powerful as $\mathcal{L}^+$ in means of expression and proof.

(i) $\Sigma \subseteq \Sigma^+$ *and* $\boldsymbol{\Phi} \subseteq \boldsymbol{\Phi}^+$.

(ii) *For every* $\Psi \subseteq \Sigma$ *and* $X \in \Sigma$, *if* $\Psi \vdash X$, *then* $\Psi \vdash^+ X$.

[2]*This has indeed been shown to be the case; cf. footnote 8*, p. 66.

To show that $\mathcal{L}^+$ is not richer than $\mathcal{L}$ we define recursively an appropriate *translation* (or *elimination*) *mapping* $\boldsymbol{G}$ from $\mathcal{L}^+$ into $\mathcal{L}$.

(iii) *$\boldsymbol{G}$ is the unique function F satisfying the conditions* (α)–(θ) (*where* $x, y \in \boldsymbol{\Upsilon}$, $A, B \in \boldsymbol{\Pi}$, *and* $X, Y \in \boldsymbol{\Phi}^+$):

- (α) $DoF = \boldsymbol{\Phi}^+$;
- (β) $F(x\mathring{\mathbf{1}}y) = x\mathring{\mathbf{1}}y$ *and* $F(x\boldsymbol{E}y) = x\boldsymbol{E}y$;
- (γ) $F(xA\boldsymbol{+}By) = F(xAy) \boldsymbol{\vee} F(xBy)$;
- (δ) $F(xA\boldsymbol{\odot}By) = \boldsymbol{\exists}_z[F(xAz) \boldsymbol{\wedge} F(zBy)]$, *where z is the first variable which is different from x and y*;
- (ε) $F(xA^{-}y) = \boldsymbol{\neg}F(xAy)$ *and* $F(xA^{\smile}y) = F(yAx)$;
- (ς) $F(A \boldsymbol{=} B) = \boldsymbol{\forall}_{xy}[F(xAy) \boldsymbol{\leftrightarrow} F(xBy)]$;
- (η) $F(X \boldsymbol{\rightarrow} Y) = FX \boldsymbol{\rightarrow} FY$ *and* $F(\boldsymbol{\neg}X) = \boldsymbol{\neg}FX$;
- (θ) $F(\boldsymbol{\forall}_x Y) = \boldsymbol{\forall}_x FY$.

In a detailed presentation, the above definition would be preceded, of course, by the proof that there exists a uniquely determined function F satisfying conditions (α)–(θ).

The next theorem states some simple properties of the function $\boldsymbol{G}$. They are easily derived from the definitions of the notions involved by induction on formulas and predicates in $\mathcal{L}^+$ or $\mathcal{L}$.

(iv) (α) *$\boldsymbol{G}$ is a recursive function.*

- (β) $\boldsymbol{G}X = X$ *if* $X \in \boldsymbol{\Phi}$.
- (γ) $\boldsymbol{\Upsilon\phi G}X = \boldsymbol{\Upsilon\phi}X$ *for every* $X \in \boldsymbol{\Phi}^+$.
- (δ) *$\boldsymbol{G}$ maps $\boldsymbol{\Phi}^+$ onto $\boldsymbol{\Phi}$, and Σ^+ onto Σ.*
- (ε) $\boldsymbol{G}X \equiv^+ X$ *for every* $X \in \boldsymbol{\Phi}^+$.
- (ς) $\boldsymbol{G}^*\Psi \equiv^+ \Psi$ *for every* $\Psi \subseteq \Sigma^+$.

In connection with (ς) recall that

$$\boldsymbol{G}^*\Psi \text{ (the } \boldsymbol{G}\text{-image of } \Psi) = \{\boldsymbol{G}X : X \in \Psi\}.$$

The main property of the function $\boldsymbol{G}$ is stated in the following theorem.

(v) *For every* $\Psi \subseteq \Sigma^+$ *and* $X \in \Sigma^+$, *we have* $\Psi \vdash^+ X$ *iff* $\boldsymbol{G}^*\Psi \vdash \boldsymbol{G}X$.

To obtain the implication from left to right in (v) we proceed by induction on sentences derivable in $\mathcal{L}^+$ from a given set Ψ, using the recursive definition of $\vdash^+$; cf. §§2.2, 1.3. More specifically, we set

$$\Omega = \{Y : Y \in \Sigma^+ \text{ and } \boldsymbol{G}^*\Psi \vdash \boldsymbol{G}Y\},$$

and we show that Ω satisfies conditions (α)–(γ) in 1.3(ii) (replacing Σ and $\boldsymbol{\Lambda}$ by Σ^+ and $\boldsymbol{\Lambda}^+$); we conclude that $X \in \Omega$ whenever $\Psi \vdash^+ X$. The implication in the opposite direction follows easily from (ii) and (iv)(ε),(ς).

In informal remarks we shall refer to (v) as the *main mapping theorem* for $\mathcal{L}^+$ and $\mathcal{L}$, and we shall extend this terminology in analogous situations to other pairs of formalisms.

(vi) $G^*\boldsymbol{\Theta\eta}^+\Psi = \boldsymbol{\Theta\eta}G^*\Psi = \boldsymbol{\Theta\eta}^+\Psi \cap \Sigma$ *for every* $\Psi \subseteq \Sigma^+$.

In fact, using (iv)(β) and (v), we easily establish the inclusions:

$$G^*\boldsymbol{\Theta\eta}^+\Psi \subseteq \boldsymbol{\Theta\eta}G^*\Psi \subseteq (\boldsymbol{\Theta\eta}^+\Psi) \cap \Sigma \subseteq G^*\boldsymbol{\Theta\eta}^+\Psi.$$

From the results in (iv) and (v) we shall now derive some simple corollaries which do not explicitly involve the auxiliary function $\boldsymbol{G}$ and are more directly related to the problem of comparing the expressive and deductive powers of the formalisms $\mathcal{L}$ and $\mathcal{L}^+$. This will give us an opportunity to clarify the precise meaning of this problem, thus suggesting a notation which can be applied in similar contexts to other formalisms as well.

A direct consequence of (iv)(δ),(ε) is

(vii) *For every* $X \in \Sigma^+$ *there is a* $Y \in \Sigma$ *such that* $X \equiv^+ Y$.

For reasons which seem intuitively clear we regard (vii) as a formal statement of the fact that the *expressive power of* $\mathcal{L}^+$ *at most equals that of* $\mathcal{L}$. Since, on the other hand, $\mathcal{L}$ is a subformalism of $\mathcal{L}^+$, the expressive power of $\mathcal{L}$ at most equals that of $\mathcal{L}^+$. To express this formally we can use the following statement, which is dual to (vii) and is a trivial consequence of (i).

(viii) *For every* $Y \in \Sigma$ *there is an* $X \in \Sigma^+$ *such that* $X \equiv^+ Y$.

By (vii) and (viii), $\mathcal{L}$ and $\mathcal{L}^+$ have equal expressive powers, or, as we shall say more frequently, are *equipollent in means of expression.* For this reason the conjunction of (vii) and (viii), or even (vii) alone, may be called the *first equipollence theorem.*

As a consequence of (iv)(β) and (v) we obtain at once

(ix) *For every* $\Psi \subseteq \Sigma$ *and* $X \in \Sigma$, *if* $\Psi \vdash^+ X$, *then* $\Psi \vdash X$.

We regard (ix) as a formal expression of the fact that the *deductive power of* $\mathcal{L}^+$ *at most equals that of* $\mathcal{L}$. The converse of (ix), i.e., the fact that the deductive power of $\mathcal{L}$ at most equals that of $\mathcal{L}^+$, is stated formally in (ii), and thus is simply one of the two conditions characterizing $\mathcal{L}$ as a subformalism of $\mathcal{L}^+$. Hence (ix) and (ii) express together the fact that $\mathcal{L}$ and $\mathcal{L}^+$ are *equipollent in means of proof.* For this reason the conjunction of (ix) and (ii), or even (ix) alone, may be called the *second equipollence theorem.*

(x) $\boldsymbol{\Theta\eta}\Psi = \boldsymbol{\Theta\eta}^+\Psi \cap \Sigma$ *for every* $\Psi \subseteq \Sigma$.

Indeed this is just another way of expressing the fact that $\mathcal{L}$ and $\mathcal{L}^+$ are equipollent in means of proof.

(xi) *If* Φ *is a theory in* $\mathcal{L}^+$, *then* $\Phi \cap \Sigma$ *is a theory in* $\mathcal{L}$, *and* $\Phi = \boldsymbol{\Theta\eta}^+(\Phi \cap \Sigma)$. *Moreover,* Φ *will satisfy one of the following six conditions:*

(α) *consistency,*
(β) *completeness,*

(γ) *finite axiomatizability,*
(δ) *decidability,*
(ε) *essential undecidability,*
(ς) *hereditary undecidability,*

relative to the formalism $\mathcal{L}^+$ iff $\Phi \cap \Sigma$ satisfies the same condition relative to the formalism $\mathcal{L}$.

The proof, using (x) and most of the results in (iv)–(vi), is straightforward.

Consider now an arbitrary system $\mathcal{S}$ (not necessarily a system of set theory) formalized in $\mathcal{L}$. Following the suggestion in §1.6 we treat $\mathcal{S}$ as a relativization of the formalism $\mathcal{L}$, obtained by replacing the notion $\vdash$ of derivability for $\mathcal{L}$ with a stronger notion $\vdash_\Phi$ of derivability relative to some base set $\Phi \subseteq \Sigma$. We correlate with $\mathcal{S}$ an extended system $\mathcal{S}^+$ which is formalized in $\mathcal{L}^+$ and has again the base Φ. If, in particular, $\mathcal{S}$ is an axiomatic system in $\mathcal{L}$ with the axiom set $\mathbf{A\xi}$ (cf. §§1.3 and 1.6), then $\mathcal{S}^+$ is the axiomatic system in $\mathcal{L}^+$ with the same axiom set, i.e., $\mathbf{A\xi}^+ = \mathbf{A\xi}$. Obviously the equipollence theorems (vii)–(ix) and (ii) continue to hold if the relations $\equiv^+$, $\vdash^+$, and $\vdash$ involved in them are relativized to $\mathcal{S}^+$ or $\mathcal{S}$, i.e., are replaced by $\equiv^+_\Phi$, $\vdash^+_\Phi$, and $\vdash_\Phi$ respectively. In this sense we say that systems $\mathcal{S}$ and $\mathcal{S}^+$ are equipollent in means of expression and proof.

The second equipollence theorem suggests the possibility of some simplification in our symbolism. It follows indeed from (ii) and (ix) that the formulas $\Psi \vdash X$ and $\Psi \vdash^+ X$ are equivalent whenever the first of them has a well-defined meaning, i.e., whenever $\Psi \subseteq \Sigma$ and $X \in \Sigma$. Hence we could refrain from using the symbol "$\vdash$" altogether and replace it everywhere by "$\vdash^+$". We prefer, however, partly for typographical reasons, to change the symbolism in the opposite direction: we extend the use of "$\vdash$" so as to cover all the situations in which we have employed "$\vdash^+$" until now. This includes, in particular, such contexts as "$\Psi \vdash^+_\Phi X$". We also make an analogous stipulation concerning "$\equiv$" and "$\equiv^+$". We may, of course, return to the original symbolism in some exceptional cases, so as to enhance the clarity of the text.

Notice that $\mathbf{\Theta\eta}\Psi$ cannot be identified with $\mathbf{\Theta\eta}^+\Psi$, even if $\Psi \subseteq \Sigma$.

2.4. The equipollence of a system with an extension

As we have said before, we are not interested in developing a general theory of formalisms. However, the notion of equipollence between two formalisms and, more generally, between two systems, is of such fundamental importance for our work that it seems proper to state some observations regarding this notion in a general and precise way. This will be the task of the remaining part of the present chapter.

The systems with which we shall concern ourselves are systems developed in interpreted formalisms in the sense of §1.6 (although most of our observations will apply to uninterpreted formalisms as well). Furthermore, the formalisms in which these systems are developed, and hence the systems themselves, satisfy postulates (FI)–(FV) in §1.6 and are semantically sound. We assume that the notion of recursiveness and related notions such as recursive enumerability have been appropriately extended to sets of sentences, and to relations between and operations on these sentences. In particular, we assume that in each formalism the set of all sentences is recursive. We do not assume, in general, that the finitary part, or even the singleton part, of the derivability relation of a formalism is recursively enumerable (cf. §1.3, where this terminology is introduced). Whenever this assumption is essential, it will be explicitly stated. However, it is worth pointing out that this assumption does indeed hold for almost all formalisms discussed in this book.

In this section we consider an arbitrary system

$$\mathcal{S}^{(1)} = \langle \Sigma^{(1)}, \vdash^{(1)}, \mathsf{RE}^{(1)}, \mathsf{MO}^{(1)} \rangle$$

and some extension of it,

$$\mathcal{S}^{(2)} = \langle \Sigma^{(2)}, \vdash^{(2)}, \mathsf{RE}^{(2)}, \mathsf{MO}^{(2)} \rangle$$

(cf. §1.6). We define:

(i) (α) *$\mathcal{S}^{(1)}$ and $\mathcal{S}^{(2)}$ are said to be equipollent in means of expression if for every $X \in \Sigma^{(2)}$ there is a $Y \in \Sigma^{(1)}$ such that $X \equiv^{(2)} Y$, and also for every $Y \in \Sigma^{(1)}$ there is an $X \in \Sigma^{(2)}$ such that $X \equiv^{(2)} Y$.*

(β) *$\mathcal{S}^{(1)}$ and $\mathcal{S}^{(2)}$ are said to be equipollent in means of proof provided, for every $\Psi \subseteq \Sigma^{(1)}$ and $X \in \Sigma^{(1)}$, we have $\Psi \vdash^{(2)} X$ iff $\Psi \vdash^{(1)} X$.*

(γ) *$\mathcal{S}^{(1)}$ and $\mathcal{S}^{(2)}$ are simply called equipollent if they are equipollent in means of both expression and proof.*

In view of the fact that $\mathcal{S}^{(1)}$ is a subsystem of $\mathcal{S}^{(2)}$ we can easily simplify the above definitions by noticing that

(ii) (α) *$\mathcal{S}^{(1)}$ is equipollent with $\mathcal{S}^{(2)}$ in means of expression if for every $X \in \Sigma^{(2)}$ there is a $Y \in \Sigma^{(1)}$ with $X \equiv^{(2)} Y$.*

(β) *$\mathcal{S}^{(1)}$ is equipollent with $\mathcal{S}^{(2)}$ in means of proof provided, for every $\Psi \subseteq \Sigma^{(1)}$ and $X \in \Sigma^{(1)}$, that $\Psi \vdash^{(2)} X$ implies $\Psi \vdash^{(1)} X$.*

The symbol "$\equiv^{(2)}$" in (i)(α) may be interpreted either syntactically or semantically. In case $\mathcal{S}^{(2)}$ is semantically complete (e.g., when $\mathcal{S}^{(2)} = \mathcal{L}^+$) these two interpretations obviously coincide. We shall frequently deal, however, with systems which are not semantically complete, and in which, therefore, the two interpretations cannot be identified. Thus we have to distinguish between syntactical and semantical equipollence in means of expression. Since $\mathcal{S}^{(2)}$ is assumed to be semantically sound, the semantical notion of equipollence is, in general, weaker than the syntactical one. One might be inclined to regard the semantical notion as more proper. In this work, however, the discussions will be based upon the syntactical interpretation of the equivalence symbol, unless explicitly stated otherwise. In many cases we shall use a semantical argument, but we shall do so only if it can be replaced by a syntactical one.

It may seem dubious whether Definition (i)(α) (either in its syntactical or semantical form) really captures the intuitive content of equipollence in means of expression. We shall try to explain what we mean, using as an example the formal languages $\mathcal{L}$ and $\mathcal{L}^+$. As we know, $\mathcal{L}$ and $\mathcal{L}^+$ satisfy (i)(α) (with $\mathcal{S}^{(1)} = \mathcal{L}$ and $\mathcal{S}^{(2)} = \mathcal{L}^+$); thus one could say that every statement which can be formulated in one of these two languages can be formulated in the other as well. However, this does not seem to imply that every notion expressible in one of these languages is expressible in the other, e.g., that every binary relation definable in $\mathcal{L}^+$ is also definable in $\mathcal{L}$. If, nevertheless, we feel that $\mathcal{L}$ and $\mathcal{L}^+$ *are* equipollent in means of expression, it is due to the fact that, in application to these two languages, (i)(α) continues to hold if X and Y are assumed to be, not necessarily sentences, but arbitrary formulas with the same free variables; this follows immediately from 2.3(iv)(γ),(ε). Extending these observations to arbitrary systems, we may infer that, to capture better the intuitive content of equipollence in means of expression, we would have to introduce a more elaborate concept of formal language, including in the list of fundamental components such notions as formulas, free variables, etc. We do not introduce such a concept, if only because of the fundamental role which is played in this book by formalisms without variables (and hence without formulas that are not sentences).

Furthermore, it may also seem dubious whether (i)(β) by itself really captures the intuitive content of equipollence in means of proof. For example, set $\mathcal{S}^{(2)} = \mathcal{L}$, let S be some fixed logically provable sentence of $\mathcal{L}$, and take $\mathcal{S}^{(1)}$ to be the subsystem of $\mathcal{L}$ whose only sentence is S. Thus $\Sigma^{(1)} = \{S\}$, and $\Psi \vdash^{(1)} X$ iff $\Psi \subseteq \Sigma^{(1)}$ (i.e., $\Psi = \varnothing$ or $\Psi = \{S\}$) and $X = S$; also $\mathsf{RE}^{(1)} = \mathsf{RE}[\mathcal{L}]$ and $\mathsf{MO}^{(1)}S = \mathsf{MO}S[\mathcal{L}]$. $\mathcal{S}^{(2)}$ is clearly an extension of $\mathcal{S}^{(1)}$ that is equipollent with $\mathcal{S}^{(1)}$ in means of proof in the sense of (i)(β). The deductive apparatus of $\mathcal{S}^{(1)}$, however, has none of the power and complexity which is possessed by the deductive apparatus of $\mathcal{S}^{(2)}$. In a situation such as this, where (i)(α) fails, we are inclined to treat (i)(β) merely as a necessary condition for the equipollence of a system $\mathcal{S}^{(2)}$ with one of its subsystems $\mathcal{S}^{(1)}$. We do not know any precise statement that would characterize equipollence in means of proof in a more adequate way. If systems $\mathcal{S}^{(1)}$ and $\mathcal{S}^{(2)}$ are (syntactically) equipollent in means of expression, i.e., satisfy (i)(α), then these doubts concerning the interpretation of (i)(β) seem to dissipate.

In establishing the equipollence of a system $\mathcal{S}^{(2)}$ with one of its subsystems $\mathcal{S}^{(1)}$, we shall usually construct a translation mapping G from $\mathcal{S}^{(2)}$ to $\mathcal{S}^{(1)}$ as was done in §2.3. A precise definition of this notion follows.

(iii) *G is a translation mapping from a system $\mathcal{S}^{(2)}$ to a subsystem $\mathcal{S}^{(1)}$ if*

(α) *G is a recursive function;*

(β) *G maps $\Sigma^{(2)}$ into $\Sigma^{(1)}$;*

(γ) *$GX = X$ for $X \in \Sigma^{(1)}$;*

(δ) *$GX \equiv^{(2)} X$ for $X \in \Sigma^{(2)}$;*

(ε) *for every $\Psi \subseteq \Sigma^{(2)}$ and $X \in \Sigma^{(2)}$, if $\Psi \vdash^{(2)} X$, then $G^*\Psi \vdash^{(1)} GX$.*

In case G satisfies conditions (β)–(ε), *but not necessarily* (α), *we shall refer to G as a generalized translation mapping.*

A useful modification of (iii) is given in the following corollary.

(iv) *For G to be a translation mapping (respectively, a generalized translation mapping) from $\mathcal{S}^{(2)}$ to $\mathcal{S}^{(1)}$, it is necessary and sufficient that G satisfy conditions* (iii)(α),(β),(γ),(ε) *(respectively,* (iii)(β),(γ),(ε)*), as well as the following converse of* (ε):

(ε') *for every $\Psi \subseteq \Sigma^{(2)}$ and $X \in \Sigma^{(2)}$, if $G^*\Psi \vdash^{(1)} GX$, then $\Psi \vdash^{(2)} X$.*

That a translation mapping or a generalized translation mapping G must satisfy (ε') follows from condition (iii)(δ) and the fact that $\mathcal{S}^{(1)}$ is a subsystem of $\mathcal{S}^{(2)}$. Suppose now that G satisfies (iii)(α),(β),(γ),(ε), or just (iii)(β),(γ),(ε), as well as condition (ε') above. For every $X \in \Sigma^{(2)}$ we have $GGX \equiv^{(1)} GX$ by (iii)(β),(γ), so $GX \equiv^{(2)} X$ by (ε'). Thus G satisfies (iii)(δ), as was to be shown.

The next corollary of (iii) shows that we can replace the notion of a translation mapping by a weaker notion.

(v) *If G satisfies conditions* (iii)(α),(β),(δ),(ε) *(respectively,* (iii)(β),(δ),(ε)*), as well as the following condition (trivially weaker than* (γ)*):*

(γ') *$GX \equiv^{(1)} X$ for $X \in \Sigma^{(1)}$,*

then there is a translation mapping (respectively, a generalized translation mapping) $\overline{G}$ from $\mathcal{S}^{(2)}$ to $\mathcal{S}^{(1)}$ which agrees with G on $\Sigma^{(2)} \sim \Sigma^{(1)}$.

Indeed we construct $\overline{G}$ by stipulating that $\overline{G}X = GX$ for $X \in \Sigma^{(2)} \sim \Sigma^{(1)}$ and $\overline{G}X = X$ for $X \in \Sigma^{(1)}$. Clearly $\overline{G}$ satisfies conditions (iii)(α)–(δ), (respectively, (iii)(β)–(δ)). To verify (iii)(ε), suppose that $\Psi \subseteq \Sigma^{(2)}$, $X \in \Sigma^{(2)}$, and $\Psi \vdash^{(2)} X$. Then $G^*\Psi \vdash^{(1)} GX$ by hypothesis. Now using (γ') it is easy to check that $GX \equiv^{(1)} \overline{G}X$ and $G^*\Psi \equiv^{(1)} \overline{G}^*\Psi$. Hence we conclude that $\overline{G}^*\Psi \vdash^{(1)} \overline{G}X$, which completes the proof of (v).

It may be noticed that (v) remains true if we replace in it condition (iii)(δ) by (iv)(ε'). The proof is a straightforward combination of the proofs of (iv) and (v).

The role of the notion of translation mappings in the discussion of equipollence is clarified by the next theorem.

(vi) *Let $\mathcal{S}^{(1)}$ be a subsystem of $\mathcal{S}^{(2)}$. Then the following two conditions are equivalent:*

(α) *$\mathcal{S}^{(1)}$ and $\mathcal{S}^{(2)}$ are equipollent in means of expression and proof;*

(β) *there is a generalized translation mapping from $\mathcal{S}^{(2)}$ to $\mathcal{S}^{(1)}$.*

In case the singleton part of $\vdash^{(2)}$ is recursively enumerable, each of these conditions is equivalent to:

(γ) *there is a translation mapping from $\mathcal{S}^{(2)}$ to $\mathcal{S}^{(1)}$.*

The proof that (β) implies (α) is straighforward and indeed completely analogous to the proof of the corresponding result for $\mathcal{L}$ and $\mathcal{L}^+$ given in 2.3(ii),(vii)–(ix).

To prove that (α) implies (β) we assume that the members of $\Sigma^{(1)}$ are arranged in a simple infinite sequence $Y_0, \ldots, Y_n, \ldots$. We define a function G from $\Sigma^{(2)}$ to $\Sigma^{(1)}$ as follows: $GX = X$ for $X \in \Sigma^{(1)}$ and $GX = Y_n$ for $X \in \Sigma^{(2)} \sim \Sigma^{(1)}$, where n is the smallest natural number such that $X \equiv^{(2)} Y_n$; that such an n exists follows from the equipollence of $\mathcal{S}^{(1)}$ and $\mathcal{S}^{(2)}$ in means of expression. Clearly G satisfies (iii)(β)–(δ). To verify part (ε) of (iii) suppose that $\Psi \subseteq \Sigma^{(2)}$, $X \in \Sigma^{(2)}$, and $\Psi \vdash^{(2)} X$. Since G satisfies (iii)(δ) we obtain $G^*\Psi \vdash^{(2)} GX$. From (i)(β) and (iii)(β) we conclude that $G^*\Psi \vdash^{(1)} GX$, as was to be shown.

The proof that (α) implies (γ) under the assumption that the singleton part of $\vdash^{(2)}$ is recursively enumerable proceeds similarly to the proof that (α) implies (β), only the construction of G is more involved. We first arrange the singleton part of $\vdash^{(2)}$ in a recursive sequence $\langle U_0, V_0\rangle, \ldots, \langle U_n, V_n\rangle, \ldots$. We then set $GX = X$ in case $X \in \Sigma^{(1)}$, and $GX = Y$ in case the following three conditions are satisfied:

(1) $X \in \Sigma^{(2)} \sim \Sigma^{(1)}$ and $Y \in \Sigma^{(1)}$;

(2) $\langle X, Y\rangle$ and $\langle Y, X\rangle$ both occur in our recursive sequence;

(3) if $Z \in \Sigma^{(1)}$, then $\langle X, Z\rangle$ and $\langle Z, X\rangle$ do not both occur before either $\langle X, Y\rangle$ or $\langle Y, X\rangle$ in our recursive sequence.

One can show without great difficulty that G is a well-defined recursive function. The verification of properties (iii)(β)–(ε) is straightforward.

Since (γ) clearly implies (β), and hence (α), this completes the proof of our theorem.

In concrete cases it often happens that the construction of translation mappings considerably simplifies the proof of the equipollence of a system $\mathcal{S}^{(2)}$ with a subsystem $\mathcal{S}^{(1)}$. In practice the set $\Sigma^{(2)}$ and the relation $\vdash^{(2)}$ are defined recursively, or are easily reduced to notions that are so defined (for example, sentences in $\mathcal{L}^+$ are characterized as special cases of formulas); this leads to an inductive procedure for proving facts about sentences and about derivability. It enables us, for example, to establish the hypothesis of (ii)(β)—and hence also equipollence in means of proof—by induction on the recursive definition of derivability; no such procedure seems to be available for a direct proof of this equipollence. Moreover, if $\mathcal{S}^{(2)}$ is semantically complete (as will most often be the case in the sequel), the hypothesis of (ii)(α)—and hence also equipollence in means of expression—may be established by a fairly simple semantic argument using induction on the recursive definition upon which the notion of a sentence is based. The efficiency of the method based on translation mappings will be confirmed by some developments in later chapters.

Several theorems in §2.3 (in particular 2.3(x),(xi)) can now be immediately generalized from the formalism $\mathcal{L}$ and its equipollent extension $\mathcal{L}^+$ to an arbitrary system $\mathcal{S}^{(1)}$ and any one of its equipollent extensions $\mathcal{S}^{(2)}$. As an example we formulate here in (vii) and (viii) the generalization of 2.3(xi).

(vii) *Let $\mathcal{S}^{(1)}$ be any system, $\mathcal{S}^{(2)}$ any equipollent extension of it, and Φ a theory in $\mathcal{S}^{(2)}$. Then $\Phi \cap \Sigma^{(1)}$ is a theory in $\mathcal{S}^{(1)}$, and $\Phi = \mathbf{\Theta}\boldsymbol{\eta}^{(2)}(\Phi \cap \Sigma^{(1)})$.*

(viii) *Under the hypothesis of* (vii), *if Φ satisfies one of the three conditions:* (α) *consistency,* (β) *completeness,* (γ) *finite axiomatizability, relative to the system $\mathcal{S}^{(2)}$, then $\Phi \cap \Sigma^{(1)}$ satisfies the same condition relative to the system $\mathcal{S}^{(1)}$, and conversely. The same applies to the following three conditions:* (δ) *decidability,* (ε) *essential undecidability,* (ς) *hereditary undecidability, provided there is a translation mapping of $\mathcal{S}^{(2)}$ to $\mathcal{S}^{(1)}$.*

In connection with the last part of (viii), notice that, by (vi), the hypotheses of (vii) ensure that there certainly will be a translation mapping from $\mathcal{S}^{(2)}$ to $\mathcal{S}^{(1)}$ in case the singleton part of $\vdash^{(2)}$ is recursively enumerable.

Theorems (vii) and (viii) imply there is a one-one correspondence between the theories of $\mathcal{S}^{(2)}$ and those of $\mathcal{S}^{(1)}$ which preserves various important properties of theories, among them (α)–(ς) (as well as many others not involved in our discussion). This correspondence F is determined by the stipulation:

$$F\Phi = \Phi \cap \Sigma^{(1)} \qquad \text{for every theory } \Phi \text{ in } \mathcal{S}^{(2)};$$

its inverse F^{-1} is defined by

$$F^{-1}\Phi = \mathbf{\Theta}\boldsymbol{\eta}^{(2)}\Phi \text{ for every theory } \Phi \text{ in } \mathcal{S}^{(1)}.$$

A basic property of the equipollence relation between a system and its extension is its transitivity whose exhaustive formulation is given in the following theorem, due to Givant.

(ix) *Let $\mathcal{S}^{(1)}$, $\mathcal{S}^{(2)}$, $\mathcal{S}^{(3)}$ be three systems such that $\mathcal{S}^{(1)}$ is a subsystem of $\mathcal{S}^{(2)}$, and $\mathcal{S}^{(2)}$ of $\mathcal{S}^{(3)}$. If any one of these systems is equipollent with the other two, then those other two are also equipollent.*

In proving this we restrict ourselves to the case when $\mathcal{S}^{(1)}$ is equipollent to $\mathcal{S}^{(2)}$ and $\mathcal{S}^{(3)}$, since the remaining cases are trivial. By (vi) there are generalized translation mappings G and H from $\mathcal{S}^{(3)}$ to $\mathcal{S}^{(1)}$ and from $\mathcal{S}^{(2)}$ to $\mathcal{S}^{(1)}$ respectively. We shall show that G satisfies conditions (β), (δ), (ε) in (iii) and condition (γ') in (v), with $\mathcal{S}^{(1)}$ and $\mathcal{S}^{(2)}$ replaced everywhere by $\mathcal{S}^{(2)}$ and $\mathcal{S}^{(3)}$ respectively. Indeed, the first three of these conditions are immediate consequences of the fact that G is a generalized translation mapping from $\mathcal{S}^{(3)}$ to $\mathcal{S}^{(1)}$. To verify (v)(γ') consider any $X \in \Sigma^{(2)}$. Then

(1) $\quad HX \equiv^{(2)} X \quad$ since (iii)(δ) holds with G replaced by H;

(2) $\quad HX \equiv^{(3)} X \quad$ by (1), since $\mathcal{S}^{(2)}$ is a subsystem of $\mathcal{S}^{(3)}$;

(3) $GHX \equiv^{(1)} GX$ by (2), since G satisfies (iii)(ε) (with $\Sigma^{(2)}$ replaced by $\Sigma^{(3)}$ and $\vdash^{(2)}$ by $\vdash^{(3)}$);

(4) $HX \equiv^{(1)} GX$ by (3), since $HX \in \Sigma^{(1)}$ and G satisfies (iii)(γ);

(5) $HX \equiv^{(2)} GX$ by (4), since $\mathcal{S}^{(1)}$ is a subsystem of $\mathcal{S}^{(2)}$;

(6) $GX \equiv^{(2)} X$ by (1), (5).

We can now apply Theorem (v) to conclude that the mapping K defined by the stipulation

$$KX = GX \quad \text{for } X \in \Sigma^{(3)} \sim \Sigma^{(2)} \quad \text{and} \quad KX = X \quad \text{for } X \in \Sigma^{(2)}$$

is a generalized translation mapping from $\mathcal{S}^{(3)}$ to $\mathcal{S}^{(2)}$. In view of (vi) this completes the proof of the theorem.

Suppose now that $\mathcal{F}^{(2)}$ is a formalism, $\mathcal{F}^{(1)}$ an equipollent subformalism of $\mathcal{F}^{(2)}$, and G a generalized translation mapping from $\mathcal{F}^{(2)}$ to $\mathcal{F}^{(1)}$. Given any system $\mathcal{S}^{(1)}$ in $\mathcal{F}^{(1)}$ with base Φ we can construct an equipollent extension $\mathcal{S}^{(2)}$ in $\mathcal{F}^{(2)}$ with the same base Φ; cf. remarks at the end of §2.3 concerning systems in $\mathcal{L}$ and $\mathcal{L}^+$. Conversely, given any system $\mathcal{S}^{(2)}$ in $\mathcal{F}^{(2)}$ with base Ψ we can construct an equipollent subsystem $\mathcal{S}^{(1)}$ in $\mathcal{F}^{(1)}$ with base $G^*\Psi$. Notice that $G^*\Psi$ can also serve as a base for $\mathcal{S}^{(2)}$. Thus, as we can conclude, the equipollence of a formalism $\mathcal{F}^{(2)}$ with one of its subformalisms $\mathcal{F}^{(1)}$ leads naturally to a one-one correspondence F between systems in $\mathcal{F}^{(2)}$ and systems in $\mathcal{F}^{(1)}$ such that, for every system $\mathcal{S}$ in $\mathcal{F}^{(2)}$, $F\mathcal{S}$ is an equipollent subsystem of $\mathcal{S}$; moreover, $\mathcal{S}$ and $F\mathcal{S}$ share a common base; see the remarks after Theorem (viii) for a closely related correspondence between theories in $\mathcal{F}^{(2)}$ and $\mathcal{F}^{(1)}$. On the other hand, as we shall see later on in this work, a system $\mathcal{S}$ may be equipollent with one of its subsystems even though the underlying formalisms are not equipollent.

We now wish to make some remarks concerning the notion of equipollence in application to systems developed in first-order formalisms. We recall that in Chapter 1 we introduced various metalogical concepts applying to the formalism $\mathcal{L}$ which can obviously be extended to arbitrary formalisms of predicate logic. Many of these are definable in terms of the four fundamental components of a formalism listed in §1.6. Among those which are not so definable, but which will be relevant for our remarks, are the concepts of a variable, a formula, a variable occurring free in a formula, and the equivalence of two formulas relative to a fixed set of sentences. The presence of these new concepts leads naturally to some stronger variants of the notions of equipollence in means of expression and of translation mapping, which seem more appropriate for predicate logic.

For the duration of the present section $\mathcal{S}^{(1)}$ and $\mathcal{S}^{(2)}$ shall denote systems developed respectively in formalisms $\mathcal{F}^{(1)}$ and $\mathcal{F}^{(2)}$ of predicate logic. In view of the semantical completeness theorem, a necessary and sufficient condition for $\mathcal{F}^{(1)}$ to be a subformalism of $\mathcal{F}^{(2)}$ (in the sense of §1.6) is that the set of nonlogical constants of $\mathcal{F}^{(1)}$ be included in that of $\mathcal{F}^{(2)}$. The system $\mathcal{S}^{(1)}$ will be a subsystem of $\mathcal{S}^{(2)}$ (in the sense of 1.6(v)) iff the theory of $\mathcal{S}^{(1)}$ (i.e., the set of sentences derivable in $\mathcal{F}^{(1)}$ from a base of $\mathcal{S}^{(1)}$) is included in the theory of $\mathcal{S}^{(2)}$. Notice that, if $\mathcal{S}^{(1)}$ is a subsystem of $\mathcal{S}^{(2)}$, then not only is every sentence of $\mathcal{S}^{(1)}$ a sentence of $\mathcal{S}^{(2)}$, but also every formula of $\mathcal{S}^{(1)}$ is a formula of $\mathcal{S}^{(2)}$.

(x) *Let $\mathcal{S}^{(1)}$ be a subsystem of $\mathcal{S}^{(2)}$. Then $\mathcal{S}^{(1)}$ and $\mathcal{S}^{(2)}$ are said to be strongly equipollent in means of expression if, for every $X \in \mathbf{\Phi}^{(2)}$, there is a $Y \in \mathbf{\Phi}^{(2)}$ such that $\Upsilon\phi X = \Upsilon\phi Y$ and $X \equiv^{(2)} Y$.*

There seems to be no need, and indeed no possibility, of an analogous strengthening of the notion of equipollence in means of proof. When saying that $\mathcal{S}^{(1)}$ and $\mathcal{S}^{(2)}$ are strongly equipollent we mean that they are strongly equipollent in means of expression, and equipollent, in the original sense, in means of proof.

(xi) *G is a strong translation mapping from a system $\mathcal{S}^{(2)}$ to a subsystem $\mathcal{S}^{(1)}$ if*

- (α) *G is a recursive function;*
- (β) *G maps $\mathbf{\Phi}^{(2)}$ into $\mathbf{\Phi}^{(1)}$, and $\Upsilon\phi GX = \Upsilon\phi X$ for $X \in \mathbf{\Phi}^{(2)}$;*
- (γ) *$GX = X$ for $X \in \mathbf{\Phi}^{(1)}$;*
- (δ) *$GX \equiv^{(2)} X$ for $X \in \mathbf{\Phi}^{(2)}$;*
- (ε) *for every $\Psi \subseteq \Sigma^{(2)}$ and $X \in \Sigma^{(2)}$, if $\Psi \vdash^{(2)} X$, then $G^*\Psi \vdash^{(1)} GX$.*

In case G satisfies conditions (β)–(ε), but not necessarily (α), we shall refer to G as a generalized strong translation mapping.

A notion that plays an important role in the study of systems developed in first-order formalisms, especially in the discussion of their equipollence, is that of a definitional extension. Before introducing this notion, we clarify what we mean by a possible definition of a nonlogical constant C in a system $\mathcal{S}$ to which C does not belong.

(xii) *Let C be a (first-order) nonlogical constant that does not occur among the symbols of a (first-order) system $\mathcal{S}$.*

(α) *If C is a relation symbol of rank n, then a possible definition of C in $\mathcal{S}$ is any sentence of the form*

$$\forall_{x_0 \cdots x_{n-1}}(C(x_0, \ldots, x_{n-1}) \leftrightarrow X),$$

where X is a formula of $\mathcal{S}$, and $x_0, \ldots, x_{n-1}$ are the distinct free variables of X.

(β) *If C is an operation symbol of rank n, then a possible definition of C in $\mathcal{S}$ is any sentence of the form*

$$\forall_{x_0 \cdots x_n}(C(x_0, \ldots, x_{n-1}) \mathring{\mathbf{1}} x_n \leftrightarrow X),$$

where X is a formula of $\mathcal{S}$, $x_0, \ldots, x_n$ are the distinct free variables of X, and the sentence

$$\forall_{x_0 \cdots x_{n-1}} \exists_{x_n} (X \wedge \forall_{x_{n+1}} (X[x_n/x_{n+1}] \rightarrow x_n \mathring{\mathbf{1}} x_{n+1}))$$

is derivable in $\mathcal{S}$ whenever x_{n+1} is a variable not occurring in X.

(Recall from §1.2 that $X[x_n/x_{n+1}]$ is the formula obtained by substituting x_{n+1} for x_n in X.)

(xiii) *A system $\mathcal{S}^{(2)}$ is said to be a (first-order) definitional extension of system $\mathcal{S}^{(1)}$ if the set $\Xi^{(2)}$ of nonlogical constants of $\mathcal{S}^{(2)}$ includes the set $\Xi^{(1)}$ of nonlogical constants of $\mathcal{S}^{(1)}$, and, moreover, for each C in $\Xi^{(2)} \sim \Xi^{(1)}$ there is a possible definition, Y_C, of C in $\mathcal{S}^{(1)}$ such that the set*

$$\Theta \cup \{Y_C : C \in \Xi^{(2)} \sim \Xi^{(1)}\}$$

is a base for $\mathcal{S}^{(2)}$, where Θ is the theory of $\mathcal{S}^{(1)}$.

It is easy to verify that in (xiii) we may equivalently replace Θ by any base for $\mathcal{S}^{(1)}$.

The strong equipollence of a first-order system with one of its (first-order) subsystems can be characterized in terms of the notion of a definitional extension, as the next theorem shows.

(xiv) *Let $\mathcal{S}^{(1)}$ be a subsystem of $\mathcal{S}^{(2)}$. Then the following three conditions are equivalent:*

(α) *$\mathcal{S}^{(1)}$ and $\mathcal{S}^{(2)}$ are strongly equipollent;*

(β) *$\mathcal{S}^{(2)}$ is a definitional extension of $\mathcal{S}^{(1)}$;*

(γ) *there is a generalized strong translation mapping from $\mathcal{S}^{(2)}$ to $\mathcal{S}^{(1)}$.*

Moreover, if $\mathcal{S}^{(2)}$ has a recursive base, then each of the above conditions is equivalent to the following one:

(δ) *there is a strong translation mapping from $\mathcal{S}^{(2)}$ to $\mathcal{S}^{(1)}$.*

To prove that (β) implies (γ) one defines by recursion on formulas a generalized strong translation mapping following the lines of our construction of the strong translation mapping $\boldsymbol{G}$ from $\mathcal{L}^+$ to $\mathcal{L}$; see §2.3. The details of the construction and the proof that the resulting mapping has the desired properties are well known, and we leave them to the reader. Also the proof that (γ) implies (α) is straightforward, and in fact analogous to the proof of the corresponding result for $\mathcal{L}$ and $\mathcal{L}^+$; see §2.3.

To prove that (α) implies (β) suppose, for $i = 1, 2$, that the system $\mathcal{S}^{(i)}$ is developed in the formalism $\mathcal{F}^{(i)}$ of predicate logic with $\Xi^{(i)}$ as its set of nonlogical constants, and let $\Theta^{(i)}$ be the theory of $\mathcal{S}^{(i)}$. Since $\mathcal{S}^{(1)}$ is a subsystem of $\mathcal{S}^{(2)}$ we have $\Xi^{(1)} \subseteq \Xi^{(2)}$. For each C in $\Xi^{(2)} \sim \Xi^{(1)}$ we readily construct a possible definition of C in $\mathcal{S}^{(1)}$. In fact, in view of the strong equipollence of $\mathcal{S}^{(1)}$ and $\mathcal{S}^{(2)}$ in means of expression, we may correlate with C a formula X_C of $\mathcal{S}^{(1)}$ such that one of the following sentences, (1), (2), is derivable in $\mathcal{S}^{(2)}$:

(1) $\forall_{x_0 \cdots x_{n-1}}(C(x_0, \ldots, x_{n-1}) \leftrightarrow X_C)$

in case C is a relation symbol of rank n, and

(2) $\forall_{x_0 \cdots x_n}(C(x_0, \ldots, x_{n-1}) \mathring{\mathbf{I}} x_n \leftrightarrow X_C)$

in case C is an operation symbol of rank n; here the distinct free variables of X_C are just $x_0, \ldots, x_{n-1}$ in the first case, and $x_0, \ldots, x_n$ in the second. Obviously, in the first case, (1) is a possible definition of C in $\mathcal{S}^{(1)}$. In the second case, using the equipollence of $\mathcal{S}^{(1)}$ and $\mathcal{S}^{(2)}$ in means of proof, it is easy to verify that (2) is a possible definition of C in $\mathcal{S}^{(1)}$.

Let Ω be the set of these possible definitions. Thus, we have

(3) $\Omega \subseteq \Theta^{(2)}$.

Moreover, from the fact that $\mathcal{S}^{(1)}$ is a subsystem of $\mathcal{S}^{(2)}$ we get

(4) $\Theta^{(1)} \subseteq \Theta^{(2)}$.

On the basis of the possible definitions in Ω we may construct a mapping H from $\mathbf{\Phi}[\mathcal{F}^{(2)}]$ to $\mathbf{\Phi}[\mathcal{F}^{(1)}]$ with the following properties:

(5) $HX \equiv_\Omega X\ [\mathcal{F}^{(2)}]$ and $\mathbf{\Upsilon}\phi HX = \mathbf{\Upsilon}\phi X$ for every $X \in \mathbf{\Phi}[\mathcal{F}^{(2)}]$.

Now (3) and (5) yield

(6) $\{HX : X \in \Theta^{(2)}\} \subseteq \Theta^{(2)}$.

Since each HX is in $\mathbf{\Phi}[\mathcal{F}^{(1)}]$, and since $\mathcal{S}^{(1)}$ is equipollent with $\mathcal{S}^{(2)}$ in means of proof, we get from (6) that

(7) $\{HX : X \in \Theta^{(2)}\} \subseteq \Theta^{(1)}$.

Therefore, using also (5), we arrive at

(8) $\Theta^{(1)} \cup \Omega \vdash \Theta^{(2)}\ [\mathcal{F}^{(2)}]$.

Together, (3), (4), and (8) show that $\Theta^{(1)} \cup \Omega$ is a base for $\mathcal{S}^{(2)}$, and hence that $\mathcal{S}^{(2)}$ is a definitional extension of $\mathcal{S}^{(1)}$.

It remains to show that (α) and (δ) are equivalent in case $\mathcal{S}^{(2)}$ has a recursive base. The proof is analogous to that of the equivalence of (α) and (γ) in (vi); the details will be omitted.

In various portions of this work we shall want to use the terms "possible definition" and "definitional extension" in application to formalisms other than those of the first order. For example, although $\mathcal{L}$ is a first-order formalism and $\mathcal{L}^+$ is not, we have suggested that $\mathcal{L}^+$ should be regarded as a definitional extension of $\mathcal{L}$, and (DI)–(DV) in §2.2 as possible definitions in $\mathcal{L}$ of the (second-order) constants $+$, $^-$, $\odot$, $^\smile$, and $=$, which belong to $\mathcal{L}^+$ but not to $\mathcal{L}$. In thus extending the usage of the term "possible definition", it seems that we can no longer consider just single sentences as was done in (xii). Indeed, since $\mathcal{L}^+$ contains no variables representing binary relations, we see no way of explicating the intended meaning of a constant such as $\odot$ in terms of the symbols of $\mathcal{L}$ unless we admit schemata as possible definitions. Thus, for example, the letters "A" and "B" in (DIII) are not variables of $\mathcal{L}^+$, but metalogical variables ranging over the predicates of $\mathcal{L}^+$. Each time we take for A and B specific predicates, we get a sentence Y in $\mathcal{L}^+$ which is an instance of (DIII) and which should be regarded as a possible definition of the corresponding predicate $A \odot B$ in terms of A and B. (If we wished to obtain a possible definition of the predicate $A \odot B$ explicitly in terms of a formula of $\mathcal{L}$, it would be necessary to replace the formulas xAz and zBy in Y by the formulas $\boldsymbol{G}(xAz)$ and $\boldsymbol{G}(zBy)$ respectively, where $\boldsymbol{G}$ is the function defined in 2.3(iii).) The intended meaning of $\odot$ is explicated by the (infinite) set of all sentences which together constitute for us the possible definition of the symbol $\odot$ itself. Under this modification of the notion of possible definition, $\mathcal{L}^+$ obviously proves to be a definitional extension of $\mathcal{L}$, and hence we conclude that $\mathcal{L}$ and $\mathcal{L}^+$ are equipollent, just as was shown in §2.3.

In several portions of the subsequent discussion we shall be confronted with a situation analogous to the one with which we have just been concerned. That is, we shall consider a formalism $\mathcal{F}$ and one of its extensions $\mathcal{F}'$, and we shall attempt to show that $\mathcal{F}'$ is a definitional extension of $\mathcal{F}$, and hence equipollent with $\mathcal{F}$ in means of expression and proof. The vocabularies of both $\mathcal{F}$ and $\mathcal{F}'$ are assumed to contain no variables of order higher than 1. The procedure outlined above can be applied with obvious modifications in case $\mathcal{F}$ is a first-order formalism and $\mathcal{F}'$ is obtained by enriching the vocabulary of $\mathcal{F}$ with new symbols that are either operation or relation symbols of second order. In other cases some different modifications may be needed. We do not attempt here to give a precise explication of a notion of possible definition which would be adequate for all situations that may arise, and we hope that the observations in the present section will help the reader to understand the nature of such modifications which will be needed.

The conclusion that a formalism $\mathcal{F}$ is equipollent with any one of its definitional extensions $\mathcal{F}'$ is based upon the tacit assumption that $\mathcal{F}$ has a sufficiently rich logical basis. If this is not the case, then the equipollence of $\mathcal{F}'$ and $\mathcal{F}$ may not be a trivial consequence of the fact that one is a definitional extension of the other, and in fact may not even be true. For a concrete example, the formalisms $\mathcal{L}_3$ and $\mathcal{L}_3^+$ as introduced in §3.7 do not seem to be equipollent in means of proof, even though $\mathcal{L}_3^+$ is a definitional extension of $\mathcal{L}_3$.[3]

---

[3]*Andréka and Németi have recently shown that this is indeed the case; cf. footnote 8*, p. 66.

2.5. The equipollence of two systems relative to a common extension

We are often interested in comparing the expressive and deductive powers of two systems $\mathcal{S}^{(1)}$ and $\mathcal{S}^{(2)}$, neither of which is a subsystem of the other. The problem of clarifying the meanings of these notions is rather involved and we shall not discuss it here in detail.

It appears that, for the purposes of the present work, the most convenient approach is to consider the equipollence of two systems $\mathcal{S}^{(1)}$ and $\mathcal{S}^{(2)}$ only in those situations in which they are treated as subsystems of a given third system $\mathcal{S}^{(3)}$; in other words, we shall not use here the binary relation of equipollence between $\mathcal{S}^{(1)}$ and $\mathcal{S}^{(2)}$, but the ternary relation of the equipollence between $\mathcal{S}^{(1)}$ and $\mathcal{S}^{(2)}$ relative to a system $\mathcal{S}^{(3)}$ which is assumed to be a common extension.

(i) *We say that $\mathcal{S}^{(1)}$ and $\mathcal{S}^{(2)}$ are equipollent relative to $\mathcal{S}^{(3)}$ in means of expression if $\mathcal{S}^{(3)}$ is a common extension of $\mathcal{S}^{(1)}$ and $\mathcal{S}^{(2)}$ that is equipollent with each of them in means of expression (in the sense of* 2.4(i)); *analogously for equipollence in means of proof.*

(ii) *In case $\mathcal{S}^{(1)}$ and $\mathcal{S}^{(2)}$ are equipollent relative to $\mathcal{S}^{(3)}$ in means of both expression and proof, then we say simply that $\mathcal{S}^{(1)}$ and $\mathcal{S}^{(2)}$ are equipollent relative to $\mathcal{S}^{(3)}$, and we refer to $\mathcal{S}^{(3)}$ as a common equipollent extension of $\mathcal{S}^{(1)}$ and $\mathcal{S}^{(2)}$.*

We formulate here some basic properties of our new notion of equipollence, beginning with its connection with the notion of equipollence discussed in the preceding section.

(iii) *Let $\mathcal{S}^{(1)}$ be a subsystem of $\mathcal{S}^{(2)}$. Then the following conditions are equivalent:*

(α) *$\mathcal{S}^{(1)}$ and $\mathcal{S}^{(2)}$ are equipollent in the sense of* 2.4(i);
(β) *$\mathcal{S}^{(1)}$ and $\mathcal{S}^{(2)}$ are equipollent relative to $\mathcal{S}^{(2)}$;*
(γ) *$\mathcal{S}^{(1)}$ and $\mathcal{S}^{(2)}$ are equipollent relative to $\mathcal{S}^{(3)}$ for some formalism $\mathcal{S}^{(3)}$.*

(iv) *The relation of equipollence between two systems relative to a given system $\mathcal{S}^{(4)}$ is an equivalence relation on the class of all subsystems of $\mathcal{S}^{(4)}$. In other words, for arbitrary subsystems $\mathcal{S}^{(1)}$, $\mathcal{S}^{(2)}$, $\mathcal{S}^{(3)}$ of $\mathcal{S}^{(4)}$ we have:*

(α) *$\mathcal{S}^{(1)}$ is equipollent with $\mathcal{S}^{(1)}$ relative to $\mathcal{S}^{(4)}$;*
(β) *if $\mathcal{S}^{(1)}$ is equipollent with $\mathcal{S}^{(2)}$ relative to $\mathcal{S}^{(4)}$, then $\mathcal{S}^{(2)}$ is so equipollent with $\mathcal{S}^{(1)}$ as well;*
(γ) *relative to $\mathcal{S}^{(4)}$, if $\mathcal{S}^{(1)}$ and $\mathcal{S}^{(2)}$ are equipollent and also $\mathcal{S}^{(2)}$ and $\mathcal{S}^{(3)}$, then $\mathcal{S}^{(1)}$ and $\mathcal{S}^{(3)}$ are equipollent as well.*

Theorem (iii) is an immediate consequence of 2.4(ix), while (iv) is trivial.

We can of course pass from the ternary relation of relative equipollence to a binary relation of equipollence by stipulating that two arbitrary systems $\mathcal{S}^{(1)}$ and $\mathcal{S}^{(2)}$ are equipollent iff there is a third system relative to which they are equipollent. By (iii) this binary notion of equipollence is of course an extension of the notion of equipollence defined in 2.4(i), and in fact the latter notion is just the restriction of the former notion to pairs $\mathcal{S}^{(1)}$ and $\mathcal{S}^{(2)}$ such that one of these systems is a subsystem of the other. It should be pointed out that the binary relation of equipollence just introduced exhibits certain properties which are hard to reconcile with the intuitive concept of equipollence. In particular, this relation is not transitive: we can construct first-order systems $\mathcal{T}^{(1)}$, $\mathcal{T}^{(2)}$, $\mathcal{T}^{(3)}$ such that $\mathcal{T}^{(1)}$ and $\mathcal{T}^{(2)}$ are equipollent relative to some first-order system $\mathcal{T}^{(4)}$, while $\mathcal{T}^{(2)}$ and $\mathcal{T}^{(3)}$ are equipollent relative to some first-order system $\mathcal{T}^{(5)}$, and yet $\mathcal{T}^{(1)}$ and $\mathcal{T}^{(3)}$ are not equipollent relative to any system. We construct $\mathcal{T}^{(1)}$ and $\mathcal{T}^{(3)}$ as certain systems formalized in $\mathcal{L}$. As bases for $\mathcal{T}^{(1)}$ and $\mathcal{T}^{(3)}$ we take respectively the sentences

$$S_1 = \forall_{xy}(x\boldsymbol{E}y \leftrightarrow x\mathring{\mathbf{1}}y) \quad \text{and} \quad S_3 = \forall_{xy}(x\boldsymbol{E}y \leftrightarrow \neg x\mathring{\mathbf{1}}y).$$

We construct $\mathcal{T}^{(2)}$ as a system in a formalism obtained from $\mathcal{L}$ by replacing everywhere the symbol $\boldsymbol{E}$ with a binary predicate symbol of a different shape, say $\boldsymbol{E}'$. As a base for $\mathcal{T}^{(2)}$ we take the sentence $S_2 = \forall_{xy}(x\boldsymbol{E}'y \leftrightarrow x\mathring{\mathbf{1}}y)$. Both $\mathcal{T}^{(4)}$ and $\mathcal{T}^{(5)}$ are constructed in the formalism of predicate logic with just two nonlogical constants, $\boldsymbol{E}$ and $\boldsymbol{E}'$. As bases for $\mathcal{T}^{(4)}$ and $\mathcal{T}^{(5)}$ we take respectively the sets $\{S_1, S_2\}$ and $\{S_2, S_3\}$. Obviously $\mathcal{T}^{(4)}$ is a definitional extension of both $\mathcal{T}^{(1)}$ and $\mathcal{T}^{(2)}$, while $\mathcal{T}^{(5)}$ is such an extension of $\mathcal{T}^{(2)}$ and $\mathcal{T}^{(3)}$. But $\mathcal{T}^{(1)}$ and $\mathcal{T}^{(3)}$ do not have a common equipollent extension. Indeed, if a system $\mathcal{S}$ were an equipollent extension of both $\mathcal{T}^{(1)}$ and $\mathcal{T}^{(3)}$, then the sentence S_3 would be logically provable in $\mathcal{S}$, since $\mathcal{T}^{(3)}$ is a subsystem of $\mathcal{S}$. But then S_3 would have to be logically provable in $\mathcal{T}^{(1)}$, since $S_3 \in \Sigma[\mathcal{T}^{(1)}]$ and $\mathcal{S}$ is equipollent with $\mathcal{T}^{(1)}$ in means of proof. Hence $\mathcal{T}^{(1)}$ would be inconsistent, which contradicts the fact that every structure $\langle U, U{\upharpoonright}Id\rangle$ is a model of $\mathcal{T}^{(1)}$.

For the subsequent discussion of the ternary relation of equipollence, we restrict our attention to formalisms of first-order logic and systems developed in these formalisms. Let $\mathcal{S}^{(1)}$, $\mathcal{S}^{(2)}$, and $\mathcal{S}^{(3)}$ be systems developed in formalisms of first-order logic. We define:

(v) (α) *$\mathcal{S}^{(1)}$ and $\mathcal{S}^{(2)}$ are first-order definitionally equivalent, or simply definitionally equivalent, relative to $\mathcal{S}^{(3)}$, if $\mathcal{S}^{(3)}$ is a definitional extension of both $\mathcal{S}^{(1)}$ and $\mathcal{S}^{(2)}$.*

(β) *$\mathcal{S}^{(1)}$ and $\mathcal{S}^{(2)}$ are (first-order) definitionally equivalent if there is a system developed in a formalism of first-order logic which is a common definitional extension of both of them.*

From 2.4(xiv) we at once get the following characterization of the definitional equivalence of two systems relative to a third system.

(vi) *The following conditions are equivalent to each other:*

(α) *$\mathcal{S}^{(1)}$ and $\mathcal{S}^{(2)}$ are definitionally equivalent relative to $\mathcal{S}^{(3)}$;*

(β) *$\mathcal{S}^{(3)}$ is an extension of $\mathcal{S}^{(1)}$ and $\mathcal{S}^{(2)}$ which is strongly equipollent with each of them;*

(γ) *$\mathcal{S}^{(3)}$ is an extension of $\mathcal{S}^{(1)}$ and $\mathcal{S}^{(2)}$, and there are generalized strong translation mappings from $\mathcal{S}^{(3)}$ to $\mathcal{S}^{(1)}$ and to $\mathcal{S}^{(2)}$.*

Moreover, if $\mathcal{S}^{(3)}$ *has a recursive base, then each of the above conditions is equivalent to the following one:*

(δ) $\mathcal{S}^{(3)}$ *is an extension of* $\mathcal{S}^{(1)}$ *and* $\mathcal{S}^{(2)}$, *and there are strong translation mappings from* $\mathcal{S}^{(3)}$ *to* $\mathcal{S}^{(1)}$ *and to* $\mathcal{S}^{(2)}$.

It is immediately seen from (vi) that if $\mathcal{S}^{(1)}$ and $\mathcal{S}^{(2)}$ are definitionally equivalent relative to $\mathcal{S}^{(3)}$, then they are also equipollent (in means of expression and proof) relative to $\mathcal{S}^{(3)}$.

The binary relation of definitional equivalence introduced in (v)(β) may seem to be a proper candidate for a binary relation of equipollence restricted to the domain of first-order systems. It is a strong notion applying to an important class of systems, and may be used to obtain, as the case may be, either positive or negative solutions to the equipollence problem concerning the systems involved. However, it shares an important defect with the binary relation of equipollence discussed in the remarks following (iv): it is not transitive (and hence not an equivalence relation in the domain of first-order systems). In fact, we actually showed above that the first-order systems $\mathcal{T}^{(1)}$ and $\mathcal{T}^{(2)}$ are definitionally equivalent, as are $\mathcal{T}^{(2)}$ and $\mathcal{T}^{(3)}$; however, $\mathcal{T}^{(1)}$ and $\mathcal{T}^{(3)}$ clearly have no common definitional extension.

There is another, weaker notion of definitional equivalence between two systems which is transitive, and hence is an equivalence relation on the class of all systems developed in first-order formalisms. To define this notion, let us agree (temporarily) to call two first-order formalisms $\mathcal{F}$ and $\mathcal{F}'$ *almost identical* if there is a one-one correspondence between the nonlogical symbols of $\mathcal{F}$ and $\mathcal{F}'$ which maps predicates, respectively operation symbols, of $\mathcal{F}$ to predicates, respectively operation symbols, of $\mathcal{F}'$ with the same rank. Such a correspondence extends in a well-determined way to a one-one correspondence F between the expressions of $\mathcal{F}$ and $\mathcal{F}'$; in particular, for every expression X, the expression FX differs from X only in the shape of its nonlogical constants. $\mathcal{S}$ being any system in $\mathcal{F}$ with base Φ, the system $\mathcal{S}'$ in $\mathcal{F}'$ with base $F^*\Phi$ is said to be *almost identical* with $\mathcal{S}$. We can now say that two first-order systems $\mathcal{S}^{(1)}$ and $\mathcal{S}^{(2)}$ are *definitionally equivalent in the wider sense* if there are systems $\mathcal{S}^{(1)\prime}$ and $\mathcal{S}^{(2)\prime}$ which are almost identical with $\mathcal{S}^{(1)}$ and $\mathcal{S}^{(2)}$ respectively, and such that $\mathcal{S}^{(1)\prime}$ and $\mathcal{S}^{(2)\prime}$ have no common nonlogical constants and have a common definitional extension.

The reader may notice that the systems $\mathcal{T}^{(1)}$ and $\mathcal{T}^{(3)}$ constructed above are definitionally equivalent in the wider sense, even though they are not definitionally equivalent in the original sense. There seem to be many interesting problems connected with the notion of definitional equivalence in the wider sense, even if we restrict ourselves to systems developed in our original formalism $\mathcal{L}$. However, this notion will not play a role in our further work.

CHAPTER 3

The Formalism $\mathcal{L}^{\times}$ without Variables and the Problem of Its Equipollence with $\mathcal{L}$

We are now approaching the main task of this work. We shall construct a new formalism, $\mathcal{L}^{\times}$, and study its relationship with the formalisms discussed in the preceding sections. Just as $\mathcal{L}$, the formalism $\mathcal{L}^{\times}$ proves to be a subformalism of $\mathcal{L}^{+}$. The formal structure of $\mathcal{L}^{\times}$ is much simpler than that of $\mathcal{L}$ and $\mathcal{L}^{+}$. In fact, the vocabulary of $\mathcal{L}^{\times}$ contains no variables, no quantifiers, and no sentential connectives. It will be shown that $\mathcal{L}^{\times}$ is poorer in means of expression and proof than either $\mathcal{L}$ or $\mathcal{L}^{+}$; it will also be shown that $\mathcal{L}^{\times}$ is equipollent in means of expression and proof with a formalism $\mathcal{L}_3$ which, roughly speaking, has been obtained from $\mathcal{L}$ by eliminating all but three variables from the vocabulary of the latter while otherwise leaving intact the formal structure of $\mathcal{L}$. On the other hand, we shall see in the next chapter that for each set-theoretical system $\mathcal{S}$ which is formalized in $\mathcal{L}$ and satisfies certain weak conditions (fulfilled by practically all systems known from the literature), an equipollent system $\mathcal{S}^{\times}$ can be constructed in $\mathcal{L}^{\times}$.

3.1. Syntactical and semantical notions of $\mathcal{L}^{\times}$

The vocabulary of $\mathcal{L}^{\times}$ contains just seven different symbols. Six of them are logical constants; these are two identity symbols, $\mathring{1}$ and $=$, and four operators, $+$, $^{-}$, $\odot$, and $^{\smile}$. The membership symbol $\boldsymbol{E}$ is again the only nonlogical constant of our formalism.

From the atomic predicates, $\mathring{1}$ and $\boldsymbol{E}$, we construct arbitrary predicates in the same way as this is done in $\mathcal{L}^{+}$; thus $\boldsymbol{\Pi}[\mathcal{L}^{\times}] = \boldsymbol{\Pi}$. The formulas of $\mathcal{L}^{\times}$ are just those formulas of $\mathcal{L}^{+}$ which contain no variables, quantifiers, or sentential connectives; in other words they are just the equations $A = B$, where $A, B \in \boldsymbol{\Pi}$. Since there are no variables in $\mathcal{L}^{\times}$, the formulas coincide with the sentences. Thus $\boldsymbol{\Sigma}^{\times}$, the set of all sentences in $\mathcal{L}^{\times}$, is the set of all equations $A = B$. (When introducing symbolic notation referring to $\mathcal{L}^{\times}$ we proceed as in the case of $\mathcal{L}^{+}$; cf. the final remarks in §2.1.)

(i) *Let* $X = (A \mathbf{=} B)$ *be a sentence in* $\Sigma^{\times}$. *We set:*

(α) X^{ℓ} *(the left side of* X*)* $= A$;

(β) X^{r} *(the right side of* X*)* $= B$;

(γ) $X^{t} = X^{\ell}$ *if* $X^{r} = \mathbf{1}$; $X^{t} = X^{\ell} \mathbf{\cdot} X^{r} \mathbf{+} X^{\ell -} \mathbf{\cdot} X^{r-}$ *otherwise;*

(δ) X *is called a tautology if* $X^{\ell} = X^{r}$, *i.e., if* A *coincides with* B.

As logical axioms of $\mathcal{L}^{\times}$ we select certain equations which can easily be shown to be logically valid in $\mathcal{L}^{+}$. Their number is infinite, but all of them are particular instances of a few simple axiom schemata.

(ii) S *is a logical axiom of* $\mathcal{L}^{\times}$, *in symbols* $S \in \mathbf{\Lambda}^{\times}$, *iff, for some* $A, B, C \in \mathbf{\Pi}$, S *coincides with one of the following equations* (BI)–(BX):

(BI) $A \mathbf{+} B \mathbf{=} B \mathbf{+} A$,

(BII) $A \mathbf{+} (B \mathbf{+} C) \mathbf{=} (A \mathbf{+} B) \mathbf{+} C$,

(BIII) $(A^{-} \mathbf{+} B)^{-} \mathbf{+} (A^{-} \mathbf{+} B^{-})^{-} \mathbf{=} A$,

(BIV) $A \odot (B \odot C) \mathbf{=} (A \odot B) \odot C$,

(BV) $(A \mathbf{+} B) \odot C \mathbf{=} A \odot C \mathbf{+} B \odot C$,

(BVI) $A \odot \mathring{\mathbf{1}} \mathbf{=} A$,

(BVII) $A^{\smile\smile} \mathbf{=} A$,

(BVIII) $(A \mathbf{+} B)^{\smile} \mathbf{=} A^{\smile} \mathbf{+} B^{\smile}$,

(BIX) $(A \odot B)^{\smile} \mathbf{=} B^{\smile} \odot A^{\smile}$,

(BX) $A^{\smile} \odot (A \odot B)^{-} \mathbf{+} B^{-} \mathbf{=} B^{-}$.

For obvious reasons, in defining derivability in $\mathcal{L}^{\times}$ we cannot follow directly the lines of 1.3(ii) and §2.2. To obtain an appropriate definition, notice that within $\mathcal{L}^{\times}$ we are confronted exclusively with deriving equations from sets of equations. Hence, in the process of derivation we can apply methods familiar from algebraic practice and formally described in equational logic; cf. for instance Tarski [**1968**]. More specifically, we make the following stipulation.

(iii) *For any given* $X \in \Sigma^{\times}$ *and* $\Psi \subseteq \Sigma^{\times}, X$ *is derivable from* Ψ *(in* $\mathcal{L}^{\times}$*), in symbols* $\Psi \vdash^{\times} X$, *iff* X *belongs to every set* $\Omega \subseteq \Sigma^{\times}$ *satisfying the following conditions:*

(α) $\Psi \subseteq \Omega$;

(β) $\mathbf{\Lambda}^{\times} \subseteq \Omega$;

(γ) Ω *contains all tautologies;*

(δ) *if* $B, C, D \in \mathbf{\Pi}$ *and* $B \mathbf{=} C$, $B \mathbf{=} D \in \Omega$, *then* $C \mathbf{=} D \in \Omega$;

(ε) *if* $B, C, D \in \mathbf{\Pi}$ *and* $B \mathbf{=} C \in \Omega$, *then* $B \mathbf{+} D \mathbf{=} C \mathbf{+} D$, $D \mathbf{+} B \mathbf{=} D \mathbf{+} C$, $B^{-} \mathbf{=} C^{-}$, $B \odot D \mathbf{=} C \odot D$, $D \odot B \mathbf{=} D \odot C$, *and* $B^{\smile} \mathbf{=} C^{\smile}$ *also belong to* Ω.

It should be pointed out that there is no essential difference between (β) and (γ) in their logical character and in their role in the definition of derivability.

Probably it would be more consistent to omit (γ) altogether and to include the tautologies in $\mathbf{\Lambda}^{\times}$ (i.e., to enrich the set of logical axioms by the schema "$A = A$"). However, our present formulation seems to be somewhat more suitable for the specific purposes of the present work.

It may be mentioned that, with the essential help of (β), we can show that part (γ) of this definition is superfluous and that two equations in the conclusion of (ε), $D + B = D + C$ and $D \odot B = D \odot C$, can be omitted. As it stands, however, the definition has the virtue that it does not depend on the specific form of the logical axioms. On the other hand, it can easily be shown that, independent of the form of the logical axioms, (δ) and (ε) together can be replaced by the following.

(iv) *If $A, B, C, D \in \mathbf{\Pi}$, if C occurs as a part in A or B, and if $A = B$, $C = D$ are in Ω, then the equation which we obtain from $A = B$ when we replace some occurrence of C by D also belongs to Ω.*

The rule of inference embodied in condition (iv) is referred to as the *rule of replacement* (*of equals by equals*). Thus, assuming we include all tautologies in the set of logical axioms (as suggested above), we can say that the rule of replacement can suffice in $\mathcal{L}^{\times}$ as the only direct rule of inference for deriving logically provable equations from the logical axioms. If, on the other hand, the original definition of derivability in $\mathcal{L}^{\times}$ is preserved, then the rule of replacement is an example of an indirect, or derivative, rule of inference.

The notions defined in terms of derivability for $\mathcal{L}$ (cf. §1.3) extend to $\mathcal{L}^{\times}$ by referring their definitions to the new notion of derivability, $\vdash^{\times}$. In particular, $\mathbf{\Theta}\boldsymbol{\eta}^{\times}\varnothing$ is the logic of $\mathcal{L}^{\times}$, i.e., the set of all sentences logically provable in $\mathcal{L}^{\times}$; it can be called the *algebraic* or *equational logic of one binary relation.* $\mathcal{L}^{\times}$ itself can be referred to as the formalism of such a logic. If we extend $\mathcal{L}^{\times}$ by admitting any number of additional nonlogical constants (which are all, like $\boldsymbol{E}$, binary atomic predicates), but without changing the structure of the formalism otherwise, then the logic of any formalism thus obtained can be referred to as an algebraic or equational logic of binary relations.

Since predicates and sentences in $\mathcal{L}^{\times}$ are also predicates and sentences in $\mathcal{L}^{+}$, we can apply to them automatically the semantical notions of denotation and truth defined for $\mathcal{L}^{+}$ (cf. §2.2). Since, on the other hand, $\mathcal{L}^{\times}$ has no formulas with free variables, the notion of satisfaction becomes trivial and useless when applied to $\mathcal{L}^{\times}$.

If we wish, we can define the basic semantical notions for $\mathcal{L}^{\times}$ directly, without referring to $\mathcal{L}^{+}$. The procedure is very simple: we first define denotation just as this was done for $\mathcal{L}^{+}$, and then we stipulate that a sentence $X \in \mathbf{\Sigma}^{\times}$ is true of a realization $\mathfrak{U}$ of $\mathcal{L}^{\times}$ iff the relations denoted respectively in $\mathfrak{U}$ by X^{ℓ} and X^{r} are identical. We shall return to the discussion of semantical notions for $\mathcal{L}^{\times}$ in Chapter 6.

3.2. Schemata of equations derivable in $\mathcal{L}^\times$

The formalism $\mathcal{L}^\times$ is probably less familiar to the reader than $\mathcal{L}$ or $\mathcal{L}^+$, and formal derivation in $\mathcal{L}^\times$ appears to be more difficult.

To give full proofs of various results stated in the subsequent discussion we would have to use a rather large number of simple metalogical lemmas to the effect that every instance of a given equational schema is logically provable in $\mathcal{L}^\times$ or is derivable from appropriate instances of other equational schemata. No method of proof is known which would be applicable to a large number of such lemmas and could be regarded as routine. (As we shall see in §3.4, the completeness theorem does not hold for $\mathcal{L}^\times$.) In this situation the following remarks are of interest, since they permit us to take advantage of some results which are available in the literature.

We can easily notice a close relationship between the set of schemata (BI)–(BX) in 3.1(ii) used to describe the logical axioms of $\mathcal{L}^\times$ and a set of postulates characterizing a class of abstract algebraic structures, namely the so-called relation algebras discussed, e.g., in Chin-Tarski [**1951**]. In fact, forgetting for a while our stipulations concerning the range of variables "A", "B",... and the meaning of the symbols occurring as logical constants in $\mathcal{L}^\times$, we see from *op. cit.*, pp. 344–346, that relation algebras can be characterized as those algebraic structures $\langle U, +, ^-, \odot, ^\smile, \mathring{1}\rangle$ with two binary operations, $+$ and $\odot$, two unary operations, $^-$ and $^\smile$, and a distinguished element $\mathring{1}$, in which the equations stated in (BI)–(BX) are identically satisfied. Clearly, we assume here that "A", "B",... range over arbitrary elements of U and that " $=$ " has the meaning of the ordinary identity symbol. (We disregard some symbolic differences between our present notation and the one in *op. cit.*, as well as a divergence in the treatment of $\mathring{1}$, which in *op. cit.* is not regarded as a distinguished element of the algebra.)

As a consequence of this observation, various mathematical results in the theory of relation algebras can be interpreted as metalogical results concerning the formalism $\mathcal{L}^\times$, and conversely. In particular, every equation which is shown to be identically satisfied in every relation algebra yields a schema of which all the particular instances (obtained by substituting predicates for variables) are sentences logically provable in $\mathcal{L}^\times$. To justify this remark, we apply the completeness theorem for equational logic due to Birkhoff [**1935**], p. 441, and we compare it with the definition of $\vdash^\times$ in 3.1(iii). For the same reason, every result by which a given equation is implied by other equations can be interpreted as a metalogical result to the effect that every sentence in $\Sigma^\times$ of a given form is derivable from some sentences of related forms (cf. Theorem 8.5(iv)).

We list below the simple lemmas needed for the subsequent discussion. They can all be obtained by interpretation of results stated in Chin-Tarski [**1951**] or at least can be easily derived from lemmas so obtained (possibly with the help of 2.1(ii)–(iv)). A, B, C, D in these lemmas are assumed to be arbitrary predicates.

(i) $\vdash^{\times} \{(A+B)^{\smile} = A^{\smile}+B^{\smile}, (A\cdot B)^{\smile} = A^{\smile}\cdot B^{\smile}, A^{-\smile} = A^{\smile -}, A^{\smile\smile} = A\}.$

(ii) $\vdash^{\times} \{\mathbf{1}^{\smile} = \mathbf{1},\ \mathbf{0}^{\smile} = \mathbf{0},\ \mathring{\mathbf{1}}^{\smile} = \mathring{\mathbf{1}},\ \mathring{\mathbf{0}}^{\smile} = \mathring{\mathbf{0}}\}.$

(iii) $\vdash^{\times} \{(A\odot B)^{-} = A^{-}\oplus B^{-},\ (A\oplus B)^{-} = A^{-}\odot B^{-},$
$(A\odot B)^{\smile} = B^{\smile}\odot A^{\smile},\ (A\oplus B)^{\smile} = B^{\smile}\oplus A^{\smile}\}.$

(iv) $\vdash^{\times} \{A\odot(B\odot C) = (A\odot B)\odot C,\ A\oplus(B\oplus C) = (A\oplus B)\oplus C\}.$

(v) $\vdash^{\times} \{A\odot(B+C) = A\odot B+A\odot C,\ (B+C)\odot A = B\odot A+C\odot A\}.$

(vi) $\vdash^{\times} \{A\oplus(B\cdot C) = (A\oplus B)\cdot(A\oplus C),\ (B\cdot C)\oplus A = (B\oplus A)\cdot(C\oplus A)\}.$

(vii) $A\leq B \vdash^{\times} \{A\odot C\leq B\odot C,\ C\odot A\leq C\odot B,\ A\oplus C\leq B\oplus C,$
$C\oplus A\leq C\oplus B,\ A^{\smile}\leq B^{\smile}\}.$

(viii) $\vdash^{\times} A\cdot C\odot(B\cdot D)\leq A\odot B\cdot(C\odot D).$

(ix) $\vdash^{\times} \{A\odot\mathring{\mathbf{1}} = A,\ \mathring{\mathbf{1}}\odot A = A,\ A\leq A\odot\mathbf{1},\ A\leq\mathbf{1}\odot A,\ \mathbf{0}\oplus A\leq A,$
$A\oplus\mathbf{0}\leq A,\ \mathbf{1}\odot\mathbf{1} = \mathbf{1},\ \mathbf{0}\oplus\mathbf{0} = \mathbf{0}\}.$

(x) $\vdash^{\times} \{A\odot C\cdot(B\odot C)^{-}\leq A\cdot B^{-}\odot C,\ C\odot A\cdot(C\odot B)^{-}\leq C\odot(A\cdot B^{-})\}.$

(xi) $\vdash^{\times} \{\mathring{\mathbf{1}}\leq A^{-}\oplus A^{\smile},\ A^{\smile}\odot A^{-}\leq\mathring{\mathbf{0}},\ \mathring{\mathbf{1}}\leq A^{-}+A^{\smile},\ A^{\smile}\cdot A^{-}\leq\mathring{\mathbf{0}}\}.$

(xii) $A\odot B\leq C^{-}\ \equiv^{\times}\ A^{\smile}\odot C\leq B^{-}\ \equiv^{\times}\ C\odot B^{\smile}\leq A^{-}.$

(xiii) $\vdash^{\times} \{\mathbf{1}\oplus A = \mathbf{1},\ A\oplus\mathbf{1} = \mathbf{1},\ \mathbf{0}\odot A = \mathbf{0},\ A\odot\mathbf{0} = \mathbf{0}\}.$

(xiv) $\vdash^{\times} \{\mathbf{1}\odot A = \mathbf{0}\oplus(\mathbf{1}\odot A),\ A\odot\mathbf{1} = (A\odot\mathbf{1})\oplus\mathbf{0},\ \mathbf{0}\oplus A = \mathbf{1}\odot(\mathbf{0}\oplus A),$
$A\oplus\mathbf{0} = (A\oplus\mathbf{0})\odot\mathbf{1}\}.$

(xv) $\vdash^{\times} \{A\odot B\cdot C\leq A\odot[B\cdot(A^{\smile}\odot C)],\ A\odot B\cdot C\leq C\odot B^{\smile}\cdot A\odot B\}.$

(xvi) $\vdash^{\times} \{\mathbf{1}\odot(A\odot B\cdot C) = \mathbf{1}\odot(A^{\smile}\odot C\cdot B),$
$\mathbf{0}\oplus(A\oplus B+C) = \mathbf{0}\oplus(A^{\smile}\oplus C+B)\}.$

(xvii) $\vdash^{\times} \{A\leq A\odot A^{\smile}\odot A,\ \mathbf{1}\odot A^{\smile}\odot\mathbf{1} = \mathbf{1}\odot A\odot\mathbf{1},\ (\mathbf{1}\odot A\odot\mathbf{1})^{\smile} = \mathbf{1}\odot A\odot\mathbf{1}\}.$

(xviii) $\vdash^{\times} A\odot(B\oplus C)\leq A\odot B\oplus C.$

(xix) $A\oplus B^{\smile} = \mathbf{1}\ \equiv^{\times}\ \mathbf{0}\oplus A+B = \mathbf{1}\ \equiv^{\times}\ \mathbf{0}\oplus A+(\mathbf{0}\oplus B) = \mathbf{1}.$

(xx) $A\odot\mathbf{1} = \mathbf{1}\ \equiv^{\times}\ \mathring{\mathbf{1}}\leq A\odot A^{\smile}.$

(xxi) $\vdash^{\times}\ A \odot B \cdot \mathring{\mathbf{1}} \odot \mathbf{1} = A \cdot B^{\smile} \odot \mathbf{1}$.

(xxii) $A \leq \mathring{\mathbf{1}} \vdash^{\times} \{A^{\smile} = A,\ \ A \odot A = A\}$.

(xxiii) $\{A \leq \mathring{\mathbf{1}},\ C \leq \mathring{\mathbf{1}}\} \vdash^{\times} \{A \cdot C \odot (B \cdot D) = A \odot B \cdot (C \odot D),$
$B \cdot D \odot (A \cdot C) = B \odot A \cdot (D \odot C)\}$.

(xxiv) $\vdash^{\times}\ (\mathring{\mathbf{1}} \cdot A \odot \mathbf{1})^{-} = \mathring{\mathbf{1}} \cdot A^{-} \odot \mathbf{1}$.

(xxv) $\{\mathring{\mathbf{1}} \leq A,\ A^{\smile} \leq A,\ A \odot A \leq A\} \vdash^{\times}\ A \odot (A \odot B)^{-} = (A \odot B)^{-}$.

(xxvi) $\{A^{\smile} \odot A \leq \mathring{\mathbf{1}},\ B^{\smile} \odot B \leq \mathring{\mathbf{1}}\} \vdash^{\times}\ (A \odot B)^{\smile} \odot (A \odot B) \leq \mathring{\mathbf{1}}$.

(xxvii) $A^{\smile} \odot A \leq \mathring{\mathbf{1}} \vdash^{\times} \{A \odot (B \cdot C) = A \odot B \cdot (A \odot C),$
$B \cdot C \odot A^{\smile} = B \odot A^{\smile} \cdot (C \odot A^{\smile})\}$.

(xxviii) $A^{\smile} \odot A \leq \mathring{\mathbf{1}} \vdash^{\times} \{A^{\smile} \odot [B \cdot (A \odot C)] = A^{\smile} \odot B \cdot C,$
$C \odot A^{\smile} \cdot B \odot A = C \cdot (B \odot A)\}$.

(xxix) $\vdash^{\times}\ [A \cdot (A^{-} \oplus \mathring{\mathbf{1}})]^{\smile} \odot [A \cdot (A^{-} \oplus \mathring{\mathbf{1}})] \leq \mathring{\mathbf{1}}$.

(xxx) $\{A^{\smile} \odot A \leq \mathring{\mathbf{1}}, A \odot \mathbf{1} = \mathbf{1}\} \vdash^{\times} \{(A \odot B)^{-} = A \odot B^{-}, (B \odot A^{\smile})^{-} = B^{-} \odot A^{\smile}\}$.

(xxxi) $\{A^{\smile} \odot A \leq \mathring{\mathbf{1}}, B^{\smile} \odot B \leq \mathring{\mathbf{1}}, A \odot \mathbf{1} \cdot (B \odot \mathbf{1}) = \mathbf{0}\} \vdash^{\times} (A + B)^{\smile} \odot (A + B) \leq \mathring{\mathbf{1}}$.

As is easily seen, the inclusions $\mathring{\mathbf{1}} \leq A$, $A^{\smile} \leq A$, $A \odot A \leq A$, and $A^{\smile} \odot A \leq \mathring{\mathbf{1}}$ respectively express the fact that the relation denoted by A is reflexive, symmetric, transitive, and functional. This helps to grasp the content of (xxii) and (xxv)–(xxxi).

It should be noticed that the same relationship which holds between the set of logical axiom schemata for $\mathcal{L}^{\times}$ and the set of postulates for relation algebras holds also between the set of the first three of these schemata, (BI)–(BIII) in §3.1, and a familiar set of postulates for Boolean algebras (treated as structures $\langle U, +, {}^{-}\rangle$) which is due to Huntington [**1933**], [**1933a**]. Hence, all the identities and implications between equations which are known from the theory of Boolean algebras yield analogous results concerning $\mathcal{L}^{\times}$. In this way we get for instance:

(xxxii) $(A = B)\ \equiv^{\times}\ (A \cdot B + A^{-} \cdot B^{-} = \mathbf{1})$ *for any* $A, B \in \mathbf{\Pi}$, *or equivalently,* $X \equiv^{\times} (X^{t} = \mathbf{1})$ *for every* $X \in \Sigma^{\times}$.

(xxxiii) $\mathbf{1} = \mathbf{0} \vdash^{\times} X$ *for every* $X \in \Sigma^{\times}$, *i.e.*, $\mathbf{1} = \mathbf{0} \vdash^{\times} \Sigma^{\times}$.

While we shall have the opportunity of referring to (i)–(xxxi), analogous lemmas obtained by interpretation of various well-known results from the theory of Boolean algebras will not be referred to here in any explicit way. When such a lemma is applied in the subsequent discussion, this fact is sometimes indicated by "BA", and in simpler cases even this indication is omitted. Also, some of the simpler lemmas among (i)–(xxxi) will be occasionally used without any explicit reference; this applies, in particular, to (i)–(iv), (vii), and (ix). The distributive laws (v), (vi), and (xxvii) can clearly be extended by induction from two predicates, B and C, to any finite sequence $\langle D_0, \dots, D_m \rangle$ of predicates, and in this extended form they will be applied in some arguments.

3.3. A deduction theorem for $\mathcal{L}^{\times}$

We shall now state some results which apply to $\mathcal{L}^{\times}$ and are closely related in content to the deduction theorems in §1.3.

(i) *For any $\Psi \subseteq \Sigma^{\times}$ and $X, Y \in \Sigma^{\times}$, we have*

$$\Psi \cup \{X\} \vdash^{\times} Y \quad \textit{iff} \quad \Psi \vdash^{\times} \mathbf{1} \odot X^{t\smile} \odot \mathbf{1} + Y^t = \mathbf{1}.$$

(ii) *For any $A, B \in \Pi$ and $\Psi \subseteq \Sigma^{\times}$, we have*

$$\Psi \cup \{A = \mathbf{1}\} \vdash^{\times} B = \mathbf{1} \quad \textit{iff} \quad \Psi \vdash^{\times} \mathbf{1} \odot A^{\smile} \odot \mathbf{1} + B = \mathbf{1}.$$

Apparently, (i) is more general than (ii), since the latter is partly restricted to equations X with $X^r = \mathbf{1}$. However, in view of 3.1(i)(γ) and 3.2(xxxii), each of the two statements can be directly derived from the other.

One half of (ii), the implication from right to left, is rather trivial; we use 3.2(xiii) and BA (in addition to some general properties of derivability mentioned in §1.6). To obtain the implication in the opposite direction we apply induction on sentences derivable in $\mathcal{L}^{\times}$ from the set $\Psi \cup \{A = \mathbf{1}\}$. In fact, we set

$$\Omega = \{X : X \in \Sigma^{\times} \text{ and } \Psi \vdash^{\times} \mathbf{1} \odot A^{\smile} \odot \mathbf{1} + X^t = \mathbf{1}\}.$$

We then show that Ω satisfies 3.1(iii)(α)–(ε), with "Ψ" replaced by "$\Psi \cup \{A = \mathbf{1}\}$", and hence contains every sentence X such that $\Psi \cup \{A = \mathbf{1}\} \vdash^{\times} X$. The argument is based primarily on BA. In addition, we use various lemmas from §3.2: 3.2(vii) in proving that $A = \mathbf{1} \in \Omega$ (a part of 3.1(iii)(α)); 3.2(i),(ii),(xvii) in proving that $B^{\smile} = C^{\smile} \in \Omega$ whenever $B = C \in \Omega$ (a part of 3.1(iii)(ε)); 3.2(iv),(vii), (x) in proving that $B \odot D = C \odot D, D \odot B = D \odot C \in \Omega$ whenever $B = C \in \Omega$ (another part of 3.1(iii)(ε)).

A close relationship between (i), (ii), and 1.3(iv) is obvious. If (ii) is referred to $\mathcal{L}^{+}$ instead of to $\mathcal{L}^{\times}$, it becomes an immediate corollary of Theorem 1.3(iv) (which extends to $\mathcal{L}^{+}$ according to the remarks preceding 2.2(i)) and Theorem 2.2(ii). In view of this relationship we refer to (i) and (ii) as the *deduction theorems for $\mathcal{L}^{\times}$*.

The last remark suggests the idea of introducing the general notion of a deduction theorem for a given formalism, and thus of singling out *formalisms (provided) with a deduction theorem.* Let $\mathcal{F}$ be any formalism for which the notion of derivability $\vdash$ has properties (FI)–(FV) listed in §1.6, and to which the notion of recursiveness has been properly extended (cf. §1.3). We assume, in particular, that the set $\Sigma[\mathcal{F}]$ is recursive and that every theory in $\mathcal{F}$ with a recursively enumerable base is itself recursively enumerable. By a *deduction theorem for* $\mathfrak{F}$ we understand any theorem of the following form:

For any $X, Y \in \Sigma[\mathfrak{F}]$ *and any* $\Psi \subseteq \Sigma[\mathfrak{F}]$, *we have*

$$\Psi \cup \{X\} \vdash Y \ [\mathfrak{F}] \quad \textit{iff} \quad \Psi \vdash (X \twoheadrightarrow Y) \ [\mathfrak{F}].$$

Here $\twoheadrightarrow$ is some appropriately chosen binary recursive operation on and to elements of $\Sigma[\mathfrak{F}]$; for instance, in the case of $\mathcal{L}^{\times}$ we can take for $\twoheadrightarrow$ the operation defined for any $X, Y \in \Sigma^{\times}$ by the formula

(iii) $X \twoheadrightarrow Y \;=\; (\mathbf{1} \odot X^{t-} \odot \mathbf{1} + Y^{t} = \mathbf{1})$,

in agreement with (i).

A deduction theorem has important consequences whenever it can be established. In particular, Theorems 1.3(vi),(vii) apply to every formalism $\mathfrak{F}$ with a deduction theorem, and the same is true of 1.3(viii) since, by (FV), the inconsistent theory $\Sigma[\mathfrak{F}]$ has a finite base. Thus, by (i) and 3.2(xxxiii), *Theorems* 1.3(vi)–(viii) *hold for* $\mathcal{L}^{\times}$.

In view of (iii), it is seen from (i) that the binary operation $\twoheadrightarrow$ plays essentially the same role in $\mathcal{L}^{\times}$ as the operation $\rightarrow$ in $\mathcal{L}$. We can also consider a unary operation $\neg\!\!\shortmid$ on sentences of $\mathcal{L}^{\times}$ defined by

(iv) $\neg\!\!\shortmid X \;=\; (\mathbf{1} \odot X^{t-} \odot \mathbf{1} = \mathbf{1})$.

This operation plays essentially the same role in $\mathcal{L}^{\times}$ as the operation $\neg$ in $\mathcal{L}$, as is seen, for instance, from the following theorem.[1]

(v) *For any set* $\Psi \subseteq \Sigma^{\times}$ *and any sentence* $X \in \Sigma^{\times}$, *the following two conditions are equivalent*:

(α) *X is incompatible with* Ψ,

(β) $\Psi \vdash^{\times} \neg\!\!\shortmid X$.

If, in addition, Ψ *is a theory in* $\mathcal{L}^{\times}$, *then the conditions* (α) *and* (β) *are also equivalent to*

(γ) $\neg\!\!\shortmid X \in \Psi$.

In fact, it is easily shown by means of 3.2(xxxiii) that (α) is equivalent to the condition $\Psi \cup \{X\} \vdash^{\times} \mathbf{1} = \mathbf{0}$, and it follows from (i), (iv) that this latter

[1] *Actually, it is not difficult to prove that every instance of Axioms (AI)–(AIII) in §1.3, with "$\rightarrow$" and "$\neg$" replaced by "$\twoheadrightarrow$" and "$\neg\!\!\shortmid$", and with "X", "Y", "Z" ranging over sentences in $\Sigma^{\times}$, is logically provable in $\mathcal{L}^{\times}$, and that *modus ponens* (formulated in terms of $\twoheadrightarrow$) is a valid rule of inference in $\mathcal{L}^{\times}$. This can be shown directly, with the help of the lemmas in §3.2. Alternately, using (i), (v), and 3.2(iii),(xiv), it is easy to verify that Axioms 1–5, 6*–10* from Tarski [**1956**], Article III, p. 32, are satisfied, with "S", "Cn", "c", and "n" replaced there by "$\Sigma^{\times}$", "$\Theta\eta^{\times}$", "$\twoheadrightarrow$", and "$\neg\!\!\shortmid$"; the desired result then follows from Theorem 3* and Axiom 7* of *ibid.* In consequence, all instances of true sentences of the sentential calculus (with "$\rightarrow$" and "$\neg$" replaced by "$\twoheadrightarrow$" and "$\neg\!\!\shortmid$", and with variables replaced by sentences of $\Sigma^{\times}$) are logically provable in $\mathcal{L}^{\times}$. This observation justifies our remark that the operations $\twoheadrightarrow$ and $\neg\!\!\shortmid$ play essentially the same role in $\mathcal{L}^{\times}$ as do the operations $\rightarrow$ and $\neg$ in $\mathcal{L}$. (However, the observation will not be used in the subsequent discussion.)

condition is equivalent to (β). In case Ψ is a theory, the equivalence of (β) and (γ) is obvious.

For later use we give here an elementary theorem based upon (v).

(vi) *Let Θ be a theory in $\mathcal{L}^\times$; let Θ', or Θ'', be respectively the set of sentences incompatible, or compatible, with Θ. If any one of the three sets $\Theta, \Theta', \Theta''$ is recursive, then the remaining two are also recursive.*

Indeed, by (iv), (v) we have for any $X \in \Sigma^\times$,

(1) $\qquad X \in \Theta'$ iff $\neg X \in \Theta$.

Primarily with the help of 3.1(i)(γ), 3.2(ix),(xiii),(xiv),(xxxii), we easily derive $X \equiv^\times \neg\neg X$, whence, using also (1),

(2) $\qquad X \in \Theta$ iff $\neg X \in \Theta'$.

Since the mapping $\neg$ is recursive, we see from (1) that the recursiveness of Θ implies that of Θ', and from (2) we derive the implication in the opposite direction. Finally, Θ'', i.e., $\Sigma^\times \sim \Theta'$, is obviously recursive iff Θ' is recursive.

3.4. The inequipollence of $\mathcal{L}^\times$ with $\mathcal{L}^+$ and $\mathcal{L}$

We turn now to the problem of comparing the formalism $\mathcal{L}^\times$ with $\mathcal{L}$ and $\mathcal{L}^+$ in regard to expressive and deductive powers.

It is easily seen that $\mathcal{L}^\times$ is a subformalism of $\mathcal{L}^+$ in the following sense.

(i) $\Sigma^\times \subseteq \Sigma^+$.

(ii) *For every $\Psi \subseteq \Sigma^\times$ and $X \in \Sigma^\times$, if $\Psi \vdash^\times X$, then $\Psi \vdash X$.*

Theorems (i) and (ii) are analogues of 2.3(i),(ii); because of the peculiar structure of $\mathcal{L}^\times$, we do not refer in (i) to arbitrary formulas. Notice that in formulating the conclusion of (ii) we apply the convention stated at the end of §2.3.

A consequence of (i) and (ii) is:

(iii) *If Θ is a theory in $\mathcal{L}^+$, then $\Theta \cap \Sigma^\times$ is a theory in $\mathcal{L}^\times$.*

Another consequence of (ii) is that the formalism $\mathcal{L}^\times$, just like $\mathcal{L}^+$ and $\mathcal{L}$, is semantically sound; cf. §1.6.

As a subformalism of $\mathcal{L}^+$, $\mathcal{L}^\times$ is at most as powerful as $\mathcal{L}^+$ in means of expression and proof. Since, as we know from Chapter 2, $\mathcal{L}^+$ is equipollent with its subformalism $\mathcal{L}$, we can also say that $\mathcal{L}^\times$ is at most as powerful as $\mathcal{L}$ (relative to $\mathcal{L}^+$); cf. §2.5. The problem naturally arises whether $\mathcal{L}^\times$ is at least as powerful as $\mathcal{L}^+$ and $\mathcal{L}$, and whether, therefore, the three formalisms are equipollent. This problem divides into two parts, the first of which concerns the means of expression and the second the means of proof.

In agreement with the stipulations in §2.4, to show that $\mathcal{L}^+$ is equipollent in means of expression with its subformalism $\mathcal{L}^\times$ we would have to prove that every sentence in $\mathcal{L}^+$ is logically equivalent with some sentence in $\mathcal{L}^\times$. It has been known for a long time, however, that this is not the case. The result is essentially due to A. Korselt, and was reported in Löwenheim [**1915**], p. 448. Using our terminology, it can be formulated as follows.

(iv) *There are sentences $X \in \Sigma^+$, and actually $X \in \Sigma$, such that $X \equiv Y$ does not hold for any $Y \in \Sigma^\times$. For instance, the sentence*

$$S = \forall_{xyz}\exists_u(\neg x\mathring{\mathbf{1}}u \wedge \neg y\mathring{\mathbf{1}}u \wedge \neg z\mathring{\mathbf{1}}u)$$

is of this kind.

The proof is not difficult.

In view of the semantical soundness of $\mathcal{L}^+$, (iv) can be interpreted not only syntactically but also semantically: there is no $Y \in \Sigma^\times$ such that S and Y have all models in common. On the other hand, it is easily seen that S is *almost semantically equivalent* with some $Y \in \Sigma^\times$ in the following sense: there is a natural number n such that for every structure $\mathfrak{U} = \langle U, E\rangle$ with $|U| \geq n$, $\mathfrak{U}$ is a model of S iff it is a model of Y. Indeed, S expresses the fact that the universe has at least 4 elements, and hence every structure $\mathfrak{U}$ with $|U| \geq 4$ is a model of S. Consequently, S is almost equivalent with every $Y \in \Sigma^\times$ which is logically provable, e.g., $Y = (\mathbf{1} = \mathbf{1})$. It is known that every $X \in \Sigma^+$ in which the nonlogical constant $\boldsymbol{E}$ does not occur is almost equivalent with some $Y \in \Sigma^\times$.

In this connection, Tarski has modified and supplemented Korselt's argument to obtain the following improvement of (iv).

(v) *The sentence*

$$S' = \forall_{xyz}\exists_u(xEu \wedge yEu \wedge zEu)$$

is not almost semantically equivalent with any $Y \in \Sigma^\times$.

The solution of the equipollence problem in its application to means of proof is also negative, as was first pointed out by Givant. To show that $\mathcal{L}^\times$ and $\mathcal{L}^+$ are not equipollent (in the sense of §2.4), it suffices to exhibit a set Ψ of sentences in $\mathcal{L}^\times$ and a sentence X in $\mathcal{L}^\times$ such that $\Psi \vdash X$ holds while $\Psi \vdash^\times X$ fails. It turns out that this can be achieved by taking the empty set for Ψ.

The problem proves to be closely related to the representation problem for relation algebras.[2] It is well known that the solution of the latter problem is negative; cf. Lyndon [**1950**], [**1956**]. A very simple solution of the problem has been presented in McKenzie [**1970**], where a nonrepresentable relation algebra with one generator and 16 different elements has been constructed. An analysis of this construction leads to the following conclusion.

[2]*See §8.3 for a brief discussion of this problem.

(vi) *There are sentences $X \in \Sigma^{\times}$ such that $\vdash X$ holds while $\vdash^{\times} X$ fails. For instance, the sentence*

$$T = \left[\mathbf{1} \odot \left(\boldsymbol{E} \oplus \boldsymbol{E} + [\boldsymbol{E}^2 + \mathring{\mathbf{1}} + (\boldsymbol{E} + \boldsymbol{E}^{\smile})^{-2} \cdot \boldsymbol{E}^{\smile -}] \cdot \boldsymbol{E}^{-} + (\boldsymbol{E} \odot \boldsymbol{E}^{\smile})^{-}\right) \odot \mathbf{1} = \mathbf{1}\right]$$

is of this kind.[3,4]

Sentence T is simpler than the one that Givant constructed directly from McKenzie's argument. (The simplification was obtained jointly by McNulty and Tarski.) Still, T is rather involved. It would be interesting to find the shortest possible sentence which could serve the same purpose.

An equivalent formulation of the first statement in (vi) is

(vii) $\boldsymbol{\Theta\eta}^{\times}\varnothing \subset \boldsymbol{\Theta\eta}^{+}\varnothing \cap \Sigma^{\times}$.

While $\boldsymbol{\Theta\eta}^{\times}\varnothing$ is the set of all logically provable sentences in $\mathcal{L}^{\times}$, $\boldsymbol{\Theta\eta}^{+}\varnothing \cap \Sigma^{\times}$ is clearly the set of all logically true sentences in $\mathcal{L}^{\times}$. Thus (vii) shows that *the formalism $\mathcal{L}^{\times}$ is not semantically complete.*

[3*] A similar result, formulated for relation algebras, was recently published in Jónsson [**1982**], Theorem 7.9.

[4*] We sketch a proof that $\vdash^{\times} T$ fails, using the terminology and theorems of §8.5 that allow us to interpret results about relation algebras as results about $\mathcal{L}^{\times}$. We begin by constructing McKenzie's relation algebra. Let $\langle A, +, ^{-}\rangle$ be a sixteen element Boolean algebra with the four atoms a, b, c, d. We define a binary operation $\odot$ and a unary operation $^{\smile}$ on these atoms via the tables

$^{\smile}$	
a	a
b	c
c	b
d	d

$\odot$	a	b	c	d
a	a	b	c	d
b	b	b	1	$b+d$
c	c	1	c	$c+d$
d	d	$b+d$	$c+d$	$a+b+c$

Both $\odot$ and $^{\smile}$ extended in the obvious ways to all of A so that the distributive laws, (Ra V) and (Ra VIII) from 8.2(i), are valid. Let $\mathfrak{A}$ be the algebra of type $\langle 2, 1, 2, 1, 0, 0\rangle$ defined by

$$\mathfrak{A} = \langle A, +, ^{-}, \odot, ^{\smile}, \mathring{1}, E\rangle,$$

where $\mathring{1} = a$ and $E = a+b = \mathring{1}+b$. It is a simple matter to check that all the axioms defining the class of relation algebras are valid in $\mathfrak{A}$. To show that T fails in $\mathfrak{A}$ it suffices, by 3.2(xiii), to show that

$$E \oplus E + [E^2 + \mathring{1} + (E+E^{\smile})^{-2} \cdot E^{\smile -}] \cdot E^{-} + (E \odot E^{\smile})^{-} = 0. \tag{1}$$

It is simple to see that

$$E \oplus E = 0 \quad \text{and} \quad (E \odot E^{\smile})^{-} = 0.$$

Since

$$(E+E^{\smile})^{-2} = \mathring{1} + b + c,$$

we get that

$$(E+E^{\smile})^{-2} \cdot E^{\smile -} = b.$$

Therefore,

$$[E^2 + \mathring{1} + (E+E^{\smile})^{-2} \cdot E^{\smile -}] \cdot E^{-} \;=\; E \cdot E^{-} \;=\; 0,$$

which verifies (1). Since the set Ξ of axioms of RA is valid in $\mathfrak{A}$, we can now conclude that

$$\Xi \nvdash T \;[\mathcal{E}^{\mathsf{EA}}]. \tag{2}$$

But

$$\boldsymbol{\Theta\eta}^{\times}\varnothing = \boldsymbol{\Theta\eta}^{\mathsf{EA}}\Xi \cap \Sigma^{\times},$$

by 8.5(iv), so the nonderivability of T in $\mathcal{L}^{\times}$ follows immediately from (2).

By (iv) and (vi), *the formalism $\mathcal{L}^\times$ is not equipollent with, and actually* (in view of (i), (ii)), *is poorer than $\mathcal{L}^+$ in means of both expression and proof.* Consequently, *$\mathcal{L}^\times$ is not equipollent with $\mathcal{L}$ relative to $\mathcal{L}^+$.* These negative results lead in a natural way to several other problems which will be discussed in the remaining part of the present chapter.

3.5. The inequipollence of extensions of $\mathcal{L}^\times$ with $\mathcal{L}^+$ and $\mathcal{L}$

The question arises whether the inequipollence of $\mathcal{L}^\times$ with $\mathcal{L}$ and $\mathcal{L}^+$ is simply due to an overly restrictive selection of both the logical operators of $\mathcal{L}^+$ and $\mathcal{L}^\times$, and the logical axiom schemata of $\mathcal{L}^\times$. Is it possible, in fact, to extend the formalisms of $\mathcal{L}^\times$ and $\mathcal{L}^+$ so as to secure their equipollence with each other and with $\mathcal{L}$, while simultaneously keeping intact their basic structural features? In discussing this problem, we shall assume that the extended formalisms $\overline{\mathcal{L}}^\times$ and $\overline{\mathcal{L}}^+$ with which we shall deal are obtained from $\mathcal{L}^\times$ and $\mathcal{L}^+$ by means of the following modifications.

(1) The vocabularies of $\mathcal{L}^\times$ and $\mathcal{L}^+$ are enriched by including finitely many new constant symbols. Just as the symbols of $\mathcal{L}^\times$ (with the exception of $=$), the new symbols are atomic predicates denoting some binary relations, or else operators denoting operations on and to binary relations, each operator having a definite rank, but not necessarily rank 1 or 2. (To simplify the formulations, we treat the atomic predicates as operators of rank 0.) Moreover, we assume that, on the basis of the semantical definitions adopted for $\overline{\mathcal{L}}^\times$ and $\overline{\mathcal{L}}^+$, and specifically on the basis of the definition of denotation (in terms of which the meanings of the new constants in each realization are specified), all these new constants prove to be logical symbols in a sense to be made precise below.

(2) The set of logical axiom schemata (BI)–(BX) of $\mathcal{L}^\times$, given in §3.1, is supplemented by finitely many new schemata which, loosely speaking, have the same structure as the original ones — i.e., all the particular instances of these schemata are equations between predicates (of $\overline{\mathcal{L}}^\times$) — and which are logically valid on the basis of the semantical definitions adopted for $\overline{\mathcal{L}}^\times$.

Conditions (1) and (2) imply some further minor changes of a rather obvious nature in the definitions of predicates and derivability given in 2.1(i) and 3.1(iii). The formal structure of $\mathcal{L}^\times$ remains otherwise unchanged. As regards $\overline{\mathcal{L}}^+$, we do not impose on it a condition analogous to (2), for example, that its set of logical axiom schemata be obtained from that of $\mathcal{L}^+$ by adding finitely many new schemata which have the same structure as Schemata (DI)–(DV). Indeed, we shall obtain certain negative results that hold even when $\overline{\mathcal{L}}^+$ is not a definitional extension of $\mathcal{L}$.

Since $\boldsymbol{E}$ is the only nonlogical constant of $\overline{\mathcal{L}}^\times$ and $\overline{\mathcal{L}}^+$, the realizations of these two formalisms are the same structures $\mathfrak{U} = \langle U, E\rangle$ that are realizations of $\mathcal{L}^\times$ and $\mathcal{L}^+$. Relative to these realizations we define the basic semantical notions applicable to $\overline{\mathcal{L}}^\times$ and $\overline{\mathcal{L}}^+$ (denotation and truth), proceeding analogously as in the case of $\mathcal{L}^\times$ and $\mathcal{L}^+$; cf. §3.1, last paragraph, and §2.2. A detailed formulation of the semantical definitions depends, of course, on the intended meaning of the new constants included in the vocabularies of $\overline{\mathcal{L}}^{\times}$ and $\overline{\mathcal{L}}^{+}$.

The assumption that these new constants are logical symbols will play an essential role in our discussion. It is based upon a conception of logical symbols which we believe was suggested for the first time in Lindenbaum–Tarski [**1936**]. (An English translation of this paper occurs as Article XIII in Tarski [**1956**]. See also Tarski [**1986**].) Due to the breadth of this conception, the results of the discussion can be expressed in a strong and general form.

Consider any nonempty set U (which can be regarded as the universe of a realization $\mathfrak{U} = \langle U, E\rangle$ of $\mathcal{L}^{\times}$). Thinking in the framework of the theory of types, we can construct from this basic universe U various derivative universes of higher types, e.g., the universe U' of all n-ary relations among elements of U, the universe U'' of p-ary operations on and to members of U', the universe U''' of q-ary relations among members of U' (where n, p, q are fixed positive integers), etc.

Every permutation P of the basic universe (i.e., every one-one mapping of U onto U) induces in a natural way a uniquely determined permutation $\tilde{P}$ of any given derivative universe $\tilde{U}$. A member M of $\tilde{U}$ is said to be invariant under P if $\tilde{P}(M) = M$.

We now stipulate:

(i) *Given a basic universe U, a member M of any derivative universe $\tilde{U}$ is said to be logical, or a logical object, if it is invariant under every permutation P of U.*

(Strictly speaking, since an object M can be a member of many derivative universes, we should use in (i) the phrase "*is said to be logical, or a logical object, as a member of $\tilde{U}$*".)

(ii) *A symbol S of the formalism $\overline{\mathcal{L}}^{\times}$ is said to be logical, or a logical constant, if, for every given realization $\mathfrak{U}$ of this formalism with the universe U, S denotes a logical object in some derivative universe $\tilde{U}$.*

On the basis of (i) one can show, for example, that for every (nonempty) U there are only four logical binary relations between elements of U: the universal relation $U \times U$, the empty relation $\emptyset$, the identity relation $U \mathbin{1} Id$, and the diversity relation $(U \times U) \cap Di$. On the other hand, if U is infinite, then there are infinitely many, and in fact nondenumerably many, logical unary, or binary, operations on and to relations between elements of U. Examples of such operations are those denoted by the operators of $\mathcal{L}^{\times}$, i.e., formation of unions, complementation, relative multiplication, and conversion.

The symbol $\mathring{1}$ in $\mathcal{L}$, $\mathcal{L}^{+}$, and $\mathcal{L}^{\times}$ is logical in the sense of (ii), and so are the symbol $=$ and the four operators in $\mathcal{L}^{+}$ and $\mathcal{L}^{\times}$. As a matter of fact, the remaining symbols in $\mathcal{L}^{+}$ and $\mathcal{L}^{\times}$ which were referred to from the start as logical constants, i.e., the implication and negation symbols and the universal quantifier, can also be subsumed under logical constants in the sense of (ii); this would require, however, some expansion and elaboration of our terminology.

It may be pointed out that the exact definitions of logical objects and symbols given in (i) and (ii) are essentially involved only in the results of the present section. Elsewhere in this work we use the terms "logical symbol", "logical notion", etc., in an informal way which seems, however, to always be compatible with the definitions given above.

The problem of whether any of the extended formalisms $\overline{\mathcal{L}}^\times$ is equipollent with $\mathcal{L}$ and $\mathcal{L}^+$ (or $\overline{\mathcal{L}}^+$) splits again into two parts concerning respectively the means of expression and those of proof.

As regards the first part, we shall show that $\overline{\mathcal{L}}^\times$ can never be equipollent with $\mathcal{L}$ or $\mathcal{L}^+$ (and hence cannot be equipollent with $\overline{\mathcal{L}}^+$) in means of expression. Notice that, in general, neither of the formalisms $\mathcal{L}^+$ and $\overline{\mathcal{L}}^\times$ is a subformalism of the other, if only for the reason that neither of the corresponding sets of sentences, Σ^+ and $\overline{\Sigma}^\times$, includes the other. Since, however, $\mathcal{L}^+$ and $\overline{\mathcal{L}}^\times$ have the same realizations, at least the problem of their semantical equipollence in means of expression has a well-defined meaning: $\mathcal{L}^+$ and $\overline{\mathcal{L}}^\times$ are so equipollent just in case for every $X \in \Sigma^+$ there is a $Y \in \overline{\Sigma}^\times$, and for every $Y \in \overline{\Sigma}^\times$ there is an $X \in \Sigma^+$, such that in all realizations of $\mathcal{L}^+$, X is true iff Y is true. An analogous definition applies to $\mathcal{L}$ and $\overline{\mathcal{L}}^\times$.

To obtain a negative solution of the problem we first introduce notations for several particular predicates in $\mathbf{\Pi}$.

(iii) $C_0 = \boldsymbol{E} \cdot \mathring{\mathbf{1}}$, $C_1 = \boldsymbol{E}^{-} \cdot \mathring{\mathbf{1}}$, $C_2 = C_0 \odot \mathbf{1} \odot C_1$, $C_3 = C_1 \odot \mathbf{1} \odot C_0$, $C_4 = C_0 \odot \mathbf{1} \odot C_0 \cdot \mathring{\mathbf{0}}$, $C_5 = C_1 \odot \mathbf{1} \odot C_1 \cdot \mathring{\mathbf{0}}$.

(iv) $D_Y = C_{k_0} + \cdots + C_{k_{n-1}}$ *whenever* $0 \le k_0 < \cdots < k_{n-1} \le 5$ *and* $Y = \{k_0, \ldots, k_{n-1}\}$ *(so that* $n \le 6$*).*

Next, using these notations, we establish three lemmas, (v)–(vii).

(v) *The following equations are all logically valid:*

(α) $D_\varnothing = \mathbf{0}$, $D_6 = \mathbf{1}$, $D_{\{0,1\}} = \mathring{\mathbf{1}}$, $D_{\{2,\ldots,5\}} = \mathring{\mathbf{0}}$, $D_{\{k\}} = C_k$ *for* $k = 0, \ldots, 5$;

(β) $D_Y + D_Z = D_{Y \cup Z}$, $D_Y \cdot D_Z = D_{Y \cap Z}$, $D_Y^- = D_{6 \sim Y}$ *for any* $Y, Z \subseteq 6$.

This follows directly from (iii), (iv), and 2.2(DI)–(DV). (Recall that $\{0, \ldots, 5\} = 6$.)

Now consider any formalism $\overline{\mathcal{L}}^\times$ of the kind described, with new logical operators $\boldsymbol{O}_0, \ldots, \boldsymbol{O}_{m-1}$, where each $\boldsymbol{O}_i$ has rank $r(i)$. Let $\overline{\mathbf{\Pi}}$ be the set of all predicates of $\overline{\mathcal{L}}^\times$. Thus we have $\boldsymbol{O}_i(A_0, \ldots, A_{r(i)-1}) \in \overline{\mathbf{\Pi}}$ whenever $A_0, \ldots, A_{r(i)-1} \in \overline{\mathbf{\Pi}}$. In the remaining part of this section we frequently use realizations $\mathfrak{U} = \langle U, E \rangle$ of $\mathcal{L}^\times$ in which $E \subseteq U \upharpoonright Id$; for brevity we shall refer to them as *Id-models.*

(vi) *For every Id-model* $\mathfrak{U}$ *and every* $A \in \overline{\mathbf{\Pi}}$ *there is a* $Y \subseteq 6$ *such that* $A = D_Y$ *holds in* $\mathfrak{U}$.

To prove this, fix an $A \in \overline{\mathbf{\Pi}}$ and let $R, S_0, \ldots, S_5$ be the relations on $\mathfrak{U}$ denoted respectively by $A, C_0, \ldots, C_5$. We have:

(1) if $S_k \cap R \neq \varnothing$, then $S_k \subseteq R$, for each $k \in 6$.

Indeed, suppose $\langle x_0, x_1 \rangle \in S_k \cap R$, and let $\langle y_0, y_1 \rangle \in S_k$. It is easy to construct an automorphism F of $\mathfrak{U}$ which maps x_i to y_i for $i = 0, 1$. To give an example, if $x_0 = y_1$ and $x_1 \neq y_0$, then we define F by stipulating:

$$Fx_0 = y_0, \quad Fy_0 = x_1, \quad Fx_1 = y_1, \text{ and } Fz = z \text{ for } z \neq x_0, x_1, y_0.$$

From $x_0 S_k x_1$ and $y_0 S_k y_1$ and (iii) we easily see that $x_0 = x_1$ iff $y_0 = y_1$, and $x_i E x_i$ iff $y_i E y_i$ for $i = 0, 1$. Using this and the fact that $\mathfrak{U}$ is an *Id*-model, one readily checks that F is a well-defined automorphism of $\mathfrak{U}$. Since R is the relation denoted by $A \in \overline{\Pi}$, it is clear that R is invariant under F, and hence from $x_0 R x_1$ we conclude $y_0 R y_1$. Now $\langle y_0, y_1 \rangle$ in S_k is arbitrary, so $S_k \subseteq R$, as was to be shown.

Let Y be the set of $k \in 6$ for which $S_k \subseteq R$. By (1) and (iv), (v), we obtain that $R = \bigcup\{S_k : k \in Y\}$, i.e., that $A \equiv D_Y$ holds in $\mathfrak{U}$, as was to be shown.

More involved is the proof of the next lemma, which presents an essential improvement of (vi).

(vii) *There is a set* $\Delta \subseteq \overline{\Pi}$ *satisfying the following conditions:*

(α) Δ *is finite;*

(β) *for every* $A \in \overline{\Pi}$ *there is a* $B \in \Delta$ *such that* $A \equiv B$ *holds in all Id-models of* $\overline{\mathcal{L}}^{\times}$.

To prove (vii) we first specify a natural number n and we construct n classes $\mathsf{K}_0, \ldots, \mathsf{K}_{n-1}$ satisfying the following conditions:

(1) $\bigcup_{j<n} \mathsf{K}_j$ is precisely the class of all *Id*-models;

(2) given $A \in \overline{\Pi}$, there is, for each $j = 0, \ldots, n-1$, a $Y \subseteq 6$ such that $A \equiv D_Y$ is in $\mathbf{\Theta}\rho\mathsf{K}_j$ (the theory of K_j).

We proceed as follows. The set $Sb6 = \{Y : Y \subseteq 6\}$ has, of course, cardinality 64, and thus can be represented in the form

(3) $Sb6 = \{S_0, \ldots, S_{63}\}$.

Put

$$M = \{\langle i, X \rangle : i < m \text{ and } X = \langle X_0, \ldots, X_{r(i)-1} \rangle \in {}^{r(i)}Sb6\}$$

(where m is the number of new logical operators $\boldsymbol{O}_i$, and $r(i)$ is the rank of $\boldsymbol{O}_i$). M is of course finite, so that we can specify n by setting

(4) $n = |{}^{M}Sb6| = 64^{|M|}$.

Consequently, the members of ${}^{M}Sb6$, i.e., the functions from M to $Sb6$, can be arranged in a finite sequence of length n, so that

(5) $\quad {}^{M}\mathfrak{Sb}6 = \{F_0, \ldots, F_{n-1}\}.$

Using the functions F_j we now define the classes K_j, $j = 0, \ldots, n-1$, by letting

(6) $\quad \mathsf{K}_j = \{\mathfrak{U}\colon \mathfrak{U}$ is an Id-model and
$\boldsymbol{O}_i(D_{X_0}, \ldots, D_{X_{r(i)-1}}) \mathbin{\boldsymbol{=}} D_{F_j(i,X)} \in \boldsymbol{\Theta}\rho\mathfrak{U}$ for all $\langle i, X\rangle \in M\}.$

With every given Id-model $\mathfrak{U}$ we can correlate a definite function F_j from (5) such that $\mathfrak{U} \in \mathsf{K}_j$. In fact for each $\langle i, X\rangle \in M$ there exists, by (vi) and (3), an S_k for which

(7) $\quad \boldsymbol{O}_i(D_{X_0}, \ldots, D_{X_{r(i)-1}}) \mathbin{\boldsymbol{=}} D_{S_k} \in \boldsymbol{\Theta}\rho\mathfrak{U}.$

The function F_j is completely determined by stipulating that $F_j(i, X) = S_k$, where k is the least number for which (7) holds. Hence, by comparing (6) and (7), we see at once that $\mathfrak{U} \in \mathsf{K}_j$. Thus the classes $\mathsf{K}_0, \ldots, \mathsf{K}_{n-1}$ satisfy condition (1).

To show that they satisfy condition (2) as well, we let Ω be the set of predicates $A \in \overline{\boldsymbol{\Pi}}$ for which the conclusion of (2) holds, and we show by induction on predicates that all members of $\overline{\boldsymbol{\Pi}}$ are in Ω. Clearly $\mathring{\boldsymbol{1}}, \boldsymbol{E} \in \Omega$ since, by (iii), (v), and (1), we have $\mathring{\boldsymbol{1}} \mathbin{\boldsymbol{=}} D_{\{0,1\}}$, $\boldsymbol{E} \mathbin{\boldsymbol{=}} D_{\{0\}} \in \boldsymbol{\Theta}\rho\mathsf{K}_j$ for $j = 0, \ldots, n-1$. Now let $\boldsymbol{O}_i$, with $0 \leq i < m$, be an arbitrary operator in $\overline{\mathcal{L}}^\times$ and $A_0, \ldots, A_{r(i)-1}$ arbitrary predicates in Ω. Given any $j = 0, \ldots, n-1$, we can find, by the definition of Ω, sets $X_k \subseteq 6$ with $k = 0, \ldots, r(i) - 1$, such that $A_k \mathbin{\boldsymbol{=}} D_{X_k} \in \boldsymbol{\Theta}\rho\mathsf{K}_j$. Therefore

$$\boldsymbol{O}_i(A_0, \ldots, A_{r(i)-1}) \mathbin{\boldsymbol{=}} \boldsymbol{O}_i(D_{X_0}, \ldots, D_{X_{r(i)-1}}) \in \boldsymbol{\Theta}\rho\mathsf{K}_j,$$

so that, by (6),

$$\boldsymbol{O}_i(A_0, \ldots, A_{r(i)-1}) \mathbin{\boldsymbol{=}} D_{F_j(i,X)} \in \boldsymbol{\Theta}\rho\mathsf{K}_j,$$

and, finally, $\boldsymbol{O}_i(A_0, \ldots, A_{r(i)-1}) \in \Omega$. This completes the induction argument. We have thus shown that $\overline{\boldsymbol{\Pi}} \subseteq \Omega$ and that, in consequence, $\mathsf{K}_0, \ldots, \mathsf{K}_{n-1}$ satisfy condition (2).

In view of (2) we can correlate with each $A \in \overline{\boldsymbol{\Pi}}$ a sequence

$$GA = \langle G_0 A, \ldots, G_{n-1} A\rangle \in {}^{n}\mathfrak{Sb}6$$

(where n is defined as in (4)), by stipulating: for each $j < n$, choose the least k such that $A \mathbin{\boldsymbol{=}} D_{S_k} \in \boldsymbol{\Theta}\rho\mathsf{K}_j$ and set $G_j A = S_k$. From this definition we see that, for any $A, B \in \overline{\boldsymbol{\Pi}}$ with $GA = GB$, we have $A \mathbin{\boldsymbol{=}} B \in \boldsymbol{\Theta}\rho\mathsf{K}_j$ for every $j < n$. Therefore, by (1), we get:

(8) $\quad$ whenever $A, B \in \overline{\boldsymbol{\Pi}}$ and $GA = GB$, then $A \mathbin{\boldsymbol{=}} B$ holds in every Id-model.

As is well known, the set $\overline{\boldsymbol{\Pi}}$ is obviously denumerable and thus can be represented as the range of an infinite sequence of type ω. For each $Y \in G^*\overline{\boldsymbol{\Pi}}$ let HY be the first predicate in this sequence such that $G(HY) = Y$. Hence, by taking $H(GA)$ for B in (8) we conclude that

(9) $A \mathbin{\boldsymbol{=}} H(GA)$ holds in all Id-models for every $A \in \overline{\mathbf{\Pi}}$.

If we now put $\Delta = H^*(G^*\overline{\mathbf{\Pi}})$ (so that Δ is the range of H), we obtain a set which satisfies the conditions of our theorem. In fact, Δ satisfies (α) since $G^*\overline{\mathbf{\Pi}}$ is included in the finite set ${}^n\mathfrak{Sb}6$, while (9) directly implies that Δ satisfies (β) as well.

We can now state the main result concerning the equipollence of $\overline{\mathcal{L}}^{\times}$ with $\mathcal{L}$ and $\mathcal{L}^{+}$ in means of expression.[5]

(viii) *Given any formalism $\overline{\mathcal{L}}^{\times}$ (of the type discussed in the beginning of the section) there is an $X \in \Sigma$, and hence also an $X \in \Sigma^{+}$, such that X is not semantically equivalent with any $Y \in \overline{\Sigma}^{\times}$. In fact, there is a natural number p such that in the infinite sequence of sentences $S_0, \ldots, S_n, \ldots$ in Σ determined by the condition*

$$S_n = \forall_{v_0 \cdots v_n} \exists_{v_{n+1}} (\neg v_0 \mathring{\mathbf{1}} v_{n+1} \wedge \cdots \wedge \neg v_n \mathring{\mathbf{1}} v_{n+1}),$$

every sentence S_n with $n \geq p$ has this property.

Indeed, let Δ be as in (vii) and assume $|\Delta| = q$. Then there are at most q distinct sentences among the S_n's which are equivalent with some sentences in $\overline{\Sigma}^{\times}$. For suppose, to the contrary, that the $q+1$ sentences $S_{n_0}, \ldots, S_{n_q}$ are, respectively, equivalent to $Y_0, \ldots, Y_q \in \overline{\Sigma}^{\times}$. For each $i = 0, \ldots, q$ there is, by (vii), a predicate $B_i \in \Delta$ for which $Y_i^t \mathbin{\boldsymbol{=}} B_i$ holds in all Id-models. Since $|\Delta| = q$, there must be distinct i, j with $0 \leq i < j \leq q$ such that B_i and B_j coincide. Thus $Y_i^t \mathbin{\boldsymbol{=}} Y_j^t$, and hence also $S_i \leftrightarrow S_j$, holds in all Id-models. But this contradicts the fact that $S_i \leftrightarrow S_j$ is false in the Id-model $\langle i+2, i+2 \upharpoonright Id \rangle$. The theorem now easily follows.

Theorem (viii) is a far-reaching generalization of Korselt's result 3.4(iv); Theorem 3.4(v) can also be generalized in an analogous way.

The first part of (viii) states in a formal way that there is no formalism $\overline{\mathcal{L}}^{\times}$ whose expressive power at least equals that of $\mathcal{L}$ or $\mathcal{L}^{+}$. Thus, *a fortiori*, $\mathcal{L}$, $\mathcal{L}^{+}$, and $\overline{\mathcal{L}}^{+}$ are not (semantically) equipollent with $\overline{\mathcal{L}}^{\times}$ in means of expression.

Some generalizations of (viii) are known. In fact, it can be shown that (viii) extends from formalisms of type $\overline{\mathcal{L}}^{\times}$ to a considerably wider class of formalisms. To obtain formalisms of this class we include in the vocabulary of $\mathcal{L}^{\times}$ any finite number of logical constants denoting relations (not necessarily binary) among elements of the universe, operations on and to such relations, as well as relations among such relations, all of them of arbitrary but well-determined finite ranks.

[5]This result was found by Tarski in 1940 and was announced in a somewhat more general form in Tarski [**1941**], p. 89.

We can even provide such formalisms with finitely many logical connectives (like those occurring in $\mathcal{L}$), as well as with finitely many variables and logical quantifiers (like the universal quantifier)—thus giving them the appearance of formal languages more similar to $\mathcal{L}^{+}$ than to $\mathcal{L}^{\times}$. (Some simple examples of such formalisms will be discussed below in §§3.7–3.10.) Theorem (viii) applies to all extensions of $\mathcal{L}^{\times}$ thus obtained. Notice that the proof of (viii) is completely independent of the way in which derivability has been defined for the formalism discussed.

The problem whether any formalism $\overline{\mathcal{L}}^{\times}$ (of the type described at the beginning of the section) can be equipollent with $\overline{\mathcal{L}}^{+}$ in means of proof is still open, even if we restrict ourselves to the case when $\overline{\mathcal{L}}^{+}$ is a definitional extension of $\mathcal{L}$ (in the same sense as is $\mathcal{L}^{+}$). In fact, the problem is even open in the special case when $\overline{\mathcal{L}}^{+}$ and $\mathcal{L}^{+}$ coincide, and $\overline{\mathcal{L}}^{\times}$ is obtained from $\mathcal{L}^{\times}$ by leaving the vocabulary unchanged and by adjoining to $\mathbf{A}^{\times}$ finitely many new schemata, all instances of which are, as before, logically valid in $\mathcal{L}^{+}$. Clearly such a formalism $\overline{\mathcal{L}}^{\times}$ is a subformalism of $\mathcal{L}^{+}$, and the notion of derivability in it is, in general, stronger than that in $\mathcal{L}^{\times}$. This special case of the problem would, of course, be solved negatively if we managed to show that

(ix) *There is a sentence $X \in \Sigma^{\times}$ for which $\vdash^{+} X$ holds while $\vdash X$ $[\overline{\mathcal{L}}^{\times}]$ fails.*[6]

We do not know whether (ix) is true. If, however, the vocabularies of the formalisms involved contained not just the nonlogical atomic predicate $\boldsymbol{E}$, but rather infinitely many nonlogical atomic predicates, then (ix) would be an easy consequence of a result of Monk [**1964**] to the effect that the equational theory of representable relation algebras is not finitely based. However, allowing infinitely many such predicates would destroy one of those essential structural features of $\mathcal{L}^{\times}$ which secure its eminently finitistic character.

3.6. $\mathcal{L}^{\times}$-expressibility

We return to the negative results stated in 3.4(iv),(v). A sentence X in $\mathcal{L}^{+}$ will be referred to as *expressible in* $\mathcal{L}^{\times}$ or $\mathcal{L}^{\times}$*-expressible*, if $X \equiv Y$ for some $Y \in \Sigma^{\times}$. We shall use analogous terminology with respect to other formalisms as well. In this case the relation $\equiv$ is to be understood, in general, as semantical equivalence. (If, however, the formalisms discussed are subformalisms of $\mathcal{L}^{+}$, then $\equiv$ can of course be equally well interpreted as syntactical equivalence in $\mathcal{L}^{+}$.) By 3.4(iv),(v) we have:

(i) *S, S' are not $\mathcal{L}^{\times}$-expressible.*

The arguments used in establishing (i) have a restricted range of application and, in general, do not work for other sentences in Σ or Σ^{+} with equally simple

[6]*This special case was recently solved negatively in Maddux [**1987**]: the set of all true equations of $\mathcal{L}^{\times}$ is not finitely schematizable. His proof proceeds by showing that the set of all equations with just one variable that are true in all representable relation algebras is not finitely based.

structures. Recently, Kwatinetz [**1981**] published a fairly general method (discovered around 1972) which enabled him to show that several simple sentences with interesting mathematical contents are not $\mathcal{L}^\times$-expressible. We give here a few examples of such sentences.

(ii) *S_0, S_1, S_2 are not $\mathcal{L}^\times$-expressible, where*

$$\begin{aligned} S_0 &= \forall_{xy}\exists_z[zEx \land zEy \land \forall_u(uEx \land uEy \rightarrow uEz)], \\ S_1 &= \forall_{xy}\exists_z\forall_u(uEz \leftrightarrow uEx \lor uEy), \\ S_2 &= \forall_{xy}[\exists_z(xEz \land yEz) \rightarrow \exists_z\forall_u(uEz \leftrightarrow u\mathring{1}x \lor u\mathring{1}y)]. \end{aligned}$$

If interpreted in a structure $\langle U, E\rangle$ where E is a partial ordering relation, S_0 obviously expresses the fact that any two elements of U have a greatest lower bound (in the order established by E). Sentences S_1 and S_2 are provable in practically all set-theoretical systems known from the literature, and frequently occur in axiom sets of these systems; S_1 is the union axiom and S_2 the restricted pair axiom. Some other related sentences have been shown not to be $\mathcal{L}^\times$-expressible by Ulf Wostner; see Wostner [**1976**].

On the other hand, we know a great many sentences in Σ which are $\mathcal{L}^\times$-expressible; by 2.3(iv)(δ),(ε) all the sentences $\boldsymbol{G}Y$ with $Y \in \Sigma^\times$ have this property.

The problem of whether a particular sentence X in Σ or Σ^+ is $\mathcal{L}^\times$-expressible may present essential difficulties. As an example of a simple sentence for which the problem is open we mention

$$S_3 = \forall_{xy}\exists_z\forall_u(uEz \leftrightarrow u\mathring{1}x \lor uEy).$$

This sentence is well known in its set-theoretical interpretation and is provable in various familiar systems of set theory.

Around 1971, Kwatinetz (settling a problem stated in Tarski [**1941**], p. 89) showed that there is no general method which enables us to solve each particular case of this problem in finitely many steps; see Kwatinetz [**1981**]. In other words,

(iii) *The set of all sentences in $\mathcal{L}$, as well as the set of all sentences in $\mathcal{L}^+$, which are $\mathcal{L}^\times$-expressible is not recursive.*

In view of this last result, the problem of finding partial criteria for $\mathcal{L}^\times$-expressibility and $\mathcal{L}^\times$-inexpressibility becomes interesting and important. This is the problem of constructing various comprehensive and simply defined recursive sets which can be shown to consist exclusively of $\mathcal{L}^\times$-expressible or $\mathcal{L}^\times$-inexpressible sentences. At present we do not know any interesting criteria for $\mathcal{L}^\times$-inexpressibility, but we know some for $\mathcal{L}^\times$-expressibility. We shall become acquainted with such criteria in the subsequent discussion; cf. §§3.10 and 4.7.

The negative results stated in 3.4(vi) give rise to some problems analogous to those which were just discussed.

While $\boldsymbol{\Theta\eta}\varnothing$ and $\boldsymbol{\Theta\eta}^+\varnothing$ are well known not to be recursive, the question arises whether the sets $\boldsymbol{\Theta\eta}^\times\varnothing$ and $\boldsymbol{\Theta\eta}^+\varnothing\cap\Sigma^\times$ are recursive. We shall see later, in §4.7, that the answer in both cases is negative. Hence, we may be interested in partial criteria for a sentence to belong or not to belong to either of the two sets, and especially to $\boldsymbol{\Theta\eta}^\times\varnothing$. In particular, it seems likely that useful results could be obtained (perhaps with the help of computers) to the effect that every sentence belonging to $\boldsymbol{\Theta\eta}^+\varnothing \cap \Sigma^\times$ and having a sufficiently simple structure (e.g., being a particular instance of a sufficiently simple schema) belongs also to $\boldsymbol{\Theta\eta}^\times\varnothing$.

No interesting results of this type are known at present. The simplicity of a sentence (or a schema) is not, of course, a precisely defined or intuitively unambiguous notion. We could agree, for instance, to take the length of a sentence (i.e., the number of occurrences of symbols) as the measure of its simplicity. In this case the results of the type proposed could lead us to the solution of a problem mentioned towards the end of §3.4, in fact, to the construction of the shortest possible sentence which belongs to $\boldsymbol{\Theta\eta}^+\varnothing \cap \Sigma^\times$ but not to $\boldsymbol{\Theta\eta}^\times\varnothing$.

3.7. The three-variable formalisms $\mathcal{L}_3$ and $\mathcal{L}_3^+$

Since $\mathcal{L}^\times$ is poorer than $\mathcal{L}$ in means of expression and proof, the problem naturally arises whether a subformalism of $\mathcal{L}$ could be constructed which would be equipollent with $\mathcal{L}^\times$ in both respects.

A possible solution of this problem is suggested by the following considerations. By 2.3(iv)(ε), for every $X \in \Sigma^\times$ there is a $Y \in \Sigma$ such that X and Y are logically equivalent (in $\mathcal{L}^+$) and hence also semantically equivalent; in fact, $Y = \boldsymbol{G}X$ is such a sentence. Moreover, from the definition of $\boldsymbol{G}$ in 2.3(iii) it is easily seen that, in case $X \in \Sigma^\times$, the sentence $\boldsymbol{G}X$ contains at most three different variables, namely $\boldsymbol{x}, \boldsymbol{y}$, and $\boldsymbol{z}$.

On the other hand, by considering various sentences $Y \in \Sigma$ satisfying this last condition, we observe that each of them can easily be transformed into an equivalent sentence $X \in \Sigma^\times$. We give here two examples:

$$\boldsymbol{\forall}_x\boldsymbol{\exists}_y\boldsymbol{\forall}_z(z\boldsymbol{E}y \leftrightarrow z\mathring{\mathbf{1}}x) \;\equiv\; \boldsymbol{E}\cdot(\mathring{\mathbf{1}} \oplus \boldsymbol{E}^-)\odot\mathbf{1} = \mathbf{1},$$

$$\boldsymbol{\exists}_z\boldsymbol{\forall}_{xy}(x\boldsymbol{E}y \wedge y\boldsymbol{E}z \rightarrow x\boldsymbol{E}z) \;\equiv\; \mathbf{1}\odot[(\boldsymbol{E}^{-\smile} \oplus \boldsymbol{E}^{-\smile} \oplus \boldsymbol{E})\cdot\mathring{\mathbf{1}}]\odot\mathbf{1} = \mathbf{1}.$$

Assume now we have succeeded in showing that our observation applies to all sentences in Σ which contain no more than three different variables. It then becomes plausible that a subformalism of $\mathcal{L}$ which is equipollent with $\mathcal{L}^\times$ can be constructed simply by restricting the number of variables available to three and by otherwise leaving the structure of $\mathcal{L}$ virtually unchanged. Such a subformalism will certainly be equipollent with $\mathcal{L}^\times$ in means of expression. We recall from §2.5 that we have decided not to discuss in this work the equipollence of two formalisms in means of proof, unless they are presented as subformalisms of a third formalism, and at least one of them proves to be equipollent with the latter. In the present case, a subformalism of $\mathcal{L}^+$ obtained by a similar restriction of the number of variables appears to be a natural candidate for this third formalism.

We shall try to realize the idea outlined above. The subformalisms of $\mathcal{L}$ and $\mathcal{L}^+$ with which we shall deal will be denoted respectively by "$\mathcal{L}_3$" and "$\mathcal{L}_3^+$".

$\mathcal{L}_3^+$ is conceived here as the formalism obtained from $\mathcal{L}_3$ in exactly the same way in which the formalism $\mathcal{L}^+$ was obtained from $\mathcal{L}$. With this in mind we shall restrict ourselves for the most part to $\mathcal{L}_3$ in this section. The extension of our observations to $\mathcal{L}_3^+$ is straightforward, and therefore in the subsequent discussion we shall use them in reference to both $\mathcal{L}_3$ and $\mathcal{L}_3^+$.

The vocabulary of $\mathcal{L}_3$ contains just the three variables $\boldsymbol{x}, \boldsymbol{y}$, and $\boldsymbol{z}$; we set $\boldsymbol{\Upsilon}_3 = \{\boldsymbol{x}, \boldsymbol{y}, \boldsymbol{z}\}$. The constants and predicates are the same as in $\mathcal{L}$. The definitions of formulas and sentences are virtually unchanged; $\boldsymbol{\Phi}_3$ and $\boldsymbol{\Sigma}_3$ are the sets of formulas and sentences in $\mathcal{L}_3$. The set $\boldsymbol{\Lambda}_3$ of logical axioms is determined by the formula $\boldsymbol{\Lambda}_3 = \boldsymbol{\Lambda} \cap \boldsymbol{\Sigma}_3$. The definition of derivability, $\vdash_3$, for $\mathcal{L}_3$ differs from the corresponding definition for $\mathcal{L}$ only in that it refers to $\boldsymbol{\Sigma}_3$ and $\boldsymbol{\Lambda}_3$, and not to $\boldsymbol{\Sigma}$ and $\boldsymbol{\Lambda}$. As the realizations of $\mathcal{L}_3$ we take just the realizations of $\mathcal{L}$. Hence, the notion of semantical consequence in $\mathcal{L}_3$ coincides with the notion of semantical consequence in $\mathcal{L}$ restricted to sentences and sets of sentences in $\boldsymbol{\Sigma}_3$.

$\mathcal{L}_3$ turns out to be a formalism with significant deductive power. Many theorem schemata which are known to be logically provable in $\mathcal{L}$ (more precisely, all instances of which are known to be so provable) and which play an important role in the development of this formalism continue to be provable in $\mathcal{L}_3$ when restricted to sentences in $\boldsymbol{\Sigma}_3$. Examples of such schemata are provided, for instance, by Lemmas 1–17, 20, and 21 in Tarski [**1965**]. To convince oneself that these lemmas can be applied to $\mathcal{L}_3$, it suffices to study the details of their proofs, and to check that, in the situation when the lemmas refer exclusively to sentences in $\mathcal{L}_3$, all the formal derivations explicitly or implicitly involved in the proofs can be carried out without using any variables different from $\boldsymbol{x}, \boldsymbol{y}, \boldsymbol{z}$. The same observation extends to Lemmas *113–*170 in Quine [**1951**]. To see this, notice that the only axiom schema in Quine's system which is not among the axiom schemata of $\mathcal{L}$ is the one formulated in *104. However, by Lemma 15 in Tarski [**1965**], *104 is logically provable in $\mathcal{L}$. Therefore, Lemmas *113–*170 hold for $\mathcal{L}$ and hence, as before, when properly restricted they hold for $\mathcal{L}_3$ as well. In the sequel, when speaking of elementary properties of $\mathcal{L}_3$ we shall have in mind the properties obtained as restrictions of the lemmas in Tarski [**1965**] and Quine [**1951**] mentioned above, and perhaps a few further properties of a related nature.

As an example of a frequently used elementary property of $\mathcal{L}_3$ which we obtain in this fashion, we state the following result.

(i) *Consider the schema*

$$\text{(R)} \qquad [X \leftrightarrow X'] \rightarrow [Y \leftrightarrow Y'],$$

where X, X', Y, Y' are formulas such that X occurs as a subformula in Y, and Y' differs from Y only in that one such occurrence of X in Y has been replaced by X'. If S is any instance of (R) *with $X, X', Y, Y' \in \boldsymbol{\Phi}_3$, then $\vdash_3 S$; hence $Y \equiv_3 Y'$ whenever $X \equiv_3 X'$.*

Indeed, by Quine's Lemma *120, Schema (R) restricted as in (i) is logically provable in $\mathcal{L}$, and the argument carries over to $\mathcal{L}_3$. (R) is called the *schema*, and (i) the *law*, *of equivalent replacement*. We shall often use (i) above without referring to it explicitly.

$\mathcal{L}_3$ and $\mathcal{L}_3^+$ have been introduced as candidates for formalisms equipollent with $\mathcal{L}^\times$. In the next chapter it will be shown that $\mathcal{L}^\times$ is adequate as a framework for the formalization of set theory, and hence in a sense for the whole of mathematics. From this point of view, $\mathcal{L}_3$ (as well as $\mathcal{L}_3^+$) proves to have serious defects as a candidate for this task.[7] Actually it seems unlikely that any two of the three formalisms $\mathcal{L}_3, \mathcal{L}_3^+$, and $\mathcal{L}^\times$ are equipollent with each other.[8]

The main basis for this conjecture is the fact that, in spite of our previous observations concerning the significant deductive power of $\mathcal{L}_3$, we see no way of establishing for this formalism certain fundamental metalogical laws that hold for $\mathcal{L}$ and are used constantly in the development of formalized mathematical theories within predicate logic. In establishing these laws for $\mathcal{L}$ we make essential use of the fact that for any formula in $\mathcal{L}$, there are variables in the vocabulary of $\mathcal{L}$ which do not occur in this formula; in the case of $\mathcal{L}_3$ such variables may not be available. For similar reasons, other such laws are not even expressible in $\mathcal{L}_3$ unless they are severely restricted or undergo a substantial reformulation.

The laws to which we have referred above deal primarily with renaming bound variables and with substituting new variables for free variables. They usually have the form of statements asserting the validity of all instances (in a given formalism) of some general schemata.

To give some examples, without stating any general law on renaming bound variables (such as *171 in Quine [**1951**]), consider the following sentence, which has the form of a biconditional:

$$\forall_{xy}\exists_z(xEz \wedge zEy) \leftrightarrow \forall_{zy}\exists_x(zEx \wedge xEy).$$

Clearly each side of this biconditional is obtained from the other by renaming bound variables. Obviously the sentence is logically provable in $\mathcal{L}$. It seems, however, quite unlikely that it is also provable in $\mathcal{L}_3$.[9] As regards laws on substitution, the situation is more involved. Consider the following *general schema of simple substitution* (i.e., substitution for single variables):

(AVI′) $$[\forall_x X \rightarrow X[x/y]],$$

where X is any formula and x, y are any variables. (Recall that $X[x/y]$ is the formula obtained from X by substituting y for x; see §1.2.) (AVI′) is, of course,

[7] *However, see the footnote, p. 143.

[8] *Recently, Andréka and Németi have communicated to us that $\mathcal{L}_3$ and $\mathcal{L}_3^+$, in this form, are not equipollent in means of proof, and that, in fact, the sentence

$$\forall_{xy}\Big(\exists_z\,[z\mathring{1}x \wedge \exists_x\,(x\mathring{1}y \wedge \exists_y[y\mathring{1}z \wedge \exists_z(xEz \wedge zEy)])] \leftrightarrow \exists_z\,[z\mathring{1}y \wedge \exists_y\,(y\mathring{1}x \wedge \exists_x[x\mathring{1}z \wedge \exists_z(xEz \wedge zEy)])]\Big)$$

is derivable in $\mathcal{L}_3^+$, but not in $\mathcal{L}_3$. On p. 68 we point out why $\mathcal{L}_3^+$ is not equipollent with $\mathcal{L}^\times$.

[9] *Recently, Németi showed that this sentence is, in fact, not derivable in $\mathcal{L}_3$.

a generalization of the axiom schema (AVI). It is well known that every sentence in $\mathcal{L}$ which is an instance of (AVI′) is logically provable. If we restrict (AVI′) to those cases when the substitution does not require any renaming of bound variables (see §1.2), then all instances of (AVI′) so restricted which are in Σ_3 are provable in $\mathcal{L}_3$. But in other cases, the formula $X[x/y]$ may contain variables different from $\boldsymbol{x}, \boldsymbol{y}$, and $\boldsymbol{z}$, so that the corresponding instance of (AVI′) is not even expressed in $\mathcal{L}_3$. For instance, if

$$X = \forall_{\boldsymbol{y}}(\boldsymbol{x}E\boldsymbol{y}) \land \forall_{\boldsymbol{x}}\exists_{\boldsymbol{z}}(\boldsymbol{x}E\boldsymbol{z}),$$

then by §1.2,

$$X[\boldsymbol{x}/\boldsymbol{y}] = \forall_{\boldsymbol{u}}(\boldsymbol{y}E\boldsymbol{u}) \land \forall_{\boldsymbol{x}}\exists_{\boldsymbol{z}}(\boldsymbol{x}E\boldsymbol{z}).$$

It proves possible, however, to reformulate the definition of substitution in such a way that the formula $X[x/y]$ in the new sense is always logically equivalent in $\mathcal{L}$ to the formula $X[x/y]$ in the old sense, and that, at the same time, $X[x/y]$ is an expression in $\mathcal{L}_3$ whenever X, x, and y are expressions in $\mathcal{L}_3$. In formulating this new definition, it is convenient for technical reasons to first treat the general case of three variable substitution, i.e., of substituting variables u_0, u_1, u_2 for distinct variables x_0, x_1, x_2 in a formula X, and then to treat simple substitution as a special case of three variable substitution. The definition is constructed by recursion on formulas, along the lines of Monk [**1971**], p. 354, as follows. Let $X \in \mathbf{\Phi}_3$ and $x_0, x_1, x_2, u_0, u_1, u_2 \in \mathbf{\Upsilon}_3$, where x_0, x_1, x_2 are distinct. In case X is atomic the substitution is performed simply by replacing everywhere in X the variables x_0, x_1, x_2 with u_0, u_1, u_2, respectively. (Of course, in this case at most two of the variables x_0, x_1, x_2 actually occur in X; if X has the form $A = B$, and thus no variable occurs in it, then the result of substitution is clearly X itself.) The recursive procedure in case X is of the form $\neg Y$ or $Y \rightarrow Z$ is obvious. Finally, if X has the form, say, $\forall_{x_1} Y$, then we set

$$X[x_0/u_0, x_1/u_1, x_2/u_2] = \forall_v(Y[x_0/u_0, x_1/v, x_2/u_2]),$$

where v is the first variable different from u_0 and u_2. The cases when X has the form $\forall_{x_0} Y$ or $\forall_{x_2} Y$ are treated analogously. (Since $\{x_0, x_1, x_2\} = \mathbf{\Upsilon}_3$, this exhausts all possible cases.)

Other substitutions applicable in $\mathcal{L}_3$ easily reduce to the one discussed above. Thus, if $x_0, x_1, u_0, u_1 \in \mathbf{\Upsilon}_3$ and $x_0 \neq x_1$, we stipulate that $X[x_0/u_0, x_1/u_1]$ is $X[x_0/u_0, x_1/u_1, x_2/x_2]$, where x_2 is the unique variable in $\mathbf{\Upsilon}_3$ different from x_0, x_1. Similarly, for any $x_0, u_0 \in \mathbf{\Upsilon}_3$ we take $X[x_0/u_0]$ to be the formula $X[x_0/u_0, x_1/x_1, x_2/x_2]$, where x_1, x_2, with $\mathit{in}\, x_1 < \mathit{in}\, x_2$, are the first two variables different from x_0. The conventions regarding the notations $X[u]$, $X[u, v]$, $X[u, v, w]$ remain the same as in §1.2.

From now on, *whenever we shall speak about substitution in a formalism with only three different variables, we shall use this notion in the sense just defined.* In particular, this applies to the schema (AVI′). It seems very likely, however, that, under this new definition, various sentences in $\mathcal{L}_3$ which are instances of (AVI′) are not logically provable in $\mathcal{L}_3$.

In addition to (AVI′), we wish to consider another schema of a similar character:

(AIX′) $$[x\mathring{1}y \to (X \to X[x/y])].$$

It is closely related to, but much stronger and more general than, (AIX). Strictly speaking, (AIX′) is not an extension of (AIX), since there are instances of the latter which are not instances of the former; for example, the sentence $\forall_{xy}(x\mathring{1}y \to (xEx \to yEx))$ is such an instance. However, all instances of (AIX) are easily derivable with the help of (AIX′). The statement asserting the validity of all instances of (AIX′) (in a given formalism) may be called the *general Leibniz law.* As in the case of (AVI′), it seems very likely that not all instances of (AIX′) are logically provable in $\mathcal{L}_3$.

In addition to, and independent of, the defects discussed so far, $\mathcal{L}_3$ and $\mathcal{L}_3^+$ exhibit a further defect of a perhaps more special character. It comes to light when one tries to establish the equipollence of $\mathcal{L}_3^+$ with $\mathcal{L}^\times$ and, specifically, to prove that $\mathcal{L}^\times$ is a subformalism of $\mathcal{L}_3^+$. Indeed, it turns out that this is not the case. One of the axiom schemata for $\mathcal{L}^\times$, namely (BIV) in 3.1(ii), has instances which are not logically provable in $\mathcal{L}_3^+$.[10] This can be established by adapting and modifying an unpublished construction due to J. C. C. McKinsey, who used it to show that the associative law for relative products, which is one of the postulates for the theory of relation algebras, is not derivable from the remaining postulates for this theory.

Instead of the axiom schema (BIV) we can consider the following *associativity schema*:

(AX) $$[\exists_z(\exists_y(X[x,y] \wedge Y[y,z]) \wedge Z[z,y]) \leftrightarrow \exists_z(X[x,z] \wedge \exists_x(Y[z,x] \wedge Z[x,y]))],$$

where X, Y, Z are three arbitrary formulas with just two free variables, x and y. While Schema (BIV) is applicable to $\mathcal{L}^\times$ and $\mathcal{L}_3^+$, Schema (AX) can be applied to $\mathcal{L}_3$ and $\mathcal{L}_3^+$. Assume for the moment that all the defects of $\mathcal{L}_3$ and $\mathcal{L}_3^+$ considered above which do not involve (BIV) have been removed. We can then show that the sets of all instances of (BIV) and (AX) are logically equivalent in $\mathcal{L}_3^+$. We can conclude that the instances of (AX) in the formalism $\mathcal{L}_3^+$ are not all logically provable in that formalism, and we can extend this conclusion to $\mathcal{L}_3$.[11]

From this discussion it appears clear that the equipollence of $\mathcal{L}_3$ and $\mathcal{L}_3^+$ with $\mathcal{L}^\times$ cannot be achieved without substantially modifying the first two formalisms, and indeed without strengthening the notion of derivability in both of them. It turns out that to achieve the desired equipollence, it suffices to strengthen the sets $\mathbf{\Lambda}_3$ and $\mathbf{\Lambda}_3^+$ of logical axioms by replacing in them the instances of (AIX) with all those of (AIX′) and by adjoining all the instances of (AX). The first modification removes all defects connected with the procedures of substitution

[10]*In Németi [**1985**] it is shown that (BIV) is not even derivable in some rather strong Q-systems (see §4.5) formalized in $\mathcal{L}_3^+$.

[11]*Henkin [**1973**] contains a proof of the following closely related result: in certain three-variable subformalisms of a first-order formalism with at least three nonlogical atomic binary predicates, there are instances of (AX) that are not logically provable.

and renaming bound variables. This enables us to carry out the proof of the equipollence of $\mathcal{L}_3$ and $\mathcal{L}_3^+$ along the lines of the equipollence proof for $\mathcal{L}$ and $\mathcal{L}^+$ in §2.3. It also enables us to carry out all steps in the proof that $\mathcal{L}^\times$ is a subformalism of $\mathcal{L}_3^+$, except for the derivation of (BIV). The second modification overcomes this last difficulty and paves the way for the full proof of the equipollence of $\mathcal{L}_3^+$ and $\mathcal{L}^\times$.

It may be noticed that, for the modified formalisms $\mathcal{L}_3$ and $\mathcal{L}_3^+$, it still remains true that $\mathcal{L}_3^+$ can be obtained from $\mathcal{L}_3$ in exactly the same way as $\mathcal{L}^+$ from $\mathcal{L}$.

The modifications just described have some negative aspects. The notion of substitution has been essentially involved in the construction of the logical axiom set $\mathbf{\Lambda}_3$ and thus, indirectly, in the definition of derivability for $\mathcal{L}_3$. In addition, the restricted number of variables available in $\mathcal{L}_3$ forces us to use a notion of substitution that is much less natural and much more complicated than the original notion used in the development of $\mathcal{L}$. In consequence, our formalization of $\mathcal{L}_3$ is deprived of a feature that is a real virtue of the formalization of $\mathcal{L}$ outlined in Chapter 1 (see §1.3), namely the elimination of substitution from that definition. As the reader will see, this leads to complications in the metamathematical study of $\mathcal{L}_3$ and, in particular, in some inductive arguments concerning the notion of derivability. Unfortunately, we see no way of formalizing $\mathcal{L}_3$ along the lines of the construction in Chapter 1.[12]

Somewhat unexpectedly, however, we do know an equivalent formalization of $\mathcal{L}_3^+$ along the lines of Chapters 1 and 2. This equivalent formalization is obtained by changing the set of axiom schemata. The new set, $\mathbf{\Lambda}_3^{+\prime}$, consists of the following schemata: (AI)–(AIX) from 1.3(i), restricted to formulas in $\mathbf{\Phi}_3^+$; (DI), (DII), (DIV), (DV) from §2.2; the schema

$$\text{(DIII}'\text{)} \qquad \forall_{xy}[xA \odot By \leftrightarrow \exists_z(xAz \wedge zBy)],$$

where $x, y, z \in \mathbf{\Upsilon}_3$ and z is the first variable different from x and y; and finally, (BIV) from 3.1(ii), instead of (AX). (In Schema (DIII′), the variables x, y from $\mathbf{\Upsilon}_3$ are arbitrary, while in the weaker schema (DIII) from §2.2, $\boldsymbol{x}, \boldsymbol{y}, \boldsymbol{z}$ are, in order, the first three variables of $\mathbf{\Upsilon}$. For uniformity we could also replace (DI), (DII), and (DIV) by corresponding stronger schemata (DI′), (DII′), and (DIV′), analogous to (DIII′).) Obviously the notion of substitution is not involved in the construction of $\mathbf{\Lambda}_3^{+\prime}$. The proof that the two formalizations are equivalent, i.e., that the notions of derivability are the same, is not obvious. In fact, the proof that all instances of (AIX′) and (AX) are logically provable on the basis of $\mathbf{\Lambda}_3^{+\prime}$ is rather involved.

The axiom set $\mathbf{\Lambda}_3^{+\prime}$ will not be referred to in our further discussion. However, we may mention that, if the modified formalism $\mathcal{L}_3^+$ were based upon $\mathbf{\Lambda}_3^{+\prime}$, then the proof of the equipollence of $\mathcal{L}_3^+$ with $\mathcal{L}^\times$ (outlined in §3.9) would become simpler, while the proof of its equipollence with the modified $\mathcal{L}_3$ (outlined in §3.8) would become more complicated.

---

[12]*See the last four paragraphs of the footnote, p. 71.

Borrowing an idea used for related purposes in Maddux [**1978a**], p. 188, we can conjecture the possibility of an equivalent axiomatization of $\mathcal{L}_3$ (and $\mathcal{L}_3^+$) which is rather simpler than the one adopted here in so far as it does not involve the complicated notion of substitution defined above. To describe this axiomatization we agree, for any given formula X and variables y, z, to denote by X_{yz} the formula obtained from X by transposing y and z, i.e., by replacing every occurrence of y (whether free or bound) with z, and every occurrence of z with y. The axiom schemata (AI)–(AVIII) remain unchanged. Instead of (AIX′) we adjoin the original schema (AIX) and also a new *transposition schema*,

$$\text{(AIX}''\text{)} \qquad [x \mathring{\mathbf{1}} y \rightarrow (X \rightarrow X_{xy})],$$

where X is any formula and x, y any variables. (AX) is replaced by a closely related schema formulated in terms of transposition instead of substitution:

$$\text{(AX}'\text{)} \qquad [\exists_z(\exists_y(X \wedge Y_{xz}) \wedge Z_{xz}) \leftrightarrow \exists_z(X_{yz} \wedge \exists_x(Y_{yz} \wedge Z))],$$

where X, Y, Z are any three formulas, each with just two free variables, $\boldsymbol{x}$ and $\boldsymbol{y}$. It is easily seen that all instances of (AIX″) and (AX′) in $\mathcal{L}_3$ are logically derivable using (AI)–(AVIII), (AIX′), and (AX) as logical axiom schemata. (In fact, the derivability of (AIX″) follows at once from Theorems 3.8(i),(iv) in the next section.) We have not given thought to the problem of whether (AIX′) is derivable from (AI)–(AIX), (AIX″), and (AX′). Even if the answer to this problem is affirmative, we are not sure that the formalism $\mathcal{L}_3$ with the modified axiom set would provide a more convenient basis for our subsequent discussion.[13]

[13]*Recently, Givant has shown that one can equipollently formalize $\mathcal{L}_3$ and $\mathcal{L}_3^+$ using (AIX″) (and (AX′)) instead of (AIX′) (and (AX)). We give here a sketch of the proof.

Let Γ be the set of all instances (in Σ_3) of Schemata (AI)–(AVIII), (AIX′), and Δ the set of all instances of Schemata (AI)–(AIX), (AIX″). By 3.8(iv), all instances in Δ of (AIX″) are derivable from Γ, and the instances of (AIX) are easily seen to be so derivable. (Schema (AX) is not involved in the proof of 3.8(iv).) Therefore, all sentences in Δ are derivable from Γ. To prove the converse it suffices to derive all instances of (AIX′) from Δ.

First of all, on the basis of Δ we can prove the theorem on alphabetic variants, 3.8(vii). Indeed, 3.8(vii) rests on 3.8(v), which, in turn, is proved by applying 3.8(iv), i.e., by using various instances of (AIX″). Next, one proves the following lemma.

(1) For any distinct $x, y, z \in \Upsilon_3$ with $\mathit{in}\, y < \mathit{in}\, x$, and for any $Y \in \Phi_3$, the formula $(\forall_x Y)[y/z]$ is an alphabetic variant of $\forall_x(Y[y/z])$.

The proof of (1) is not difficult. It uses the fact that any such Y must be a Boolean combination of atomic and universally quantified formulas. Combining (1) and the theorem on alphabetic variants we obtain:

(2) For any distinct $x, y, z \in \Upsilon_3$ with $\mathit{in}\, y < \mathit{in}\, x$, and for any $Y \in \Phi_3$, the formula $(\forall_x Y)[y/z] \leftrightarrow \forall_x(Y[y/z])$ is derivable from Δ.

The final step is to show, for every $X \in \Phi_3$ and every $x, y \in \Upsilon_3$, that

(3) $[x \mathring{\mathbf{1}} y \rightarrow (X \leftrightarrow X[x/y])]$ is derivable from Δ.

The proof of (3) is by induction on formulas. Most of the steps are rather straightforward and use only some lemmas from Tarski [**1965**]. Of course the key cases are when X is of the form $\forall_u Y$ and $x \neq y$. If $u = x$, then $X[x/y] = X$ and (3) is trivial. If $u = y$, then

$$X[x/y] = (\forall_y Y)[x/y] = (\forall_y Y)[x/y, y/x] = X_{xy},$$

To conclude, we wish to point out that the text of this section contains several statements, the wording of which makes it clear that they are to be treated merely as conjectures. We are inclined to believe that the conjectures are correct, but we have made no attempt to demonstrate them. These conjectures have some interest in their own right, but their validity does not substantially influence our further discussion.

---

by 3.8(i),(ii), so that (3) is an easy consequence of (AIX″). If $u \neq x, y$ and $\mathit{in}\, u < \mathit{in}\, x$, then $X[x/y] = \forall_u(Y[x/y])$, so that (3) follows from the induction hypothesis and Lemma 11 in Tarski [**1965**]. Finally, if $u \neq x, y$ and $\mathit{in}\, x < \mathit{in}\, u$, then (3) follows from the induction hypothesis and (2) above.

The fact that we can use Δ as a logical base for $\mathcal{L}_3$ simplifies the statement and proof of 3.9(iv), as well as the proof of 3.9(v). Indeed, it appears that parts $(\alpha_5), (\alpha_6)$, and (β) can be dropped entirely from 3.9(iv), and that the proof of 3.9(v) in case $X \in \Delta$ is an instance of (AIX″) or (AX′) requires the consideration of many fewer cases than for the corresponding schemata (AIX′) and (AX).

Quite recently, Andréka and Németi have informed us that they have found a set Ξ of axioms for $\mathcal{L}_3$ in which the notion of substitution is not directly involved. The idea behind their axiomatization is the observation of Tarski [**1965**], p. 62, that the formula $\exists_x(x\mathring{\mathbf{1}}y \wedge X)$ can serve as an adequate substitute for the formula $X[x/y]$ in the development of first-order logic. (See also Tarski [**1951**] and Henkin–Tarski [**1961**], p. 86.)

To define Ξ it is helpful to introduce some abbreviations. For any formulas $X, Y \in \Phi_3$ in which the variable z does not occur free, we set

$$X \circledast Y = \exists_z[\exists_y(y\mathring{\mathbf{1}}z \wedge X) \wedge \exists_x(x\mathring{\mathbf{1}}z \wedge Y)],$$
$$X^\circ = \exists_z(z\mathring{\mathbf{1}}y \wedge \exists_y[y\mathring{\mathbf{1}}x \wedge \exists_x(x\mathring{\mathbf{1}}z \wedge X)]).$$

The formulas $X \circledast Y$ and X° are respectively equivalent to the formulas $\exists_z(X[x,z] \wedge Y[z,y])$ and $X[y,x]$ in $\mathcal{L}$, and thus can play the role of the "relative product" of two formulas X and Y, and of the "converse" of a formula X (cf. Henkin–Tarski [**1961**], pp. 86, 103). The definition of X° is slightly complicated owing to the difficulties involved in using Tarski's idea to define certain simultaneous substitutions.

The set Ξ consists of all instances in $\mathcal{L}_3$ of the following thirteen schemata: (AI)–(AIX), the schema

$$[x\mathring{\mathbf{1}}y \rightarrow (\exists_x(x\mathring{\mathbf{1}}y \wedge X) \rightarrow X)]$$

(for $X, Y \in \Phi_3$ and $x, y \in \Upsilon_3$), which, in view of the observations made above, is closely related to Schema (AIX′), and the schemata

$$[(X \circledast Y) \circledast Z \leftrightarrow X \circledast (Y \circledast Z)],$$
$$[(X \circledast Y)^\circ \leftrightarrow Y^\circ \circledast X^\circ],$$
$$[X^{\circ\circ} \leftrightarrow X],$$

where X and Y are arbitrary formulas in Φ_3 in which the variable z does not occur free.

Although this axiomatization shows that, in principle, we can avoid the notion of substitution in defining the set of logical axioms of $\mathcal{L}_3$, it does not seem to be a particularly simple axiomatization. It is true that if we used it as the basis for our discussion in this and the next section, then Lemma 3.9(iv) would become superfluous. However, it is not clear whether the proof of 3.9(v) would become substantially simpler than, say, its proof using the axiom set Δ mentioned above. It would be quite interesting to find a simple and natural axiomatization of $\mathcal{L}_3$ (and of the formalism obtained from $\mathcal{L}_3$ by deleting all instances of Schema (AX)) that does not involve the notion of substitution.

3.8. The equipollence of $\mathcal{L}_3$ and $\mathcal{L}_3^+$

Henceforth, *by* $\mathcal{L}_3$ *and* $\mathcal{L}_3^+$ *we shall always understand the formalisms obtained from those described in the first part of* §3.7 *by replacing* (AIX) *with* (AIX′) *and adding* (AX) *in the definitions of* $\mathbf{\Lambda}_3$ *and* $\mathbf{\Lambda}_3^+$. Our main task now is to establish the equipollence (in means of expression and proof) of the three formalisms $\mathcal{L}_3, \mathcal{L}_3^+$, and $\mathcal{L}^{\times}$, and, more specifically, to show that $\mathcal{L}_3$ and $\mathcal{L}^{\times}$ have $\mathcal{L}_3^+$ as their common equipollent extension.

To carry out this task we shall need to know that certain basic metalogical theorems which are known to hold for $\mathcal{L}$, especially those of the types discussed in §3.7, can be extended to $\mathcal{L}_3$ (and $\mathcal{L}_3^+$). A simple theorem of this character, the law of equivalent replacement, was stated, with a hint of its proof, in 3.7(i). Two further theorems of this kind, concerning Schemata (AIX′) and (AX), have become available simply because all instances of these schemata have now been included in $\mathbf{\Lambda}_3$. Another such theorem, in fact the general law of simple substitution (to the effect that all sentences in Σ_3 which are instances of (AVI′) are logically provable in $\mathcal{L}_3$) admits now a straightforward proof along the lines of the proof of Lemma 14 in Tarski [**1965**].

Among the remaining theorems of the same type, the most important for our purposes is the *law of renaming bound variables*, which we are going to take up now. We begin with a series of lemmas. The first two of them do not involve the notion of derivability in the formalism $\mathcal{L}_3$, but are merely consequences of the definition of substitution adopted in this formalism.

(i) *Let P be any permutation of $\Upsilon_3 = \{\boldsymbol{x}, \boldsymbol{y}, \boldsymbol{z}\}$. Then for every $X \in \mathbf{\Phi}_3$, the formula $X[P\boldsymbol{x}, P\boldsymbol{y}, P\boldsymbol{z}]$ coincides with the formula obtained from X by simultaneously replacing everywhere $\boldsymbol{x}, \boldsymbol{y}$, and $\boldsymbol{z}$ with $P\boldsymbol{x}, P\boldsymbol{y}$, and $P\boldsymbol{z}$ respectively.*

This can be proved by a straightforward induction on formulas.

(ii) *If $\Upsilon_3 = \{u, v, w\}$ and X is a formula in $\mathbf{\Phi}_3$ beginning with $\forall_u$, then*

(α) $X[u/v, v/u] = X[v/u]$,

(β) $X[v/w, w/v] = X[v/u][w/v][u/w]$.

This theorem obviously expresses the fact that a simultaneous substitution for two variables reduces to a series of simple substitutions, i.e., substitutions for single variables. To prove it, let $X = \forall_u Z$. Applying our definition of substitution and (i) above, we get

$$X[v/u] = (\forall_u Z)[u/u, v/u, w/w] = \forall_v (Z[u/v, v/u, w/w]) = X[u/v, v/u],$$

which gives (α). Applying (α) three times, and using (i), we obtain (β).

(iii) *For every $X \in \mathbf{\Phi}_3$ and $v, w \in \Upsilon_3$ we have* $\vdash_3 [v\mathbf{\dot{1}}w \rightarrow (X \leftrightarrow X[v/w])]$.

This is easily derived by applying (AIX′), first to the given formula X, and then to $\neg X$.

(iv) *For every* $X \in \mathbf{\Phi}_3$ *and* $v, w \in \Upsilon_3$ *we have* $\vdash_3 [v\mathring{\mathbf{1}}w \rightarrow (X \leftrightarrow X[v/w, w/v])]$.

When $v = w$, (iv) is trivial, in view of (i). Assume, therefore, that $v \neq w$, and proceed by induction on formulas. Let Ω be the set of all formulas X in $\mathbf{\Phi}_3$ that satisfy the conclusion of (iv).

In case X is atomic we set

(1) $\quad X' = X[v/w, w/v]$

and observe that

(2) $\quad X[v/w] = X'[v/w]$.

By (iii) we get

(3) $\quad \vdash_3 \{[v\mathring{\mathbf{1}}w \rightarrow (X \leftrightarrow X[v/w])],\ [v\mathring{\mathbf{1}}w \rightarrow (X' \leftrightarrow X'[v/w])]\}$.

Clearly, (1)–(3) imply

(4) $\quad X \in \Omega$ whenever $X \in \mathbf{\Phi}_3$ and X is atomic.

We obviously have

(5) $\quad \neg Y,\ Y \rightarrow Z \in \Omega$ whenever $Y, Z \in \Omega$.

Consider still the case when X begins with $\forall_u$, where u is any variable in Υ_3. If $u = v$ or $u = w$, then the desired conclusion follows easily from (ii)(α) and (iii). Suppose now that $u \neq v, w$, so that u, v, and w are all distinct. By three consecutive applications of (iii) we arrive at

$$\vdash_3 [u\mathring{\mathbf{1}}v \wedge v\mathring{\mathbf{1}}w \rightarrow (X \leftrightarrow X[v/u][w/v][u/w])],$$

and hence, by (ii)(β), at

(6) $\quad \vdash_3 [u\mathring{\mathbf{1}}v \wedge v\mathring{\mathbf{1}}w \rightarrow (X \leftrightarrow X[v/w, w/v])]$.

Since X begins with $\forall_u$, we have $u \notin \Upsilon\phi X$, and hence, by (i), $u \notin \Upsilon\phi X[v/w, w/v]$. Thus, from (6) we conclude that

$$\vdash_3 [\exists_u(u\mathring{\mathbf{1}}v) \wedge v\mathring{\mathbf{1}}w \rightarrow (X \leftrightarrow X[v/w, w/v])],$$

and consequently, by (AVIII),

$$\vdash_3 [v\mathring{\mathbf{1}}w \rightarrow (X \leftrightarrow X[v/w, w/v])].$$

This last formula shows that

(7) $\quad X \in \Omega$ whenever $X \in \mathbf{\Phi}_3$ and X begins with $\forall$.

The conclusion follows directly from (4), (5), (7).

(v) *For any $X \in \mathbf{\Phi}_3$, if P is a permutation of $\mathbf{\Upsilon}_3$ which is the identity when restricted to $\mathbf{\Upsilon}\phi X$, then $X \equiv_3 X[P\boldsymbol{x}, P\boldsymbol{y}, P\boldsymbol{z}]$.*

We first consider the case when P leaves at least one variable fixed, say $\boldsymbol{x}$, so that $P\boldsymbol{x} = \boldsymbol{x}$. In view of (i), the conclusion obviously holds if P is the identity on all three variables. We can therefore restrict ourselves to the case when $P\boldsymbol{x} = \boldsymbol{x}$, $P\boldsymbol{y} = \boldsymbol{z}$, and $P\boldsymbol{z} = \boldsymbol{y}$, and thus to the case when P is a transposition. It follows then by the hypothesis of (v) that $\boldsymbol{y}, \boldsymbol{z}$ occur free neither in X nor, by (i), in $X[\boldsymbol{y}/\boldsymbol{z}, \boldsymbol{z}/\boldsymbol{y}]$. Since by (iv) we have

$$\vdash_3 [\boldsymbol{y}\mathring{\mathbf{1}}\boldsymbol{z} \rightarrow (X \leftrightarrow X[\boldsymbol{y}/\boldsymbol{z}, \boldsymbol{z}/\boldsymbol{y}])],$$

we can conclude

$$\vdash_3 [\exists_{\boldsymbol{yz}}(\boldsymbol{y}\mathring{\mathbf{1}}\boldsymbol{z}) \rightarrow (X \leftrightarrow X[\boldsymbol{y}/\boldsymbol{z}, \boldsymbol{z}/\boldsymbol{y}])],$$

whence

$$X \equiv_3 X[\boldsymbol{y}/\boldsymbol{z}, \boldsymbol{z}/\boldsymbol{y}].$$

A moment's reflection suffices to see that this is just the conclusion of our theorem in the case under consideration.

We still have to consider the case when P leaves no variable fixed. Hence, by hypothesis, $\mathbf{\Upsilon}\phi X = \emptyset$, i.e., $X \in \Sigma_3$. Also, as is well known, in this case P is a composition of two transpositions, say Q and R, so that $P = Q \circ R$. Now, as has been shown above, the conclusion of our theorem holds for transpositions, so we have

$$X \equiv_3 X[R\boldsymbol{x}, R\boldsymbol{y}, R\boldsymbol{z}],$$

and

$$X[R\boldsymbol{x}, R\boldsymbol{y}, R\boldsymbol{z}] \equiv_3 (X[R\boldsymbol{x}, R\boldsymbol{y}, R\boldsymbol{z}])[Q\boldsymbol{x}, Q\boldsymbol{y}, Q\boldsymbol{z}],$$

whence

$$X \equiv_3 (X[R\boldsymbol{x}, R\boldsymbol{y}, R\boldsymbol{z}])[Q\boldsymbol{x}, Q\boldsymbol{y}, Q\boldsymbol{z}].$$

Since $P = Q \circ R$, it is clearly seen from (i) that the last formula yields directly our conclusion, and the proof is complete.

To formulate the law of renaming bound variables we need the notion of an alphabetic variant of a given formula X. (The term "alphabetic variant" originates with Quine [**1951**], p. 111.) This notion is defined as follows.

(vi) *Let $X, Y \in \mathbf{\Phi}_3$.*

(α) *Y is called a direct alphabetic variant of X if there is a permutation P of $\mathbf{\Upsilon}_3$ such that P is the identity when restricted to $\mathbf{\Upsilon}\phi X$ and $Y = X[P\boldsymbol{x}, P\boldsymbol{y}, P\boldsymbol{z}]$.*

(β) *In general, Y is called an alphabetic variant of X if there is a finite sequence $\langle Z_0, \ldots, Z_m\rangle$ of formulas in $\mathbf{\Phi}_3$ such that $X = Z_0, Y = Z_m$, and, for every i with $0 \leq i < m$, the formula Z_{i+1} is obtained from Z_i by replacing a single occurrence of some subformula of Z_i with some direct alphabetic variant of this subformula.*

This definition of the notion of an alphabetic variant of X differs from the ones implicitly or explicitly used in the literature in that its formulation does not necessitate the use of new variables not occurring in X.

The law on renaming bound variables can now be formulated in the following way.

(vii) *If $X, Y \in \mathbf{\Phi}_3$ and Y is an alphabetic variant of X, then $X \equiv_3 Y$.*

In case Y is a direct alphabetic variant of X, Theorem (vii) is an immediate consequence of (v), (vi)(α). This extends to arbitrary alphabetic variants by a straightforward induction based upon (vi)(β) and using 3.7(i).

We take up the proof of the equipollence of $\mathcal{L}_3$ and $\mathcal{L}_3^+$. We shall frequently use here the fact that various results stated in Chapter 2, especially in §2.3, continue to hold if the notions involved in these results that refer to the formalism $\mathcal{L}$ or $\mathcal{L}^+$ are now referred to $\mathcal{L}_3$ or $\mathcal{L}_3^+$; more specifically, this means that symbols like "$\mathbf{\Phi}$", "$\mathbf{\Phi}^+$", "$\vdash$", "$\vdash^+$", etc., are now provided with a subscript "3". For brevity, we shall refer to statements obtained in this fashion as *appropriately modified* statements.

We see at once that the appropriately modified statements 2.3(i),(ii) hold.

(viii) (α) *$\Sigma_3 \subseteq \Sigma_3^+$ and $\mathbf{\Phi}_3 \subseteq \mathbf{\Phi}_3^+$.*
(β) *For every $\Psi \subseteq \Sigma_3$ and $X \in \Sigma_3$, if $\Psi \vdash_3 X$, then $\Psi \vdash_3^+ X$.*

Thus, $\mathcal{L}_3$ is a subformalism of $\mathcal{L}_3^+$.

As a translation mapping from $\mathcal{L}_3^+$ to $\mathcal{L}_3$ we shall use the translation mapping $\mathbf{G}$ defined as in 2.3(iii), but restricted in its domain to $\mathbf{\Phi}_3^+$; we shall use the same symbol "$\mathbf{G}$" (without any subscript) to denote this restricted mapping. It is obvious that the appropriately modified results 2.3(iv)(α)–(γ) are valid for the restricted mapping $\mathbf{G}$.

This applies also to 2.3(iv)(δ) if we recall Definition 2.3(iii) and notice, in particular, the exact wording of 2.3(iii)(δ),(ς). As regards 2.3(iv)(ε), suppose that a detailed proof of this theorem for $\mathcal{L}$ and $\mathcal{L}^+$ is available and that this proof has a purely syntactical (proof-theoretical) character, avoiding any application of the semantical completeness theorem. There is no difficulty in supplying such a proof on the basis of the definitions of $\mathbf{\Lambda}$ and $\mathbf{\Lambda}^+$ given in §§1.3 and 2.2, and then to modify it so that it applies to $\mathcal{L}_3$ and $\mathcal{L}_3^+$. We proceed by induction on predicates and formulas, and we use some of the lemmas in Tarski [**1965**] and Quine [**1951**] which, according to the remarks in §3.7, extend to the formalisms presently discussed; we also use the law on renaming bound variables, (vii). The fact that the appropriately modified 2.3(iv)(ε) holds in $\mathcal{L}_3^+$ obviously implies that the same is true of 2.3(iv)(ς). Consequently, the appropriately modified 2.3(iv) is valid.

(ix) *The restricted mapping $\mathbf{G}$ has the following properties.*

(α) *$\mathbf{G}$ is a recursive function.*

(β) *$GX = X$ if $X \in \mathbf{\Phi}_3$.*
(γ) *$\Upsilon\phi(GX) = \Upsilon\phi X$ for every $X \in \mathbf{\Phi}_3^+$.*
(δ) *G maps $\mathbf{\Phi}_3^+$ onto $\mathbf{\Phi}_3$, and Σ_3^+ onto Σ_3.*
(ε) *$GX \equiv_3^+ X$ for every $X \in \mathbf{\Phi}_3^+$.*
(ς) *$G^*\Psi \equiv_3^+ \Psi$ for every $\Psi \subseteq \Sigma_3^+$.*

To proceed further we need the following lemma.

(x) *$G(X[u_0, u_1, u_2]) \equiv_3 (GX)[u_0, u_1, u_2]$ for all $X \in \mathbf{\Phi}_3^+$ and $u_0, u_1, u_2 \in \Upsilon_3$.*

This is proved by induction on formulas; the case of atomic formulas requires a preliminary induction on predicates and makes use of (vii).

We can now establish the main mapping theorem for $\mathcal{L}_3$ and $\mathcal{L}_3^+$ by showing that

(xi) *For every $\Psi \subseteq \Sigma_3^+$ and $X \in \Sigma_3^+$, we have $\Psi \vdash_3^+ X$ iff $G^*\Psi \vdash_3 GX$.*

The proof of this is entirely analogous to that of 2.3(v). Some attention should be given to that portion of the proof in which we deal with Axiom Schemata (AIX′) and (AX), and show that, for every $Y \in \Sigma_3^+$ which is an instance of one of these two schemata, we have $\vdash_3 GY$. We use here in an essential way Lemma (x).

From (ix)(β) and (xi) we easily derive the appropriately modified statement 2.3(vi), just as in §2.3.

As was pointed out in 2.4(vi), the main mapping theorem (together with such statements as 2.3(i),(iv)(β),(δ),(ε)) implies directly the proper equipollence theorems 2.3(ii),(vii)–(ix), and this applies not only to $\mathcal{L}$ and $\mathcal{L}^+$, but also to any formalisms $\mathcal{F}$ and $\mathcal{F}^+$. Thus, in particular, the proper equipollence theorems hold for $\mathcal{L}_3$ and $\mathcal{L}_3^+$.

(xii) (α) *For every $X \in \Sigma_3$ there is a $Y \in \Sigma_3^+$ such that $X \equiv_3^+ Y$, and for every $X \in \Sigma_3^+$ there is a $Y \in \Sigma_3$ such that $X \equiv_3^+ Y$.*
(β) *For every $\Psi \subseteq \Sigma_3$ and $X \in \Sigma_3$, we have $\Psi \vdash_3^+ X$ iff $\Psi \vdash_3 X$.*

3.9. The equipollence of $\mathcal{L}^\times$ and $\mathcal{L}_3^+$

We now take up the problem of the equipollence of $\mathcal{L}^\times$ and $\mathcal{L}_3^+$. We shall still refer here to the argument in §2.3 as a paradigm for proofs of equipollence between a formalism and one of its subformalisms. Since, however, $\mathcal{L}^\times$ contains no variables, the notion of a formula (with free variables) is not applicable to it, and in consequence, certain portions of §2.3 in which this notion occurs will be disregarded.

First of all, $\mathcal{L}^\times$ is a subformalism of $\mathcal{L}_3^+$ (in the sense of §1.6).

(i) *$\Sigma^\times \subseteq \Sigma_3^+$; more specifically, $\Sigma^\times$ is the set of all sentences of Σ_3^+ that are atomic.*

(ii) *For every* $\Psi \subseteq \Sigma^{\times}$ *and* $X \in \Sigma^{\times}$, *if* $\Psi \vdash^{\times} X$, *then* $\Psi \vdash_3^+ X$.

The proof of (i) is obvious. That of (ii) is also rather straightforward; we make essential use of Axioms (DI)–(DV) in §2.2 and apply to them several times the law on renaming bound variables, 3.8(vii).

We now construct a function $\boldsymbol{H}$ on $\boldsymbol{\Phi}_3^+$ which, when restricted to $\boldsymbol{\Sigma}_3^+$, proves to be an appropriate translation mapping from $\mathcal{L}_3^+$ to $\mathcal{L}^{\times}$. Its definition is complicated and must be formulated with care. We give here enough hints for constructing such a definition, without formulating it precisely in all details.

$\boldsymbol{H}$ is defined by recursion on formulas, and maps $\boldsymbol{\Phi}_3^+$ into the set of all quantifier-free formulas in $\boldsymbol{\Phi}_3^+$. At each recursive passage we distinguish several cases dependent on whether the number of variables occurring free in the formulas involved is $0, 1, 2$, or 3. The definition of $\boldsymbol{H}$ is designed so that we always have $\boldsymbol{\Upsilon\phi H}X = \boldsymbol{\Upsilon\phi}X$ and $\boldsymbol{H}X \equiv_3^+ X$; in case $|\boldsymbol{\Upsilon\phi}X| < 3$, $\boldsymbol{H}X$ is an atomic formula; in case $|\boldsymbol{\Upsilon\phi}X| = 3$, i.e., $\boldsymbol{\Upsilon\phi}X = \boldsymbol{\Upsilon}_3 = \{\boldsymbol{x}, \boldsymbol{y}, \boldsymbol{z}\}$, $\boldsymbol{H}X$ is a conjunction of a finite sequence of formulas $U_0, \ldots, U_n$, where each U_k is of a special form,

$$U_k \;=\; \boldsymbol{x}A_k\boldsymbol{y} \vee \boldsymbol{x}B_k\boldsymbol{z} \vee \boldsymbol{y}C_k\boldsymbol{z},$$

with appropriately chosen predicates A_k, B_k, C_k. Formulas of this special form will be referred to in this section as canonical formulas.

The definition of $\boldsymbol{H}X$ for an atomic X runs as follows: if $X = uAv$ (where $u, v \in \boldsymbol{\Upsilon}_3$ and $A \in \boldsymbol{\Pi}$), then $\boldsymbol{H}X = X$ in case $\boldsymbol{in}\, u \leq \boldsymbol{in}\, v$, and $\boldsymbol{H}X = vA^{\smile}u$ otherwise; if $X = (A = B)$ (with $A, B \in \boldsymbol{\Pi}$), then $\boldsymbol{H}X = X$.

Assume $X = \neg Y$; $\boldsymbol{H}X$ is defined in terms of $\boldsymbol{H}Y$. If $|\boldsymbol{\Upsilon\phi}Y| = 3$ and $\boldsymbol{H}Y$ is a conjunction of n canonical formulas, then $\boldsymbol{H}X$ proves to be a conjunction of 3^n canonical formulas; $\boldsymbol{H}X$ is constructed from $\boldsymbol{H}Y$ by performing on $\neg HY$ certain simple transformations based primarily on some well-known laws of sentential logic (De Morgan law, distributive, commutative, and associative laws); the order in which these transformations are performed is unambiguously determined. For example, let $n = 2$, so that

$$\boldsymbol{H}Y \;=\; (\boldsymbol{x}A_0\boldsymbol{y} \vee \boldsymbol{x}B_0\boldsymbol{z} \vee \boldsymbol{y}C_0\boldsymbol{z}) \wedge (\boldsymbol{x}A_1\boldsymbol{y} \vee \boldsymbol{x}B_1\boldsymbol{z} \vee \boldsymbol{y}C_1\boldsymbol{z}).$$

Then we set

$$\boldsymbol{H}X \;=\; U_0 \wedge \cdots \wedge U_8,$$

where

$$\begin{aligned} U_0 \;&=\; \boldsymbol{x}A_0^- + A_1^-\boldsymbol{y} \vee \boldsymbol{x}\mathbf{0}\boldsymbol{z} \vee \boldsymbol{y}\mathbf{0}\boldsymbol{z}, \\ U_1 \;&=\; \boldsymbol{x}A_0^-\boldsymbol{y} \vee \boldsymbol{x}B_1^-\boldsymbol{z} \vee \boldsymbol{y}\mathbf{0}\boldsymbol{z}, \\ &\;\;\vdots \\ U_8 \;&=\; \boldsymbol{x}\mathbf{0}\boldsymbol{y} \vee \boldsymbol{x}\mathbf{0}\boldsymbol{z} \vee \boldsymbol{y}C_0^- + C_1^-\boldsymbol{z}. \end{aligned}$$

If $|\boldsymbol{\Upsilon\phi}Y|$ is 1 or 2 and $\boldsymbol{H}Y = uAv$, set $\boldsymbol{H}X = uA^-v$. If $|\boldsymbol{\Upsilon\phi}Y| = 0$, so that $\boldsymbol{H}Y \in \Sigma^{\times}$, set $\boldsymbol{H}X = [\mathbf{1} \odot (\boldsymbol{H}Y)^{t-} \odot \mathbf{1} = \mathbf{1}]$.

Now assume $X = Y \to Z$. This part of the definition is the most involved in that we have many cases. Each case is determined by the ordered pair of two sets, $\boldsymbol{\Upsilon}\phi Y$ and $\boldsymbol{\Upsilon}\phi Z$. Hence, in principle, we have to consider 64 different cases (however, some of them differ very little from one another). Out of these 64 cases, only six will be examined here.

For typographical convenience we set $Y' = \neg Y$ and notice that $\boldsymbol{\Upsilon}\phi Y' = \boldsymbol{\Upsilon}\phi Y$. $\boldsymbol{H}X$ can be expressed in terms of $\boldsymbol{H}Y'$ and $\boldsymbol{H}Z$. In our first four cases we assume that $\boldsymbol{\Upsilon}\phi Z = \{\boldsymbol{x}, \boldsymbol{y}, \boldsymbol{z}\}$ and that $\boldsymbol{H}Z$ is a conjunction of q canonical formulas; in actual examination of these cases we take, for simplicity, $q = 1$, and in fact we let

$$(1) \qquad \boldsymbol{H}Z = \boldsymbol{x}D\boldsymbol{y} \vee \boldsymbol{x}E\boldsymbol{z} \vee \boldsymbol{y}F\boldsymbol{z}.$$

We consider the cases when $\boldsymbol{\Upsilon}\phi Y'$ coincides with one of the four sets $\{\boldsymbol{x}, \boldsymbol{y}, \boldsymbol{z}\}$, $\{\boldsymbol{x}, \boldsymbol{z}\}$, $\{\boldsymbol{y}\}$, $\varnothing$. $\boldsymbol{H}X$ is constructed by transforming $\boldsymbol{H}Y' \vee \boldsymbol{H}Z$ in a way similar to the way in which $\neg\boldsymbol{H}Y$ was previously transformed.

Case 1. $\boldsymbol{\Upsilon}\phi Y' = \{\boldsymbol{x}, \boldsymbol{y}, \boldsymbol{z}\}$ and $\boldsymbol{H}Y'$ is a conjunction of p canonical formulas; then $\boldsymbol{H}X$ is a conjunction of $p \cdot q$ such formulas. (The fact that, by the preceding part of our definition, p must be of the form 3^n plays no role in our discussion.) For simplicity take $p = 3$, and in fact let

$$(2) \qquad \begin{aligned} \boldsymbol{H}Y' = {} & (\boldsymbol{x}A_0\boldsymbol{y} \vee \boldsymbol{x}B_0\boldsymbol{z} \vee \boldsymbol{y}C_0\boldsymbol{z}) \wedge (\boldsymbol{x}A_1\boldsymbol{y} \vee \boldsymbol{x}B_1\boldsymbol{z} \vee \boldsymbol{y}C_1\boldsymbol{z}) \wedge \\ & (\boldsymbol{x}A_2\boldsymbol{y} \vee \boldsymbol{x}B_2\boldsymbol{z} \vee \boldsymbol{y}C_2\boldsymbol{z}). \end{aligned}$$

Since we want $\boldsymbol{H}X$ to be logically equivalent to $\neg\boldsymbol{H}Y \vee \boldsymbol{H}Z$, and hence to $\boldsymbol{H}Y' \vee \boldsymbol{H}Z$, we set

$$\begin{aligned} \boldsymbol{H}X = {} & (\boldsymbol{x}A_0 + D\boldsymbol{y} \vee \boldsymbol{x}B_0 + E\boldsymbol{z} \vee \boldsymbol{y}C_0 + F\boldsymbol{z}) \wedge \\ & (\boldsymbol{x}A_1 + D\boldsymbol{y} \vee \boldsymbol{x}B_1 + E\boldsymbol{z} \vee \boldsymbol{y}C_1 + F\boldsymbol{z}) \wedge \\ & (\boldsymbol{x}A_2 + D\boldsymbol{y} \vee \boldsymbol{x}B_2 + E\boldsymbol{z} \vee \boldsymbol{y}C_2 + F\boldsymbol{z}). \end{aligned}$$

Case 2. $\boldsymbol{\Upsilon}\phi Y' = \{\boldsymbol{x}, \boldsymbol{z}\}$; say $\boldsymbol{H}Y' = \boldsymbol{x}A\boldsymbol{z}$. Then we set

$$\boldsymbol{H}X = \boldsymbol{x}D\boldsymbol{y} \vee \boldsymbol{x}A + E\boldsymbol{z} \vee \boldsymbol{y}F\boldsymbol{z}.$$

Case 3. $\boldsymbol{\Upsilon}\phi Y' = \{\boldsymbol{y}\}$; say $\boldsymbol{H}Y' = \boldsymbol{y}A\boldsymbol{y}$. Then

$$\boldsymbol{H}X = \boldsymbol{x}(\mathbf{1} \odot (A \cdot \mathring{\mathbf{1}}) + D)\boldsymbol{y} \vee \boldsymbol{x}E\boldsymbol{z} \vee \boldsymbol{y}(A \cdot \mathring{\mathbf{1}} \odot \mathbf{1} + F)\boldsymbol{z}.$$

Case 4. $\boldsymbol{\Upsilon}\phi Y' = \varnothing$; say $\boldsymbol{H}Y' = (A = B)$. Then

$$\begin{aligned} \boldsymbol{H}X = {} & \boldsymbol{x}[\mathbf{0} \oplus (A \cdot B + A^- \cdot B^-) \oplus \mathbf{0} + D]\boldsymbol{y} \vee \\ & \boldsymbol{x}[\mathbf{0} \oplus (A \cdot B + A^- \cdot B^-) \oplus \mathbf{0} + E]\boldsymbol{z} \vee \\ & \boldsymbol{y}[\mathbf{0} \oplus (A \cdot B + A^- \cdot B^-) \oplus \mathbf{0} + F]\boldsymbol{z}. \end{aligned}$$

We also wish to consider two cases when both Y' and Z have less than three free variables.

Case 5. $\Upsilon\phi Y' = \{\boldsymbol{y}\}$ and $\Upsilon\phi Z = \{\boldsymbol{x}, \boldsymbol{z}\}$; say $\boldsymbol{H}Y' = \boldsymbol{y}A\boldsymbol{y}$ and $\boldsymbol{H}Z = \boldsymbol{x}E\boldsymbol{z}$. We then set

$$\boldsymbol{H}X = \boldsymbol{x}\mathbf{1}\odot(A\cdot\mathring{\mathbf{1}})\boldsymbol{y} \vee \boldsymbol{x}E\boldsymbol{z} \vee \boldsymbol{y}A\cdot\mathring{\mathbf{1}}\odot\mathbf{1}\boldsymbol{z}.$$

Case 6. $\Upsilon\phi Y' = \{\boldsymbol{z}\}$ and $\Upsilon\phi Z = \{\boldsymbol{x}, \boldsymbol{z}\}$; say $\boldsymbol{H}Y' = \boldsymbol{z}A\boldsymbol{z}$ and $\boldsymbol{H}Z = \boldsymbol{x}E\boldsymbol{z}$. Then

$$\boldsymbol{H}X = \boldsymbol{x}[\mathbf{1}\odot(A\cdot\mathring{\mathbf{1}})+E]\boldsymbol{z}.$$

We still have to define $\boldsymbol{H}X$ in case $X = \forall_u Y$. If $u \notin \Upsilon\phi Y$, we set $\boldsymbol{H}X = \boldsymbol{H}Y$. Assume now that $u \in \Upsilon\phi Y$. Then $|\Upsilon\phi X| < 3$ and $\boldsymbol{H}X$ must be an atomic formula. Consider the case $|\Upsilon\phi Y| = 3$, so that $\boldsymbol{H}Y$ has the form

$$\boldsymbol{H}Y = (\boldsymbol{x}A_0\boldsymbol{y} \vee \boldsymbol{x}B_0\boldsymbol{z} \vee \boldsymbol{y}C_0\boldsymbol{z}) \wedge \cdots \wedge (\boldsymbol{x}A_n\boldsymbol{y} \vee \boldsymbol{x}B_n\boldsymbol{z} \vee \boldsymbol{y}C_n\boldsymbol{z}).$$

We have $u \in \Upsilon_3$; let, e.g., $u = \boldsymbol{y}$. We then set

$$\boldsymbol{H}X = \boldsymbol{x}(A_0 \oplus C_0 + B_0) \cdot \cdots \cdot (A_n \oplus C_n + B_n)\boldsymbol{z}.$$

In the case $|\Upsilon\phi Y| = 2$, assume, e.g., $\boldsymbol{H}Y = \boldsymbol{x}A\boldsymbol{y}$ and $u = \boldsymbol{y}$; set $\boldsymbol{H}X = \boldsymbol{x}A\oplus\mathbf{0}\boldsymbol{x}$. In case $|\Upsilon\phi Y| = 1$ and $\boldsymbol{H}Y = uAu$, set $\boldsymbol{H}X = (A+\mathring{\mathbf{0}} = \mathbf{1})$.

We hope the above outline gives an adequate idea of the definition of $\boldsymbol{H}$. We now state formally several properties of $\boldsymbol{H}$.

(iii) (α) *$\boldsymbol{H}$ is a recursive function.*

(β) *$\boldsymbol{H}X = X$ for every $X \in \Sigma^{\times}$.*

(γ) *Assume $X \in \boldsymbol{\Phi}_3^+$.*

(γ_1) *If $|\Upsilon\phi X| = 3$, then there exists a uniquely determined number $n \in \omega$ and three uniquely determined sequences $A, B, C \in {}^{n+1}\Pi$ such that*

$$\boldsymbol{H}X = (\boldsymbol{x}A_0\boldsymbol{y} \vee \boldsymbol{x}B_0\boldsymbol{z} \vee \boldsymbol{y}C_0\boldsymbol{z}) \wedge \cdots \wedge (\boldsymbol{x}A_n\boldsymbol{y} \vee \boldsymbol{x}B_n\boldsymbol{z} \vee \boldsymbol{y}C_n\boldsymbol{z}).$$

(γ_2) *If $1 \le |\Upsilon\phi X| \le 2$, then there exist a uniquely determined $A \in \Pi$ and uniquely determined $u, v \in \Upsilon_3$, with $\mathit{in}\, u \le \mathit{in}\, v$ and $\{u, v\} = \Upsilon\phi X$, such that $\boldsymbol{H}X = uAv$.*

(γ_3) *If $|\Upsilon\phi X| = 0$, then there exist uniquely determined $A, B \in \Pi$ such that $\boldsymbol{H}X = (A = B)$.*

(δ) (δ_1) *$\boldsymbol{H}$ maps the set $\boldsymbol{\Phi}_3^+$ into the set of quantifier-free formulas in $\boldsymbol{\Phi}_3^+$, and the set Σ_3^+ onto $\Sigma^{\times}$.*

(δ_2) *$\Upsilon\phi\boldsymbol{H}X = \Upsilon\phi X$ for every $X \in \boldsymbol{\Phi}_3^+$.*

(ε) *$\boldsymbol{H}X \equiv_3^+ X$ for every $X \in \boldsymbol{\Phi}_3^+$.*

Parts (α) and (β) follow immediately from the definition of $\boldsymbol{H}$. (γ) is established by an easy induction on formulas in $\boldsymbol{\Phi}_3^+$. (δ) is an obvious consequence of (γ), at one point with the help of (β). The proof of (ε), again by induction on formulas in $\boldsymbol{\Phi}_3^+$ and with the help of some elementary properties of $\mathcal{L}_3^+$, is straightforward.

From (δ) and (ε) we see that $\boldsymbol{H}$ maps every formula in $\mathcal{L}_3^+$ to a logically equivalent formula containing no quantifiers. Thus, using frequently employed terminology, we can say that $\boldsymbol{H}$ permits us to eliminate quantifiers from any given formula in $\mathcal{L}_3^+$.

Before proceeding further, we state three preliminary results, (iv)–(vi), that will be involved in our discussion. The proofs of these results do not present any essential difficulties. Certain portions of the arguments require a familiarity with the fundamental equational laws of $\mathcal{L}^\times$ that are listed in §3.2, and some skill in deriving further laws of a similar nature. The proofs of (iv) and (v) are long and tedious. These are typical "proofs by cases", and the number of cases involved is so large that a complete presentation of the arguments would certainly wear out the reader.[14] Therefore, in each proof we shall restrict ourselves to illustrating the argument by means of two examples of average difficulty. Only (vi) will be proved in full.

The content of (iv) can be approximately described in the following way: given any formula X and a substitution instance $Y = X[r, s, t]$, then Lemma (iv), in combination with (iii)(γ), shows how $\boldsymbol{H}Y$ is determined by $\boldsymbol{H}X$ and the variables r, s, and t. The formulation of (iv) is quite involved. To make it somewhat less cumbersome we introduce auxiliary notation.

Let X be any given formula in $\mathbf{\Phi}_3^+$. If $|\mathbf{\Upsilon}\phi X| = 3$, then we shall use "$n^{(X)}$", "$A^{(X)}$", "$B^{(X)}$", "$C^{(X)}$" to denote respectively the uniquely determined number n and sequences A, B, C that satisfy the conclusion of (γ_1) in (iii). Analogously, if $1 \leq |\mathbf{\Upsilon}\phi X| \leq 2$, then we shall use "$A^{(X)}$", "$u^{(X)}$", "$v^{(X)}$" to denote respectively the uniquely determined predicate A and variables u, v that satisfy the conclusion of (γ_2) in (iii). (An analogous notation in case $|\mathbf{\Upsilon}\phi X| = 0$ will not be needed.)

(iv) *Let $X \in \mathbf{\Phi}_3^+$ and $r, s, t \in \mathbf{\Upsilon}_3$. We set $Y = X[r, s, t]$.*

(α) *If $|\mathbf{\Upsilon}\phi X| = 3$ and $|\{r, s, t\}| = 3$, then $n^{(Y)} = n^{(X)}$, and there is a permutation P of $\{0, \ldots, n^{(X)}\}$ such that for each $i = 0, \ldots, n^{(X)}$ the following conclusions hold*:

$$(\alpha_1) \vdash^\times \{A_i^{(Y)} \mathbin{\pmb{=}} A_{Pi}^{(X)},\ B_i^{(Y)} \mathbin{\pmb{=}} B_{Pi}^{(X)},\ C_i^{(Y)} \mathbin{\pmb{=}} C_{Pi}^{(X)}\} \quad \textit{in case } \langle r, s, t\rangle = \langle \boldsymbol{x}, \boldsymbol{y}, \boldsymbol{z}\rangle,$$

$$(\alpha_2) \vdash^\times \{A_i^{(Y)} \mathbin{\pmb{=}} (A_{Pi}^{(X)})^\smile,\ B_i^{(Y)} \mathbin{\pmb{=}} C_{Pi}^{(X)},\ C_i^{(Y)} \mathbin{\pmb{=}} B_{Pi}^{(X)}\} \quad \textit{in case } \langle r, s, t\rangle = \langle \boldsymbol{y}, \boldsymbol{x}, \boldsymbol{z}\rangle,$$

$$(\alpha_3) \vdash^\times \{A_i^{(Y)} \mathbin{\pmb{=}} B_{Pi}^{(X)},\ B_i^{(Y)} \mathbin{\pmb{=}} A_{Pi}^{(X)},\ C_i^{(Y)} \mathbin{\pmb{=}} (C_{Pi}^{(X)})^\smile\} \quad \textit{in case } \langle r, s, t\rangle = \langle \boldsymbol{x}, \boldsymbol{z}, \boldsymbol{y}\rangle,$$

$$(\alpha_4) \vdash^\times \{A_i^{(Y)} \mathbin{\pmb{=}} (C_{Pi}^{(X)})^\smile,\ B_i^{(Y)} \mathbin{\pmb{=}} (B_{Pi}^{(X)})^\smile,\ C_i^{(Y)} \mathbin{\pmb{=}} (A_{Pi}^{(X)})^\smile\} \quad \textit{in case } \langle r, s, t\rangle = \langle \boldsymbol{z}, \boldsymbol{y}, \boldsymbol{x}\rangle,$$

$$(\alpha_5) \vdash^\times \{A_i^{(Y)} \mathbin{\pmb{=}} (B_{Pi}^{(X)})^\smile,\ B_i^{(Y)} \mathbin{\pmb{=}} (C_{Pi}^{(X)})^\smile,\ C_i^{(Y)} \mathbin{\pmb{=}} A_{Pi}^{(X)}\} \quad \textit{in case } \langle r, s, t\rangle = \langle \boldsymbol{y}, \boldsymbol{z}, \boldsymbol{x}\rangle,$$

[14]*See the fourth paragraph of the footnote, pp. 70–71.

$(\alpha_6) \vdash^{\times} \{A_i^{(Y)} = C_{Pi}^{(X)},\ B_i^{(Y)} = (A_{Pi}^{(X)})^{\smile},\ C_i^{(Y)} = (B_{Pi}^{(X)})^{\smile}\}$
in case $\langle r, s, t\rangle = \langle \boldsymbol{z}, \boldsymbol{x}, \boldsymbol{y}\rangle$.

(β) *If* $|\Upsilon\phi X| = 3$ *and* $1 \leq |\{r, s, t\}| \leq 2$, *we set*

$$E = (A_0^{(X)} \cdot \mathring{\mathbf{1}} \odot \mathbf{1} + B_0^{(X)} + C_0^{(X)}) \cdot \cdots \cdot (A_{n(X)}^{(X)} \cdot \mathring{\mathbf{1}} \odot \mathbf{1} + B_{n(X)}^{(X)} + C_{n(X)}^{(X)}),$$

$$F = (A_0^{(X)} + B_0^{(X)} \cdot \mathring{\mathbf{1}} \odot \mathbf{1} + (C_0^{(X)})^{\smile}) \cdot \cdots \cdot (A_{n(X)}^{(X)} + B_{n(X)}^{(X)} \cdot \mathring{\mathbf{1}} \odot \mathbf{1} + (C_{n(X)}^{(X)})^{\smile}),$$

$$G = (A_0^{(X)} + B_0^{(X)} + \mathbf{1} \odot (C_0^{(X)} \cdot \mathring{\mathbf{1}})) \cdot \cdots \cdot (A_{n(X)}^{(X)} + B_{n(X)}^{(X)} + \mathbf{1} \odot (C_{n(X)}^{(X)} \cdot \mathring{\mathbf{1}})),$$

$$I = (A_0^{(X)} + B_0^{(X)} + C_0^{(X)}) \cdot \cdots \cdot (A_{n(X)}^{(X)} + B_{n(X)}^{(X)} + C_{n(X)}^{(X)}),$$

and we conclude

$(\beta_1) \vdash^{\times} A^{(Y)} = E$ *in case* $\mathbf{in}\, r = \mathbf{in}\, s < \mathbf{in}\, t$,
$(\beta_2) \vdash^{\times} A^{(Y)} = E^{\smile}$ *in case* $\mathbf{in}\, r = \mathbf{in}\, s > \mathbf{in}\, t$,
$(\beta_3) \vdash^{\times} A^{(Y)} = F$ *in case* $\mathbf{in}\, r = \mathbf{in}\, t < \mathbf{in}\, s$,
$(\beta_4) \vdash^{\times} A^{(Y)} = F^{\smile}$ *in case* $\mathbf{in}\, r = \mathbf{in}\, t > \mathbf{in}\, s$,
$(\beta_5) \vdash^{\times} A^{(Y)} = G$ *in case* $\mathbf{in}\, r < \mathbf{in}\, s = \mathbf{in}\, t$,
$(\beta_6) \vdash^{\times} A^{(Y)} = G^{\smile}$ *in case* $\mathbf{in}\, r > \mathbf{in}\, s = \mathbf{in}\, t$,
$(\beta_7) \vdash^{\times} A^{(Y)} \cdot \mathring{\mathbf{1}} = I \cdot \mathring{\mathbf{1}}$ *in case* $\mathbf{in}\, r = \mathbf{in}\, s = \mathbf{in}\, t$.

(γ) *If* $|\Upsilon\phi X| = 2$ *and* $u^{(Y)} \neq v^{(Y)}$, *then we have*

$(\gamma_1) \vdash^{\times} A^{(Y)} = A^{(X)}$ *in those cases when* $\langle u^{(X)}, v^{(X)}\rangle$ *and* $\langle u^{(Y)}, v^{(Y)}\rangle$ *coincide respectively with* $\langle \boldsymbol{x}, \boldsymbol{y}\rangle$ *and* $\langle r, s\rangle$, *or with* $\langle \boldsymbol{x}, \boldsymbol{z}\rangle$ *and* $\langle r, t\rangle$, *or with* $\langle \boldsymbol{y}, \boldsymbol{z}\rangle$ *and* $\langle s, t\rangle$;
$(\gamma_2) \vdash^{\times} A^{(Y)} = (A^{(X)})^{\smile}$ *in the remaining cases.*

(δ) *If* $1 \leq |\Upsilon\phi X| \leq 2$ *and* $u^{(Y)} = v^{(Y)}$, *then 1* $\vdash^{\times} A^{(Y)} \cdot \mathring{\mathbf{1}} = A^{(X)} \cdot \mathring{\mathbf{1}}$.

(ε) *If* $|\Upsilon\phi X| = 0$, *then* $\boldsymbol{H}Y \equiv^{\times} \boldsymbol{H}X$.

A detailed proof proceeds by induction on formulas X. Thus, the proof divides into four parts according as (a) X is atomic, or X has one of the forms (b) $X = \neg Z$, (c) $X = W \rightarrow Z$, or (d) $X = \forall_z Z$. In each of these four parts we must consider all the particular cases of (iv) specified by which variables occur free in X and which variables are taken for r, s, and t—thus cases (α_1)–(α_6), (β_1)–(β_7), (γ_1), (γ_2), (δ), and (ε) described in the statement of (iv). As mentioned above, we shall illustrate the method of argumentation with only two examples.

Let Ω be the set of formulas X in $\mathbf{\Phi}_3^+$ such that, for any choice of $r, s, t \in \Upsilon_3$, all the conclusions of (iv) hold. Our first example belongs to part (b) of the proof. Thus, the proof of this part is based on the following inductive premise:

(1) $\quad X = \neg Z$ and $Z \in \Omega$;

our task is to show that (1) implies

(2) $\quad X \in \Omega$.

To specify the particular case which we wish to treat, we let

(3) $|\boldsymbol{\Upsilon}\phi X| = 3,$

(4) $\langle r, s, t\rangle = \langle \boldsymbol{z}, \boldsymbol{x}, \boldsymbol{y}\rangle;$

hence we are dealing with the case (α_6) in (iv).

Clearly, by (1) and (3) we have

(5) $|\boldsymbol{\Upsilon}\phi Z| = 3;$

from (4) and the definition of Y we get

(6) $Y = X[\boldsymbol{z}, \boldsymbol{x}, \boldsymbol{y}].$

Setting

(7) $V = Z[\boldsymbol{z}, \boldsymbol{x}, \boldsymbol{y}],$

we obtain

(8) $Y = \neg V$

by (1), (6), (7), and the definition of substitution in §3.7. For notational simplicity we let

(9) $n^{(Z)} = 0.$

By (4), (5), (7), (9), we see that our premise $Z \in \Omega$ in (1) reduces to

(10) $n^{(V)} = n^{(Z)} = 0,$

(11) there is a permutation Q of $\{0\}$ such that (α_6), with P, X, Y replaced respectively by Q, Z, V, holds.

Since Q is obviously the identity permutation, we can rewrite (11) as

(12) $\vdash^{\times} \{A_0^{(V)} \mathbin{\boldsymbol{=}} C_0^{(Z)},\ B_0^{(V)} \mathbin{\boldsymbol{=}} (A_0^{(Z)})^{\smile},\ C_0^{(V)} \mathbin{\boldsymbol{=}} (B_0^{(Z)})^{\smile}\}.$

In view of the definition of $\boldsymbol{H}$ outlined above, we see from (1), (8), and (9) that

(13) $n^{(X)} = n^{(Y)} = 2,$

(14) $A_0^{(X)} = (A_0^{(Z)})^{-},\ B_1^{(X)} = (B_0^{(Z)})^{-},\ C_2^{(X)} = (C_0^{(Z)})^{-}$, and
$A_1^{(X)} = A_2^{(X)} = B_0^{(X)} = B_2^{(X)} = C_0^{(X)} = C_1^{(X)} = \mathbf{0},$

(15) the formulas in (14) continue to hold if X and Z are replaced everywhere by Y and V.

Let P be the permutation of $\{0, 1, 2\}$ given by $P0 = 2, P1 = 0$, and $P2 = 1$. Using (12), (14), (15) we then easily check with the help of 3.2(i),(ii) that

$$(16) \qquad \vdash^{\times} \{A_i^{(Y)} = C_{Pi}^{(X)},\ B_i^{(Y)} = (A_{Pi}^{(X)})^{\smile},\ C_i^{(Y)} = (B_{Pi}^{(X)})^{\smile}\} \ \text{ for } \ i = 0, 1, 2.$$

From (16) we see that (α_6) holds; hence, in view of (13), all the conclusions of (iv) that are relevant for our particular case as specified by (3) and (4) do hold for X. (The remaining parts of (iv), namely (α_1)–(α_5), (β_1)–(β_7), (γ_1), (γ_2), (δ), (ε), hold vacuously since the premises involved in these parts are incompatible with (3), (4).) Thus, we arrive at (2), and this was just our task.

Our second example belongs to part (c) of the proof. Thus, we assume as the inductive premise

$$(17) \qquad X = W \rightarrow Z \ \text{ and } \ W, Z \in \Omega;$$

our task is again to derive (2) from (17). It turns out that part (c) involves a complication which does not occur in other parts of the proof. To avoid inessential distractions we choose for our example the same particular case of (iv) as was chosen for our first example, i.e., (α_6). Thus, we assume (3) and (4). However, because our inductive premise in part (c) involves two formulas, W and Z, and not just one as in parts (b) and (d), in the discussion of any particular case in this part we need to know not only which variables occur free in X, but also which of them occur free in W and which in Z. In other words, within a particular case we may have to consider a large number of subcases (indeed, 27 subcases under assumption (3)). For our example we select one of these subcases, namely the one specified by the condition

$$(18) \qquad \Upsilon\phi W = \{\boldsymbol{x}, \boldsymbol{y}\} \ \text{ and } \ \Upsilon\phi Z = \{\boldsymbol{z}\}.$$

Clearly, (6) continues to hold. Setting

$$(19) \qquad U = W[\boldsymbol{z}, \boldsymbol{x}, \boldsymbol{y}] \ \text{ and } \ V = Z[\boldsymbol{z}, \boldsymbol{x}, \boldsymbol{y}],$$

we get

$$(20) \qquad Y = U \rightarrow V,$$

$$(21) \qquad \Upsilon\phi U = \{\boldsymbol{x}, \boldsymbol{z}\} \ \text{ and } \ \Upsilon\phi V = \{\boldsymbol{y}\},$$

by (17), (6), (19) and the definition of substitution in §3.7.

From (4), (18), (19), (21) we see that the second part of (17) reduces to

$$(22) \qquad \vdash^{\times} \{A^{(U)} = (A^{(W)})^{\smile},\ A^{(V)} \cdot \mathring{1} = A^{(Z)} \cdot \mathring{1}\}.$$

In view of the definition of $\boldsymbol{H}$, we get from (17), (18), (20), (21) that

(23) $\quad n^{(X)} = n^{(Y)} = 0,$

(24) $\quad A_0^{(X)} = (A^{(W)})^{-}, \quad B_0^{(X)} = \mathbf{1} \odot (A^{(Z)} \cdot \mathring{\mathbf{1}}), \quad C_0^{(X)} = \mathbf{1} \odot (A^{(Z)} \cdot \mathring{\mathbf{1}}),$

(25) $\quad A_0^{(Y)} = \mathbf{1} \odot (A^{(V)} \cdot \mathring{\mathbf{1}}), \quad B_0^{(Y)} = (A^{(U)})^{-}, \quad C_0^{(Y)} = A^{(V)} \cdot \mathring{\mathbf{1}} \odot \mathbf{1}.$

Using (22), (24), (25) and 3.2(i),(ii),(iii),(xxii) we easily conclude that

(26) $\quad \vdash^{\times} \{A_0^{(Y)} = C_0^{(X)},\ B_0^{(Y)} = (A_0^{(X)})^{\smile},\ C_0^{(Y)} = (B_0^{(X)})^{\smile}\}.$

From (26) we see that (α_6) holds, provided we take P to be the identity permutation on $\{0\}$. Hence, in view of (23), all relevant conclusions of (iv) hold for X, and we arrive at (2). This completes our discussion of (iv).

(v) $\vdash^{\times} \boldsymbol{H}X$ *for every* $X \in \mathbf{\Lambda}_3^{+}$.

In fact, we first recall that $\mathbf{\Lambda}_3^{+}$ consists of all sentences of $\mathbf{\Sigma}_3^{+}$ which are instances of the schemata (AI)–(AVIII), (AIX′), (AX), and (DI)–(DV) (cf. §§1.3, 3.7, and 2.2). In principle, a separate proof is needed for each of these schemata. Given a particular instance $X \in \mathbf{\Sigma}_3^{+}$ of one of these schemata, we first compute $\boldsymbol{H}X$ on the basis of the definition of $\boldsymbol{H}$ (this usually requires distinguishing many cases). Then we show that $\vdash^{\times} \boldsymbol{H}X$. To this end we apply some of the lemmas listed in §3.2; we also use implicitly a number of analogous laws of a Boolean algebraic character which are not listed in §3.2. Moreover, in the cases of (AIX′) and (AX) we must use Lemma (iv).

As in the proof of Lemma (iv), we shall illustrate the methods of argumentation by considering two examples. For our first example we take a particular instance X of Schema (AIV). Thus X has the form

(1) $\quad X = [\forall_u \forall_v Z \rightarrow \forall_v \forall_u Z],$

where $u, v \in \mathbf{\Upsilon}_3$ and $Z \in \mathbf{\Phi}_3^{+}$. To specify the particular example which we shall consider, we make the additional assumptions: $|\mathbf{\Upsilon}\phi Z| = 3$ (if $|\mathbf{\Upsilon}\phi Z| < 3$, the argument is quite simple) and $u = \boldsymbol{y}, v = \boldsymbol{x}$. Thus, by ($\gamma_1$) in (iii),

(2) $\quad \boldsymbol{H}Z \ = \ (\boldsymbol{x}A_0\boldsymbol{y} \vee \boldsymbol{x}B_0\boldsymbol{z} \vee \boldsymbol{y}C_0\boldsymbol{z}) \wedge \cdots \wedge (\boldsymbol{x}A_n\boldsymbol{y} \vee \boldsymbol{x}B_n\boldsymbol{z} \vee \boldsymbol{y}C_n\boldsymbol{z})$

for some $n \in \omega$ and some $A, B, C \in {}^{n+1}\mathbf{\Pi}$, and by (1),

(3) $\quad X \ = \ \forall_{\boldsymbol{z}}(\forall_{\boldsymbol{y}}\forall_{\boldsymbol{x}} Z \rightarrow \forall_{\boldsymbol{x}}\forall_{\boldsymbol{y}} Z).$

Setting

(4) $\quad D = (A_0^{\smile} \oplus B_0 + C_0) \cdot \dots \cdot (A_n^{\smile} \oplus B_n + C_n),$
$\quad\quad E = (A_0 \oplus C_0 + B_0) \cdot \dots \cdot (A_n \oplus C_n + B_n),$

we obtain by the definition of $\boldsymbol{H}$ and (2),

$$\boldsymbol{H}(\forall_{\boldsymbol{y}}\forall_{\boldsymbol{x}}Z) = \boldsymbol{z}\mathbf{0}\oplus D\boldsymbol{z}, \quad \boldsymbol{H}(\forall_{\boldsymbol{x}}\forall_{\boldsymbol{y}}Z) = \boldsymbol{z}\mathbf{0}\oplus E\boldsymbol{z},$$

and hence by (3),

(5) $\quad \boldsymbol{H}X = [(\mathbf{0}\oplus D)^{-} + (\mathbf{0}\oplus E) + \mathring{\mathbf{0}} = \mathbf{1}].$

By 3.2(vi) we get

(6) $\quad \vdash^{\times} \mathbf{0}\oplus D = [\mathbf{0}\oplus(A_0^{\smile}\oplus B_0 + C_0)]\cdot\cdots\cdot[\mathbf{0}\oplus(A_n^{\smile}\oplus B_n + C_n)],$

(7) $\quad \vdash^{\times} \mathbf{0}\oplus E = [\mathbf{0}\oplus(A_0\oplus C_0 + B_0)]\cdot\cdots\cdot[\mathbf{0}\oplus(A_n\oplus C_n + B_n)].$

Now 3.2(xvi) yields

$$\vdash^{\times} \mathbf{0}\oplus(A_k^{\smile}\oplus B_k + C_k) = \mathbf{0}\oplus(A_k\oplus C_k + B_k),$$

for $k = 0, \ldots, n$, whence by (6) and (7),

(8) $\quad \vdash^{\times} \mathbf{0}\oplus D = \mathbf{0}\oplus E.$

From (5) and (8) we conclude at once with the help of BA that

(9) $\quad \vdash^{\times} \boldsymbol{H}X,$

and this is just what we wanted to prove.

For our second example we take a particular instance X of Schema (AIX$'$).[15] Thus, let

(10) $\quad X = [u\mathring{\mathbf{1}}v \rightarrow (Z \rightarrow Z[u/v])],$

where $u, v \in \boldsymbol{\Upsilon}_3$ and $Z \in \boldsymbol{\Phi}_3^+$. Again we assume $|\boldsymbol{\Upsilon}\phi Z| = 3$ and $u = \boldsymbol{y}, v = \boldsymbol{x}$. Thus, by (10),

(11) $\quad X = \forall_{\boldsymbol{x}}\forall_{\boldsymbol{y}}\forall_{\boldsymbol{z}}[\boldsymbol{y}\mathring{\mathbf{1}}\boldsymbol{x} \rightarrow (Z \rightarrow Z[\boldsymbol{y}/\boldsymbol{x}])].$

From $(\gamma_1), (\gamma_2)$ in (iii), and from the definition of $\boldsymbol{H}$, we see that

(12) $\quad \boldsymbol{H}(\boldsymbol{y}\mathring{\mathbf{1}}\boldsymbol{x}) = \boldsymbol{x}\mathring{\mathbf{1}}^{\smile}\boldsymbol{y},$

(13) $\quad \boldsymbol{H}Z = (\boldsymbol{x}A_0\boldsymbol{y} \vee \boldsymbol{x}B_0\boldsymbol{z} \vee \boldsymbol{y}C_0\boldsymbol{z}) \wedge \cdots \wedge (\boldsymbol{x}A_n\boldsymbol{y} \vee \boldsymbol{x}B_n\boldsymbol{z} \vee \boldsymbol{y}C_n\boldsymbol{z})$

for some $n \in \omega$ and some $A, B, C \in {}^{n+1}\boldsymbol{\Pi}$, and

(14) $\quad \boldsymbol{H}(Z[\boldsymbol{y}/\boldsymbol{x}]) = \boldsymbol{x}F\boldsymbol{z}$

for some $F \in \boldsymbol{\Pi}$. For notational simplicity we assume $n = 0$, so that (13) becomes

[15]*See the fourth paragraph of the footnote, pp. 70–71.

(15) $\quad \boldsymbol{H}Z \;=\; \boldsymbol{x}A_0\boldsymbol{y} \vee \boldsymbol{x}B_0\boldsymbol{z} \vee \boldsymbol{y}C_0\boldsymbol{z}.$

Using the definition of $\boldsymbol{H}$, we obtain step-by-step:

(16) $\quad \boldsymbol{H}(Z \to Z[\boldsymbol{y}/\boldsymbol{x}]) = (\boldsymbol{x}A_0^{-}\boldsymbol{y} \vee \boldsymbol{x}\mathbf{0}\mathbf{+}F\boldsymbol{z} \vee \boldsymbol{y}\mathbf{0}\boldsymbol{z}) \wedge (\boldsymbol{x}\mathbf{0}\boldsymbol{y} \vee \boldsymbol{x}B_0^{-}\mathbf{+}F\boldsymbol{z} \vee \boldsymbol{y}\mathbf{0}\boldsymbol{z}) \wedge (\boldsymbol{x}\mathbf{0}\boldsymbol{y} \vee \boldsymbol{x}\mathbf{0}\mathbf{+}F\boldsymbol{z} \vee \boldsymbol{y}C_0^{-}\boldsymbol{z})$

by (14), (15);

(17) $\quad \boldsymbol{H}(\boldsymbol{y}\mathring{\boldsymbol{\imath}}\boldsymbol{x} \to (Z \to Z[\boldsymbol{y}/\boldsymbol{x}])) \;=\; (\boldsymbol{x}\mathring{\boldsymbol{\imath}}^{\smile -}\mathbf{+}A_0^{-}\boldsymbol{y} \vee \boldsymbol{x}\mathbf{0}\mathbf{+}F\boldsymbol{z} \vee \boldsymbol{y}\mathbf{0}\boldsymbol{z}) \wedge (\boldsymbol{x}\mathring{\boldsymbol{\imath}}^{\smile -}\mathbf{+}\mathbf{0}\boldsymbol{y} \vee \boldsymbol{x}B_0^{-}\mathbf{+}F\boldsymbol{z} \vee \boldsymbol{y}\mathbf{0}\boldsymbol{z}) \wedge (\boldsymbol{x}\mathring{\boldsymbol{\imath}}^{\smile -}\mathbf{+}\mathbf{0}\boldsymbol{y} \vee \boldsymbol{x}\mathbf{0}\mathbf{+}F\boldsymbol{z} \vee \boldsymbol{y}C_0^{-}\boldsymbol{z})$

by (12), (16);

(18) $\quad \boldsymbol{H}X = \big([\mathring{\boldsymbol{\imath}}^{\smile -}\mathbf{+}A_0^{-}\mathbf{+}(\mathbf{0}\mathbf{+}F\oplus\mathbf{0}^{\smile})]\cdot[\mathring{\boldsymbol{\imath}}^{\smile -}\mathbf{+}\mathbf{0}\mathbf{+}(B_0^{-}\mathbf{+}F\oplus\mathbf{0}^{\smile})] \cdot[\mathring{\boldsymbol{\imath}}^{\smile -}\mathbf{+}\mathbf{0}\mathbf{+}(\mathbf{0}\mathbf{+}F\oplus C_0^{-\smile})]\oplus\mathbf{0}\mathbf{+}\mathring{\mathbf{0}}=\mathbf{1}\big)$

by (11), (17).

We now use Lemma (iv), with $X, Y, \langle r, s, t\rangle$ replaced respectively by Z, $Z[\boldsymbol{y}/\boldsymbol{x}]$, $\langle \boldsymbol{x}, \boldsymbol{x}, \boldsymbol{z}\rangle$. Since $|\Upsilon\phi Z| = 3$, $|\Upsilon\phi Z[\boldsymbol{y}/\boldsymbol{x}]| = 2$, and $\boldsymbol{in}\,\boldsymbol{x} = \boldsymbol{in}\,\boldsymbol{x} < \boldsymbol{in}\,\boldsymbol{z}$, we see that case ($\beta_1$) of (iv) applies. In view of (14) and (15) we therefore have

(19) $\quad \vdash^{\times}\; F \mathbf{=} A_0\cdot\mathring{\boldsymbol{\imath}}\odot\mathbf{1}\mathbf{+}B_0\mathbf{+}C_0.$

Setting

(20) $\quad G = \mathring{\boldsymbol{\imath}}^{\smile -}\mathbf{+}A_0^{-}\mathbf{+}(\mathbf{0}\mathbf{+}F\oplus\mathbf{0}^{\smile}),$

we obtain successively:

(21) $\quad \vdash^{\times}\; G \mathbf{=} (\mathring{\boldsymbol{\imath}}\cdot A_0)^{-}\mathbf{+}(F\oplus\mathbf{0})$ by (20), 3.2(ii);

(22) $\quad \vdash^{\times}\; G \mathbf{\geq} (\mathring{\boldsymbol{\imath}}\cdot A_0)^{-}\mathbf{+}(A_0\cdot\mathring{\boldsymbol{\imath}}\odot\mathbf{1}\oplus\mathbf{0})$ by (19), (21), 3.2(vii);

(23) $\quad \vdash^{\times}\; G \mathbf{\geq} (\mathring{\boldsymbol{\imath}}\cdot A_0)^{-}\mathbf{+}A_0\cdot\mathring{\boldsymbol{\imath}}\odot\mathbf{1}$ by (22), 3.2(xiv);

(24) $\quad \vdash^{\times}\; G \mathbf{\geq} (\mathring{\boldsymbol{\imath}}\cdot A_0)^{-}\mathbf{+}A_0\cdot\mathring{\boldsymbol{\imath}}$ by (23), 3.2(ix);

(25) $\quad \vdash^{\times}\; G \mathbf{=} \mathbf{1}$ by (24), BA.

In an analogous fashion we get

(26) $\quad \vdash^{\times}\; \mathring{\boldsymbol{\imath}}^{\smile -}\mathbf{+}\mathbf{0}\mathbf{+}(B_0^{-}\mathbf{+}F\oplus\mathbf{0}^{\smile}) \mathbf{=} \mathbf{1},$

(27) $\quad \vdash^{\times}\; \mathring{\boldsymbol{\imath}}^{\smile -}\mathbf{+}\mathbf{0}\mathbf{+}(\mathbf{0}\mathbf{+}F\oplus C_0^{-\smile}) \mathbf{=} \mathbf{1}.$

(To obtain (27) we use, in particular, 3.2(xi).) From (18), (20), (25)–(27), and 3.2(xiii) we easily arrive at (9). With this, our discussion of (v) has been completed.

(vi) *If* $X, Y \in \Sigma_3^+$, *then* $\{\boldsymbol{H}X,\ \boldsymbol{H}(X \to Y)\} \vdash^\times\ \boldsymbol{H}Y$.

Indeed, by (iii)(δ) we have $\boldsymbol{H}X, \boldsymbol{H}Y \in \Sigma^\times$, and hence by 3.2(xxxii),

(1) $\boldsymbol{H}X \equiv^\times [(\boldsymbol{H}X)^t = \mathbf{1}], \quad \boldsymbol{H}Y \equiv^\times [(\boldsymbol{H}Y)^t = \mathbf{1}].$

On the other hand, we obtain from the definition of $\boldsymbol{H}$,

(2) $\boldsymbol{H}(X \to Y) \;=\; [\mathbf{1} \odot (\boldsymbol{H}X)^{t-} \odot \mathbf{1} + (\boldsymbol{H}Y)^t = \mathbf{1}].$

By 3.2(xiii) and BA, (1) and (2) imply at once the conclusion of (vi).

We come to the main mapping theorem for $\mathcal{L}^\times$ and $\mathcal{L}_3^+$.

(vii) *For every* $\Psi \subseteq \Sigma_3^+$ *and* $X \in \Sigma_3^+$, *we have* $\Psi \vdash_3^+ X$ *iff* $\boldsymbol{H}^*\Psi \vdash^\times \boldsymbol{H}X$.

The proof is entirely analogous to that of 2.3(v). In establishing the implication from left to right we make essential use of (v), (vi); in the opposite direction we use (ii), (iii)(ε).

We can now use the observations in §§2.3 and 2.4 to derive from (vii) (with the help of (i), (iii)(β),(δ)) the proper equipollence theorems for $\mathcal{L}^\times$ and $\mathcal{L}_3^+$.

(viii) *For every* $X \in \Sigma_3^+$ *there is a* $Y \in \Sigma^\times$, *and for every* $Y \in \Sigma^\times$ *there is an* $X \in \Sigma_3^+$, *such that* $X \equiv_3^+ Y$.

(ix) *For every* $\Psi \subseteq \Sigma^\times$ *and* $X \in \Sigma^\times$, *we have* $\Psi \vdash_3^+ X$ *iff* $\Psi \vdash^\times X$.

We can summarize less formally the main results obtained in §3.8 and the present section by saying that *the formalisms* $\mathcal{L}_3$ *and* $\mathcal{L}^\times$ *have the formalism* $\mathcal{L}_3^+$ *as their common equipollent extension, and hence* $\mathcal{L}_3$ *and* $\mathcal{L}^\times$ (*treated as subformalisms of* $\mathcal{L}_3^+$) *are equipollent in means of expression and proof.*

The construction used here to establish these equipollence results has clearly some serious defects, if only from the point of view of mathematical elegance. Actually, this applies to the proof of the equipollence of $\mathcal{L}_3^+$ and $\mathcal{L}^\times$. The splintered character of the definition of the translation mapping $\boldsymbol{H}$, with its many cases, is a principal cause of the fragmented nature of certain portions of the argument; the involved notion of substitution (which we have to use because of the restricted number of variables in our formalisms) is another detrimental factor. As a final result, the construction is so cumbersome in some of its parts—culminating in the proofs of (iv) and (v)—that we did not even attempt to present them in full. A different construction that would remove most of the present defects would be very desirable indeed.

As easy corollaries of the equipollence results we get the following.

(x) $\mathbf{\Theta}\eta_3\Psi = \mathbf{\Theta}\eta_3^+\Psi \cap \Sigma_3$ *for every* $\Psi \subseteq \Sigma_3$.

(xi) $\mathbf{\Theta}\eta^{\times}\Psi = \mathbf{\Theta}\eta_3^+\Psi \cap \Sigma^{\times}$ *for every* $\Psi \subseteq \Sigma^{\times}$.

(xii) *If* Φ *is a theory in* $\mathcal{L}_3^+$, *then* $\Phi \cap \Sigma_3$ *is a theory in* $\mathcal{L}_3$ *and* $\Phi \cap \Sigma^{\times}$ *a theory in* $\mathcal{L}^{\times}$, *and* $\Phi = \mathbf{\Theta}\eta_3^+(\Phi \cap \Sigma_3) = \mathbf{\Theta}\eta_3^+(\Phi \cap \Sigma^{\times})$. *If, moreover, any of the theories* Φ, $\Phi \cap \Sigma_3$, *or* $\Phi \cap \Sigma^{\times}$ *satisfies one of the conditions* (α)–(ς) *in* 2.3(xi), *then the remaining two theories also satisfy this condition.*

To prove (x)–(xii) we argue in exactly the same way as we did in deriving 2.3(x),(xi); compare also 2.4(vii),(viii), and the subsequent remarks.

An immediate consequence of the equipollence results is that the formalisms $\mathcal{L}_3$ and $\mathcal{L}_3^+$, just as $\mathcal{L}^{\times}$, are not equipollent with $\mathcal{L}$ and $\mathcal{L}^+$, and indeed are actually poorer than them in means of both expression and proof. In fact, the sentences S, S' of $\mathcal{L}$ given in 3.4(iv),(v) are not logically equivalent with any sentence of $\mathcal{L}_3$. The sentence T of $\mathcal{L}_3^+$ given in 3.4(vi) is logically provable in $\mathcal{L}^+$, but not in $\mathcal{L}_3^+$, and hence the correlated sentence $\boldsymbol{G}T$ is provable in $\mathcal{L}$, but not in $\mathcal{L}_3$. From this last observation it follows also that $\mathcal{L}_3$ and $\mathcal{L}_3^+$ are not semantically complete.

Nevertheless, it will be seen from our results in the next chapter that $\mathcal{L}_3$ can serve as an adequate framework for formalizing the whole of set theory, and hence, in essence, the whole of mathematics.

It seems appropriate to conclude this section with some remarks of a historical nature. The proof of the equipollence of $\mathcal{L}_3, \mathcal{L}_3^+$, and $\mathcal{L}^{\times}$ was first outlined by Tarski in essentially the following form. For $\mathcal{L}_3$ and $\mathcal{L}_3^+$ he took the formalisms described in the first part of §3.7, but enriched with appropriate associativity schemata. As translation mappings he used the same mappings $\boldsymbol{G}$ and $\boldsymbol{H}$ that were used above. In presenting his argument (which followed the lines of the development given above) he used the assumption that various basic metalogical laws holding for $\mathcal{L}$ and $\mathcal{L}^+$ can be extended to $\mathcal{L}_3$ and $\mathcal{L}_3^+$ by analyzing carefully their original proofs; however, he treated this assumption as a "working hypothesis" which required careful checking.[16]

In working out a detailed presentation of Tarski's argument, Givant realized that the "working hypothesis" was unwarranted, and might well prove to be false if the formalisms $\mathcal{L}_3$ and $\mathcal{L}_3^+$ were left unchanged. He suggested, therefore, a rather radical extension of these formalisms consisting in the replacement of (AIX) by (AIX′) on the respective lists of logical axiom schemata. As a result of this modification the notion of substitution becomes essentially involved in the description of $\mathcal{L}_3$ and $\mathcal{L}_3^+$. For obvious reasons the original definition of substitution given in §1.2 is not suitable in this context, and he suggested using the variant given in Monk [**1971**] instead; to secure the usefulness of this notion of substitution, Givant proved a special theorem, (iv), which establishes a connection between substitution and the translation mapping $\boldsymbol{H}$. In consequence,

[16]*Cf. Henkin–Tarski [**1961**], p. 73.

he succeeded in obtaining a full and exact proof of the equipollence results. This is just the proof that has been sketched here in the last few sections. (Givant also observed that, while the modification of $\mathcal{L}_3$ discussed above seems to be essential, in the case of $\mathcal{L}_3^+$ a far less radical one is adequate for our purposes; we have in mind the modification briefly mentioned at the end of §3.7.)

The reader will see in the next section that Givant's idea can be used to obtain an interesting and reasonable formalization of logics $\mathcal{L}_n$ with n variables, for $n > 3$.

3.10. Subformalisms of $\mathcal{L}$ and $\mathcal{L}^+$ with finitely many variables

In this final section of the chapter we shall concern ourselves with some problems that do not belong to the main stream of our discussion, but are closely related to it.

The inclusion of the associativity schema (AX) in the lists of logical axiom schemata for $\mathcal{L}_3$ and $\mathcal{L}_3^+$ was motivated exclusively by our desire to construct subformalisms of $\mathcal{L}$ and $\mathcal{L}^+$ that are equipollent with $\mathcal{L}^\times$ (cf. §3.7). We may therefore consider the (*standardized*) *formalisms* $\mathcal{L}s_3$ and $\mathcal{L}s_3^+$ obtained from $\mathcal{L}_3$ and $\mathcal{L}_3^+$ by deleting (AX) in both $\mathbf{\Lambda}_3$ and $\mathbf{\Lambda}_3^+$. These standardized formalisms are undoubtedly more natural and more interesting in their own right than $\mathcal{L}_3$ and $\mathcal{L}_3^+$. Clearly, they differ from $\mathcal{L}_3$ and $\mathcal{L}_3^+$ as originally described in §3.7 only in that (AIX) has been replaced with (AIX$'$). By analyzing the arguments in §3.8 we easily see that $\mathcal{L}s_3$ is an equipollent subformalism of $\mathcal{L}s_3^+$.

Two problems naturally arise concerning $\mathcal{L}s_3$ and $\mathcal{L}s_3^+$. Assume that $\mathcal{L}o^\times$ is the formalism obtained from $\mathcal{L}^\times$ by removing (BIV) from the list of logical axiom schemata in §3.1. The first problem is whether $\mathcal{L}o^\times$ is equipollent with the formalisms $\mathcal{L}s_3$ and $\mathcal{L}s_3^+$. From a result of Maddux [**1978a**], Chapter 11, it follows that the solution of this problem is negative. Maddux has shown that the formalism $\mathcal{L}w^\times$ obtained from $\mathcal{L}^\times$ by weakening the associative law for $\odot$, and in fact by replacing it with a special case,

(BIV$'$) $$(A\odot B)\odot\mathbf{1} = A\odot(B\odot\mathbf{1}),$$

is equipollent with $\mathcal{L}s_3^+$, and that (BIV$'$) cannot be derived from the remaining axiom schemata of $\mathcal{L}w^\times$. Thus, $\mathcal{L}o^\times$ is weaker than $\mathcal{L}w^\times$, and hence also weaker than $\mathcal{L}s_3$ and $\mathcal{L}s_3^+$, in means of proof. Maddux has also shown that (BIV$'$) can be replaced on the list of axiom schemata for $\mathcal{L}w^\times$ with a still more special case of the associative law, namely with

$$(A\odot\mathbf{1})\odot\mathbf{1} = A\odot(\mathbf{1}\odot\mathbf{1})$$

or, equivalently, with

$$(A\odot\mathbf{1})\odot\mathbf{1} = A\odot\mathbf{1}.$$

Finally, he has proved that (BIV) cannot be derived in $\mathcal{L}w^\times$, so that $\mathcal{L}w^\times$ is actually weaker than $\mathcal{L}^\times$ in means of proof.

The second problem, somewhat related to the first, but still open, is more essential for our main purposes. We should like to know whether the formalism

$\mathcal{L}w^\times$ provides an adequate basis upon which our fundamental results (to be presented in the next chapter) could be reconstructed and our main goal, a formalization of set theory without the use of variables, could be achieved.[17] Even if the answer to this question is affirmative, the proof of it may be a hard task. The development of the logic and metalogic of $\mathcal{L}^\times$, to the extent that this is needed for our purposes, is not easy even in the presence of the full associative law, and would probably become much more involved if this law were not available.

By 3.9(viii), for every $X \in \Sigma_3^+$ there is a $Y \in \Sigma^\times$ which is logically equivalent with X in $\mathcal{L}_3^+$. Since $\mathcal{L}_3^+$ is obviously a subformalism of $\mathcal{L}^+$, the sentences X and Y are *a fortiori* logically equivalent in $\mathcal{L}^+$. Thus, using the terminology introduced in §3.6, we obtain

(i) *Every sentence in Σ_3^+, and hence every sentence in Σ_3, is $\mathcal{L}^\times$-expressible.*

The sets Σ_3 and Σ_3^+ are obviously recursive, so (i) provides partial criteria for $\mathcal{L}^\times$-expressibility of sentences in $\mathcal{L}$ and $\mathcal{L}^+$. We shall show how these criteria can be widened. To this end we single out some new sets of formulas and sentences in $\mathcal{L}$ and $\mathcal{L}^+$. Let $\boldsymbol{\Phi}_{(3)}$ (to be distinguished from $\boldsymbol{\Phi}_3$!) be the set of all $X \in \boldsymbol{\Phi}$ such that $|\boldsymbol{\Upsilon}\phi Y| \leq 3$ for every subformula Y of X. The sets $\Sigma_{(3)}$, $\boldsymbol{\Phi}_{(3)}^+$, and $\Sigma_{(3)}^+$ are defined analogously. Notice that these definitions impose no restrictions either on the total number of variables occurring in formulas involved or on the indices of variables occurring free. We can prove the following.

(ii) *For every $X \in \boldsymbol{\Phi}_{(3)}^+$, and hence also for every $X \in \boldsymbol{\Phi}_{(3)}$, there is a quantifier-free $Y \in \boldsymbol{\Phi}^+$ such that $X \equiv Y$ and $\boldsymbol{\Upsilon}\phi X = \boldsymbol{\Upsilon}\phi Y$; if, moreover, $|\boldsymbol{\Upsilon}\phi X| \leq 2$, then Y is atomic.*

To show this, we define a function $\boldsymbol{H}'$ which is closely related to the function $\boldsymbol{H}$ introduced in §3.9 (and actually coincides with $\boldsymbol{H}$ when its domain is restricted to $\boldsymbol{\Phi}_3^+$). $\boldsymbol{H}'$ is defined by recursion on $\boldsymbol{\Phi}_{(3)}^+$; recursion is warranted by the obvious fact that all the subformulas of a formula in $\boldsymbol{\Phi}_{(3)}^+$ belong themselves to $\boldsymbol{\Phi}_{(3)}^+$. A detailed definition of $\boldsymbol{H}'$ differs from the definition of $\boldsymbol{H}$ only in that the variables involved are not assumed to coincide with $\boldsymbol{x}$, $\boldsymbol{y}$, or $\boldsymbol{z}$. For example, by a canonical formula we now understand any formula of the form

$$uAv \vee uBw \vee vCw,$$

where $u, v, w \in \boldsymbol{\Upsilon}$ and $\mathbf{in}\, u < \mathbf{in}\, v < \mathbf{in}\, w$. The proof that $\boldsymbol{H}'$ has the properties strictly analogous to 3.9(iii)(γ),(δ),(ε), with $\equiv_3^+$ replaced by $\equiv^+$, presents no difficulty, and (ii) is an immediate consequence of these properties. Thus, we see that the mapping $\boldsymbol{H}'$ permits us to eliminate quantifiers from any formula in $\boldsymbol{\Phi}_{(3)}^+$—just as $\boldsymbol{H}$ does with formulas in $\boldsymbol{\Phi}_3^+$ (see §3.9).

[17]*Recently, in Németi [**1985**], it has been shown that the answer to this question is affirmative: $\mathcal{L}w^\times$ provides an adequate basis for the formalization of set theory (see the footnote, p. 143). On the other hand, in Németi [**1987**] it is proved that $\mathcal{L}o^\times$ does not provide an adequate basis for such a formalization (see footnote 6*, p. 138).

As a particular case of (ii) we obtain

(iii) *Every sentence in* $\mathbf{\Sigma}^+_{(3)}$, *and hence every sentence in* $\mathbf{\Sigma}_{(3)}$, *is* $\mathcal{L}^\times$*-expressible.*

Since the sets $\mathbf{\Sigma}_{(3)}$ and $\mathbf{\Sigma}^+_{(3)}$ are recursive, (iii) provides new partial criteria for $\mathcal{L}^\times$-expressibility which are wider than those in (i).

In the next theorem, (iv), by supplementing (i) and (iii) we obtain new full characterizations of $\mathcal{L}^\times$-expressibility (i.e., necessary and sufficient conditions, but not necessarily of a recursive character). In formulating (iv) we find it convenient to speak not of the sets $\mathbf{\Sigma}_{(3)}$ and $\mathbf{\Sigma}^+_{(3)}$, but of the corresponding formalisms $\mathcal{L}_{(3)}$ and $\mathcal{L}^+_{(3)}$. These two formalisms are, respectively, assumed to be subformalisms of $\mathcal{L}$ and $\mathcal{L}^+$ whose sets of sentences coincide with $\mathbf{\Sigma}_{(3)}$ and $\mathbf{\Sigma}^+_{(3)}$. Further information concerning $\mathcal{L}_{(3)}$ and $\mathcal{L}^+_{(3)}$ (such as the definitions of derivability) is not needed for understanding (iv) and will not be given here.

(iv) *If any sentence in* $\mathcal{L}^+$ (*or in* $\mathcal{L}$) *is expressible in one of the five formalisms* $\mathcal{L}^\times, \mathcal{L}_3, \mathcal{L}^+_3, \mathcal{L}_{(3)}, \mathcal{L}^+_{(3)}$, *then it is expressible in the remaining four.*

In fact, the passages from $\mathcal{L}^\times$ to $\mathcal{L}^+_3$, from $\mathcal{L}_3$ to $\mathcal{L}_{(3)}$, and from $\mathcal{L}_{(3)}$ to $\mathcal{L}^+_{(3)}$ are obvious; to pass from $\mathcal{L}^+_3$ to $\mathcal{L}_3$ we use 3.8(ix)(δ),(ε), and to pass from $\mathcal{L}^+_{(3)}$ to $\mathcal{L}^\times$ we use (iii).

An analogous result of a more general character, concerning formulas and not just sentences, can be established; however, $\mathcal{L}^\times$-expressibility must be replaced by expressibility in terms of a quantifier-free formula of $\mathcal{L}^+_3$.

It may be interesting to observe that, by (iv), $\mathcal{L}^\times$-expressibility and $\mathcal{L}_3$-expressibility are equivalent properties (of sentences in $\mathcal{L}$), even though the formalisms $\mathcal{L}^\times$ and $\mathcal{L}_3$ differ so radically from each other in their formal structures. In this connection we may note the following reformulation of the result of Kwatinetz [**1981**] stated in 3.6(iii).

(v) *The set of all sentences in* $\mathcal{L}$ (*or* $\mathcal{L}^+$) *which are* $\mathcal{L}_3$*-expressible is not recursive.*

To conclude, we should like to point out that, following the pattern of §§3.7 and 3.8, formalisms $\mathcal{L}_n$ and $\mathcal{L}^+_n$ can be constructed for every integer $n > 3$. The study of these formalisms may have some intrinsic value for finitistic trends that are present in contemporary foundational research. We shall not concern ourselves with formalisms $\mathcal{L}_n$ and $\mathcal{L}^+_n$ for $n = 1$ or 2.

To construct $\mathcal{L}_n$ and $\mathcal{L}^+_n$ for $n > 3$, we first define $\mathbf{\Phi}_n$ and $\mathbf{\Phi}^+_n$ to be respectively the sets of formulas in $\mathbf{\Phi}$ and $\mathbf{\Phi}^+$ in which no variables different from $\mathbf{v}_0, \ldots, \mathbf{v}_{n-1}$ occur, and we set $\mathbf{\Sigma}_n = \mathbf{\Phi}_n \cap \mathbf{\Sigma}$, $\mathbf{\Sigma}^+_n = \mathbf{\Phi}^+_n \cap \mathbf{\Sigma}^+$. We consider the sets of logical axiom schemata for $\mathcal{L}$ and $\mathcal{L}^+$, with (AIX) replaced by Schema (AIX′) from §3.7; the sets of all instances of these schemata that belong to $\mathbf{\Sigma}_n$ and $\mathbf{\Sigma}^+_n$, respectively, are taken as $\mathbf{\Lambda}_n$ and $\mathbf{\Lambda}^+_n$, the sets of logical axioms for $\mathcal{L}_n$ and $\mathcal{L}^+_n$.

One can convince oneself without great difficulty that various important metalogical laws concerning $\mathcal{L}$ and $\mathcal{L}^+$ can be extended to the formalisms $\mathcal{L}_n$ and $\mathcal{L}_n^+$ just described. We have in mind such laws as the general law of substitution and the general Leibniz law—both of them for multivariable, and not just simple, substitution—as well as the law of renaming bound variables. (For the case $n = 3$, cf. §§3.7 and 3.8.) In particular, notice that, in opposition to the case $n = 3$, the associativity schemata (AX) from §3.7 has not been used in the construction of $\mathbf{\Lambda}_n$ and $\mathbf{\Lambda}_n^+$ for $n > 3$. The reason is that all instances of this schema turn out to be easily derivable from $\mathbf{\Lambda}_n$ and $\mathbf{\Lambda}_n^+$ (by means of *modus ponens* alone). From this remark and the description of $\mathcal{L}_n$ and $\mathcal{L}_n^+$, it is obvious that, for any $n, m \in \omega$ with $3 \leq n \leq m$, $\mathcal{L}_n$ and $\mathcal{L}_n^+$ are respectively subformalisms of $\mathcal{L}_m$ and $\mathcal{L}_m^+$.

The formalisms just described lead, in a natural way, to the extension of problems previously discussed for $\mathcal{L}_3$ and $\mathcal{L}^\times$ to formalisms $\mathcal{L}_n$ for $n > 3$. We shall give here a short survey of some results obtained so far in this direction, and of open problems that seem to us interesting.

We consider first the problems of expressibility. Notice that, in view of (iv), the results of 3.4(iv),(v) can now be reformulated as

(vi) *There are sentences in $\mathcal{L}_4$, e.g., sentences S, S' in* 3.4(iv),(v), *which are not $\mathcal{L}_3$-expressible.*

The question naturally arises whether (vi) remains true if $\mathcal{L}_3$ and $\mathcal{L}_4$ are respectively replaced by $\mathcal{L}_n$ and $\mathcal{L}_{n+1}$, where n is an arbitrary positive integer ≥ 4. It has been shown that this is indeed the case; see Kwatinetz [**1981**]. The fact that for every $n \geq 4$, there are sentences in $\mathcal{L}$ (not necessarily in $\mathcal{L}_{n+1}$) which are not $\mathcal{L}_n$-expressible is also a simple consequence of Theorem 3.5(viii) in its general form. (Compare here the remarks after 3.5(viii), where formalisms $\mathcal{L}_n$ and their extensions $\mathcal{L}_n^+$ are implicitly referred to.)

By the results of Kwatinetz [**1981**], Theorem (v) remains true when applied to arbitrary formalisms $\mathcal{L}_n$ with $n \geq 4$. On the other hand, it seems to be an open problem whether, for any given $n \geq 3$, the set of sentences in $\mathcal{L}_{n+1}$ that are $\mathcal{L}_n$-expressible is recursive.

We turn now to the problems of provability. It will be seen that here there are still many unanswered questions. As already noted at the end of the last section, there are sentences in $\mathbf{\Sigma}_3$ that are valid, but not provable in $\mathcal{L}_3$, i.e.,

$$\mathbf{\Theta}\boldsymbol{\eta}_3\varnothing \neq \mathbf{\Theta}\boldsymbol{\eta}\varnothing \cap \mathbf{\Sigma}_3.$$

Does this formula continue to hold when we replace $\mathbf{\Theta}\boldsymbol{\eta}_3\varnothing$ by $\mathbf{\Theta}\boldsymbol{\eta}_n\varnothing \cap \mathbf{\Sigma}_3$ for $n \geq 4$, i.e., do we have the inequality

$$\mathbf{\Theta}\boldsymbol{\eta}_n\varnothing \cap \mathbf{\Sigma}_3 \neq \mathbf{\Theta}\boldsymbol{\eta}\varnothing \cap \mathbf{\Sigma}_3$$

for every $n \geq 4$? Maddux [**1978a**] has shown that

$$\mathbf{\Theta}\boldsymbol{\eta}_3\varnothing = \mathbf{\Theta}\boldsymbol{\eta}_4\varnothing \cap \mathbf{\Sigma}_3,$$

so that the inequality holds when $n = 4$. It is an open problem whether it continues to hold for every $n > 4$. Related to this problem is the one of determining for which values of n the inequality

$$\mathbf{\Theta}\eta_n\varnothing \cap \Sigma_3 \neq \mathbf{\Theta}\eta_{n+1}\varnothing \cap \Sigma_3$$

holds. For $n = 3$ it fails by the above result of Maddux. On the other hand, it holds for $n = 4$; indeed the sentence ***GT*** mentioned at the end of the last section is in $\mathbf{\Theta}\eta_5\varnothing \cap \Sigma_3$, but not in $\mathbf{\Theta}\eta_3\varnothing$. Therefore, in view of Maddux's result, ***GT*** is not in $\mathbf{\Theta}\eta_4\varnothing \cap \Sigma_3$. Again, it is not known whether this inequality holds for $n > 4$.[18]

Finally, it will follow from our results in §4.7 (cf. remarks after 4.7(vi)) that $\mathbf{\Theta}\eta_n^+\varnothing \cap \Sigma^\times$ is not recursive for any $n \geq 3$. Therefore we can infer from 3.9(xii) and the equipollence of $\mathcal{L}_n$ with $\mathcal{L}_n^+$ that $\mathbf{\Theta}\eta_n\varnothing \cap \Sigma_3$ is not recursive for any $n > 3$. We do not know whether the set $(\mathbf{\Theta}\eta_{n+1}\varnothing \sim \mathbf{\Theta}\eta_n\varnothing) \cap \Sigma_3$ is recursive for any $n > 3$.

We may consider problems of a still more general character, involving two or even three varying indices; for instance, the problem for which triples $\langle k, m, n\rangle$ with $k \leq n < m$ the inequality

$$\mathbf{\Theta}\eta_n\varnothing \cap \Sigma_k \neq \mathbf{\Theta}\eta_m\varnothing \cap \Sigma_k$$

holds. However, it is easily seen that, for any fixed n, if this inequality holds in the particular case when $k = 3$ and $m = n + 1$, then it holds for all k and m with $k \leq n < m$.

To conclude, we mention that Maddux actually established the results attributed to him in this section for a somewhat different axiomatization of $\mathcal{L}_n$; cf. Maddux [**1978a**]. However, he has communicated to us that his methods and results can also be applied to our particular axiomatization of $\mathcal{L}_n$.

[18]*Maddux [**1987**] has recently shown that

$$\mathbf{\Theta}\eta_n\varnothing \cap \Sigma_3 \neq \mathbf{\Theta}\eta\varnothing \cap \Sigma_3$$

for all $n > 4$, and that

$$\mathbf{\Theta}\eta_n\varnothing \cap \Sigma_3 \neq \mathbf{\Theta}\eta_{n+1}\varnothing \cap \Sigma_3$$

for infinitely many n.

CHAPTER 4

The Relative Equipollence of $\mathcal{L}$ and $\mathcal{L}^{\times}$, and the Formalization of Set Theory in $\mathcal{L}^{\times}$

From the results in §3.4 we know that $\mathcal{L}^{\times}$ is poorer than $\mathcal{L}^{+}$ and $\mathcal{L}$ in means of expression and proof. In this chapter we shall be able, however, to establish a kind of relative equipollence between $\mathcal{L}^{+}$ and $\mathcal{L}^{\times}$ (and hence also between $\mathcal{L}$ and $\mathcal{L}^{\times}$). More specifically, we shall single out a certain set Σ of sentences in $\mathcal{L}^{\times}$ and we shall show that, for every sentence X in Σ, the formalisms $\mathcal{L}^{+}$ and $\mathcal{L}^{\times}$ when relativized to X (or, more properly, to $\{X\}$, cf. §1.6) turn out to be equipollent in means of expression and proof. This means that every sentence in $\mathcal{L}^{+}$ is equivalent with some sentence in $\mathcal{L}^{\times}$ relative to X, and whenever a sentence Y in $\mathcal{L}^{\times}$ is derivable in $\mathcal{L}^{+}$ from a set Ψ of sentences in $\mathcal{L}^{\times}$, then it is also derivable in $\mathcal{L}^{\times}$ from Ψ relative to X.

Sentences forming the set Σ have a rather specialized character, and therefore the results obtained may not seem very interesting from a general metalogical point of view. We shall see, however, that some sentences of Σ occur as provable sentences in practically all systems of set theory known from the literature, when formalized in $\mathcal{L}^{+}$; this applies also to systems formalized in $\mathcal{L}$, except that we have to consider, instead of sentences of Σ (which are not sentences in the language $\mathcal{L}$), some logically equivalent sentences formulated in $\mathcal{L}$. As a consequence, our general results, which establish a relative equipollence between $\mathcal{L}$— or $\mathcal{L}^{+}$— and $\mathcal{L}^{\times}$, carry with them the full equipollence of every such set-theoretical system $\mathcal{S}$ formalized in $\mathcal{L}$, or system $\mathcal{S}^{+}$ formalized in $\mathcal{L}^{+}$, with an appropriately constructed system $\mathcal{S}^{\times}$ in $\mathcal{L}^{\times}$.[1]

4.1. Conjugated quasiprojections and sentences Q_{AB}

The set Σ referred to above consists of all sentences $\boldsymbol{Q}_{AB}$ correlated with arbitrary ordered pairs of predicates $A, B \in \boldsymbol{\Pi}$ by the following formula:

[1]This result was explicitly stated in Tarski [**1953**]. It was implicitly mentioned in Chin [**1948**], p. 2, as well as in Chin–Tarski [**1951**], pp. 341–343, with references to Tarski's seminar at the University of California at Berkeley in 1945, where the result was presented and discussed.

(i) $\boldsymbol{Q}_{AB} = [((A^\smile \odot A + B^\smile \odot B)^- + \mathring{1}) \cdot (A^\smile \odot B) = 1]$.

To grasp the meaning of $\boldsymbol{Q}_{AB}$, notice some obvious consequences of (i), where A, B are, as before, arbitrary predicates.

(ii) $\boldsymbol{Q}_{AB} \equiv^\times \{A^\smile \odot A \leq \mathring{1},\ B^\smile \odot B \leq \mathring{1},\ A^\smile \odot B = 1\}$.

(iii) $\boldsymbol{Q}_{AB} \equiv \forall_{xyz}[(xAy \wedge xAz) \vee (xBy \wedge xBz) \rightarrow y\mathring{1}z] \wedge \forall_{xy}\exists_z(zAx \wedge zBy)$.

In derivations within $\mathcal{L}^\times$ we shall refer more frequently to (ii) than (i).

From (iii) it is clearly seen that $\boldsymbol{Q}_{AB}$ expresses the following facts: the binary relations F and G (between elements of a set U) which are respectively denoted by A and B in a given realization $\langle U, E\rangle$ of our formal language are functions; moreover, for any $x, y \in U$ there is a $z \in U$ such that $Fz = x$ and $Gz = y$. Two relations F and G with these properties will be referred to as *conjugated quasiprojections* or simply *quasiprojections* (on a set U). Thus, $\langle U, E\rangle$ is a model of $\boldsymbol{Q}_{AB}$ iff the two relations denoted in $\langle U, E\rangle$ by A and B are quasiprojections on U.

A familiar example of conjugated quasiprojections is provided by the ordinary projections defined over ordered couples. Consider a (nonempty) set U satisfying the condition: $U \times U \subseteq U$, i.e., $\langle x, y\rangle \in U$ whenever $x, y \in U$. By *projections over ordered couples* we understand, as usual, the two functions F and G whose common domain is $U \times U$ and which correlate with every ordered pair $z = \langle x, y\rangle$ of $U \times U$ its first term $Fz = x$ and its second term $Gz = y$. F and G can be jointly referred to as *conjugated projections over ordered pairs* of elements of U. The conjugated projections over ordered pairs are obviously conjugated quasiprojections in our sense. Notice the following two simple sentences in $\mathcal{L}^\times$ which express certain properties of conjugated projections that do not apply in general to conjugated quasiprojections.

(iv) $(A \odot A^\smile) \cdot (B \odot B^\smile) \leq \mathring{1}, \quad A \odot 1 = B \odot 1$.

As the reader may notice from the later discussion, the fact that quasiprojections are not required to satisfy the above sentences will considerably facilitate our presentation.

The problem naturally arises of characterizing those sets U on which conjugated quasiprojections exist.

(v) *In order that there exist two conjugated quasiprojections on a set U, it is necessary and sufficient that U have at most one element or be infinite.*

In fact, from the definition of quasiprojections we can conclude that the existence of quasiprojections on a set U is equivalent to the condition $|U \times U| \leq |U|$. On the other hand, it is well known that the latter condition holds iff U is infinite or has at most one element. These two observations lead at once to (v). (The proof of sufficiency requires the axiom of choice.)

We establish some consequences of $\boldsymbol{Q}_{AB}$ in $\mathcal{L}^\times$ which will be used in various proofs in this chapter.

(vi) $Q_{AB} \equiv^{\times} Q_{BA}$ *for every* $A, B \in \Pi$.

This is an immediate consequence of (ii) and 3.2(i)–(iii).

(vii) $Q_{AB} \vdash^{\times} \{\mathbf{1} \odot A = \mathbf{1},\ \mathbf{1} \odot B = \mathbf{1},\ A^{\smile} \odot \mathbf{1} = \mathbf{1},$
$B^{\smile} \odot \mathbf{1} = \mathbf{1},\ A^{\smile} \odot A = \mathring{\mathbf{1}},\ B^{\smile} \odot B = \mathring{\mathbf{1}}\}$

for every $A, B \in \Pi$.

To prove this on the basis of (ii) we notice that, by 3.2(vii) and BA, we have

$$Q_{AB} \vdash^{\times} \{A^{\smile} \odot \mathbf{1} = \mathbf{1},\ \mathbf{1} \odot B = \mathbf{1}\},$$

and therefore, by (vi) above,

$$Q_{AB} \vdash^{\times} \{B^{\smile} \odot \mathbf{1} = \mathbf{1},\ \mathbf{1} \odot A = \mathbf{1}\}.$$

Hence, finally, by 3.2(xx),(i) we obtain

$$Q_{AB} \vdash^{\times} \{A^{\smile} \odot A = \mathring{\mathbf{1}},\ B^{\smile} \odot B = \mathring{\mathbf{1}}\}.$$

The proof of the next theorem is the first long and rather involved derivation in this chapter within the formalism $\mathcal{L}^{\times}$. In connection with such derivations, the reader may recall the closing remarks of §3.2. The particular proof of this theorem presented below is due to Maddux.

(viii) $Q_{AB} \vdash^{\times} (C_0 \odot D_0) \cdot (C_1 \odot D_1) =$
$[(C_0 \odot A^{\smile}) \cdot (C_1 \odot B^{\smile})] \odot [(A \odot D_0) \cdot (B \odot D_1)]$

for every $A, B, C_0, C_1, D_0, D_1 \in \Pi$.

Let C_0, C_1, D_0, D_1 be arbitrary predicates and set

(1) $L = C_0 \odot D_0 \cdot (C_1 \odot D_1)$.

We obtain successively:

(2) $\vdash^{\times} C_0 \odot A^{\smile} \cdot (C_1 \odot B^{\smile}) \odot [A \odot D_0 \cdot (B \odot D_1)]$
$\mathrel{\underline{\leq}} C_0 \odot A^{\smile} \odot A \odot D_0 \cdot (C_1 \odot B^{\smile} \odot B \odot D_1)$
by 3.2(viii);

(3) $Q_{AB} \vdash^{\times} C_0 \odot A^{\smile} \cdot (C_1 \odot B^{\smile}) \odot [A \odot D_0 \cdot (B \odot D_1)] \mathrel{\underline{\leq}} L$
by (1), (2), (ii), 3.2(vii),(ix).

To derive the reverse inclusion it proves convenient to use the following abbreviations:

(4) $R = C_0 \odot B^{\smile} \cdot A^{\smile}$, $\quad S = B \odot D_0 \cdot [A \odot C_1 \odot D_1]$,
$T = B \odot A^{\smile} \cdot (A \odot C_1 \odot B^{\smile})$.

Proceeding step-by-step, we obtain:

(5) $\vdash^{\times} \{R^{\smile} \leq A,\ T^{\smile} \leq A \odot B^{\smile}\}$ by (4), 3.2(i),(iii);

(6) $\boldsymbol{Q}_{AB} \vdash^{\times} C_0 = A^{\smile} \odot B \cdot C_0$ by (ii);

(7) $\boldsymbol{Q}_{AB} \vdash^{\times} C_0 \leq C_0 \odot B^{\smile} \cdot A^{\smile} \odot B$ by (6), 3.2(xv);

(8) $\boldsymbol{Q}_{AB} \vdash^{\times} L \leq R \odot (B \odot D_0) \cdot (C_1 \odot D_1)$ by (1), (4), (7), 3.2(vii);

(9) $\boldsymbol{Q}_{AB} \vdash^{\times} L \leq R \odot \left(B \odot D_0 \cdot [R^{\smile} \odot (C_1 \odot D_1)]\right)$ by (8), 3.2(xv);

(10) $\boldsymbol{Q}_{AB} \vdash^{\times} L \leq R \odot S$ by (4), (5), (9), 3.2(vii);

(11) $\vdash^{\times} S \leq A \odot C_1 \odot D_1 \cdot (B \odot \mathbf{1})$ by (4), 3.2(vii);

(12) $\vdash^{\times} S \leq B \odot \mathbf{1} \odot D_1^{\smile} \cdot (A \odot C_1) \odot D_1$ by (11), 3.2(xv);

(13) $\boldsymbol{Q}_{AB} \vdash^{\times} S \leq B \odot A^{\smile} \odot B \cdot (A \odot C_1) \odot D_1$ by (12), (ii), 3.2(vii);

(14) $\boldsymbol{Q}_{AB} \vdash^{\times} S \leq A \odot C_1 \odot B^{\smile} \cdot (B \odot A^{\smile}) \odot B \odot D_1$ by (13), 3.2(xv) (with A replaced by $B \odot A^{\smile}$ and C by $A \odot C_1$), 3.2(vii);

(15) $\boldsymbol{Q}_{AB} \vdash^{\times} S \leq T \odot (B \odot D_1) \cdot (B \odot D_0)$ by (4), (14), BA;

(16) $\boldsymbol{Q}_{AB} \vdash^{\times} S \leq T \odot \left(B \odot D_1 \cdot [T^{\smile} \odot (B \odot D_0)]\right)$ by (15), 3.2(xv);

(17) $\boldsymbol{Q}_{AB} \vdash^{\times} S \leq T \odot [B \odot D_1 \cdot (A \odot D_0)]$ by (5), (16), (ii), 3.2(vii);

(18) $\boldsymbol{Q}_{AB} \vdash^{\times} L \leq R \odot T \odot [A \odot D_0 \cdot (B \odot D_1)]$ by (10), (17), 3.2(vii);

(19) $\vdash^{\times} C_0 \odot B^{\smile} \cdot A^{\smile} \odot [B \odot A^{\smile} \cdot (A \odot C_1 \odot B^{\smile})]$
$\leq (C_0 \odot B^{\smile} \odot B \odot A^{\smile}) \cdot (A^{\smile} \odot A \odot C_1 \odot B^{\smile})$
by 3.2(viii);

(20) $\boldsymbol{Q}_{AB} \vdash^{\times} R \odot T \leq C_0 \odot A^{\smile} \cdot (C_1 \odot B^{\smile})$ by (4), (19), (ii), 3.2(vii);

(21) $\boldsymbol{Q}_{AB} \vdash^{\times} L \leq C_0 \odot A^{\smile} \cdot (C_1 \odot B^{\smile}) \odot [A \odot D_0 \cdot (B \odot D_1)]$
by (18), (20), 3.2(vii).

From (1), (3), and (21) we conclude that (viii) holds.

(ix) *For any* $A, B, C, D \in \mathbf{\Pi}$ *we have*

(α) $\boldsymbol{Q}_{AB} \vdash^{\times} A^{\smile} \odot (A \odot C \odot B^{\smile} \cdot \mathring{\mathbf{i}} \cdot D) \odot B = A^{\smile} \odot (\mathring{\mathbf{i}} \cdot D) \odot B \cdot C,$

(β) $\boldsymbol{Q}_{AB} \vdash^{\times} A^{\smile} \odot (A \odot C \odot B^{\smile} \cdot \mathring{\mathbf{i}}) \odot B = C,$

(γ) $\boldsymbol{Q}_{AB} \vdash^{\times} \mathbf{1} \odot (A \odot C \odot B^{\smile} \cdot \mathring{\mathbf{i}}) \odot \mathbf{1} = \mathbf{1} \odot C \odot \mathbf{1}.$

In fact, applying successively both parts of 3.2(xxviii) we get

$$\boldsymbol{Q}_{AB} \vdash^{\times} A^{\smile} \odot (A \odot C \odot B^{\smile} \cdot \mathring{\mathbf{i}} \cdot D) \odot B = A^{\smile} \odot (\mathring{\mathbf{i}} \cdot D) \cdot (C \odot B^{\smile}) \odot B,$$

$$\boldsymbol{Q}_{AB} \vdash^{\times} A^{\smile} \odot (\mathring{\mathbf{i}} \cdot D) \cdot (C \odot B^{\smile}) \odot B = A^{\smile} \odot (\mathring{\mathbf{i}} \cdot D) \odot B \cdot C,$$

which yields (α). Taking $\mathbf{1}$ for D in (α) we obtain (β), with the help of (ii) and 3.2(ix).

By using repeatedly 3.2(vii), we derive from (β):

$$\boldsymbol{Q}_{AB} \vdash^{\times} C \leq \mathbf{1} \odot (A \odot C \odot B^{\smile} \cdot \mathring{\mathbf{i}}) \odot \mathbf{1},$$

and we obtain directly

$$\vdash^{\times} \mathbf{1} \odot (A \odot C \odot B^{\smile} \cdot \mathring{\mathbf{i}}) \odot \mathbf{1} \leq \mathbf{1} \odot C \odot \mathbf{1}.$$

These two formulas easily lead to (γ).

The following consequences of $\boldsymbol{Q}_{AB}$ are given mainly because of their intrinsic interest.[2]

(x) *For any* $A, B, C, D \in \mathbf{\Pi}$ *we have*

(α) $\boldsymbol{Q}_{AB} \vdash^{\times} C \odot D = \mathbf{1} \odot (D^{\smile} \odot A^{\smile} \cdot B^{\smile}) \cdot (C \odot A^{\smile}) \odot B,$

(β) $\boldsymbol{Q}_{AB} \vdash^{\times} C \odot D = \mathbf{1} \odot (D \odot B^{\smile} \cdot A^{\smile}) \cdot (C \odot A^{\smile}) \odot B,$

(γ) $\boldsymbol{Q}_{AB} \vdash^{\times} C^{\smile} = \mathbf{1} \odot (C \odot A^{\smile} \cdot B^{\smile}) \cdot A^{\smile} \odot B.$

We first derive (α):

(1) $\boldsymbol{Q}_{AB} \vdash^{\times} C \odot D = C \odot (A^{\smile} \odot B \cdot D)$ by (ii);

(2) $\boldsymbol{Q}_{AB} \vdash^{\times} C \odot D \leq C \odot A^{\smile} \odot (A \odot D \cdot B) \cdot \mathbf{1}$ by (1), 3.2(xv), BA;

(3) $\boldsymbol{Q}_{AB} \vdash^{\times} C \odot D \leq \mathbf{1} \odot (D^{\smile} \odot A^{\smile} \cdot B^{\smile}) \cdot (C \odot A^{\smile}) \odot (A \odot D \cdot B)$
by (2), 3.2(xv) (with B replaced by $A \odot D \cdot B$, A by $C \odot A^{\smile}$, and C by $\mathbf{1}$);

(4) $\boldsymbol{Q}_{AB} \vdash^{\times} C \odot D \leq \mathbf{1} \odot (D^{\smile} \odot A^{\smile} \cdot B^{\smile}) \cdot (C \odot A^{\smile}) \odot B$
by (3), 3.2(vii);

(5) $\boldsymbol{Q}_{AB} \vdash^{\times} \mathbf{1} \odot (D^{\smile} \odot A^{\smile} \cdot B^{\smile}) \cdot (C \odot A^{\smile})$
$\leq C \odot A^{\smile} \odot (A \odot D \cdot B) \cdot \mathbf{1} \odot (D^{\smile} \odot A^{\smile} \cdot B^{\smile})$
by 3.2(xv);

---

[2] *See the remark at the end of the second paragraph of §8.6 to the effect that ORA's can be construed as algebras of type $\langle 2, 1, 1, 0, 0 \rangle$.

(6) $\quad \boldsymbol{Q}_{AB} \vdash^\times \mathbf{1} \odot (D^\smile \odot A^\smile \cdot B^\smile) \cdot (C \odot A^\smile) \odot B \le C \odot A^\smile \odot A \odot D \odot B^\smile \odot B$ by (5), 3.2(vii);

(7) $\quad \boldsymbol{Q}_{AB} \vdash^\times \mathbf{1} \odot (D^\smile \odot A^\smile \cdot B^\smile) \cdot (C \odot A^\smile) \odot B \le C \odot D$ by (6), (ii), 3.2(vii),(ix).

Steps (4) and (7) immediately yield (α). From (α) we get (β) at once by applying 3.2(xvi). To derive (γ) we replace C and D in (α) with $\mathring{\mathbf{1}}$ and $C^\smile$.

In formulating and establishing the main results of this chapter we shall replace given predicates A and B by predicates $A^\square$ and $B^\square$ such that from $\boldsymbol{Q}_{AB}$ we can derive not only $\boldsymbol{Q}_{A^\square B^\square}$, but also $A^\square \odot \mathbf{1} = \mathbf{1}$ and $B^\square \odot \mathbf{1} = \mathbf{1}$. To this end we stipulate:

(xi) $C^\square = C + \mathring{\mathbf{1}} \cdot (C^- \oplus \mathbf{0})$ *for every* $C \in \boldsymbol{\Pi}$.

Notice that if C denotes a function F in a given realization $\langle U, E\rangle$, then $C^\square$ denotes the function on U which coincides with F on DoF and with the identity function on $U \sim DoF$.

(xii) $\vdash^\times A^\square \odot \mathbf{1} = \mathbf{1}$ *and* $\boldsymbol{Q}_{AB} \vdash^\times \boldsymbol{Q}_{A^\square B^\square}$ *for every* $A, B \in \boldsymbol{\Pi}$.

Indeed we obtain successively:

(1) $\quad \vdash^\times \mathring{\mathbf{1}} \cdot (A^- \oplus \mathbf{0}) \odot \mathbf{1} = [\mathring{\mathbf{1}} \cdot (A \odot \mathbf{1}) \odot \mathbf{1}]^-$ by 3.2(xxiv);

(2) $\quad \vdash^\times \mathring{\mathbf{1}} \cdot (A^- \oplus \mathbf{0}) \odot \mathbf{1} = (A \odot \mathbf{1})^-$ by (1), 3.2(xxi);

(3) $\quad \vdash^\times A^\square \odot \mathbf{1} = \mathbf{1}$ by (xi), (2), 3.2(v);

(4) $\quad \vdash^\times A \odot \mathbf{1} \cdot [\mathring{\mathbf{1}} \cdot (A^- \oplus \mathbf{0}) \odot \mathbf{1}] = \mathbf{0}$ by (2);

(5) $\quad \vdash^\times [\mathring{\mathbf{1}} \cdot (A^- \oplus \mathbf{0})]^\smile \odot [\mathring{\mathbf{1}} \cdot (A^- \oplus \mathbf{0})] \le \mathring{\mathbf{1}}$ by 3.2(ii),(vii),(ix);

(6) $\quad \boldsymbol{Q}_{AB} \vdash^\times A^{\square\smile} \odot A^\square \le \mathring{\mathbf{1}}$ by (4), (5), (ii), (xi), 3.2(xxxi);

(7) $\quad \boldsymbol{Q}_{AB} \vdash^\times B^{\square\smile} \odot B^\square \le \mathring{\mathbf{1}}$ by (vi), (6);

(8) $\quad \boldsymbol{Q}_{AB} \vdash^\times \mathbf{1} = A^{\square\smile} \odot B^\square$ by (ii), (xi), 3.2(vii).

Theorem (xii) follows immediately from (3), (6), (7), and (8), by (ii).

4.2. Systems of conjugated quasiprojections and systems of predicates $\boldsymbol{P}_{AB}$

In this and the next section we consider two arbitrary predicates $A, B \in \boldsymbol{\Pi}$, assumed to be fixed throughout the discussion. In terms of A and B we define an infinite sequence of predicates

$$\boldsymbol{P}_{AB} = \langle \boldsymbol{P}^{(0)}_{AB}, \boldsymbol{P}^{(1)}_{AB}, \ldots, \boldsymbol{P}^{(m)}_{AB}, \ldots\rangle$$

by stipulating

(i) $\boldsymbol{P}_{AB}^{(m)} = A^m \odot B$ *for each* $m \in \omega$.

The significance of these predicates under the assumption of $\boldsymbol{Q}_{AB}$ is brought out in the following statements.

(ii) $\boldsymbol{Q}_{AB} \vdash \forall_{v_0 v_1 v_2}(v_0 \boldsymbol{P}_{AB}^{(m)} v_1 \wedge v_0 \boldsymbol{P}_{AB}^{(m)} v_2 \rightarrow v_1 \mathring{1} v_2)$ *for each* $m \in \omega$.

(iii) $\boldsymbol{Q}_{AB} \vdash \forall_{v_0 \cdots v_m} \exists_{v_{m+1}} (v_{m+1} \boldsymbol{P}_{AB}^{(k_0)} v_0 \wedge \cdots \wedge v_{m+1} \boldsymbol{P}_{AB}^{(k_m)} v_m)$ *for each* $m \in \omega$ *and each sequence* $\langle k_0, \ldots, k_m \rangle$ *of distinct natural numbers.*

(iv) *If* $C = \boldsymbol{P}_{AB}^{(m)}$, $D = \boldsymbol{P}_{AB}^{(n)}$, *and* $m, n \in \omega$ *with* $m \neq n$, *then* $\boldsymbol{Q}_{AB} \vdash^{\times} \boldsymbol{Q}_{CD}$.

Theorem (ii) is a straightforward consequence of (i) and 4.1(iii). The proof of (iii) can easily be reduced to the special case when $k_i = i$ for $i = 0, \ldots, m$. To prove the special case one proceeds by induction on m, using 4.1(iii); as usual, the completeness theorem for $\mathcal{L}^+$ may be used to simplify the argument. Theorem (iv) may be derived from (i) and 4.1(ii),(vii), with the help of some elementary lemmas from §3.2, by a double induction on m and n.

In connection with (ii) and (iii) we may generalize the notion of (a pair of) conjugated quasiprojections: for each $m \in \omega$ a sequence $\langle H_0, \ldots, H_m \rangle$ of binary relations between elements of a set U will be referred to as a *sequence of* $m+1$ *conjugated quasiprojections* (on U) just in case all the relations $H_0, \ldots, H_m$ are functions, and for any elements $y_0, \ldots, y_m$ in U there is an x in U such that $H_0 x = y_0, \ldots, H_m x = y_m$. A familiar example of such a sequence is provided by the projections defined over ordered $(m+1)$-tuples of elements of U (presuming, of course, that $\langle y_0, \ldots, y_m \rangle \in U$ whenever $y_0, \ldots, y_m \in U$). From (ii) and (iii) we conclude that, in case A and B are assumed to denote two conjugated quasiprojections, the corresponding relations denoted by $\boldsymbol{P}_{AB}^{(k_0)}, \ldots, \boldsymbol{P}_{AB}^{(k_m)}$ form a sequence of $m+1$ conjugated quasiprojections.

It follows trivially from (iv) that 4.1(vii)–(x) continue to hold if we replace all occurrences of A and B on the right-hand side of $\vdash^{\times}$ by $\boldsymbol{P}_{AB}^{(m)}$ and $\boldsymbol{P}_{AB}^{(n)}$ respectively, whenever $m, n \in \omega$ and $m \neq n$. We thus get, in particular,

(v) $\boldsymbol{Q}_{AB} \vdash^{\times} \{\mathbf{1} \odot \boldsymbol{P}_{AB}^{(m)} = \mathbf{1}, \boldsymbol{P}_{AB}^{(m)\smile} \odot \mathbf{1} = \mathbf{1}, \boldsymbol{P}_{AB}^{(m)\smile} \odot \boldsymbol{P}_{AB}^{(m)} = \mathring{\mathbf{1}}\}$ *for every* $m \in \omega$.

Theorem 4.1(viii) can now be generalized as follows.

(vi) *Let* $\langle k_0, \ldots, k_m \rangle$ *be any finite sequence of distinct natural numbers. Then*

$$\boldsymbol{Q}_{AB} \vdash^{\times} (C_0 \odot D_0) \cdot \cdots \cdot (C_m \odot D_m) =$$
$$[(C_0 \odot \boldsymbol{P}_{AB}^{(k_0)\smile}) \cdot \cdots \cdot (C_m \odot \boldsymbol{P}_{AB}^{(k_m)\smile})] \odot [(\boldsymbol{P}_{AB}^{(k_0)} \odot D_0) \cdot \cdots \cdot (\boldsymbol{P}_{AB}^{(k_m)} \odot D_m)]$$

for arbitrary sequences $\langle C_0, \ldots, C_m \rangle$ *and* $\langle D_0, \ldots, D_m \rangle$ *of predicates.*

We first establish (vi) under the hypothesis that $k_i = i$ for $i = 0, \ldots, m$. The argument proceeds by induction on m. In case $m = 0$ the proof reduces to showing that

$$\boldsymbol{Q}_{AB} \vdash^{\times} C_0 \odot D_0 = C_0 \odot B^{\smile} \odot (B \odot D_0).$$

This however follows at once from 4.1(vii).

Assume now that the conclusion holds for a given m in ω, and let $\langle C_0, \ldots, C_{m+1}\rangle$ and $\langle D_0, \ldots, D_{m+1}\rangle$ be any two sequences of predicates. We introduce two abbreviations:

$$S = (C_1 \odot \boldsymbol{P}_{AB}^{(0)\smile}) \cdot \cdots \cdot (C_{m+1} \odot \boldsymbol{P}_{AB}^{(m)\smile}),$$
$$T = (\boldsymbol{P}_{AB}^{(0)} \odot D_1) \cdot \cdots \cdot (\boldsymbol{P}_{AB}^{(m)} \odot D_{m+1}).$$

By applying 3.2(xxvii) (extended to finite sequences) we get

$$(1) \qquad \boldsymbol{Q}_{AB} \vdash^{\times} S \odot A^{\smile} = (C_1 \odot \boldsymbol{P}_{AB}^{(0)\smile} \odot A^{\smile}) \cdot \cdots \cdot (C_{m+1} \odot \boldsymbol{P}_{AB}^{(m)\smile} \odot A^{\smile}).$$

Using (i) and 3.2(iii) we also have

$$(2) \qquad \vdash^{\times} \boldsymbol{P}_{AB}^{(k)\smile} \odot A^{\smile} = \boldsymbol{P}_{AB}^{(k+1)\smile} \quad \text{for} \quad k = 0, \ldots, m.$$

From (1) and (2) we therefore obtain

$$(3) \qquad \boldsymbol{Q}_{AB} \vdash^{\times} S \odot A^{\smile} = (C_1 \odot \boldsymbol{P}_{AB}^{(1)\smile}) \cdot \cdots \cdot (C_{m+1} \odot \boldsymbol{P}_{AB}^{(m+1)\smile}).$$

In a completely analogous fashion we can derive

$$(4) \qquad \boldsymbol{Q}_{AB} \vdash^{\times} A \odot T = (\boldsymbol{P}_{AB}^{(1)} \odot D_1) \cdot \cdots \cdot (\boldsymbol{P}_{AB}^{(m+1)} \odot D_{m+1}).$$

Now by the induction hypothesis applied to the two sequences $\langle C_1, \ldots, C_{m+1}\rangle$ and $\langle D_1, \ldots, D_{m+1}\rangle$ we see that

$$(5) \qquad \boldsymbol{Q}_{AB} \vdash^{\times} (C_1 \odot D_1) \cdot \cdots \cdot (C_{m+1} \odot D_{m+1}) = S \odot T$$

and therefore

$$(6) \qquad \boldsymbol{Q}_{AB} \vdash^{\times} (C_0 \odot D_0) \cdot \cdots \cdot (C_{m+1} \odot D_{m+1}) = C_0 \odot D_0 \cdot (S \odot T).$$

From 4.1(vi),(viii) we obtain

$$(7) \qquad \boldsymbol{Q}_{AB} \vdash^{\times} C_0 \odot D_0 \cdot (S \odot T) = C_0 \odot B^{\smile} \cdot (S \odot A^{\smile}) \odot (B \odot D_0 \cdot (A \odot T)).$$

Putting (3), (4), (6), and (7) together, we arrive at

$$\boldsymbol{Q}_{AB} \vdash^{\times} (C_0 \odot D_0) \cdot \cdots \cdot (C_{m+1} \odot D_{m+1}) =$$
$$[(C_0 \odot \boldsymbol{P}_{AB}^{(0)\smile}) \cdot \cdots \cdot (C_{m+1} \odot \boldsymbol{P}_{AB}^{(m+1)\smile})] \odot [(\boldsymbol{P}_{AB}^{(0)} \odot D_0) \cdot \cdots \cdot (\boldsymbol{P}_{AB}^{(m+1)} \odot D_{m+1})].$$

The conclusion continues to hold for $m + 1$. This completes the proof of (vi) under the special hypothesis.

In the general case let n be the largest of the natural numbers $k_0, \ldots, k_m$. Given two sequences $\langle C_0, \ldots, C_m\rangle$ and $\langle D_0, \ldots, D_m\rangle$ of predicates, we define new sequences $\langle C'_0, \ldots, C'_n\rangle$ and $\langle D'_0, \ldots, D'_n\rangle$ by stipulating that

(8) $C'_{k_i} = C_i$ and $D'_{k_i} = D_i$ for $i = 0, \ldots, m$;

(9) $C'_j = D'_j = \mathbf{1}$ in case $0 \leq j \leq n$ and $j \neq k_0, \ldots, k_m$.

By the special case of our theorem which was established above, we have

$$\begin{aligned} \boldsymbol{Q}_{AB} \vdash^{\times} & (C'_0 \odot D'_0) \cdot \cdots \cdot (C'_n \odot D'_n) \\ & = [(C'_0 \odot \boldsymbol{P}^{(0)\smile}_{AB}) \cdot \cdots \cdot (C'_n \odot \boldsymbol{P}^{(n)\smile}_{AB})] \odot [(\boldsymbol{P}^{(0)}_{AB} \odot D'_0) \cdot \cdots \cdot (\boldsymbol{P}^{(n)}_{AB} \odot D'_n)]. \end{aligned}$$

From this, and by (9) and 3.2(vii),(ix), we obtain

$$\begin{aligned} (10)\ \boldsymbol{Q}_{AB} \vdash^{\times} & (C'_{k_0} \odot D'_{k_0}) \cdot \cdots \cdot (C'_{k_m} \odot D'_{k_m}) \leq \\ & [(C'_{k_0} \odot \boldsymbol{P}^{(k_0)\smile}_{AB}) \cdot \cdots \cdot (C'_{k_m} \odot \boldsymbol{P}^{(k_m)\smile}_{AB})] \odot [(\boldsymbol{P}^{(k_0)}_{AB} \cdot D'_{k_0}) \cdot \cdots \cdot (\boldsymbol{P}^{(k_m)}_{AB} \odot D'_{k_m})]. \end{aligned}$$

On the other hand, using (iv), 3.2(viii), and 4.1(vii), we derive by induction on m:

$$\begin{aligned} (11)\ \boldsymbol{Q}_{AB} \vdash^{\times} & [(C'_{k_0} \odot \boldsymbol{P}^{(k_0)\smile}_{AB}) \cdot \cdots \cdot (C'_{k_m} \odot \boldsymbol{P}^{(k_m)\smile}_{AB})] \\ & \odot [(\boldsymbol{P}^{(k_0)}_{AB} \odot D'_{k_0}) \cdot \cdots \cdot (\boldsymbol{P}^{(k_m)}_{AB} \odot D'_{k_m})] \leq (C'_{k_0} \odot D'_{k_0}) \cdot \cdots \cdot (C'_{k_m} \odot D'_{k_m}). \end{aligned}$$

From (8), (10), and (11) we obtain at once the desired conclusion.

The results which will be stated in the subsequent part of this section are of some interest in their own right, but will not be applied until Chapter 7.

From (vi) we obtain the following corollary.

(vii) *Let $m, n \in \omega$ with $n \leq m$ and let $\langle k_0, \ldots, k_m\rangle$ be an arbitrary sequence of distinct natural numbers. Then for any predicates $D_0, \ldots, D_m$ we have*

$$\begin{aligned} (\alpha)\ \boldsymbol{Q}_{AB} \vdash^{\times} & [(\mathbf{1} \odot D_0) \cdot \cdots \cdot (\mathbf{1} \odot D_{n-1})] \cdot D_n \cdot [(\mathbf{1} \odot D_{n+1}) \cdot \cdots \cdot (\mathbf{1} \odot D_m)] \\ & = \boldsymbol{P}^{(k_n)\smile}_{AB} \odot [(\boldsymbol{P}^{(k_0)}_{AB} \odot D_0) \cdot \cdots \cdot (\boldsymbol{P}^{(k_m)}_{AB} \odot D_m)]; \\ (\beta)\ \boldsymbol{Q}_{AB} \vdash^{\times} & [(D_0 \odot \mathbf{1}) \cdot \cdots \cdot (D_{n-1} \odot \mathbf{1})] \cdot D_n \cdot [(D_{n+1} \odot \mathbf{1}) \cdot \cdots \cdot (D_m \odot \mathbf{1})] \\ & = [(D_0 \odot \boldsymbol{P}^{(k_0)\smile}_{AB}) \cdot \cdots \cdot (D_m \odot \boldsymbol{P}^{(k_m)\smile}_{AB})] \odot \boldsymbol{P}^{(k_n)}_{AB}. \end{aligned}$$

Indeed,

$$\begin{aligned} (1)\quad \boldsymbol{Q}_{AB} \vdash^{\times} & [(\mathbf{1} \odot D_0) \cdot \cdots \cdot (\mathbf{1} \odot D_{n-1})] \cdot (\mathring{\mathbf{1}} \odot D_n) \cdot [(\mathbf{1} \odot D_{n+1}) \cdot \cdots \cdot (\mathbf{1} \odot D_m)] \\ & = ([(\mathbf{1} \odot \boldsymbol{P}^{(k_0)\smile}_{AB}) \cdot \cdots \cdot (\mathbf{1} \odot \boldsymbol{P}^{(k_{n-1})\smile}_{AB})] \cdot (\mathring{\mathbf{1}} \odot \boldsymbol{P}^{(k_n)\smile}_{AB}) \\ & \qquad \cdot [(\mathbf{1} \odot \boldsymbol{P}^{(k_{n+1})\smile}_{AB}) \cdot \cdots \cdot (\mathbf{1} \odot \boldsymbol{P}^{(k_m)\smile}_{AB})]) \\ & \quad \odot [(\boldsymbol{P}^{(k_0)}_{AB} \odot D_0) \cdot \cdots \cdot (\boldsymbol{P}^{(k_m)}_{AB} \odot D_m)] \text{ by (vi);} \end{aligned}$$

(2) $\boldsymbol{Q}_{AB} \vdash^\times [(\mathbf{1}\odot D_0)\cdot\dots\cdot(\mathbf{1}\odot D_{n-1})]\cdot D_n\cdot[(\mathbf{1}\odot D_{n+1})\cdot\dots\cdot(\mathbf{1}\odot D_m)]$
$\mathrel{\underline{\le}} \boldsymbol{P}_{AB}^{(k_n)\smile}\odot[(\boldsymbol{P}_{AB}^{(k_0)}\odot D_0)\cdot\dots\cdot(\boldsymbol{P}_{AB}^{(k_m)}\odot D_m)]$
by (1) and 3.2(vii),(ix);

(3) $\vdash^\times \boldsymbol{P}_{AB}^{(k_n)\smile}\odot[(\boldsymbol{P}_{AB}^{(k_0)}\odot D_0)\cdot\dots\cdot(\boldsymbol{P}_{AB}^{(k_m)}\odot D_m)]$
$\mathrel{\underline{\le}} (\boldsymbol{P}_{AB}^{(k_n)\smile}\odot\boldsymbol{P}_{AB}^{(k_0)}\odot D_0)\cdot\dots\cdot(\boldsymbol{P}_{AB}^{(k_n)\smile}\odot\boldsymbol{P}_{AB}^{(k_m)}\odot D_m)$
by BA and 3.2(viii);

(4) $\boldsymbol{Q}_{AB} \vdash^\times \boldsymbol{P}_{AB}^{(k_n)\smile}\odot[(\boldsymbol{P}_{AB}^{(k_0)}\odot D_0)\cdot\dots\cdot(\boldsymbol{P}_{AB}^{(k_m)}\odot D_m)]$
$\mathrel{\underline{\le}} [(\mathbf{1}\odot D_0)\cdot\dots\cdot(\mathbf{1}\odot D_{n-1})]\cdot D_n\cdot[(\mathbf{1}\odot D_{n+1})\cdot\dots\cdot(\mathbf{1}\odot D_m)]$
by (3), (iv) and 4.1(ii).

Part (α) follows at once from (2) and (4). We easily obtain (β) from (α) by replacing $D_0,\dots,D_m$ in (α) with $D_0^\smile,\dots,D_m^\smile$, and applying 3.2(i)–(iii).

(viii) *Let $m,n\in\omega$ with $n\le m$, and let $\langle k_0,\dots,k_m\rangle$ be any sequence of distinct natural numbers. Given arbitrary predicates $C_0,\dots,C_m$, set*

$$D = (\boldsymbol{P}_{AB}^{(k_0)}\odot C_0\odot\boldsymbol{P}_{AB}^{(k_0)\smile})\cdot\dots\cdot(\boldsymbol{P}_{AB}^{(k_m)}\odot C_m\odot\boldsymbol{P}_{AB}^{(k_m)\smile})$$

and

$$E = [(\mathbf{1}\odot C_0\odot\mathbf{1})\cdot\dots\cdot(\mathbf{1}\odot C_{n-1}\odot\mathbf{1})]\cdot[(\mathbf{1}\odot C_{n+1}\odot\mathbf{1})\cdot\dots\cdot(\mathbf{1}\odot C_m\odot\mathbf{1})].$$

Then

(α) $\vdash^\times_{\boldsymbol{Q}_{AB}} \boldsymbol{P}_{AB}^{(k_n)\smile}\odot D\odot\boldsymbol{P}_{AB}^{(k_n)} = C_n\cdot E;$

(β) $\{\mathbf{1}\odot C_i\odot\mathbf{1}=\mathbf{1}\colon i\le m,\ i\ne n\}\vdash^\times_{\boldsymbol{Q}_{AB}} \boldsymbol{P}_{AB}^{(k_n)\smile}\odot D\odot\boldsymbol{P}_{AB}^{(k_n)} = C_n;$

(γ) $\{\neg C_i=\mathbf{0}\colon i\le m,\ i\ne n\}\vdash_{\boldsymbol{Q}_{AB}} \boldsymbol{P}_{AB}^{(k_n)\smile}\odot D\odot\boldsymbol{P}_{AB}^{(k_n)} = C_n.$

To prove (α) we first apply (vii)(α), with D_i replaced by $C_i\odot\boldsymbol{P}_{AB}^{(k_i)\smile}$ for $i=0,\dots,m$, and obtain

$$\begin{aligned}\boldsymbol{Q}_{AB} \vdash^\times\ & \boldsymbol{P}_{AB}^{(k_n)\smile}\odot D\\ &= [(\mathbf{1}\odot C_0\odot\boldsymbol{P}_{AB}^{(k_0)\smile})\cdot\dots\cdot(\mathbf{1}\odot C_{n-1}\odot\boldsymbol{P}_{AB}^{(k_{n-1})\smile})]\cdot(C_n\odot\boldsymbol{P}_{AB}^{(k_n)\smile})\\ &\quad\cdot[(\mathbf{1}\odot C_{n+1}\odot\boldsymbol{P}_{AB}^{(k_{n+1})\smile})\cdot\dots\cdot(\mathbf{1}\odot C_m\odot\boldsymbol{P}_{AB}^{(k_m)\smile})].\end{aligned}$$

From this we arrive directly at (α) by applying (vii)(β), with D_i replaced by $\mathbf{1}\odot C_i$ for $i=0,\dots,m$ and $i\ne n$, and by C_i for $i=n$. (β) is an immediate consequence of (α), and (γ) follows at once from (β) with the help of 2.2(iii).

With "$\vdash^\times$" replaced by "$\vdash$" in (α) and (β), Theorem (viii) can be somewhat generalized. To this end, with any sequence $\langle A_0,\dots,A_m\rangle$ of predicates ($m\in\omega$), we correlate the sentence $\boldsymbol{Q}(A_0,\dots,A_m)$ defined as follows:

(ix) $\boldsymbol{Q}(A_0, \ldots, A_m) = S_0 \wedge \cdots \wedge S_m \wedge T$, *where*

$$S_k = \forall_{v_0 v_1 v_2}(v_0 A_k v_1 \wedge v_0 A_k v_2 \rightarrow v_1 \mathring{\mathbf{1}} v_2) \ \textit{for} \ k = 0, \ldots, m, \ \textit{and}$$

$$T = \forall_{v_0 \cdots v_m} \exists_{v_{m+1}}(v_{m+1} A_0 v_0 \wedge \cdots \wedge v_{m+1} A_m v_m).$$

Thus $\boldsymbol{Q}(A_0, \ldots, A_m)$ expresses the fact that relations denoted by $A_0, \ldots, A_m$ (in any realization of $\mathcal{L}^+$) are $m+1$ conjugated quasiprojections; cf. (ii), (iii), and the subsequent remarks. It then turns out that Theorem (viii) (with "$\vdash^{\times}$" replaced by "$\vdash$") continues to hold if we change everywhere $\boldsymbol{Q}_{AB}, \boldsymbol{P}_{AB}^{(k_0)}, \ldots, \boldsymbol{P}_{AB}^{(k_m)}$ to $\boldsymbol{Q}(A_0, \ldots, A_m), A_0, \ldots, A_m$, respectively. The changes needed in the proof are obvious.

Theorem (viii) can clearly be formulated in set-theoretical terms. We state this formulation explicitly for the generalized form of (viii) just indicated.

(x) *Given any $m \in \omega$, let $\langle H_0, \ldots, H_m \rangle$ be a sequence of $m+1$ conjugated quasiprojections on a given set U; let O be the $(m+1)$-ary operation from and to (binary) relations on U defined by the condition*

$$O(R_0, \ldots, R_m) = \bigcap\{H_k | R_k | H_k^{-1} \colon k \leq m\}.$$

In case $R_0, \ldots, R_m$ are nonempty, we then have

$$R_n = H_n^{-1} | O(R_0, \ldots, R_m) | H_n \quad \textit{for} \quad n = 0, \ldots, m.$$

In other words, when restricted to nonempty relations on U, O has the fundamental property of the operation of forming ordered $(m+1)$-tuples, i.e., $O(R_0, \ldots, R_m)$ uniquely determines each of its arguments $R_0, \ldots, R_m$. (x) can be proved directly by an elementary set-theoretical argument.

As was observed by Givant, some theorems can be established which (with one trivial exception) improve (viii) and (x). We state these in (xi) and (xii) below. In the case of (xii) it is easy to explain in what this improvement consists: it turns out that if the operation O from (x) is replaced by a more elaborately constructed operation O', then the restriction to nonempty relations $R_0, \ldots, R_m$ may be deleted provided only that $|U| \geq 2$. (However, in the exceptional case $|U| = 1$, (xii) can easily be seen to fail.)

(xi) *Let $m, n \in \omega$ with $n \leq m$ and let $\langle k_0, \ldots, k_m \rangle$ be any sequence of distinct natural numbers. Given arbitrary predicates $C_0, \ldots, C_m$, set*

$$D' = [\boldsymbol{P}_{AB}^{(k_0)} \odot (\mathring{\mathbf{0}} + A \odot C_0 \odot B^{\smile}) \odot \boldsymbol{P}_{AB}^{(k_0)\smile}] \cdot \cdots \cdot [\boldsymbol{P}_{AB}^{(k_m)} \odot (\mathring{\mathbf{0}} + A \odot C_m \odot B^{\smile}) \odot \boldsymbol{P}_{AB}^{(k_m)\smile}]$$

and let E be as in (viii). *Then*

(α) $\quad \boldsymbol{Q}_{AB} \vdash^{\times} A^{\smile} \odot (\boldsymbol{P}_{AB}^{(k_n)\smile} \odot D' \odot \boldsymbol{P}_{AB}^{(k_n)} \cdot \mathring{\mathbf{1}}) \odot B = C_n \cdot (\mathbf{1} \odot \mathring{\mathbf{0}} \odot \mathbf{1} + E)$;

(β) $\mathbf{1} \odot \mathring{\mathbf{0}} \odot \mathbf{1} = \mathbf{1} \vdash^{\times} A^{\smile} \odot (\boldsymbol{P}_{AB}^{(k_n)\smile} \odot D' \odot \boldsymbol{P}_{AB}^{(k_n)} \cdot \mathring{\mathbf{1}}) \odot B = C_n$;

(γ) $\quad \neg\mathring{\mathbf{0}} = \mathbf{0} \vdash A^{\smile} \odot (\boldsymbol{P}_{AB}^{(k_n)\smile} \odot D' \odot \boldsymbol{P}_{AB}^{(k_n)} \cdot \mathring{\mathbf{1}}) \odot B = C_n$.

The proof of (α) is based on (viii) and proceeds as follows.

(1) $\boldsymbol{Q}_{AB} \vdash^\times ([\mathbf{1}\odot(\mathring{\mathbf{0}}+A\odot C_0\odot B^\smile)\odot\mathbf{1}]\cdot\cdots\cdot[\mathbf{1}\odot(\mathring{\mathbf{0}}+A\odot C_{n-1}\odot B^\smile)\odot\mathbf{1}])$
$\cdot([\mathbf{1}\odot(\mathring{\mathbf{0}}+A\odot C_{n+1}\odot B^\smile)\odot\mathbf{1}]\cdot\cdots\cdot[\mathbf{1}\odot(\mathring{\mathbf{0}}+A\odot C_m\odot B^\smile)\odot\mathbf{1}])$
$=\mathbf{1}\odot\mathring{\mathbf{0}}\odot\mathbf{1}+E$ by BA, 3.2(v), and 4.1(vii);

(2) $\boldsymbol{Q}_{AB} \vdash^\times \boldsymbol{P}_{AB}^{(k_n)\smile}\odot D'\odot\boldsymbol{P}_{AB}^{(k_n)} = (\mathring{\mathbf{0}}+A\odot C_n\odot B^\smile)\cdot(\mathbf{1}\odot\mathring{\mathbf{0}}\odot\mathbf{1}+E)$
by (1) and (viii)(α);

(3) $\boldsymbol{Q}_{AB} \vdash^\times A^\smile\odot(\boldsymbol{P}_{AB}^{(k_n)\smile}\odot D'\odot\boldsymbol{P}_{AB}^{(k_n)}\cdot\mathring{\mathbf{1}})\odot B$
$= C_n\cdot\,(A^\smile\odot[(\mathbf{1}\odot\mathring{\mathbf{0}}\odot\mathbf{1}+E)\cdot\mathring{\mathbf{1}}]\odot B)$
by (2), BA, 4.1(ix)(α);

(4) $\vdash^\times \mathbf{1}\odot(\mathbf{1}\odot\mathring{\mathbf{0}}\odot\mathbf{1}+E)\odot\mathbf{1}=\mathbf{1}\odot\mathring{\mathbf{0}}\odot\mathbf{1}+E$
by the definition of E and 3.2(vii)–(ix);

(5) $\vdash^\times A^\smile\odot[A\odot(\mathbf{1}\odot\mathring{\mathbf{0}}\odot\mathbf{1}+E)\odot B^\smile\cdot\mathring{\mathbf{1}}]\odot B$
$\leq A^\smile\odot[(\mathbf{1}\odot\mathring{\mathbf{0}}\odot\mathbf{1}+E)\cdot\mathring{\mathbf{1}}]\odot B$
by (4), 3.2(vii);

(6) $\boldsymbol{Q}_{AB} \vdash^\times \mathbf{1}\odot\mathring{\mathbf{0}}\odot\mathbf{1}+E\leq A^\smile\odot[(\mathbf{1}\odot\mathring{\mathbf{0}}\odot\mathbf{1}+E)\cdot\mathring{\mathbf{1}}]\odot B$
by (5), 4.1(ii),(ix)(β);

(7) $\boldsymbol{Q}_{AB} \vdash^\times \mathbf{1}\odot\mathring{\mathbf{0}}\odot\mathbf{1}+E= A^\smile\odot[(\mathbf{1}\odot\mathring{\mathbf{0}}\odot\mathbf{1}+E)\cdot\mathring{\mathbf{1}}]\odot B$
by (4), (6), and 3.2(vii);

(8) $\boldsymbol{Q}_{AB} \vdash^\times A^\smile\odot(\boldsymbol{P}_{AB}^{(k_n)\smile}\odot D'\odot\boldsymbol{P}_{AB}^{(k_n)}\cdot\mathring{\mathbf{1}})\odot B = C_n\cdot(\mathbf{1}\odot\mathring{\mathbf{0}}\odot\mathbf{1}+E)$
by (3) and (7).

Part (β) is an immediate consequence of (α), and (γ) follows from (β) with the help of 2.2(iii).

With some changes, the remarks following (viii) apply to (xi) as well.

(xii) *Given any $m\in\omega$, let $\langle H_0,\dots,H_m\rangle$ be a sequence of $m+1$ conjugated quasiprojections on a given set U, and F,G any pair of conjugated quasiprojections on U; let O' be the $(m+1)$-ary operation from and to relations on U defined by the condition*

$$O'(R_0,\dots,R_m)=\bigcap\{H_k|(Di\cup F|R_k|G^{-1})|H_k^{-1}:\,k\leq m\}.$$

If U has at least two elements, then for any relations $R_0,\dots,R_m$ we have

$$R_n=F^{-1}|\,[(H_n^{-1}|O'(R_0,\dots,R_m)|H_n)\cap Id]\,|G\quad\textit{for}\quad n=0,\dots,m.$$

Theorem (xii) is derived from (x) in much the same way as (xi) from (viii).

For later purposes we state a property of the predicates $\boldsymbol{P}^{(m)}_{A^\square B^\square}$, where $A^\square$ and $B^\square$ are the predicates obtained respectively from A and B according to 4.1(xi). It is established by induction on m, with the help of 4.1(vi),(xii).

(xiii) $\vdash^\times \{\boldsymbol{P}^{(m)}_{A^\square B^\square} \boldsymbol{\odot} \mathbf{1} = \mathbf{1},\ \mathbf{1} \boldsymbol{\odot} \boldsymbol{P}^{(m)\smile}_{A^\square B^\square} = \mathbf{1}\}$ *for every* $m \in \omega$.

4.3. Historical remarks regarding the translation mapping from $\mathcal{L}^+$ to $\mathcal{L}^\times$

In establishing the relative equipollence of $\mathcal{L}^+$ and $\mathcal{L}^\times$, i.e., the equipollence of the systems obtained by relativizing the formalisms $\mathcal{L}^+$ and $\mathcal{L}^\times$ to any given sentence $\boldsymbol{Q}_{AB}$, we shall apply the same general method which was used to establish the equipollence of $\mathcal{L}^+$ and $\mathcal{L}$ in Chapter 2 and of $\mathcal{L}_3^+$ and $\mathcal{L}^\times$ in Chapter 3.[3] As was stated in §3.4, $\mathcal{L}^\times$ is a subformalism of $\mathcal{L}^+$, and hence its expressive and deductive powers at most equal those of $\mathcal{L}^+$. To show that, conversely, the expressive and deductive powers of $\mathcal{L}^+$ at most equal those of $\mathcal{L}^\times$ relative to any given sentence $\boldsymbol{Q}_{AB}$, we shall use again an appropriately constructed translation mapping. Actually, we shall construct not a single mapping, but a whole system of mappings $\boldsymbol{K}_{AB}$ indexed (just as the system of sentences $\boldsymbol{Q}_{AB}$) by arbitrary pairs of predicates A, B. Each $\boldsymbol{K}_{AB}$ is defined on the set Σ^+ and maps this set onto $\Sigma^\times$.

Recall that to define the translation mapping from Σ^+ to Σ in Chapter 2 we first constructed a certain translation mapping $\boldsymbol{G}$ from $\boldsymbol{\Phi}^+$ to $\boldsymbol{\Phi}$ with the property that, for each $X \in \boldsymbol{\Phi}^+$, X and $\boldsymbol{G}X$ have the same free variables; then we took the restriction of $\boldsymbol{G}$ to Σ^+. This approach cannot be utilized in constructing the system of mappings $\boldsymbol{K}_{AB}$ because $\mathcal{L}^\times$ does not contain formulas with free variables.

In his original construction Tarski first defined an auxiliary system of mappings $\boldsymbol{L}_{AB}$ with the following properties.

(i) $\boldsymbol{L}_{AB}$ *is a recursive function from* $\boldsymbol{\Phi}^+$ *into* $\boldsymbol{\Phi}^+$.

(ii) *For every* $X \in \boldsymbol{\Phi}^+$ *with canonical sequence* $\langle x_0, \ldots, x_{m-1}\rangle$ *there exist a variable* u *and predicates* $G_0, \ldots, G_{m-1}, H$ *(all of them uniquely determined) such that*

$$\boldsymbol{L}_{AB}X = \boldsymbol{\forall}_u(x_0 G_0 u \boldsymbol{\vee} \cdots \boldsymbol{\vee} x_{m-1} G_{m-1} u \boldsymbol{\vee} uHu) \quad \text{and} \quad \vdash^\times \mathbf{0} \boldsymbol{\oplus} H = H.$$

(iii) *In particular, for every* $X \in \Sigma^+$ *there is a uniquely determined* $H \in \boldsymbol{\Pi}$ *such that*

$$\boldsymbol{L}_{AB}X = \boldsymbol{\forall}_{\boldsymbol{x}}(\boldsymbol{x}H\boldsymbol{x}) \quad \text{and} \quad \vdash^\times \mathbf{0} \boldsymbol{\oplus} H = H.$$

(iv) $\boldsymbol{\Upsilon\phi}(\boldsymbol{L}_{AB}X) = \boldsymbol{\Upsilon\phi}X$ *for every* $X \in \boldsymbol{\Phi}^+$.

(v) $\boldsymbol{L}_{AB}X \equiv_{\boldsymbol{Q}_{AB}} X$ *for every* $X \in \boldsymbol{\Phi}^+$.

As a framework for later remarks, and perhaps for some historical interest, we state here the definition of $\boldsymbol{L}_{AB}$ explicitly (although, as the reader will notice, the notion so defined is not involved in our further development).

[3]*Thus, the proof we shall give may be regarded as a syntactical proof. In the footnote on p. 244 we discuss briefly a semantical proof essentially due to Maddux.

(vi) *For any* $A, B \in \mathbf{\Pi}$, $\boldsymbol{L}_{AB}$ *is the unique function* F *satisfying conditions* (α)–(η) *given below. In these conditions* $X, Y, Z \in \mathbf{\Phi}^+$; m, n, p, i, j, k *are natural numbers;* $\langle x_0, \ldots, x_{m-1}\rangle$, $\langle y_0, \ldots, y_{n-1}\rangle$, $\langle z_0, \ldots, z_{p-1}\rangle$ *are respectively the canonical sequences of* X, Y, Z; u *is the first variable such that* $u \notin \mathbf{\Upsilon}\phi X$; $r, s \in \mathbf{\Upsilon}$; $C, D, H, G_0, \ldots, G_{m-1}, I_0, \ldots, I_{n-1}, J_0, \ldots, J_{p-1} \in \mathbf{\Pi}$.

(α) $DoF = \mathbf{\Phi}^+$.

(β) *If* $X = rCs$, *then*

$$\begin{aligned} FX &= \forall_u(rCu \vee s\mathring{\mathbf{0}}u \vee u\mathbf{0}u) && \textit{in case } \mathit{in}\, r < \mathit{in}\, s,\\ FX &= \forall_u(sC^{\smile}u \vee r\mathring{\mathbf{0}}u \vee u\mathbf{0}u) && \textit{in case } \mathit{in}\, r > \mathit{in}\, s,\\ FX &= \forall_u(rC + \mathring{\mathbf{0}}u \vee u\mathbf{0}u) && \textit{in case } \mathit{in}\, r = \mathit{in}\, s. \end{aligned}$$

(γ) *If* $X = (C = D)$, *then* $FX = \forall_x(x\mathbf{0} \oplus X^t x)$.

(δ) *If*

$$\begin{aligned} X &= \neg Y,\\ FY &= \forall_u(x_0 G_0 u \vee \cdots \vee x_{m-1}G_{m-1}u \vee uCu), \quad \textit{and}\\ H &= \mathbf{1} \odot (C^{-} \cdot \mathring{\mathbf{1}}) \odot [(G_0^{\smile -} \odot (A^0 \odot B)^{\smile}) \cdot \cdots \cdot (G_{m-1}^{\smile -} \odot (A^{m-1} \odot B)^{\smile})], \end{aligned}$$

then

$$FX = \forall_u[x_0(A^0 \odot B)^{\smile -}u \vee \cdots \vee x_{m-1}(A^{m-1} \odot B)^{\smile -}u \vee uHu].$$

(ε) *If*

$$\begin{aligned} X &= Y \to Z,\\ F(\neg Y) &= \forall_r(y_0 I_0 r \vee \cdots \vee y_{n-1}I_{n-1}r \vee rCr),\\ FZ &= \forall_s(z_0 J_0 s \vee \cdots \vee z_{p-1}J_{p-1}s \vee sDs), \quad \textit{and}\\ H &= (\mathbf{0} \oplus (C + \mathring{\mathbf{0}}) \oplus A^{\smile -}) + (\mathbf{0} \oplus (D + \mathring{\mathbf{0}}) \oplus B^{\smile -}), \end{aligned}$$

then

$$FX = \forall_u(x_0 G_0 u \vee \cdots \vee x_{m-1}G_{m-1}u \vee uHu),$$

where for each $k < m$

$$\begin{aligned} &G_k = I_i \odot A^{\smile} + J_j \odot B^{\smile} \quad \textit{in case } x_k = y_i = z_j \textit{ for some } i < n,\ j < p,\\ &G_k = I_i \odot A^{\smile} \quad \textit{in case } x_k = y_i \neq z_0, \ldots, z_{p-1} \textit{ for some } i < n,\\ &G_k = J_j \odot B^{\smile} \quad \textit{in case } x_k = z_j \neq y_0, \ldots, y_{n-1} \textit{ for some } j < p. \end{aligned}$$

(η) *If*

$$\begin{aligned} X &= \forall_r Y \quad \textit{and}\\ FY &= \forall_s(y_0 I_0 s \vee \cdots \vee y_{n-1}I_{n-1}s \vee sCs), \end{aligned}$$

then

$$\begin{aligned} FX = \forall_u[&y_0 I_0 u \vee \cdots \vee y_{k-1}I_{k-1}u \vee y_{k+1}I_{k+1}u \vee \cdots\\ &\vee y_{n-1}I_{n-1}u \vee u(\mathbf{0} \oplus I_k + C)u]\\ &\textit{in case } r = y_k,\ k < n, \quad \textit{and} \end{aligned}$$

$$FX = FY \quad \textit{in case} \quad r \neq y_0, \ldots, y_{n-1}.$$

A consequence of (ii) and (v) is that, on the basis of the sentence $\boldsymbol{Q}_{AB}$, we can partly eliminate quantifiers from any given formula X in $\boldsymbol{\Phi}^+$; more specifically, we can construct another formula in $\boldsymbol{\Phi}^+$, namely $\boldsymbol{L}_{AB}X$, which is equivalent with X under $\boldsymbol{Q}_{AB}$ and which has only one occurrence of a quantifier. Hence, it easily follows that in case $|\boldsymbol{\Upsilon}\phi X| \leq 2$, the quantifiers can be eliminated entirely; e.g., if $m = 2$ in (ii), then

$$X \equiv_{\boldsymbol{Q}_{AB}} x_0(G_0 \boldsymbol{\oplus} (\mathring{\boldsymbol{0}} \boldsymbol{+} H) \boldsymbol{\oplus} G_1^{\smile})x_1.$$

Obviously, throughout the whole discussion we could use, instead of universally quantified disjunctive formulas, the dual existentially quantified conjunctive formulas.

The definition of a system of mappings $\boldsymbol{K}_{AB}$ in terms of the mapping $\boldsymbol{L}_{AB}$ is simple.

(vii) *For any $A, B \in \boldsymbol{\Pi}$, $\boldsymbol{K}_{AB}$ is the unique function F such that*

(α) $DoF = \Sigma^+$,

(β) $FX = X$ *in case* $X \in \Sigma^\times$,

(γ) $FX = (H \boldsymbol{=} \boldsymbol{1})$ *in case* $X \in \Sigma^+ \sim \Sigma^\times$ *and H is the uniquely determined predicate for which* $\boldsymbol{L}_{AB}X = \boldsymbol{\forall}_{\boldsymbol{x}}(\boldsymbol{x}H\boldsymbol{x})$.

(We could simplify this definition by removing condition (β) and extending (γ) to all $X \in \Sigma^\times$. Such a change would exert only a minimal influence on the present discussion, but it would necessitate the use of a different notion of translation mapping than the one introduced in 2.4(iii); cf. 2.4(v).)

On the other hand, the proof (by induction on sentences derivable in $\mathcal{L}^+$) that $\boldsymbol{K}_{AB}$ has the desired property, namely

(viii) *For every $\Psi \subseteq \Sigma^+$ and every $X \in \Sigma^+$, if $\Psi \vdash^+ X$, then*

$$\boldsymbol{K}^*_{AB}\Psi \vdash^\times_{\boldsymbol{Q}_{AB}} \boldsymbol{K}_{AB}X,$$

turns out to be more complicated than one would expect. Curiously enough, one of the most involved parts of the proof is where we show that $\boldsymbol{K}_{AB}X$ is derivable from $\boldsymbol{Q}_{AB}$ whenever X is a logical axiom of $\mathcal{L}^+$ of the form (AI), i.e., X has the form of a sentential syllogism (cf. §1.3).

There is another construction of translation mappings which leads to some simplification of both the basic definitions and the proofs of the fundamental results. This construction was discovered by Monk around 1960 (but was never published and was not known to the authors) and was rediscovered in 1974 by Maddux in a slightly modified form. With their permission we shall use the new construction as a base for the subsequent discussion, and in fact we shall present it in the form given by Maddux. In particular, the specific proof of (viii) that we shall give in the next section is essentially due to Maddux.

To describe the main difference between the original and the new constructions (disregarding inessential details), we should point out that in both cases we can define by recursion a function $\boldsymbol{L}_{AB}$ satisfying conditions (i)–(v) above. In the case of the original construction, the predicates $G_0, \ldots, G_{m-1}, H$ occurring in $\boldsymbol{L}_{AB}X$ according to (ii) are rather involved expressions which depend on the whole structure of X. In contrast to this, the predicates $G_0, \ldots, G_{m-1}$ in the new construction depend only on the set $\Upsilon\phi X$, and in fact they are the predicates $\boldsymbol{P}_{AB}^{(k_0)\smile-}, \ldots, \boldsymbol{P}_{AB}^{(k_{m-1})\smile-}$, where the k_i's are the indices of the corresponding variables x_i in the canonical sequence of X. As regards the new definition of H, it is somewhat more involved than the old one in case X is either an atomic or a universally quantified formula, but rather simpler in the remaining cases. Altogether, the new definition of $\boldsymbol{L}_{AB}$ appears to be considerably less involved than the old one.

A further simplification in the new construction results from the observation that, for every formula X with specified free variables, $\boldsymbol{L}_{AB}X$ is now uniquely determined by the predicate H. We also know from (vii) that the translation mapping $\boldsymbol{K}_{AB}$ can actually be defined in terms of H. Since the mapping $\boldsymbol{L}_{AB}$ plays only an auxiliary role in the construction, it can now be eliminated entirely, provided we define by recursion a mapping $\boldsymbol{M}_{AB}$ which correlates with any given formula X the predicate $H = \boldsymbol{M}_{AB}X$. This is essentially the course we shall follow.

If for any reason it should prove convenient, we can obviously reintroduce $\boldsymbol{L}_{AB}$ by setting explicitly

(ix) $\boldsymbol{L}_{AB}X = \forall_u(x_0\boldsymbol{P}_{AB}^{(k_0)\smile-}u \lor \cdots \lor x_{m-1}\boldsymbol{P}_{AB}^{(k_{m-1})\smile-}u \lor u\boldsymbol{M}_{AB}Xu)$ *for every* $X \in \Phi^+$, *where* $\langle x_0, \ldots, x_{m-1}\rangle$ *is the canonical sequence of* X *and* $k_i = \mathbf{in}\, x_i$ *for* $i = 0, \ldots, m-1$.

4.4. Proof of the main mapping theorem for $\mathcal{L}^{\times}$ and $\mathcal{L}^{+}$

We begin by introducing some notation. We consider two arbitrary predicates A and B, which are regarded as fixed throughout this section. For abbreviation we set

(i) $\boldsymbol{P}_n = \boldsymbol{P}_{A^{\square}B^{\square}}^{(n)}$ *for every* $n \in \omega$.

(Notice that in (i) we have used $A^{\square}, B^{\square}$, and not merely A, B.)

For further reference we restate 4.2(iii)–(vi),(xiii), using 4.1(xii) and the abbreviation in (i).

(ii) $\boldsymbol{Q}_{AB} \vdash \forall_{v_0\cdots v_m}\exists_{v_{m+1}}(v_{m+1}\boldsymbol{P}_{k_0}v_0 \land \cdots \land v_{m+1}\boldsymbol{P}_{k_m}v_m)$ *for every sequence* $\langle k_0, \ldots, k_m\rangle$ *of distinct natural numbers.*

(iii) $\boldsymbol{Q}_{AB} \vdash^{\times} \{\boldsymbol{P}_n^{\smile} \odot \boldsymbol{P}_n = \mathring{1},\ \boldsymbol{P}_n^{\smile} \odot \boldsymbol{P}_m = 1,\ \boldsymbol{P}_n \odot 1 = 1,\ 1 \odot \boldsymbol{P}_n = 1,$
$\boldsymbol{P}_n^{\smile} \odot 1 = 1,\ 1 \odot \boldsymbol{P}_n^{\smile} = 1\}$ *for all* $n, m \in \omega$ *with* $n \neq m$.

(iv) $\boldsymbol{Q}_{AB} \vdash^{\times} (C_0 \odot D_0) \cdot \cdots \cdot (C_m \odot D_m)$
$= [(C_0 \odot \boldsymbol{P}_{k_0}^{\smile}) \cdot \cdots \cdot (C_m \odot \boldsymbol{P}_{k_m}^{\smile})] \odot [(\boldsymbol{P}_{k_0} \odot D_0) \cdot \cdots \cdot (\boldsymbol{P}_{k_m} \odot D_m)]$

for every sequence $\langle k_0, \ldots, k_m\rangle$ *of distinct natural numbers and for any two sequences* $\langle C_0, \ldots, C_m\rangle, \langle D_0, \ldots, D_m\rangle$ *of predicates.*

Every finite set $I \subseteq \omega$ determines, of course, a unique strictly increasing sequence of integers $\langle i_0, \ldots, i_{m-1} \rangle$ with range I. We correlate with I a predicate $\boldsymbol{V}_{(I)}$ by setting

(v) $\boldsymbol{V}_{(I)} = (\boldsymbol{P}_{i_0} \odot \boldsymbol{P}_{i_0}^{\smile}) \cdot \cdots \cdot (\boldsymbol{P}_{i_{m-1}} \odot \boldsymbol{P}_{i_{m-1}}^{\smile})$.

Some useful consequences of results in §4.2 are listed in the following theorem.

(vi) *Let I, J be any finite subsets of ω. Then*

(α) $\vdash^\times \boldsymbol{V}_{(J)} \leq \boldsymbol{V}_{(I)}$ *whenever* $I \subseteq J$,

(β) $\boldsymbol{Q}_{AB} \vdash^\times \mathring{\mathbf{1}} \leq \boldsymbol{V}_{(J)}$,

(γ) $\vdash^\times \boldsymbol{V}_{(I)}^{\smile} = \boldsymbol{V}_{(I)}$,

(δ) $\boldsymbol{Q}_{AB} \vdash^\times \boldsymbol{V}_{(I)} \odot \boldsymbol{V}_{(I)} \leq \boldsymbol{V}_{(I)}$,

(ε) $\boldsymbol{Q}_{AB} \vdash^\times \boldsymbol{V}_{(I)} \odot \boldsymbol{V}_{(J)} = \boldsymbol{V}_{(I \cap J)}$,

(ς) $\boldsymbol{Q}_{AB} \vdash^\times \boldsymbol{V}_{(I)}^{-} \oplus \boldsymbol{V}_{(J)}^{-} = \boldsymbol{V}_{(I \cap J)}^{-}$.

The proof of (α)–(γ) is based on (v) and elementary parts of §3.2; in the case of (β) we also use (iii) and 3.2(xx).

The proof of (ε) is more involved. Let m be the least natural number such that $i < m$ for $i \in I \cup J$, and recall that $m = \{0, \ldots, m-1\}$. We define two sequences $\langle C_0, \ldots, C_{m-1} \rangle$ and $\langle D_0, \ldots, D_{m-1} \rangle$ of predicates by stipulating:

$$C_i = \boldsymbol{P}_i \quad \text{for } i \in I \text{ and } C_i = \mathbf{1} \text{ for } i \in m \sim I;$$
$$D_j = \boldsymbol{P}_j^{\smile} \quad \text{for } j \in J \text{ and } D_j = \mathbf{1} \text{ for } j \in m \sim J.$$

From this, by (iii) and 3.2(ix), we clearly obtain

(1) $\boldsymbol{Q}_{AB} \vdash^\times C_i \odot \boldsymbol{P}_i^{\smile} = \mathbf{1}$ for $i \in m \sim I$,

(2) $\boldsymbol{Q}_{AB} \vdash^\times \boldsymbol{P}_j \odot D_j = \mathbf{1}$ for $j \in m \sim J$,

(3) $\boldsymbol{Q}_{AB} \vdash^\times C_l \odot D_l = \mathbf{1}$ for $l \in m \sim I \cap J$.

With the help of (v), we get from (1)–(3) respectively:

(4) $\boldsymbol{Q}_{AB} \vdash^\times \boldsymbol{V}_{(I)} = (C_0 \odot \boldsymbol{P}_0^{\smile}) \cdot \cdots \cdot (C_{m-1} \odot \boldsymbol{P}_{m-1}^{\smile})$,

(5) $\boldsymbol{Q}_{AB} \vdash^\times \boldsymbol{V}_{(J)} = (\boldsymbol{P}_0 \odot D_0) \cdot \cdots \cdot (\boldsymbol{P}_{m-1} \odot D_{m-1})$,

(6) $\boldsymbol{Q}_{AB} \vdash^\times \boldsymbol{V}_{(I \cap J)} = (C_0 \odot D_0) \cdot \cdots \cdot (C_{m-1} \odot D_{m-1})$.

Now (4)–(6), together with (iv), immediately yield (ε).

Finally, both (δ) and (ς) follow directly from (ε), and the proof of (α)–(ς) is complete.

Together, statements (vi)(β)–(δ) express the fact that in any realization of $\mathcal{L}^+$ which satisfies $\boldsymbol{Q}_{AB}$, every relation denoted by a predicate of the form $\boldsymbol{V}_{(I)}$ (I a finite subset of ω) is an equivalence relation.

The definition of $\boldsymbol{M}_{AB}$, constructed by recursion on formulas, runs as follows.

(vii) *$\boldsymbol{M}_{AB}$ is the unique function F satisfying conditions (α)–(ς) below for arbitrary $C, D \in \boldsymbol{\Pi}$, $X, Y \in \boldsymbol{\Phi}^{+}$, and $i, j \in \omega$:*

(α) $DoF = \boldsymbol{\Phi}^{+}$;
(β) $F(\boldsymbol{v}_i C \boldsymbol{v}_j) = \mathring{\mathbf{1}} \cdot (\boldsymbol{P}_i \odot C \odot \boldsymbol{P}_j^{\smile}) \odot \mathbf{1}$;
(γ) $F(C = D) = \mathbf{0} \oplus \mathring{\mathbf{1}} \cdot [\boldsymbol{P}_0 \odot (C \cdot D + C^{-} \cdot D^{-}) \odot \boldsymbol{P}_1^{\smile}] \odot \mathbf{1}$;
(δ) $F(\neg X) = (FX)^{-}$;
(ε) $F(X \rightarrow Y) = (FX)^{-} + FY$;
(ς) $F(\forall_{\boldsymbol{v}_i} X) = \boldsymbol{V}_{(I)}^{-} \oplus FX$, *where* $I = \boldsymbol{in}^{*}\, \Upsilon\phi(\forall_{\boldsymbol{v}_i} X)$.

(Recall that $\boldsymbol{in}^{*}\, \Upsilon\phi\forall_{\boldsymbol{v}_i} X = \{\boldsymbol{in}\, x : x \in \Upsilon\phi\forall_{\boldsymbol{v}_i} X\}$.)

We list various consequences of this definition.

(viii) *$\boldsymbol{M}_{AB}$ is a recursive function mapping $\boldsymbol{\Phi}^{+}$ into $\boldsymbol{\Pi}$.*

(ix) *For every $X, Y \in \boldsymbol{\Phi}^{+}$ and every $i \in \omega$ we have:*

(α) $\vdash^{\times} \boldsymbol{M}_{AB}(X \vee Y) = \boldsymbol{M}_{AB}X + \boldsymbol{M}_{AB}Y$;
(β) $\vdash^{\times} \boldsymbol{M}_{AB}(X \wedge Y) = \boldsymbol{M}_{AB}X \cdot \boldsymbol{M}_{AB}Y$;
(γ) $\vdash^{\times} \boldsymbol{M}_{AB}(X \leftrightarrow Y) = (\boldsymbol{M}_{AB}X = \boldsymbol{M}_{AB}Y)^{t}$;
(δ) $\vdash^{\times} \boldsymbol{M}_{AB}(\exists_{\boldsymbol{v}_i} X) = \boldsymbol{V}_{(I)} \odot \boldsymbol{M}_{AB}X$, *where* $I = \boldsymbol{in}^{*}\, \Upsilon\phi(\exists_{\boldsymbol{v}_i} X)$;
(ε) $\boldsymbol{M}_{AB}(X \rightarrow Y) = \mathbf{1} \equiv^{\times} \boldsymbol{M}_{AB}X \leq \boldsymbol{M}_{AB}Y$;
(ς) $\boldsymbol{M}_{AB}(X \leftrightarrow Y) = \mathbf{1} \equiv^{\times} \boldsymbol{M}_{AB}X = \boldsymbol{M}_{AB}Y$.

The proofs of (viii) and (ix) are obvious.

(x) $\vdash^{\times} \{\boldsymbol{M}_{AB}X = (\boldsymbol{M}_{AB}X) \oplus \mathbf{0},\ \boldsymbol{M}_{AB}X = (\boldsymbol{M}_{AB}X) \odot \mathbf{1}\}$ *for every $X \in \boldsymbol{\Phi}^{+}$.*

The proof is straightforward and proceeds by induction on formulas; we use 3.2(xiv) and notice that, as a simple consequence of it,

$$(A \odot \mathbf{1} = A) \equiv^{\times} (A \oplus \mathbf{0} = A) \text{ for every } A \in \boldsymbol{\Pi}.$$

(xi) *Let $\mathfrak{U} = \langle U, E\rangle$ be a model of $\boldsymbol{Q}_{AB}$ and X any formula with canonical sequence $\langle \boldsymbol{v}_{k_0}, \dots, \boldsymbol{v}_{k_{m-1}}\rangle$. Let R and $H_0, \dots, H_{m-1}$ be respectively the relations denoted by $\boldsymbol{M}_{AB}X$ and $\boldsymbol{P}_{k_0}, \dots, \boldsymbol{P}_{k_{m-1}}$ in $\mathfrak{U}$. Then for every $u \in U$, we have uRu iff the sequence $\langle H_0 u, \dots, H_{m-1}u\rangle$ satisfies X in $\mathfrak{U}$.*

To understand (xi) notice that, by the hypotheses and (iii), all the relations H_n are functions with domain U (whence the $H_n u$'s are function values). The straightforward proof of (xi) proceeds by induction on formulas, using the definition of $\boldsymbol{M}_{AB}$, the definition of satisfaction, and (ii), (iii), (x).

In view of the semantical completeness of $\mathcal{L}^{+}$, Lemma (xi), together with (ii) and (iii), leads directly to the following purely syntactical result.

(xii) *Let $X \in \boldsymbol{\Phi}^{+}$, let $\langle x_0, \dots, x_{m-1}\rangle$ be any sequence of variables such that $\Upsilon\phi X \subseteq \{x_0, \dots, x_{m-1}\}$, and suppose u is a variable different from all x_i's. Setting $k_i = \boldsymbol{in}\, x_i$ for $i = 0, \dots, m-1$, we obtain:*

(α) $$X \equiv_{\boldsymbol{Q}_{AB}} \forall_u (x_0 \boldsymbol{P}_{k_0}^{\smile -} u \vee \cdots \vee x_{m-1} \boldsymbol{P}_{k_{m-1}}^{\smile -} u \vee u \boldsymbol{M}_{AB} X u);$$

$$(\alpha') \qquad X \equiv_{\boldsymbol{Q}_{AB}} \exists_u (x_0 \boldsymbol{P}_{k_0}^{\smile} u \wedge \cdots \wedge x_{m-1} \boldsymbol{P}_{k_{m-1}}^{\smile} u \wedge u \boldsymbol{M}_{AB} X u);$$

$$(\beta) \quad u\boldsymbol{M}_{AB}Xu \equiv_{\boldsymbol{Q}_{AB}} \forall_{x_0 \cdots x_{m-1}} (x_0 \boldsymbol{P}_{k_0}^{\smile -} u \vee \cdots \vee x_{m-1} \boldsymbol{P}_{k_{m-1}}^{\smile -} u \vee X);$$

$$(\beta') \quad u\boldsymbol{M}_{AB}Xu \equiv_{\boldsymbol{Q}_{AB}} \exists_{x_0 \cdots x_{m-1}} (x_0 \boldsymbol{P}_{k_0}^{\smile} u \wedge \cdots \wedge x_{m-1} \boldsymbol{P}_{k_{m-1}}^{\smile} u \wedge X).$$

Theorem (xii) throws light on the role played by the mapping in our discussion. We know from (viii) that $\boldsymbol{M}_{AB}$ is a mapping which correlates with any formula X in $\mathcal{L}^+$ a predicate $\boldsymbol{M}_{AB}X$. (xii)(β) shows us how $\boldsymbol{M}_{AB}X$ can be explicitly constructed (more precisely, explicitly defined relative to $\boldsymbol{Q}_{AB}$) in terms of X. (We use here the fact that, in view of (x), the formula $u\boldsymbol{M}_{AB}Xu$ in (xii)(β) may be replaced by $u\boldsymbol{M}_{AB}Xv$, where v is any variable different from u.) Conversely, (xii)(α) shows us how X can be explicitly constructed in terms of $\boldsymbol{M}_{AB}X$. In this sense the predicates $\boldsymbol{M}_{AB}X$ may be said to *represent*, or to be (*binary*) *representatives of*, formulas X with any number of free variables.

It may be mentioned here that the idea of constructing, for any formula, a predicate which may serve as its binary representative is not new. In particular, it was known to Tarski; the proofs of results announced in Tarski [**1954**], [**1954a**], which will be discussed in Chapter 7 below, are based on it.

(xiii) *For every $X \in \boldsymbol{\Phi}^+$ with $\Upsilon\phi X = \{\boldsymbol{x}, \boldsymbol{y}\}$ there is a predicate C such that $X \equiv_{\boldsymbol{Q}_{AB}} \boldsymbol{x}C\boldsymbol{y}$; in fact, for C we can take either*

$$\boldsymbol{P}_0^{\smile -} \oplus (\mathring{0} + \boldsymbol{M}_{AB}X) \oplus \boldsymbol{P}_1^{-} \quad \textit{or} \quad \boldsymbol{P}_0^{\smile} \odot (\mathring{1} \cdot \boldsymbol{M}_{AB}X) \odot \boldsymbol{P}_1.$$

This is a direct corollary of (xii)(α),(α').

(xiv) $X \equiv_{\boldsymbol{Q}_{AB}} (\boldsymbol{M}_{AB}X = \mathbf{1})$ *for every* $X \in \Sigma^+$.

Indeed, by (xii)(α) we have $X \equiv_{\boldsymbol{Q}_{AB}} (\mathring{1} \leq \boldsymbol{M}_{AB}X)$, while (x) easily implies $(\mathring{1} \leq \boldsymbol{M}_{AB}X) \equiv_{\boldsymbol{Q}_{AB}} (\boldsymbol{M}_{AB}X = \mathbf{1})$.

Theorem (xii)(α),(α'), as well as Theorems (xiii) and (xiv), are clearly results concerning the role of $\boldsymbol{M}_{AB}$ in eliminating quantifiers on the basis of $\boldsymbol{Q}_{AB}$ (cf. the remarks following 4.3(vi)).

(xv) *For every $X, Y \in \boldsymbol{\Phi}^+$, we have $X \equiv_{\boldsymbol{Q}_{AB}} Y$ iff $\boldsymbol{Q}_{AB} \vdash \boldsymbol{M}_{AB}X = \boldsymbol{M}_{AB}Y$.*

Indeed, (xv) is a simple consequence of (xii) and (x).

By (xv), two formulas X and Y are equivalent relative to $\boldsymbol{Q}_{AB}$ iff their binary representatives, $\boldsymbol{M}_{AB}X$ and $\boldsymbol{M}_{AB}Y$, are so equivalent (in the sense that $\boldsymbol{Q}_{AB} \vdash \boldsymbol{M}_{AB}X = \boldsymbol{M}_{AB}Y$); as a consequence, the many-one mapping $\boldsymbol{M}_{AB}$ from formulas to predicates induces a one-one mapping from equivalence classes of formulas to predicates.

In this connection the following observation, due to Givant, may be of some interest. It is possible to restrict the domain of the mapping $\boldsymbol{M}_{AB}$, i.e., the set $\boldsymbol{\Phi}^+$, in an inessential way so that this mapping itself (and not just the mapping induced on equivalence classes) is one-one. In fact, let us agree to say that X is a *formula without useless quantifiers*, for brevity a *w.u.q. formula*, if X has no subformula of the form $\forall_x Y$ with $x \notin \Upsilon\phi Y$. By deleting from any given formula X all useless quantifier expressions $\forall_x$ we obtain a uniquely determined w.u.q. formula X' which is, of course, logically equivalent to X. Using this terminology we get:

(xvi) *The mapping $\boldsymbol{M}_{AB}$ with the domain restricted to the set of w.u.q. formulas is one-one. In other words, for any w.u.q. formulas X and Y, we have $X = Y$ if (and only if) $\boldsymbol{M}_{AB}X = \boldsymbol{M}_{AB}Y$.*

This can be proved by induction on formulas, using the recursive definition of $\boldsymbol{M}_{AB}$.

We shall now study some deeper properties of the mapping from $\mathcal{L}^+$ to $\mathcal{L}^{\times}$ which correlates with every X in Σ^+ the sentence $\boldsymbol{M}_{AB}X = \mathbf{1}$ in $\Sigma^{\times}$. It should be clear from various remarks in §4.3 that this mapping is closely related to the translation mapping $\boldsymbol{K}_{AB}$. Almost all the results established in (xvii)–(xxxi) below concern the derivability in $\mathcal{L}^{\times}$ of sentences of the form $\boldsymbol{M}_{AB}X = \mathbf{1}$ under various assumptions on X. As will be seen, these results will lead directly to the main mapping theorem for $\mathcal{L}^+$ and $\mathcal{L}^{\times}$.

We begin with a theorem closely related to (xiv).

(xvii) $X \equiv^{\times}_{\boldsymbol{Q}_{AB}} (\boldsymbol{M}_{AB}X = \mathbf{1})$ *for all* $X \in \Sigma^{\times}$.

To prove this, we use the notation of 3.1(i)(γ) and set, as an abbreviation,

(1) $C = \boldsymbol{P}_0 \odot X^t \odot \boldsymbol{P}_1^{\smile}$.

We obtain successively:

(2) $\boldsymbol{Q}_{AB} \vdash^{\times} \boldsymbol{M}_{AB}X \leq (\boldsymbol{P}_0^{\smile -} \oplus \mathring{\mathbf{1}} \cdot C \odot \mathbf{1}) \cdot (\boldsymbol{P}_0^{\smile} \odot \boldsymbol{P}_1)$ by (1), (vii)(γ), (iii);

(3) $\boldsymbol{Q}_{AB} \vdash^{\times} \boldsymbol{M}_{AB}X \leq \boldsymbol{P}_0^{\smile} \odot (\mathring{\mathbf{1}} \cdot C \odot \mathbf{1} \cdot \boldsymbol{P}_1)$ by (2), 3.2(iii),(x);

(4) $\vdash^{\times} \mathring{\mathbf{1}} \cdot C \odot \mathbf{1} \cdot (\mathring{\mathbf{1}} \odot \boldsymbol{P}_1) = \mathring{\mathbf{1}} \cdot C \odot \boldsymbol{P}_1$ by 3.2(xxiii);

(5) $\boldsymbol{Q}_{AB} \vdash^{\times} \boldsymbol{M}_{AB}X \leq \boldsymbol{P}_0^{\smile} \odot (\mathring{\mathbf{1}} \cdot C) \odot \boldsymbol{P}_1$ by (3), (4);

(6) $\boldsymbol{Q}_{AB} \vdash^{\times} \boldsymbol{M}_{AB}X \leq X^t$ by (1), (5), (iii), 4.1(ix)(β);

(7) $\boldsymbol{M}_{AB}X = \mathbf{1} \vdash^{\times}_{\boldsymbol{Q}_{AB}} X$ by (6), 3.2(xxxii);

(8) $X \vdash^{\times} \boldsymbol{M}_{AB}X = \mathbf{1}$ by (vii)(γ), (iii), and 3.2(xxxii),(xiii).

In view of (7) and (8), the proof is complete.

The next theorem can be compared to (x).

(xviii) $\boldsymbol{Q}_{AB} \vdash^\times \boldsymbol{M}_{AB}[X] = \mathbf{0} \oplus \boldsymbol{M}_{AB}X$ *for every* $X \in \boldsymbol{\Phi}^+$.

In fact, we first show that in case X is a sentence of the form $\forall_{x_0 \cdots x_m} Y$, then we have

$$\boldsymbol{Q}_{AB} \vdash^\times \boldsymbol{M}_{AB}X = \mathbf{0} \oplus \boldsymbol{M}_{AB}Y.$$

The proof uses (vii)(ς), (vi)(ς), and the fact that $\boldsymbol{V}_{(\varnothing)} = \mathbf{1}$ by (v). With the help of this result we establish the theorem by induction on formulas; we apply here the definition of $\boldsymbol{M}_{AB}$ in (vii) and the equivalence $(\mathbf{0} \oplus C = C) \equiv^\times (\mathbf{1} \odot C = C)$ for every $C \in \boldsymbol{\Pi}$, which is a simple consequence of 3.2(xiv).

From (xviii), using 3.2(ix),(xiii), we immediately obtain:

(xix) $(\boldsymbol{M}_{AB}[X] = \mathbf{1}) \equiv^\times_{\boldsymbol{Q}_{AB}} (\boldsymbol{M}_{AB}X = \mathbf{1})$ *for every* $X \in \boldsymbol{\Phi}^+$.

Our next task is to show that

$$\boldsymbol{Q}_{AB} \vdash^\times \boldsymbol{M}_{AB}S = \mathbf{1}$$

for every logical axiom S of $\mathcal{L}^+$. By 1.3(i) and the remarks at the beginning of §2.2, the set $\boldsymbol{\Lambda}^+$ of these axioms is divided into fourteen mutually exclusive groups, each of which consists of all instances of one of the schemata (AI)–(AIX) and (DI)–(DV). In principle each group requires a separate argument and may be conveniently treated in a separate lemma. However, in view of the similarity of some of the arguments, we find it possible to occasionally treat two or more groups in the same lemma. In this series of lemmas two auxiliary results of a different character, (xxii) and (xxvi), will be proved separately so as to make them available for the subsequent discussion.

(xx) *If S is a sentence of one of the forms* (AI)–(AIII), *then*

$$\boldsymbol{Q}_{AB} \vdash^\times \boldsymbol{M}_{AB}S = \mathbf{1}.$$

This is actually an immediate consequence of (xix) and (vii)(δ),(ε). For example, suppose S is an instance of (AII), i.e.,

$$S = [(\neg X \rightarrow X) \rightarrow X]$$

for some $X \in \boldsymbol{\Phi}^+$. By (vii)($\delta$),($\varepsilon$) and BA we obviously have

$$\vdash^\times \boldsymbol{M}_{AB}((\neg X \rightarrow X) \rightarrow X) = \mathbf{1}.$$

Hence, by (xix) we obtain directly the conclusion.

The argument just completed would be more involved if we did not use Theorem (xix). This theorem will frequently be used in an analogous way in the subsequent lemmas.

(xxi) *If S is a sentence of the form* (AIV), *then* $\boldsymbol{Q}_{AB} \vdash^{\times} \boldsymbol{M}_{AB}S = \mathbf{1}$.

Assume $S = [\forall_x \forall_y X \rightarrow \forall_y \forall_x X]$, where $X \in \boldsymbol{\Phi}^+$ and $x, y \in \boldsymbol{\Upsilon}$. Let $i = \boldsymbol{in}\, x$, $j = \boldsymbol{in}\, y$, and $J = \boldsymbol{in}^* \boldsymbol{\Upsilon}\phi X$. We get:

(1) $\boldsymbol{Q}_{AB} \vdash^{\times} (\boldsymbol{V}^{-}_{(J \sim \{i,j\})} \oplus \boldsymbol{V}^{-}_{(J \sim \{j\})} \oplus \boldsymbol{M}_{AB}X)^{-} = (\boldsymbol{V}^{-}_{(J \sim \{i,j\})} \oplus \boldsymbol{M}_{AB}X)^{-}$ by (vi)(ς);

(2) $\boldsymbol{Q}_{AB} \vdash^{\times} \boldsymbol{V}^{-}_{(J \sim \{i,j\})} \oplus \boldsymbol{V}^{-}_{(J \sim \{i\})} \oplus \boldsymbol{M}_{AB}X = \boldsymbol{V}^{-}_{(J \sim \{i,j\})} \oplus \boldsymbol{M}_{AB}X$ by (vi)(ς);

(3) $\boldsymbol{Q}_{AB} \vdash^{\times} \boldsymbol{M}_{AB}(\forall_x \forall_y X \rightarrow \forall_y \forall_x X) = \mathbf{1}$ by (1), (2), and (vii)(ε),(ς).

In view of (xix), the desired conclusion follows directly from (3).

(xxii) *Let $X \in \boldsymbol{\Phi}^+$ and $I = \boldsymbol{in}^* \boldsymbol{\Upsilon}\phi X$. We then have*

$$\boldsymbol{Q}_{AB} \vdash^{\times} \{\boldsymbol{V}_{(I)} \odot \boldsymbol{M}_{AB}X = \boldsymbol{M}_{AB}X,\ \boldsymbol{V}^{-}_{(I)} \oplus \boldsymbol{M}_{AB}X = \boldsymbol{M}_{AB}X\}.$$

To prove this, set

(1) $\Gamma = \{Y : Y \in \boldsymbol{\Phi}^+$ and $\boldsymbol{Q}_{AB} \vdash^{\times} \boldsymbol{V}_{(J)} \odot \boldsymbol{M}_{AB}Y = \boldsymbol{M}_{AB}Y$, where $J = \boldsymbol{in}^* \boldsymbol{\Upsilon}\phi Y\}$.

By induction on formulas we show that $X \in \Gamma$ for every $X \in \boldsymbol{\Phi}^+$.

Consider first the case when $X = xCy$, with $C \in \boldsymbol{\Pi}$. Let $i = \boldsymbol{in}\, x$, $j = \boldsymbol{in}\, y$, and set

(2) $S_i = \boldsymbol{P}_i \odot \boldsymbol{P}_i^{\smile}, \quad S_j = \boldsymbol{P}_j \odot \boldsymbol{P}_j^{\smile}, \quad T = \boldsymbol{P}_i \odot C \odot \boldsymbol{P}_j^{\smile}.$

Since by hypothesis $I = \{i, j\}$, we have

(3) $\vdash^{\times} \{\boldsymbol{M}_{AB}X = \mathring{\mathbf{1}} \cdot T \odot \mathbf{1},\ \boldsymbol{V}_{(I)} \odot \boldsymbol{M}_{AB}X = S_i \cdot S_j \odot (\mathring{\mathbf{1}} \cdot T) \odot \mathbf{1}\}$ by (v), (vii)(β).

We get step-by-step:

(4) $\boldsymbol{Q}_{AB} \vdash^{\times} \{S_i \odot T = T,\ T \odot S_j^{\smile} = T\}$ by (2), (iii);

(5) $\boldsymbol{Q}_{AB} \vdash^{\times} S_i \cdot S_j \odot (T \cdot \mathring{\mathbf{1}}) \leq T \cdot (\mathring{\mathbf{1}} \odot S_j)$ by 3.2(viii) and (4);

(6) $\vdash^{\times} T \cdot (\mathring{\mathbf{1}} \odot S_j) \leq T \odot S_j^{\smile} \cdot \mathring{\mathbf{1}} \odot S_j$ by 3.2(xv);

(7) $\boldsymbol{Q}_{AB} \vdash^{\times} S_i \cdot S_j \odot (T \cdot \mathring{\mathbf{1}}) \odot \mathbf{1} \leq T \cdot \mathring{\mathbf{1}} \odot \mathbf{1}$ by (4)–(6), 3.2(vii);

(8) $\boldsymbol{Q}_{AB} \vdash^{\times} T \cdot \mathring{\mathbf{1}} \odot \mathbf{1} \leq S_i \cdot S_j \odot (T \cdot \mathring{\mathbf{1}}) \odot \mathbf{1}$ by (2), (v), (vi)(β).

In view of (3), (7), and (8) we conclude by (1) that $X \in \Gamma$ in the case considered.

Secondly, assume that $X = (C \mathbf{=} D)$ with $C, D \in \mathbf{\Pi}$. Then $I = \varnothing$ and therefore $\boldsymbol{V}_{(I)} = \mathbf{1}$, so that we obtain again $X \in \Gamma$, using (vii)(γ) and 3.2(xiv).

Next, we take up the case when $X = \neg Y$ and $Y \in \Gamma$. Since in this case $I = \boldsymbol{in}^* \Upsilon\phi X = \boldsymbol{in}^* \Upsilon\phi Y$, we have, by (1),

(9) $\boldsymbol{Q}_{AB} \vdash^\times \boldsymbol{V}_{(I)} \odot \boldsymbol{M}_{AB} Y \mathbf{=} \boldsymbol{M}_{AB} Y.$

Continuing, we get:

(10) $\boldsymbol{Q}_{AB} \vdash^\times \boldsymbol{V}_{(I)} \odot (\boldsymbol{M}_{AB} Y)^- \mathbf{=} \boldsymbol{V}_{(I)} \odot (\boldsymbol{V}_{(I)} \odot \boldsymbol{M}_{AB} Y)^-$ by (9);

(11) $\boldsymbol{Q}_{AB} \vdash^\times \boldsymbol{V}_{(I)} \odot (\boldsymbol{M}_{AB} Y)^- \mathbf{=} (\boldsymbol{V}_{(I)} \odot \boldsymbol{M}_{AB} Y)^-$ by (10), (vi)(β)–(δ), and 3.2(xxv);

(12) $\boldsymbol{Q}_{AB} \vdash^\times \boldsymbol{V}_{(I)} \odot (\boldsymbol{M}_{AB} Y)^- \mathbf{=} (\boldsymbol{M}_{AB} Y)^-$ by (9), (11).

In view of (vii)(δ), we see from (12) and (1) that $X \in \Gamma$.

The argument in the case when $X = Y \rightarrow Z$ with $Y, Z \in \Gamma$ is analogous to that in the preceding case, but uses also (vii)(ε) and (vi)(ε). We also proceed analogously in the case when $X = \forall_x Y$, applying (vi)(β)–(δ), 3.2(xxv) (without even using the assumption that $Y \in \Gamma$).

By combining all the cases above we infer by (1) that

$$\boldsymbol{Q}_{AB} \vdash^\times \boldsymbol{V}_{(I)} \odot \boldsymbol{M}_{AB} X \mathbf{=} \boldsymbol{M}_{AB} X \quad \text{for every} \quad X \in \mathbf{\Phi}^+;$$

by (vii)(δ) and 3.2(iii), this easily implies

$$\boldsymbol{Q}_{AB} \vdash^\times \boldsymbol{V}^-_{(I)} \oplus \boldsymbol{M}_{AB} X \mathbf{=} \boldsymbol{M}_{AB} X \quad \text{for every} \quad X \in \mathbf{\Phi}^+,$$

and the proof of (xxii) is completed.

(xxiii) *If S is a sentence of the form* (AV), *then* $\boldsymbol{Q}_{AB} \vdash^\times \boldsymbol{M}_{AB} S \mathbf{=} \mathbf{1}$.

Assume that

$$S = [\forall_x (X \rightarrow Y) \rightarrow (\forall_x X \rightarrow \forall_x Y)].$$

Set $I = \boldsymbol{in}^* \Upsilon\phi\forall_x X$ and $J = \boldsymbol{in}^* \Upsilon\phi\forall_x Y$. Obviously

(1) $I \cup J = \boldsymbol{in}^* \Upsilon\phi\forall_x (X \rightarrow Y) = \boldsymbol{in}^* \Upsilon\phi(\forall_x X \rightarrow \forall_x Y).$

We obtain:

(2) $\vdash^\times \boldsymbol{V}^-_{(I)} \oplus \boldsymbol{M}_{AB} X \mathbf{\le} \boldsymbol{V}^-_{(I\cup J)} \oplus (\boldsymbol{M}_{AB} X \mathbf{+} \boldsymbol{M}_{AB} Y)$ by (vi)(α), 3.2(vii);

(3) $\vdash^\times (\boldsymbol{V}^-_{(I\cup J)} \oplus [(\boldsymbol{M}_{AB} X)^- \mathbf{+} \boldsymbol{M}_{AB} Y]) \cdot (\boldsymbol{V}^-_{(I)} \oplus \boldsymbol{M}_{AB} X)$
$\mathbf{\le} \boldsymbol{V}^-_{(I\cup J)} \oplus ([(\boldsymbol{M}_{AB} X)^- \mathbf{+} \boldsymbol{M}_{AB} Y] \cdot (\boldsymbol{M}_{AB} X \mathbf{+} \boldsymbol{M}_{AB} Y))$ by (2), 3.2(vi);

(4) $\vdash^\times (V^-_{(I\cup J)} \oplus M_{AB}Y)^-$
$$\leq (V^-_{(I\cup J)} \oplus [(M_{AB}X)^- + M_{AB}Y])^- + (V^-_{(I)} \oplus M_{AB}X)^-$$
by (3), BA.

Setting now $K = in^* \Upsilon\phi Y$, so that $(I \cup J) \cap K = J$, we derive:

(5) $Q_{AB} \vdash^\times V^-_{(I\cup J)} \oplus V^-_{(K)} \oplus M_{AB}Y = V^-_{(J)} \oplus M_{AB}Y$ by (vi)(ς);

(6) $Q_{AB} \vdash^\times V^-_{(I\cup J)} \oplus M_{AB}Y = V^-_{(J)} \oplus M_{AB}Y$ by (5) and (xxii);

(7) $Q_{AB} \vdash^\times M_{AB}(\forall_x(X \to Y) \to (\forall_x X \to \forall_x Y)) = 1$ by (vii)(ε),(ς), and (1), (4), (6), BA.

From (7) and (xix) we arrive directly at the conclusion.

(xxiv) *If S is a sentence of the form* (AVI) *or* (AVII), *then*
$$Q_{AB} \vdash^\times M_{AB}S = 1.$$

In both cases the argument, based upon (xix) and (vii)(ε),(ς), is straightforward. When dealing with (AVI) we use also (vi)(β), while in the case of (AVII) we apply Lemma (xxii) above.

(xxv) *If S is a sentence of the form* (AVIII), *then* $Q_{AB} \vdash^\times M_{AB}S = 1$.

Suppose $S = [\neg\forall_x(\neg x\mathring{1}y)]$, where x, y are distinct variables. Let $i = in\, x$ and $j = in\, y$. We have

(1) $Q_{AB} \vdash^\times \mathring{1} \leq P_j^\smile \odot (P_i \cdot P_j)$ by (iii) and 3.2(xv) (with $P_j^\smile$, P_i, $\mathring{1}$ taken for A, B, C respectively);

(2) $Q_{AB} \vdash^\times P_j \odot 1 \leq P_j \odot P_j^\smile \odot (P_i \cdot P_j) \odot 1$ by (1), 3.2(vii);

(3) $Q_{AB} \vdash^\times 1 = P_j \odot P_j^\smile \odot [\mathring{1} \cdot (P_i \odot P_j^\smile)] \odot 1$ by (2), (iii), 3.2(xxi);

(4) $Q_{AB} \vdash^\times 1 = M_{AB}(\neg\forall_x(\neg x\mathring{1}y))$ by (3), (vii)(β),(δ),(ς).

By (xix), step (4) yields at once the conclusion.

(xxvi) $Q_{AB} \vdash^\times [\mathring{1} \cdot (P_i \odot C \odot P_k^\smile) \odot 1]^- = \mathring{1} \cdot (P_i \odot C^- \odot P_k^\smile) \odot 1$ *for every* $C \in \Pi$ *and* $i, k \in \omega$.

Indeed from 3.2(xxiv) we get
$$\vdash^\times [\mathring{1} \cdot (P_i \odot C \odot P_k^\smile) \odot 1]^- = \mathring{1} \cdot (P_i \odot C \odot P_k^\smile)^- \odot 1.$$

Hence, using (iii) and 3.2(xxx), we easily arrive at (xxvi).

(xxvii) *If S is a sentence of the form* (AIX), *then* $\boldsymbol{Q}_{AB} \vdash^\times \boldsymbol{M}_{AB}S = \mathbf{1}$.

Assume that

(1) $S = [x\mathring{\mathbf{1}}y \rightarrow (xCz \rightarrow yCz)]$,

where $C \in \boldsymbol{\Pi}$ and $x, y, z \in \boldsymbol{\Upsilon}$. Let $i = \mathit{in}\, x$, $j = \mathit{in}\, y$, and $k = \mathit{in}\, z$. We get consecutively:

(2) $\vdash^\times \mathring{\mathbf{1}} \cdot (P_i \odot P_j^\smile) \leq P_i \cdot (\mathring{\mathbf{1}} \odot P_j) \odot P_j^\smile$ by 3.2(xv);

(3) $\vdash^\times \mathring{\mathbf{1}} \cdot (P_i \odot P_j^\smile) \leq P_i \cdot P_j \odot [(C^- + C) \odot P_k^\smile]$ by (2), (iii) (with $C^- + C$ replacing $\mathbf{1}$), 3.2(vii);

(4) $\vdash^\times \mathring{\mathbf{1}} \cdot (P_i \odot P_j^\smile) \leq P_i \odot C^- \odot P_k^\smile + P_j \odot C \odot P_k^\smile$ by (3), 3.2(v),(vii);

(5) $\vdash^\times \mathring{\mathbf{1}} \cdot (P_i \odot P_j^\smile) \odot \mathbf{1} \leq \mathring{\mathbf{1}} \cdot (P_i \odot C^- \odot P_k^\smile) \odot \mathbf{1} + \mathring{\mathbf{1}} \cdot (P_j \odot C \odot P_k^\smile) \odot \mathbf{1}$ by (4), BA, 3.2(v);

(6) $\vdash^\times \boldsymbol{M}_{AB}(x\mathring{\mathbf{1}}y) \leq \boldsymbol{M}_{AB}(xCz \rightarrow yCz)$ by (5), (xxvi) and (vii)(β),(ε).

From (6), (xix) and (ix)(ε), we obtain the conclusion of our lemma under the assumption (1).

By hypothesis, S is an instance of (AIX), and hence it is either of the form (1), or of the form obtained from (1) by replacing xCz with zCx and yCz with zCy, or, finally, of the form obtained from (1) by replacing both xCz and yCz with $C = D$ ($C, D \in \boldsymbol{\Pi}$). The first case was settled above; the second can be dealt with in an entirely analogous manner. The proof in the third case is quite simple; one uses (xix), (vii)(ε), (ix)(ε), and BA.

We turn now to the axiom schemata (DI)–(DV).

(xxviii) *If S is a sentence of one of the forms* (DI), (DII), (DIV), *then*

$$\boldsymbol{Q}_{AB} \vdash^\times \boldsymbol{M}_{AB}S = \mathbf{1}.$$

The proof of (xxviii) is based upon (xix) and (ix)(ς). In addition, one needs (vii)(β), (ix)(α), and 3.2(v) in the case of Schema (DI), (xxvi) and (vii)(β),(δ) in the case of (DII), and (vii)(β) as well as 3.2(i),(iii),(xxii) in the case of (DIV). The details are straightforward.

(xxix) *If S is a sentence of the form* (DIII), *then* $\boldsymbol{Q}_{AB} \vdash^\times \boldsymbol{M}_{AB}S = \mathbf{1}$.

By hypothesis

$$S = [\boldsymbol{x}C \odot D\boldsymbol{y} \leftrightarrow \exists_{\boldsymbol{z}}(\boldsymbol{x}C\boldsymbol{z} \wedge \boldsymbol{z}D\boldsymbol{y})],$$

for some $C, D \in \mathbf{\Pi}$. Set

(1) $E = \boldsymbol{M}_{AB}(\boldsymbol{x}C \odot D\boldsymbol{y})$.

In successive steps we obtain:

(2) $\vdash^\times E = \boldsymbol{P}_0 \odot C \cdot (\boldsymbol{P}_1 \odot D^\smile) \odot \mathbf{1}$ by (1), (vii)(β), 3.2(xxi) (taking $\boldsymbol{P}_0 \odot C$ and $D \odot \boldsymbol{P}_1^\smile$ for A and B);

(3) $\boldsymbol{Q}_{AB} \vdash^\times E = \boldsymbol{P}_0 \odot \mathbf{1} \cdot (\boldsymbol{P}_1 \odot \mathbf{1}) \cdot [\boldsymbol{P}_0 \odot C \cdot (\boldsymbol{P}_1 \odot D^\smile) \odot \mathbf{1}]$ by (2), (iii);

(4) $\boldsymbol{Q}_{AB} \vdash^\times E = \boldsymbol{P}_0 \odot \boldsymbol{P}_0^\smile \cdot (\boldsymbol{P}_1 \odot \boldsymbol{P}_1^\smile) \cdot [\boldsymbol{P}_0 \odot C \cdot (\boldsymbol{P}_1 \odot D^\smile) \odot \boldsymbol{P}_2^\smile] \odot \mathbf{1}$ by (3), (iv), (iii);

(5) $\boldsymbol{Q}_{AB} \vdash^\times E = \boldsymbol{P}_0 \odot \boldsymbol{P}_0^\smile \cdot (\boldsymbol{P}_0 \odot C \odot \boldsymbol{P}_2^\smile) \cdot (\boldsymbol{P}_1 \odot \boldsymbol{P}_1^\smile) \cdot (\boldsymbol{P}_1 \odot D^\smile \odot \boldsymbol{P}_2^\smile) \odot \mathbf{1}$ by (4), (iii), 3.2(xxvii);

(6) $\boldsymbol{Q}_{AB} \vdash^\times \boldsymbol{P}_0 \odot \boldsymbol{P}_0^\smile \cdot (\boldsymbol{P}_0 \odot C \odot \boldsymbol{P}_2^\smile) = \boldsymbol{P}_0 \odot [\boldsymbol{P}_0^\smile \odot \mathring{\mathbf{1}} \cdot (C \odot \boldsymbol{P}_2^\smile)]$ by (iii), 3.2(xxvii);

(7) $\boldsymbol{Q}_{AB} \vdash^\times \boldsymbol{P}_0^\smile \odot \mathring{\mathbf{1}} \cdot (C \odot \boldsymbol{P}_2^\smile) = \boldsymbol{P}_0^\smile \odot [\mathring{\mathbf{1}} \cdot (\boldsymbol{P}_0 \odot C \odot \boldsymbol{P}_2^\smile)]$ by (iii), 3.2(xxviii);

(8) $\boldsymbol{Q}_{AB} \vdash^\times \boldsymbol{P}_0 \odot \boldsymbol{P}_0^\smile \cdot (\boldsymbol{P}_0 \odot C \odot \boldsymbol{P}_2^\smile) = \boldsymbol{P}_0 \odot \boldsymbol{P}_0^\smile \odot [\mathring{\mathbf{1}} \cdot (\boldsymbol{P}_0 \odot C \odot \boldsymbol{P}_2^\smile)]$ by (6), (7).

By a fully symmetric argument we get statements (6′)–(8′) differing from (6)–(8) only in that the subscript 0 is replaced everywhere with the subscript 1 and C with $D^\smile$. From (8′), by virtue of 3.2(xxii), we obtain

(9) $\boldsymbol{Q}_{AB} \vdash^\times \boldsymbol{P}_1 \odot \boldsymbol{P}_1^\smile \cdot (\boldsymbol{P}_1 \odot D^\smile \odot \boldsymbol{P}_2^\smile) = \boldsymbol{P}_1 \odot \boldsymbol{P}_1^\smile \odot [\mathring{\mathbf{1}} \cdot (\boldsymbol{P}_2 \odot D \odot \boldsymbol{P}_1^\smile)]$.

We continue:

(10) $\boldsymbol{Q}_{AB} \vdash^\times E = \boldsymbol{P}_0 \odot \boldsymbol{P}_0^\smile \odot [\mathring{\mathbf{1}} \cdot (\boldsymbol{P}_0 \odot C \odot \boldsymbol{P}_2^\smile)] \cdot (\boldsymbol{P}_1 \odot \boldsymbol{P}_1^\smile \odot [\mathring{\mathbf{1}} \cdot (\boldsymbol{P}_2 \odot D \odot \boldsymbol{P}_1^\smile)]) \odot \mathbf{1}$ by (5), (8), (9);

(11) $\boldsymbol{Q}_{AB} \vdash^\times E = \boldsymbol{P}_0 \odot \boldsymbol{P}_0^\smile \cdot (\boldsymbol{P}_1 \odot \boldsymbol{P}_1^\smile) \odot [\mathring{\mathbf{1}} \cdot (\boldsymbol{P}_0 \odot C \odot \boldsymbol{P}_2^\smile) \cdot \mathring{\mathbf{1}} \cdot (\boldsymbol{P}_2 \odot D \odot \boldsymbol{P}_1^\smile)] \odot \mathbf{1}$ by (10), 3.2(xxiii);

(12) $\quad \boldsymbol{Q}_{AB} \vdash^\times \boldsymbol{M}_{AB}(\boldsymbol{x}C \odot D\boldsymbol{y}) = \boldsymbol{M}_{AB}(\boldsymbol{\exists}_z(\boldsymbol{x}Cz \wedge zD\boldsymbol{y}))$
by (1), (11), 3.2(xxiii), and (v), (vii)(β), (ix)(β),(δ).

The desired conclusion follows directly from (12) by means of (xix) and (ix)(ς).

(xxx) *If S is a sentence of the form* (DV), *then* $\boldsymbol{Q}_{AB} \vdash^\times \boldsymbol{M}_{AB}S = \mathbf{1}$.

Suppose

$$S = (C = D \leftrightarrow \boldsymbol{\forall}_{xy}(\boldsymbol{x}C\boldsymbol{y} \leftrightarrow \boldsymbol{x}D\boldsymbol{y})),$$

where $C, D \in \boldsymbol{\Pi}$. Set

(1) $\quad E = \mathring{\mathbf{1}} \cdot [\boldsymbol{P}_0 \odot (C \cdot D + C^- \cdot D^-) \odot \boldsymbol{P}_1^\smile] \odot \mathbf{1},$

(2) $\quad F_0 = \boldsymbol{P}_0 \odot C \odot \boldsymbol{P}_1^\smile, \quad F_1 = \boldsymbol{P}_0 \odot C^- \odot \boldsymbol{P}_1^\smile,$

(3) $\quad G_0 = \boldsymbol{P}_0 \odot D \odot \boldsymbol{P}_1^\smile, \quad G_1 = \boldsymbol{P}_0 \odot D^- \odot \boldsymbol{P}_1^\smile.$

We have:

(4) $\quad \boldsymbol{M}_{AB}(C = D) = \mathbf{0} \oplus E$ by (1), (vii)(γ);

(5) $\quad \boldsymbol{Q}_{AB} \vdash^\times \boldsymbol{M}_{AB}(\boldsymbol{\forall}_{xy}(\boldsymbol{x}C\boldsymbol{y} \leftrightarrow \boldsymbol{x}D\boldsymbol{y}))$
$\quad = \mathbf{0} \oplus [(\mathring{\mathbf{1}} \cdot F_0 \odot \mathbf{1}) \cdot (\mathring{\mathbf{1}} \cdot G_0 \odot \mathbf{1}) + (\mathring{\mathbf{1}} \cdot F_1 \odot \mathbf{1}) \cdot (\mathring{\mathbf{1}} \cdot G_1 \odot \mathbf{1})]$
by (2), (3), (xviii), (xxvi), (vii)(β), (ix)(γ);

(6) $\quad \boldsymbol{Q}_{AB} \vdash^\times E = \mathring{\mathbf{1}} \cdot (F_0 \cdot G_0 + F_1 \cdot G_1) \odot \mathbf{1}$ by (1)–(3), (iii), 3.2(v),(xxvii);

(7) $\quad \boldsymbol{Q}_{AB} \vdash^\times E = (\mathring{\mathbf{1}} \cdot F_0 \odot \mathbf{1}) \cdot (\mathring{\mathbf{1}} \cdot G_0 \odot \mathbf{1}) + (\mathring{\mathbf{1}} \cdot F_1 \odot \mathbf{1}) \cdot (\mathring{\mathbf{1}} \cdot G_1 \odot \mathbf{1})$
by (6), 3.2(v),(xxiii);

(8) $\quad \boldsymbol{Q}_{AB} \vdash^\times \boldsymbol{M}_{AB}(C = D) = \boldsymbol{M}_{AB}(\boldsymbol{\forall}_{xy}(\boldsymbol{x}C\boldsymbol{y} \leftrightarrow \boldsymbol{x}D\boldsymbol{y}))$
by (4), (5), (7).

The conclusion follows from (8) by virtue of (ix)(ς).

We have thus completed the proof that $\boldsymbol{Q}_{AB} \vdash^\times \boldsymbol{M}_{AB}S = \mathbf{1}$ for every $S \in \boldsymbol{\Lambda}^+$. The next result is a preliminary form of the main mapping theorem.

(xxxi) *For every $\Psi \subseteq \boldsymbol{\Sigma}^+$ and $X \in \boldsymbol{\Sigma}^+$, in order that $\Psi \vdash_{\boldsymbol{Q}_{AB}} X$ it is necessary and sufficient that*

$$\{\boldsymbol{M}_{AB}Y = \mathbf{1} : Y \in \Psi\} \vdash^\times_{\boldsymbol{Q}_{AB}} \boldsymbol{M}_{AB}X = \mathbf{1}.$$

In fact, the sufficiency of the condition is an obvious consequence of (xiv). To prove its necessity, consider the set Ω of all sentences $X \in \Sigma^+$ such that, for a fixed $\Psi \subseteq \Sigma^+$,

$$\{\boldsymbol{M}_{AB}Y = \mathbf{1}: Y \in \Psi\} \vdash^\times_{\boldsymbol{Q}_{AB}} \boldsymbol{M}_{AB}X = \mathbf{1}.$$

Obviously $\Psi \cup \{\boldsymbol{Q}_{AB}\} \subseteq \Omega$. By (xx), (xxi), (xxiii)–(xxv), and (xxvii)–(xxx) we have $\boldsymbol{\Lambda}^+ \subseteq \Omega$. From (ix)($\varepsilon$) we easily see that $X \in \Omega$ whenever $Z, Z \twoheadrightarrow X \in \Omega$. Hence, by induction on sentences derivable in $\mathcal{L}^+$ from $\Psi \cup \{\boldsymbol{Q}_{AB}\}$, we conclude that Ω contains every sentence X for which $\Psi \vdash_{\boldsymbol{Q}_{AB}} X$. This completes the proof.

We now give the definition of the system of translation mappings $\boldsymbol{K}_{AB}$ (indexed by arbitrary pairs of predicates A, B) along the lines discussed in §4.3. We still consider A, B as fixed. From (xxxi) one could expect that $\boldsymbol{K}_{AB}$ would be determined by the condition $\boldsymbol{K}_{AB}X = (\boldsymbol{M}_{AB}X = \mathbf{1})$ for every $X \in \Sigma^+$. For reasons mentioned after 4.3(vii) we adopt here a slightly more involved definition.

(xxxii) *$\boldsymbol{K}_{AB}$ is the unique function F such that*

(α) $DoF = \Sigma^+$,

(β) $FX = X$ *in case* $X \in \Sigma^\times$,

(γ) $FX = (\boldsymbol{M}_{AB}X = \mathbf{1})$ *in case* $X \in \Sigma^+ \sim \Sigma^\times$.

The following properties of $\boldsymbol{K}_{AB}$ easily follow from the above definition.

(xxxiii) (α) *$\boldsymbol{K}_{AB}$ is a recursive function.*

(β) *$\boldsymbol{K}_{AB}$ maps Σ^+ onto $\Sigma^\times$.*

(γ) *$\boldsymbol{K}_{AB}X = X$ iff $X \in \Sigma^\times$.*

(δ) *$\boldsymbol{K}_{AB}X \equiv_{\boldsymbol{Q}_{AB}} X$ for every $X \in \Sigma^+$, and hence $\boldsymbol{K}^*_{AB}\Psi \equiv_{\boldsymbol{Q}_{AB}} \Psi$ for every $\Psi \subseteq \Sigma^+$.*

(ε) *$X \equiv \{\boldsymbol{K}_{AB}X, \boldsymbol{Q}_{AB}\} \equiv^\times [(\boldsymbol{K}_{AB}X)^t \cdot \boldsymbol{Q}^\ell_{AB} = \mathbf{1}]$ for every $X \in \Sigma^+$ such that $X \vdash \boldsymbol{Q}_{AB}$.*

We use here various properties of $\boldsymbol{M}_{AB}$, in particular (viii), (xiv). Notice that (ε) is a direct consequence of (β) and (δ).

As is easily seen from (xvii), (ε) continues to hold if $(\boldsymbol{K}_{AB}X)^t$ is replaced by $\boldsymbol{M}_{AB}X$.

The main mapping theorem follows.

(xxxiv) *For every $\Psi \subseteq \Sigma^+$ and every $X \in \Sigma^+$, we have*

$$\Psi \vdash^+_{\boldsymbol{Q}_{AB}} X \quad \textit{iff} \quad \boldsymbol{K}^*_{AB}\Psi \vdash^\times_{\boldsymbol{Q}_{AB}} \boldsymbol{K}_{AB}X.$$

In view of (xvii) and the definition of $\boldsymbol{K}_{AB}$, this theorem is a direct consequence of (xxxi).

A variant of (xxxiv), easily derivable from it with the help of (xxxiii)(β),(γ), is:

(xxxv) *If* $\boldsymbol{Q}_{AB} \in \Psi \subseteq \Sigma^+$, *then* $\boldsymbol{K}^*_{AB}\boldsymbol{\Theta\eta}^+\Psi = \boldsymbol{\Theta\eta}^\times \boldsymbol{K}^*_{AB}\Psi = \boldsymbol{\Theta\eta}^+\Psi \cap \Sigma^\times$.

Theorem (xxxiv) differs in its form from the earlier mapping theorems, 2.3(v) (for $\mathcal{L}^+$ and $\mathcal{L}$) and 3.9(vii) (for $\mathcal{L}_3^+$ and $\mathcal{L}^\times$), only in that the derivability relation has been relativized to a given sentence $\boldsymbol{Q}_{AB}$. If, instead of the formalisms $\mathcal{L}^+$ and $\mathcal{L}^\times$, we consider in (xxxiv) the corresponding systems $\mathcal{S}^+$ and $\mathcal{S}^\times$ obtained by including $\boldsymbol{Q}_{AB}$ in the logical bases of the two systems, then the difference vanishes completely.

This remark permits us to apply the observations in the latter part of §2.3 and in §2.4 to the present situation. Thus, as easy consequences of (xxxiv) (using also (xxxiii)(β),(γ)) we get both equipollence theorems for $\mathcal{L}^+$ and $\mathcal{L}^\times$.

(xxxvi) *For every* $X \in \Sigma^+$ *there is a* $Y \in \Sigma^\times$, *and also for every* $Y \in \Sigma^\times$ *there is an* $X \in \Sigma^+$, *such that* $X \equiv_{\boldsymbol{Q}_{AB}} Y$.

(xxxvii) *For every* $\Psi \subseteq \Sigma^\times$ *and* $X \in \Sigma^\times$, *we have* $\Psi \vdash^+_{\boldsymbol{Q}_{AB}} X$ *iff* $\Psi \vdash^\times_{\boldsymbol{Q}_{AB}} X$.

Theorem (xxxvi) is of course a direct corollary of (xxxiii)(δ). Since, however, (xxxiii)(δ) is easily derivable from (xxxiv), we can treat both (xxxvi) and (xxxvii) as consequences of (xxxiv) (cf. 2.4(iv)).

By (xxxvi) and (xxxvii), *the formalisms* $\mathcal{L}^+$ *and* $\mathcal{L}^\times$ *are equipollent* (*in means of expression and proof*) *relative to any given sentence* $\boldsymbol{Q}_{AB}$. *Hence, the same relationship holds between* $\mathcal{L}$ *and* $\mathcal{L}^\times$ *as well.*

We state here some consequences of (xxxvi) and (xxxvii) analogous in formulations and proofs to the results in 2.3(x),(xi) (cf. 2.4(vii),(viii)).

(xxxviii) *If* $\boldsymbol{Q}_{AB} \in \boldsymbol{\Theta\eta}^\times\Psi \subseteq \Sigma^\times$, *then* $\boldsymbol{\Theta\eta}^\times\Psi = \boldsymbol{\Theta\eta}^+\Psi \cap \Sigma^\times$.

(xxxix) *If* Φ *is a theory in* $\mathcal{L}^+$ *with* $\boldsymbol{Q}_{AB} \in \Phi$, *then* $\Phi \cap \Sigma^\times$ *is a theory in* $\mathcal{L}^\times$ *and* $\Phi = \boldsymbol{\Theta\eta}^+(\Phi \cap \Sigma^\times)$. *If, moreover,* Φ *or* $\Phi \cap \Sigma^\times$ *satisfies one of the conditions* (α)–(ε) *in* 2.3(xi), *then both* Φ *and* $\Phi \cap \Sigma^\times$ *satisfy this condition.*

Notice that Theorem (xxxix) contains no reference to the condition (ς) in 2.3(xi). However, we can supplement (xxxix) by a condition which is weaker than (ς), namely "*is hereditarily undecidable relative to* $\boldsymbol{Q}_{AB}$"; we stipulate that a theory satisfies this condition iff all its subtheories which contain $\boldsymbol{Q}_{AB}$ are undecidable.

On the other hand, we are frequently confronted with a situation when one of the theories Φ and $\Phi \cap \Sigma^\times$ satisfying the hypothesis of (xxxix) is shown to be hereditarily undecidable as a consequence of the fact that it is undecidable and finitely based. In such a situation we can use the part of (xxxix) referring to 2.3(xi)(γ),(δ) in order to show that both theories Φ and $\Phi \cap \Sigma^\times$ are finitely based and undecidable. Since both $\mathcal{L}^+$ and $\mathcal{L}^\times$ are formalisms with deduction theorems, we then conclude that both Φ and $\Phi \cap \Sigma^\times$ are hereditarily undecidable; cf. the paragraph following formula (iii) in §3.3.

Although, as we know, the formalism $\mathcal{L}^\times$ is semantically incomplete, it becomes semantically complete when relativized to the sentence $\boldsymbol{Q}_{AB}$. Specifically, as a direct consequence of (xxxvii) (or of (xxxviii)) and the semantic completeness of $\mathcal{L}^+$ it follows that:

(xl) *For every* $\Psi \subseteq \Sigma^{\times}$ *and* $X \in \Sigma^{\times}$, *we have* $\Psi \vdash^{\times}_{\boldsymbol{Q}_{AB}} X$ *iff* $\Psi \models_{\boldsymbol{Q}_{AB}} X$.

In the next theorem, an immediate corollary of (xl), we no longer consider the predicates A, B in $\mathbf{\Pi}$ to be fixed.

(xli) *Every sentence in* $\Sigma^{\times}$ *which is logically true is derivable in* $\mathcal{L}^{\times}$ *from every sentence* $\boldsymbol{Q}_{AB}$ *with* $A, B \in \mathbf{\Pi}$.

The converse of (xli) is not true. For instance, the sentence $\mathring{\mathbf{1}} \oplus \mathring{\mathbf{1}} \leq \mathring{\mathbf{1}}$ is derivable from every sentence $\boldsymbol{Q}_{AB}$, but is not logically true. However, the problem is open whether the converse of (xli) holds if we restrict ourselves to sentences in which $\mathring{\mathbf{1}}$ does not occur.[4] Some results providing partial affirmative solutions of this restricted problem are known, but they apply exclusively to the case when $\mathcal{L}^{\times}$ is provided with more than one nonlogical constant. In particular, the solution proves to be affirmative if $\mathcal{L}^{\times}$ is supplied with infinitely many such constants.

4.5. The construction of equipollent $\mathcal{Q}$-systems in $\mathcal{L}^{\times}$

The fundamental equipollence results stated in the final part of §4.4 have important implications for the problem of developing set theory within the formalism $\mathcal{L}^{\times}$.

Let $\mathcal{S}$ be a system (not necessarily a system of set theory) formalized in $\mathcal{L}$, and let $\mathcal{S}^+$ be the correlated system formalized in $\mathcal{L}^+$. Thus, the sets of (nonlogical) axioms in $\mathcal{S}$ and $\mathcal{S}^+$ coincide:

$$\mathbf{A}\boldsymbol{\xi}[\mathcal{S}] = \mathbf{A}\boldsymbol{\xi}[\mathcal{S}^+] = \mathbf{A}\boldsymbol{\xi}.$$

We are interested in the question of whether it is possible to construct a system $\mathcal{S}^{\times}$ formalized in $\mathcal{L}^{\times}$ that is equipollent with $\mathcal{S}$ and $\mathcal{S}^+$. The answer is, in general, negative. It is affirmative, however, if $\mathcal{S}$ and $\mathcal{S}^+$ are subjected to an additional assumption; the systems satisfying this assumption will be referred to as $\mathcal{Q}$-*systems*, or *quasi-projectional systems*. In view of our fundamental results, the definition of $\mathcal{Q}$-systems can be formulated as follows.

---

[4]*Andréka and Németi have shown (in a private communication) that this restricted converse of (xli) is also false. Indeed, taking X to be the implication

$$\boldsymbol{E} \odot \mathbf{1} = \boldsymbol{E} \rightarrow \mathbf{1} \odot \boldsymbol{E} = \boldsymbol{E}$$

(where $\mathbf{1}$ is redefined as $\boldsymbol{E} + \boldsymbol{E}^{-}$), they show that

(1) $\models_{\boldsymbol{Q}_{AB}} X$ for any $A, B \in \mathbf{\Pi}$, but not $\models X$.

Now X is equivalent to an equation Y (cf. 2.2(i)–(v)), for example, to the equation

$$\mathbf{1} \odot (\boldsymbol{E} \odot \mathbf{1} \cdot \boldsymbol{E}^{-}) \odot \mathbf{1} + (\mathbf{0} \oplus \boldsymbol{E}^{-}) + \boldsymbol{E} = \mathbf{1},$$

and so (1) holds with X replaced by Y. In view of (xl) we conclude that

$$\vdash^{\times}_{\boldsymbol{Q}_{AB}} Y \text{ for any } A, B \in \mathbf{\Pi}, \text{ but not } \models Y.$$

(i) *Let* $\mathcal{S}$ *and* $\mathcal{S}^+$ *be any systems formalized respectively in* $\mathcal{L}$ *and* $\mathcal{L}^+$, *with the common axiom set* $\mathbf{A}\boldsymbol{\xi}$. *Then*

(α) $\mathcal{S}^+$ *is a* $\mathfrak{Q}$*-system iff there are* $A, B \in \mathbf{\Pi}$ *such that* $\mathbf{A}\boldsymbol{\xi} \vdash^+ \boldsymbol{Q}_{AB}$;

(β) $\mathcal{S}$ *is a* $\mathfrak{Q}$*-system iff* $\mathcal{S}^+$ *is a* $\mathfrak{Q}$*-system or, equivalently, iff there are predicates* A, B *in* $\mathcal{L}^+$ *such that* $\mathbf{A}\boldsymbol{\xi} \vdash \boldsymbol{G}(\boldsymbol{Q}_{AB})$.

The function $\boldsymbol{G}$ in (β) is, of course, the translation mapping from $\mathcal{L}^+$ to $\mathcal{L}$ constructed in 2.3(iii).

We can also give a characterization of $\mathfrak{Q}$-systems in $\mathcal{L}$ which is expressed exclusively in terms referring to $\mathcal{L}$. (In formulating this characterization we use a notation for substitutions in a formula that was introduced in §3.7.)

(ii) *Given any system* $\mathcal{S}$ *in* $\mathcal{L}$ *with the axiom set* $\mathbf{A}\boldsymbol{\xi}$, *for* $\mathcal{S}$ *to be a* $\mathfrak{Q}$*-system it is necessary and sufficient that there exist formulas* $D, E \in \mathbf{\Phi}$ *satisfying the following conditions*:

(α) $\boldsymbol{\Upsilon}\phi D = \boldsymbol{\Upsilon}\phi E = \{\boldsymbol{x}, \boldsymbol{y}\}$;

(β) $\mathbf{A}\boldsymbol{\xi} \vdash \{\forall_{\boldsymbol{xyz}}(D[\boldsymbol{x},\boldsymbol{y}] \wedge D[\boldsymbol{x},\boldsymbol{z}] \rightarrow \boldsymbol{y}\mathring{\mathbf{1}}\boldsymbol{z}),\ \forall_{\boldsymbol{xyz}}(E[\boldsymbol{x},\boldsymbol{y}] \wedge E[\boldsymbol{x},\boldsymbol{z}] \rightarrow \boldsymbol{y}\mathring{\mathbf{1}}\boldsymbol{z})\}$;

(γ) $\mathbf{A}\boldsymbol{\xi} \vdash \forall_{\boldsymbol{xy}}\exists_{\boldsymbol{z}}(D[\boldsymbol{z},\boldsymbol{x}] \wedge E[\boldsymbol{z},\boldsymbol{y}])$;

(δ) D *and* E *contain no variables different from* $\boldsymbol{x}, \boldsymbol{y}, \boldsymbol{z}$.

In fact, if $\mathcal{S}$ is such a $\mathfrak{Q}$-system, then, by (i) above, there are $A, B \in \mathbf{\Pi}$ such that

(1) $$\mathbf{A}\boldsymbol{\xi} \vdash^+ \boldsymbol{Q}_{AB} \quad\text{and}\quad \mathbf{A}\boldsymbol{\xi} \vdash \boldsymbol{G}(\boldsymbol{Q}_{AB}).$$

We set

$$D = \boldsymbol{G}(\boldsymbol{x}A\boldsymbol{y}), \quad E = \boldsymbol{G}(\boldsymbol{x}B\boldsymbol{y}),$$

and, using 3.8(ix) and 4.1(iii), we easily show that the formulas D, E so defined satisfy the conditions (α)–(δ) above. If, conversely, the formulas $D, E \in \mathbf{\Phi}$ satisfying (α)–(δ) are given, we apply 3.9(iii)(γ),(ε) to derive the existence of $A, B \in \mathbf{\Pi}$ for which (1) holds.

Let $\mathcal{S}$ and $\mathcal{S}^+$ be any $\mathfrak{Q}$-systems with the axiom set $\mathbf{A}\boldsymbol{\xi}$, and let A and B be two predicates for which $\mathbf{A}\boldsymbol{\xi} \vdash^+ \boldsymbol{Q}_{AB}$. We correlate with $\mathcal{S}$ and $\mathcal{S}^+$ a system $\mathcal{S}^{\times}$ which is formalized in $\mathcal{L}^{\times}$ and whose axiom set $\mathbf{A}\boldsymbol{\xi}^{\times}$ is determined by the formula

(iii) $\mathbf{A}\boldsymbol{\xi}^{\times} = \boldsymbol{K}^*_{AB}\mathbf{A}\boldsymbol{\xi} \cup \{\boldsymbol{Q}_{AB}\}$.

Obviously $\mathbf{A}\boldsymbol{\xi}^{\times} \vdash^{\times} \boldsymbol{Q}_{AB}$, and therefore $\mathcal{S}^{\times}$ is referred to as a $\mathfrak{Q}$-system as well. Strictly speaking, $\mathcal{S}^{\times}$ is not uniquely determined by $\mathcal{S}$, since there are many different pairs $\langle A, B\rangle$ for which $\mathbf{A}\boldsymbol{\xi} \vdash^+ \boldsymbol{Q}_{AB}$ holds. In this connection notice, however, the following theorem, which is an easy consequence of 4.4(xxxv).

(iv) *If* $A, B, A', B' \in \mathbf{\Pi}$ *and* $\mathbf{A}\boldsymbol{\xi} \vdash^+ \{\boldsymbol{Q}_{AB}, \boldsymbol{Q}_{A'B'}\}$, *then*

$$\boldsymbol{\Theta}\boldsymbol{\eta}^{\times}(\boldsymbol{K}^*_{AB}\mathbf{A}\boldsymbol{\xi} \cup \{\boldsymbol{Q}_{AB}\}) = \boldsymbol{\Theta}\boldsymbol{\eta}^{\times}(\boldsymbol{K}^*_{A'B'}\mathbf{A}\boldsymbol{\xi} \cup \{\boldsymbol{Q}_{A'B'}\})$$

and hence

$$\boldsymbol{K}^*_{AB}\mathbf{A\xi} \cup \{\boldsymbol{Q}_{AB}\} \equiv^\times \boldsymbol{K}^*_{A'B'}\mathbf{A\xi} \cup \{\boldsymbol{Q}_{A'B'}\}.$$

Thus $\mathcal{S}^\times$ does not depend on the choice of a pair $\langle A, B\rangle$ in any essential way; it is true that the axiom set of $\mathcal{S}^\times$ varies with the choice of $\langle A, B\rangle$, but the theory of $\mathcal{S}^\times$ (in the sense of the penultimate paragraph of §1.3) remains unchanged.

Moreover, notice that the set $\boldsymbol{\Pi}$ of predicates, and hence also the set $\boldsymbol{\Pi}\times\boldsymbol{\Pi}$ of ordered pairs of predicates, can easily be arranged in simple infinite sequences. Therefore we can effectively correlate with every $\mathcal{Q}$-system $\mathcal{S}$ (and, more specifically, with its axiom set $\mathbf{A\xi}$) a well-defined pair $\langle A_0, B_0\rangle$ for which $\mathbf{A\xi} \vdash^+ \boldsymbol{Q}_{A_0B_0}$ holds. (The question of whether and in what sense this correlation is recursive will not be discussed here.) We could then stipulate that the pair $\langle A_0, B_0\rangle$ will always be used in defining the axiom set $\mathbf{A\xi}^\times$ of $\mathcal{S}^\times$, thus making this set uniquely determined.

In view of these last remarks, or even in view of (iv) alone, we shall speak of $\mathcal{S}^\times$ as a well-defined system formalized in $\mathcal{L}^\times$ and correlated with $\mathcal{S}$ or $\mathcal{S}^+$.

By specializing (and partly repeating) the results in §2.3 and in the latter part of §4.4, we obtain the following statement, which sums up the main properties of the translation mappings used in the discussion of the formalisms $\mathcal{L}$, $\mathcal{L}^+$, $\mathcal{L}^\times$, and describes the resulting equipollence of the correlated systems $\mathcal{S}$, $\mathcal{S}^+$, $\mathcal{S}^\times$. (In formulating this statement we do not use the stipulations made at the end of §2.3.)

(v) *Let $\mathcal{S}$ be any $\mathcal{Q}$-system in $\mathcal{L}$ with the axiom set $\mathbf{A\xi}$; let $\mathcal{S}^+$ be the correlated system in $\mathcal{L}^+$ with the same axiom set, and $\mathcal{S}^\times$ the correlated system in $\mathcal{L}^\times$ with the axiom set $\mathbf{A\xi}^\times$.*

(α) $\mathbf{A\xi} \equiv^+ \mathbf{A\xi}^\times$.

(β) *There are two recursive functions, G and K, with the following properties.*

($\beta 1$) *G maps the set $\boldsymbol{\Sigma}^+$ onto, and its subset $\boldsymbol{\Sigma}^\times$ into, $\boldsymbol{\Sigma}$ (and, moreover, it maps $\boldsymbol{\Phi}^+$ onto $\boldsymbol{\Phi}$); K maps the set $\boldsymbol{\Sigma}^+$ onto, and its subset $\boldsymbol{\Sigma}$ into, $\boldsymbol{\Sigma}^\times$.*

($\beta 2$) *$GX = X$ for every $X \in \boldsymbol{\Sigma}$ (and, indeed, for every $X \in \boldsymbol{\Phi}$); $KX = X$ for every $X \in \boldsymbol{\Sigma}^\times$.*

($\beta 3$) *$GX \equiv^+_{\mathbf{A\xi}} X$ for every $X \in \boldsymbol{\Sigma}^+$ (and, indeed, $GX \equiv^+ X$ and $\boldsymbol{\Upsilon}\phi GX = \boldsymbol{\Upsilon}\phi X$ for every $X \in \boldsymbol{\Phi}^+$); $KX \equiv^+_{\mathbf{A\xi}} X$ for every $X \in \boldsymbol{\Sigma}^+$.*

($\beta 4$) *For every $\Psi \subseteq \boldsymbol{\Sigma}^+$ and $X \in \boldsymbol{\Sigma}^+$ the following three conditions are equivalent:*

$$\Psi \vdash^+_{\mathbf{A\xi}} X, \quad G^*\Psi \vdash_{\mathbf{A\xi}} GX, \quad K^*\Psi \vdash^\times_{\mathbf{A\xi}^\times} KX.$$

For G and K we can take the functions $\boldsymbol{G}$ and $\boldsymbol{K}_{AB}$ respectively defined in 2.3(iii) and 4.4(xxxii), A and B being any two predicates for which $\mathbf{A\xi} \vdash^+ \boldsymbol{Q}_{AB}$.

(γ) *Consequently*:

(γ1) *For every $X \in \Sigma^+$ (or $\mathbf{\Phi}^+$) there is a $Y \in \Sigma$ (or $\mathbf{\Phi}$) and, trivially, for every $Y \in \Sigma$ (or $\mathbf{\Phi}$) there is an $X \in \Sigma^+$ (or $\mathbf{\Phi}^+$) such that $X \equiv^+_{\mathbf{A}\xi} Y$ (and $\mathbf{\Upsilon}\phi X = \mathbf{\Upsilon}\phi Y$); similarly with "$\Sigma$" replaced by "$\Sigma^\times$", leaving "$\Sigma^+$" untouched.*

(γ2) *For every $\Psi \subseteq \Sigma$ and $X \in \Sigma$, we have*

$$\Psi \vdash^+_{\mathbf{A}\xi} X \quad \textit{iff} \quad \Psi \vdash_{\mathbf{A}\xi} X;$$

similarly, for every $\Psi \subseteq \Sigma^\times$ and $X \in \Sigma^\times$, we have

$$\Psi \vdash^+_{\mathbf{A}\xi} X \quad \textit{iff} \quad \Psi \vdash^\times_{\mathbf{A}\xi^\times} X.$$

(γ3) $\mathbf{\Theta}\eta\mathbf{A}\xi = \mathbf{\Theta}\eta^+\mathbf{A}\xi \cap \Sigma, \quad \mathbf{\Theta}\eta^\times\mathbf{A}\xi^\times = \mathbf{\Theta}\eta^+\mathbf{A}\xi \cap \Sigma^\times.$

By (v)(γ1),(γ2), the correlated systems $\mathcal{S}^+$ and $\mathcal{S}$, as well as $\mathcal{S}^+$ and $\mathcal{S}^\times$, are equipollent in means of expression and proof, and so are the systems $\mathcal{S}$ and $\mathcal{S}^\times$, treated as subsystems of $\mathcal{S}^+$.

Our equipollence theorems carry with them various corollaries concerning the correlated $\mathcal{Q}$-systems $\mathcal{S}$, $\mathcal{S}^+$, and $\mathcal{S}^\times$. In particular (using the terminology of §1.4) we have:

(vi) *Let $\mathcal{T}$ be any of the correlated $\mathcal{Q}$-systems $\mathcal{S}$, $\mathcal{S}^+$, and $\mathcal{S}^\times$. Then $\mathcal{T}$ is semantically complete, i.e., the relations $\vdash [\mathcal{T}]$ and $\models [\mathcal{T}]$ coincide.*

In fact, the semantical completeness of $\mathcal{S}$ and $\mathcal{S}^+$ does not depend on the fact that they are $\mathcal{Q}$-systems; it is an immediate consequence of the completeness theorems for $\mathcal{L}$ and $\mathcal{L}^+$ (cf. 1.4(i) and the relevant remarks in §2.2). The semantical completeness of $\mathcal{S}^\times$ follows at once from 4.4(xl).

4.6. The formalizability of systems of set theory in $\mathcal{L}^\times$

In this section we shall attempt to convince the reader that, in general, the systems of set theory which are known from the literature and are (or can be) formalized in predicate logic appear to be $\mathcal{Q}$-systems in the sense of 4.5(i); hence, by the results in 4.5(v), for each such system an equipollent system can be constructed which is formalized in $\mathcal{L}^\times$.

There are, however, exceptions to the foregoing remarks. Two systems of set theory are known to the authors which have not been shown to be $\mathcal{Q}$-systems, and may very well prove not to be equipollent with any systems in $\mathcal{L}^\times$. These exceptional systems will be briefly discussed at the end of this section, but otherwise their existence will be disregarded in our remarks and assertions.

When speaking of systems known from the literature, we have in mind not only systems which have been actually studied or developed to any extent, but also those which potentially appear to be of interest to contemporary researchers in the foundations of set theory. (Of course, it is not excluded that some systems studied in the literature will not be accounted for in this work—simply because the authors are not acquainted with them.)

Set-theoretical systems are usually classified into those *excluding individuals* and those *admitting individuals.* Loosely speaking, systems of the first kind are ones whose set-theoretical models consist solely of classes, and thus contain only one object which has no elements, namely the empty class. In contrast, a system of the second kind has some models in which appear, besides classes, also individuals, thus objects having no elements but differing from the empty class. We restrict ourselves here to systems in which the proposition stating the existence of an empty class can be expressed and proved; all systems known from the literature seem to be of this kind. Such a system excludes or admits individuals depending on whether or not the following sentence is derivable in it:

$$U_1 = \boldsymbol{\forall}_{xy}[\boldsymbol{\forall}_z(\neg zEx \wedge \neg zEy) \rightarrow x\mathring{1}y].$$

(Compare, however, the latter part of this section for a different conception of individuals.)

Another (and probably more significant) classification of set-theoretical systems is the one into systems *excluding* or *admitting proper classes.* In this case, a system is of the first kind if all the classes in its set-theoretical models are sets, i.e., are elements of other members of the universe. On the other hand, a system of the second kind has set-theoretical models which also contain proper classes, i.e., classes that are not sets. A characteristic feature of a system without proper classes is the provability of the sentence

$$U_2 = \boldsymbol{\forall}_x\boldsymbol{\exists}_y(xEy).$$

For various metamathematical studies of set theory, another, much deeper and subtler, classification of set-theoretical systems is of essential importance. This classification concerns a very comprehensive category of systems, and consists in ascribing to each system involved a definite ordinal called the *rank* of the system; see Tarski [**1956b**] and Montague–Vaught [**1959**]. In terms of this classification, systems excluding, or admitting, proper classes appear simply as systems whose rank is, or is not, a limit ordinal. It seems that the notion of the rank of a set-theoretical system (as opposed to that of the rank of a set within a given set-theoretical system, or within a particular model of such a system) has not yet been extensively studied. At any rate, this notion will not be relevant for our discussion.

We shall first concern ourselves with set-theoretical systems excluding individuals; they are much more frequently discussed in contemporary foundational research than systems with individuals.

We begin with systems excluding, in addition to individuals, also proper classes. The best known examples of such systems are the systems of Zermelo and Zermelo–Fraenkel, together with all their variants obtained by removing some statements from their axiom sets, for instance, the axiom of choice or the axiom of infinity, or else by adding some new axioms, such as the well-foundedness axiom or various strong axioms of infinity.

It should be emphasized that, when speaking here of Zermelo and Zermelo–Fraenkel systems, we always have in mind modern versions of these systems in which certain not precisely defined terms occurring in the original versions ("definite properties" in Zermelo's axiom of subset construction; "functions" in Fraenkel's replacement axiom) have been eliminated. Only due to this elimination (suggested for the first time in Skolem [**1923**]) can the two systems serve as subjects of a rigorous metamathematical discussion. In addition, both systems in their original versions—as opposed to those which are generally used at present and in which we are now interested—appear as systems admitting individuals. It is obvious that modern versions preserving this feature of the original systems can be conceived and constructed as well.

In all known systems without proper classes the following simple sentence $\boldsymbol{P}$ proves to be valid:

(i) $\boldsymbol{P} = \forall_{xy}\exists_z\forall_u(\boldsymbol{uEz} \leftrightarrow \boldsymbol{u\mathring{1}x} \lor \boldsymbol{u\mathring{1}y})$.

This sentence is often included in the axiom set of a system $\mathcal{S}$ and is referred to as the *pair axiom* (or the *axiom of unordered pairs*). $\boldsymbol{P}$ obviously implies the sentence U_2 above.

The following elementary result is relevant for our purposes.

(ii) *There are* $A, B \in \boldsymbol{\Pi}$ *such that* $\boldsymbol{P} \equiv \boldsymbol{Q}_{AB} \equiv (A^{\smile} \odot B = \mathbf{1})$. *Such* A, B *can be obtained by setting*

$$(\alpha)\quad D = \boldsymbol{E}^{\smile} \odot [\boldsymbol{E}^{\smile} \cdot (\boldsymbol{E}^{\smile -} \oplus \mathring{1})], \qquad F = \boldsymbol{E}^{\smile} \odot \boldsymbol{E}^{\smile},$$

$$(\beta)\quad A = D \cdot (D^{-} \oplus \mathring{1}), \qquad B = F \cdot (F^{-} + A \oplus \mathring{1}).$$

The proof is rather straightforward and will be left to the reader. It is advisable to begin by "decoding" the set-theoretical content of the predicates A and B, which makes it possible to carry through the argument within the formalism $\mathcal{L}$, using 4.1(iii). The proof that $\boldsymbol{P} \vdash (A^{\smile} \odot B = \mathbf{1})$ is based on Kuratowski's classical construction of an ordered pair. In carrying out a detailed proof, the reader should keep in mind that the argument cannot be based on any special set-theoretical laws (different from $\boldsymbol{P}$), and, in particular, not on the axiom of extensionality.

It may be interesting to notice that the second part of the conclusion of (ii) can be improved in the following way.

(iii) *Let* A *and* B *be predicates defined as in* (ii)(β) *in terms of arbitrary predicates* D, F. *Then* $\boldsymbol{Q}_{AB} \equiv^\times (A^{\smile} \odot B = \mathbf{1})$.

This improvement of (ii) (which is not essential for our further discussion) is derived from 4.1(ii) by showing that the sentences $A^{\smile} \odot A \leq \mathring{1}$ and $B^{\smile} \odot B \leq \mathring{1}$ are logically provable in $\mathcal{L}^\times$. Indeed we obtain successively:

(1) $\quad G^{\smile} \odot H \leq \mathring{1} \;\equiv^\times\; (G \odot \mathring{0} \cdot H = \mathbf{0}) \;\equiv^\times\; (H \odot \mathring{0} \cdot G = \mathbf{0})$ for any $G, H \in \boldsymbol{\Pi}$ by 3.2(xii),(ii);

(2) $\vdash^\times \{A^\smile \odot A \leq \mathring{1},\ A \odot \mathring{0} \cdot A = 0\}$ by (ii)(β), 3.2(xxix), (1);

(3) $\vdash^\times B \odot \mathring{0} \cdot B = B \cdot A^- \odot \mathring{0} \cdot B + B \cdot A \odot \mathring{0} \cdot B \cdot A^- + B \cdot A \odot \mathring{0} \cdot B \cdot A$ by 3.2(v), BA;

(4) $\vdash^\times B \odot \mathring{0} \cdot B \leq B \cdot A^- \odot \mathring{0} \cdot B + B \odot \mathring{0} \cdot B \cdot A^- + A \odot \mathring{0} \cdot A$ by (3), 3.2(vii);

(5) $\vdash^\times B \cdot A^- \odot \mathring{0} \cdot B \leq F \cdot A^- \odot \mathring{0} \cdot (F^- + A \oplus \mathring{1})$ by (ii)(β), 3.2(vii);

(6) $\vdash^\times \{B \cdot A^- \odot \mathring{0} \cdot B = 0,\ B \odot \mathring{0} \cdot B \cdot A^- = 0\}$ by (5), 3.2(iii), 2.1(iii), (1);

(7) $\vdash^\times \{B \odot \mathring{0} \cdot B = 0,\ B^\smile \odot B \leq \mathring{1}\}$ by (2), (4), (6), (1).

The conclusion of (iii) follows directly from (2) and (7).

It may be mentioned that, although the relations denoted by the predicates A, B in (ii) are related to conjugated projections over ordered couples, they by no means coincide with the latter and (treated as sets of ordered couples) are actually more comprehensive. In particular, the sentences in 4.1(iv) do not hold for these specific A, B. The construction of predicates A', B' which denote precisely the conjugated projections over ordered couples (in every set-theoretical realization $\langle U, E\rangle$ in which $\boldsymbol{P}$ holds) is known; A', B' have, however, a much more complex structure than A, B.

An immediate corollary of (ii) is

(iv) *Every system $\mathcal{S}$ (in $\mathcal{L}$) in which* $\mathbf{A}\boldsymbol{\xi} \vdash \boldsymbol{P}$ *is a $\mathfrak{Q}$-system.*

Hence, indeed, all known set-theoretical systems excluding proper classes (and individuals) are $\mathfrak{Q}$-systems, and the results in §4.5 apply to each of them.

We shall now consider set-theoretical systems (excluding individuals) which admit proper classes. The best known systems of this kind are those discussed in Bernays [**1937**]–[**1954**] and Morse [**1965**], together with all their variants; the system published in Vopěnka–Hájek [**1972**] also belongs here. Bernays' system in its original version is not, strictly speaking, formalized within the ordinary first-order predicate logic, since it is provided with two different kinds of variables. However, the variant of this system presented in Gödel [**1940**] can clearly be formalized in first-order logic, using one kind of variable, and hence is better suited to our discussion. To this end it suffices to eliminate from the vocabulary of Gödel [**1940**] the two unary predicates denoting respectively the properties of being a set and a class, and this can easily be done. In the subsequent discussion, when speaking of the Bernays–Gödel system, we have in mind Gödel's variant of Bernays' system with the modification just mentioned. As regards Morse's system, its original version exhibits some very essential peculiarities (the full identification of predicates and formulas) that seem to preclude the possibility of adequately formalizing the system in conventional predicate logic, and seriously complicate the problem of direct application of our results (cf. the latter part of §5.4). Here, however, we simply disregard these peculiarities and refer to what could be called the conventional version of Morse's system as presented in Kelley [**1955**] or Monk [**1969**]. By an unpublished result of Tarski (discussed further in §5.4 below), the two versions of Morse's system are equipollent in means of expression and proof; therefore the results established here for the conventional version apply to the original version as well. From the equipollence of the two versions of Morse's system it follows, in particular, that the consistency of either one of them implies the consistency of the other (cf. Morse [**1965**], p. xxiii).

The common feature of all systems discussed above is that in each of them the sentence U_2 fails to hold, and in fact its negation is provable. Hence, the pair axiom $\boldsymbol{P}$ also fails, and Theorem (iv) is not applicable. The restricted pair axiom S_2 formulated in 3.6(ii) does hold in all these systems, but this alone does not suffice for our purposes. However, we list below a few simple sentences which also hold in the systems involved, and the set (or the conjunction) of which adequately replaces the pair axiom in the discussion of these systems:

$$\begin{aligned}
T_1 &= \forall_x[\exists_y(xEy) \rightarrow \exists_y\forall_z(zEy \leftrightarrow z\mathring{1}x)],\\
T_2 &= \exists_x[\forall_z(\neg zEx) \wedge \exists_y(xEy)],\\
T_3 &= \forall_{xy}\exists_z\forall_u(uEz \leftrightarrow uEx \vee uEy),\\
T_4 &= \forall_x\exists_y\forall_z[zEy \leftrightarrow \exists_u(uEx \wedge \forall_w(wEz \leftrightarrow w\mathring{1}u))],\\
T_5 &= \forall_x\exists_y\forall_z\left[zEy \leftrightarrow \exists_u\left(uEx \wedge \forall_w[wEz \leftrightarrow w\mathring{1}u \vee \forall_s(\neg sEw)]\right)\right],\\
T_6 &= \forall_{xy}[\forall_z(zEx \leftrightarrow zEy) \rightarrow x\mathring{1}y].
\end{aligned}$$

The mathematical content of these sentences is clear. In a set-theoretical realization of $\mathcal{L}$, T_1 expresses the fact that for every set (but not for every class) x the singleton $\{x\}$ exists; it can be called the restricted singleton axiom, and it is a particular case of the restricted pair axiom. T_2 secures the existence of the empty set. T_3 is the union axiom S_1 stated in 3.6(ii). T_4 and T_5 express respectively the facts that for every class x there exists a class y consisting of all singletons $\{u\}$ of elements u of x, or of all pairs $\{u, \varnothing\}$ correlated with such elements u. Finally, T_6 is the well-known extensionality axiom.

We can now state the following theorem.

(v) *There are $A, B \in \boldsymbol{\Pi}$ such that $\{T_1, \ldots, T_6\} \vdash \boldsymbol{Q}_{AB}$. Such A, B can be obtained by setting*

$$\begin{aligned}
C &= \boldsymbol{E}^{\smile} \odot [\boldsymbol{E}^{\smile} \cdot (\boldsymbol{E}^{\smile -} \oplus \mathring{1})], & D &= \boldsymbol{E}^{\smile} \odot (\boldsymbol{E}^{\smile} \odot \mathring{0} \cdot \boldsymbol{E}^{\smile}) \odot \boldsymbol{E}^{\smile},\\
A &= (C \oplus \boldsymbol{E}^{-}) \cdot (C^{-} \oplus \boldsymbol{E}), & B &= (D \oplus \boldsymbol{E}^{-}) \cdot (D^{-} \oplus \boldsymbol{E}).
\end{aligned}$$

By 4.1(ii), to derive $\boldsymbol{Q}_{AB}$ from $\{T_1, \ldots, T_6\}$ amounts to deriving three sentences: $A^{\smile} \odot A \leq \mathring{1}$, $B^{\smile} \odot B \leq \mathring{1}$, and $A^{\smile} \odot B = \mathbf{1}$. The first two sentences are direct consequences of the extensionality axiom T_6; in checking this we can disregard the meaning of the auxiliary predicates C, D. The derivation of the third sentence reduces to showing that, in every model $\langle U, E\rangle$ of $\{T_1, \ldots, T_6\}$, for any given $r, s \in U$ there is a $t \in U$ such that $Ft = r$ and $Gt = s$, where F and G are the functional relations denoted by A and B. To simplify the formulation, we assume that $\langle U, E\rangle$ is a set-theoretical model and we use set-theoretical notation. After having carefully decoded the meanings of A and B the construction runs as follows. By applying T_4 we obtain from the given r the class r' of all singletons $\{u\}$ with $u \in r$; similarly, we obtain from s the class s' of all singletons $\{u\}$ with $u \in s$. Next, we apply T_5 to obtain from s' the class s'' of all pairs $\{\{u\}, \varnothing\}$ with $u \in s$. Finally, by virtue of T_3, we obtain the class $t = r' \cup s''$. The remaining part of the proof involves essentially (though not exclusively) T_1 and T_2; the argument is straightforward but requires some care.

Several variants of (v) are known in which some of the sentences $T_1 - T_6$ are replaced by other, usually related sentences, and the definitions of A and B are modified. For example, leaving the sentences T_2 and T_6 unchanged, we can replace T_1, T_3, T_4, T_5 by two sentences, one of which is S_2 from 3.6(ii), and the second runs as follows:

$$\forall_{xy}\exists_z\forall_u[uEz \leftrightarrow \exists_{qr}(qEx \land rEy \land \forall_w(wEu \leftrightarrow w\mathring{1}q \lor w\mathring{1}r))].$$

We leave to the reader the construction of the appropriate predicates A, B (which, at least under the method of construction used by Tarski, are more complex than in (v)) and the proof of the resulting variant of (v).

It may be noticed that the class t constructed above from given classes r and s has the essential properties of an ordered pair $\langle r, s\rangle$, i.e., t exists for any two given members of the universe, r and s, and it determines uniquely r as its first term and s as its second term. (In addition, t is a set whenever r and s are sets.) Thus, this construction can be used as a definition of an ordered pair in those set-theoretical systems which admit proper classes and in which, therefore, the classical construction (based upon $\boldsymbol{P}$) does not work. Several other, nonequivalent constructions are known which can serve the same purpose, and each of them yields a variant of (v). The procedure does not have, however, a mechanical character: a "good" definition of ordered pairs does not lead automatically to the construction of appropriate predicates A and B in the formalism $\mathcal{L}^{+}$.

An immediate consequence of (v) is:

(vi) *Every system $\mathbb{S}$ (in $\mathcal{L}$) such that* $\mathbf{A}\boldsymbol{\xi} \vdash \{T_1, \ldots, T_6\}$ *is a $\mathfrak{Q}$-system.*

As was mentioned above, sentences $T_1, \ldots, T_6$ are valid in all set-theoretical systems admitting proper classes which are known from the literature. This proves to apply also to set-theoretical systems without proper classes, discussed previously. Thus, on the basis of (vi) alone we can claim that all familiar set-theoretical systems (without individuals) are $\mathfrak{Q}$-systems.

In turn we take up the discussion of the systems of set theory admitting individuals.

As we know from the initial remarks of this section, individuals are objects which are not sets, but which share with the empty set the property of not having any elements. On the other hand, we assume that every individual, just as every set, is a member of some class. (An individual which is not a member of any class does not belong to the field of the membership relation; thus, loosely speaking, it would have no set-theoretical connections with any other object in the universe of discourse, and hence would be useless in the development of set theory.) More generally, individuals are used in forming new classes in the same way in which sets are used. For instance, the singleton of any given individual and the pair consisting of two individuals or of any individual and a set are assumed as a rule to exist.

It appears to be impossible to differentiate between individuals and the empty set in terms of the membership relation alone. Hence, to formalize a set-theoretical system with individuals, we introduce in the vocabulary an additional nonlogical constant, e.g., an individual constant denoting the empty set or a unary predicate denoting the property of being an individual. For our present discussion, however, we wish specifically to have a binary (atomic) predicate as the new constant. We choose for this purpose, e.g., the predicate $\boldsymbol{I}$ denoting the binary relation which holds between two members of the universe iff they are identical individuals. Thus the formula $\boldsymbol{xIx}$ expresses the fact that the object represented by $\boldsymbol{x}$ is an individual. As was pointed out at the end of §1.5, the inclusion of a new binary predicate does not affect the results obtained in our earlier discussion.

As examples of set-theoretical systems with individuals but without proper classes we may mention the original (but modernized) versions of the systems of Zermelo and Zermelo–Fraenkel; for such a version of the Zermelo–Fraenkel system see, e.g., Suppes [**1960**]. In all familiar systems of this kind the pair axiom $\boldsymbol{P}$ is known to be valid. Hence, Theorems (ii), (iv), and their consequences apply to all such systems.

Examples of systems with both individuals and proper classes can be obtained as variants of systems without individuals, such as Bernays' or Morse's system. However, Theorems (v) and (vi) in the present formulations are not applicable to systems with individuals. This is seen if only from the fact that the extensionality axiom T_6 implies directly the sentence U_1, and hence is not valid in any system in which the existence of both an empty set and at least one individual can be established.

To obtain applicable results, we replace sentence T_2 by three sentences, T_2', T_2'', and T_2'''; we also replace T_5, T_6 by the related sentences T_5', T_6':

$$\begin{aligned} T_2' &= \forall_{\boldsymbol{x}}[\forall_{\boldsymbol{z}}(\neg \boldsymbol{zEx}) \rightarrow \exists_{\boldsymbol{y}}(\boldsymbol{xEy})], \\ T_2'' &= \exists_{\boldsymbol{x}}[\neg \boldsymbol{xIx} \wedge \forall_{\boldsymbol{z}}(\neg \boldsymbol{zEx})], \\ T_2''' &= \forall_{\boldsymbol{x}}[\boldsymbol{xIx} \rightarrow \forall_{\boldsymbol{z}}(\neg \boldsymbol{zEx})], \end{aligned}$$

$$T_5' = \forall_x \exists_y \forall_z \left[zEy \leftrightarrow \exists_u \left(uEx \wedge \forall_w [wEz \leftrightarrow w\mathring{I}u \vee (\neg wIw \wedge \forall_s(\neg sEw))] \right) \right],$$
$$T_6' = \forall_{xy} [\neg xIx \wedge \neg yIy \wedge \forall_z (zEx \leftrightarrow zEy) \rightarrow x\mathring{I}y].$$

All the sentences of the set $\Theta = \{T_1, T_2', T_2'', T_2''', T_3, T_4, T_5', T_6'\}$ prove to be valid in each of the set-theoretical systems with individuals mentioned above (both in those which exclude and those which admit proper classes). Furthermore, we can show that, for appropriately chosen predicates A and B, Θ implies $\boldsymbol{Q}_{AB}$; the construction of A, B and the proof of the implication are analogous to those in (v), but are more complicated in details. A direct consequence of this result is that every system $\mathcal{S}$ which is developed in the extended formalism obtained from $\mathcal{L}$ by adjoining the predicate $\boldsymbol{I}$, and for which $\mathbf{A}\xi \vdash \Theta$, is a $\mathcal{Q}$-system and is therefore equipollent with the correlated system $\mathcal{S}^\times$ in the corresponding extended formalism obtained from $\mathcal{L}^\times$.

There is still another conception of individuals embodied in some systems of set theory: an object is regarded as an individual if it contains itself as the only element, i.e., is its own singleton. Thus, the fact that the object represented by the variable $\boldsymbol{x}$ is an individual is expressed by the formula

$$\forall_y (yEx \leftrightarrow y\mathring{I}x).$$

Two well-known set-theoretical systems which admit this kind of individuals are the systems described in Quine [**1937**] and Quine [**1951**]; they will be referred to respectively as $\mathcal{Q}_1$ and $\mathcal{Q}_2$.

One virtue of this conception is that it agrees with the intuition of geometers, who may find it hard to reconcile themselves with the assertion that a point is something different from a one-point set. Another virtue is that individuals in the new sense are classes—since they contain elements—that can be distinguished from other classes in terms of the membership predicate $\boldsymbol{E}$ alone. As a consequence, the formalization of systems embodying this conception does not require the introduction of a new nonlogical constant and can be carried through in the original language $\mathcal{L}$.

Indeed, for the purposes of the present discussion, set-theoretical systems based upon the new conception of individuals can be treated as systems without individuals. This applies, in particular, to systems $\mathcal{Q}_1$ and $\mathcal{Q}_2$. In $\mathcal{Q}_1$ proper classes are excluded, and Axiom $\boldsymbol{P}$ is valid; in $\mathcal{Q}_2$ proper classes are admitted, and $\boldsymbol{P}$ fails. However, the sentences $T_1 - T_6$ hold in both $\mathcal{Q}_1$ and $\mathcal{Q}_2$. Hence, $\mathcal{Q}_1$ is a $\mathcal{Q}$-system by virtue of either (iv) or (vi), and $\mathcal{Q}_2$ is a $\mathcal{Q}$-system by (vi); consequently, the main equipollence results established in the present chapter apply to both of these systems.

As is well known, systems $\mathcal{Q}_1$ and $\mathcal{Q}_2$ differ essentially from other familiar systems of set theory with regard to their sets of axioms and provable sentences; the difference in the approach to individuals does not play any significant role in this context. Without elaborating here on these questions, we should like to mention that set-theoretical systems can be constructed which admit individuals (in the new sense of this term), but which at the same time are closely related to various systems mentioned in our earlier discussion, such as the systems of Zermelo–Fraenkel, Bernays, and Morse, and actually they can be regarded as variants of the latter. Their main peculiarity is that the well-foundedness axiom, whenever it is included in any of these systems, appears there in a somewhat weaker form, in which it does not imply the nonexistence of individuals. Naturally, these systems prove again to be Ω-systems.

Some additional observations concerning systems which represent the second conception of individuals will be found in Chapter 7.

To complete our survey of set-theoretical systems (formalized in the predicate logic $\mathcal{L}$) that are of interest for contemporary foundational research, we wish to point out two systems to which our equipollence results seem not to apply. These are systems described respectively in Mostowski [**1939**] and Ackermann [**1956**]. We shall refer to them (exclusively in the present section) as systems $\mathcal{M}$ and $\mathcal{A}$.

The first of these systems is regarded by its author as a variant of Bernays' system, and in fact as one admitting individuals (in our original conception); cf. *op. cit.*, p. 208, the first paragraph, and p. 205, Definition II. Thus, disregarding certain discrepancies between our notation and that in *op. cit.*, it seems that we can treat $\mathcal{M}$ as one of the systems discussed above. Just as Bernays' system, $\mathcal{M}$ proves then to be a system admitting proper classes; in fact, the pair axiom $\boldsymbol{P}$ can be refuted in $\mathcal{M}$, and (iv) is not applicable. On the other hand, the extensionality axiom, in both its stronger form, T_6, and its weaker form, T_6', seems not to be provable in $\mathcal{M}$ either, so that we cannot apply (vi) or its subsequently described variant to $\mathcal{M}$; in the form in which the extensionality axiom appears in $\mathcal{M}$ (*op. cit*, p. 205, Axiom 2), it does not concern proper classes and is therefore inadequate for our purposes. Thus, we see no way of showing that $\mathcal{M}$ is a Ω-system.

The last statement applies to the system $\mathcal{A}$ as well. The formalization of $\mathcal{A}$ requires the introduction of a special predicate $\boldsymbol{M}$ whose intuitive meaning can hardly be explained in our terminology. System $\mathcal{A}$ is highly incomplete with respect to sentences in which this predicate does not occur. In particular, neither the sentence $\boldsymbol{P}$ nor any of the sentences T_1, T_3, T_4, T_5 appears to be provable in $\mathcal{A}$.

As a consequence of our discussion it seems that—with these two exceptions—*all the set-theoretical systems familiar from the literature are Ω-systems, and therefore each of them can be equipollently formalized either in the simple language $\mathcal{L}^{\times}$ or in one of its extensions obtained by adjoining a new nonlogical atomic binary predicate $\boldsymbol{I}$—thus in a language whose vocabulary contains no variables, quantifiers, or sentential connectives.*

4.7. Problems of expressibility and decidability in $\mathcal{L}^{\times}$

In this section we discuss some applications of the results in §§4.4–4.6, primarily to questions of expressibility and decidability.

As we recall, the notion of $\mathcal{L}^\times$-expressibility was dealt with in §§3.6 and 3.10. To obtain further results regarding this notion, we shall only use theorems of §§4.4 and 4.6 that concern equipollence in means of expression—thus, results which are much more straightforward than, say, the mapping theorem, 4.4(xxxiv), or the second equipollence theorem, 4.4(xxxvii).

Any one of the theorems (xxxiii)(δ), (xxxiii)(ε), or (xxxvi) in §4.4 yields the following corollary.

(i) *For every $X \in \Sigma^+$, if $X \vdash \boldsymbol{Q}_{AB}$ for some $A, B \in \boldsymbol{\Pi}$, then X is $\mathcal{L}^\times$-expressible.*

Corollary (i) provides new partial criteria for $\mathcal{L}^\times$-expressibility; cf. §3.6. For instance, let Ψ be the set of all sentences X of the form $X = Y \wedge \boldsymbol{Q}_{AB}$ for some $Y \in \Sigma^+$ and $A, B \in \boldsymbol{\Pi}$. Ψ is clearly recursive, and by (i) all members of Ψ are $\mathcal{L}^\times$-expressible. We can obtain many other related criteria by taking, instead of $\boldsymbol{Q}_{AB}$, any system of sentences R_{AB} with a recursive range, provided we can show that $R_{AB} \vdash \boldsymbol{Q}_{AB}$ for some $A, B \in \boldsymbol{\Pi}$. All these criteria are deeper than those given in 3.10(i),(iii), but are also of a more specialized character. Indeed, we are rarely confronted with sentences X which are exactly of the form $X = Y \wedge \boldsymbol{Q}_{AB}$, i.e., which belong to the set Ψ defined above. More often we come across sentences which prove to satisfy the hypothesis, and hence also the conclusion, of (i), although the connection with $\boldsymbol{Q}_{AB}$ is not reflected in the form of these sentences.

As we shall see below, we can establish in this way the $\mathcal{L}^\times$-expressibility of a fairly large number of sentences in Σ with interesting mathematical contents. However, the structure of the equivalent sentences in $\Sigma^\times$ as described in 4.4(xxxiii)(ε) is as a rule quite involved, and we do not know any simpler method of constructing such sentences. The same is true, to an even greater extent, if we attempt to construct equivalent sentences, not in $\Sigma^\times$, but in Σ_3 or $\Sigma_{(3)}$.

Actually, 4.4(xxxiii)(ε) provides the only known method of establishing the $\mathcal{L}^\times$-expressibility of a sentence $X \in \Sigma$, except for those (rather trivial) cases when X obviously satisfies one of the criteria provided in 3.10(iv). The method will be extended somewhat by certain observations that will be found at the end of §6.3.

The range of application of (i) is increased by the fact that every sentence which satisfies the hypothesis of (i) is not only $\mathcal{L}^\times$-expressible, but also *essentially $\mathcal{L}^\times$-expressible* in the sense that all logically stronger sentences are also $\mathcal{L}^\times$-expressible (since they obviously satisfy the hypothesis of (i)). Examples of sentences are known which are $\mathcal{L}^\times$-expressible but not essentially $\mathcal{L}^\times$-expressible. A general result to this effect is:

(ii) *If a sentence $X \in \Sigma^+$ is essentially $\mathcal{L}^\times$-expressible, then $\neg X$ is $\mathcal{L}^\times$-expressible, but not essentially $\mathcal{L}^\times$-expressible.*

Indeed, since X is $\mathcal{L}^\times$-expressible by hypothesis, so is $\neg X$ by 2.2(i),(iii). Consider now an arbitrary $Y \in \Sigma^+$ which is not $\mathcal{L}^\times$-expressible (cf. 3.4(iv),(v)).

Obviously $Y \wedge X \vdash X$, whence $Y \wedge X$ is $\mathcal{L}^{\times}$-expressible. Similarly, if $\neg X$ were essentially $\mathcal{L}^{\times}$-expressible, then $Y \wedge \neg X$ would be $\mathcal{L}^{\times}$-expressible. Therefore, by 2.2(i),(iv) the sentence

$$(Y \wedge X) \vee (Y \wedge \neg X),$$

and hence also Y itself, would be $\mathcal{L}^{\times}$-expressible, which contradicts our assumption on Y.

By (ii), a concrete example of a sentence that is $\mathcal{L}^{\times}$-expressible, but not essentially $\mathcal{L}^{\times}$-expressible, is provided by $\neg \boldsymbol{Q}_{AB}$, where A, B are any predicates.

From 4.6(ii) it is seen directly that the sentence $\boldsymbol{P}$ is $\mathcal{L}^{\times}$-expressible. By (i) the same applies to all sentences in $\mathcal{L}^{+}$ which are logically stronger than $\boldsymbol{P}$, for instance to all sentences of the form

$$\forall_{x_0 \cdots x_{n-1}} \exists_z \forall_u (u\boldsymbol{E}z \leftrightarrow u\mathring{\mathbf{1}}x_0 \vee \cdots \vee u\mathring{\mathbf{1}}x_{n-1})$$

(where $n = 3, 4, \ldots$, and $x_0, \ldots, x_{n-1}, z, u$ are distinct variables). These are sentences that assert the existence of unordered n-tuples.

It may be interesting to compare $\boldsymbol{P}$ with sentences S_1 and S_3 in §3.6. The structure of the three sentences is virtually the same; S_3 is obtained from S_1 by changing $\boldsymbol{E}$ to $\mathring{\mathbf{1}}$ in one place, and $\boldsymbol{P}$ is obtained in exactly the same way from S_3. While $\boldsymbol{P}$ is $\mathcal{L}^{\times}$-expressible, S_1 is not, and for S_3 the problem is open.

If we use merely the fact that $\boldsymbol{P} \vdash \boldsymbol{Q}_{AB}$, then a sentence in $\mathcal{L}^{\times}$ logically equivalent with $\boldsymbol{P}$ can be obtained by means of the general method based upon 4.4(xxxiii)(ε); this will be, in fact, the sentence $(\boldsymbol{K}_{AB}\boldsymbol{P})^t \cdot (\boldsymbol{Q}_{AB})^{\ell} = \mathbf{1}$. As was pointed out above, sentences thus obtained are as a rule very involved. In the particular case of the sentence $\boldsymbol{P}$ and of appropriately chosen predicates A and B, Theorem 4.6(ii) provides us with a much simpler sentence which serves the same purpose, namely with $A^{\smile} \odot B = \mathbf{1}$. However, even this sentence is not quite simple. It may be instructive to give here an explicit formulation of this sentence by writing it exclusively in terms of the atomic predicates $\boldsymbol{E}$ and $\mathring{\mathbf{1}}$. Actually, we formulate here not $A^{\smile} \odot B = \mathbf{1}$, but a slight simplification of the equivalent sentence $B^{\smile} \odot A = \mathbf{1}$:

$$\boldsymbol{E} \odot \boldsymbol{E} \cdot \left[\mathring{\mathbf{1}} \oplus \left(\boldsymbol{E}^{-} \oplus \boldsymbol{E}^{-} + (\mathring{\mathbf{1}} \oplus \boldsymbol{E}^{-}) \cdot \boldsymbol{E} \odot \boldsymbol{E} \cdot [\mathring{\mathbf{1}} \oplus (\mathring{\mathbf{1}}^{-} \odot \boldsymbol{E} + \boldsymbol{E}^{-}) \oplus \boldsymbol{E}^{-}]\right)\right]$$
$$\odot \left(\boldsymbol{E}^{\smile} \odot [\boldsymbol{E}^{\smile} \cdot (\boldsymbol{E}^{\smile -} \oplus \mathring{\mathbf{1}})] \cdot [\boldsymbol{E}^{\smile -} \oplus (\boldsymbol{E}^{\smile -} + \boldsymbol{E}^{\smile} \odot \mathring{\mathbf{1}}^{-}) \oplus \mathring{\mathbf{1}}]\right) = \mathbf{1}.$$

By analyzing the proof of 3.8(ix)(ε) and Definition 2.3(iii), we obtain a sentence in $\mathcal{L}_3$ which is equivalent to $A^{\smile} \odot B = \mathbf{1}$ and hence also to $\boldsymbol{P}$ (obviously $\boldsymbol{P}$ is a sentence in $\mathcal{L}_4$). This is, in fact, the sentence $\boldsymbol{G}(B^{\smile} \odot A = \mathbf{1})$; it is longer than $A^{\smile} \odot B = \mathbf{1}$ and about eight times as long as $\boldsymbol{P}$. It would be interesting to find the shortest possible sentences in $\mathcal{L}^{\times}$ and $\mathcal{L}_3$ logically equivalent with $\boldsymbol{P}$.

Theorem 4.6(iv) can be used to obtain the solution of some problems mentioned in the penultimate paragraph of §3.6. We first notice the following.

(iii) *There is a $\mathfrak{Q}$-system $\mathcal{S}$ (in $\mathcal{L}$) such that $\mathbf{A}\xi$ is finite while $\mathbf{\Theta\eta A}\xi$ is undecidable and, in fact, hereditarily and essentially undecidable. The same applies to the correlated systems $\mathcal{S}^+$ (in $\mathcal{L}^+$) and $\mathcal{S}^\times$ (in $\mathcal{L}^\times$).*

In fact, by a result in Vaught [**1962**], p. 21 (which improves an earlier result of Szmielew–Tarski [**1952**]; see also Tarski–Mostowski–Robinson [**1953**], p. 34), a simple example of a system $\mathcal{S}$ with all the desired properties is obtained by setting

(iv) $\mathbf{A}\xi = \{\exists_x \forall_y(\neg yEx),\ \forall_{xy}\exists_z\forall_u(uEz \leftrightarrow uEx \vee u\mathring{1}y)\}.$

In particular, it is almost obvious that $\mathbf{A}\xi \vdash \boldsymbol{P}$, and hence, by 4.6(iv), $\mathcal{S}$ is a $\mathfrak{Q}$-system.

(In view of this last remark, the conjunction of the two sentences of $\mathbf{A}\xi$ is $\mathcal{L}^\times$-expressible. Obviously, the first of these sentences is also $\mathcal{L}^\times$-expressible. Recall that for the second sentence the problem is open; cf. §3.6.)

From the construction of the correlated systems $\mathcal{S}^+$ and $\mathcal{S}^\times$ it follows immediately that they are $\mathfrak{Q}$-systems as well, and that $\mathbf{\Theta\eta}^+\mathbf{A}\xi$ and $\mathbf{\Theta\eta}^\times\mathbf{A}\xi^\times$ are finitely based theories. By 2.3(x),(xi)(ε),(ς) we conclude that $\mathbf{\Theta\eta}^+\mathbf{A}\xi$ is hereditarily and essentially undecidable. Applying, in addition, 4.4(xxxix) and the remarks following 4.4(xxxix), we obtain the same conclusions for $\mathbf{\Theta\eta}^\times\mathbf{A}\xi^\times$.

The following corollary of (iii) provides the solution of the decision problems stated near the end of §3.6.[5]

(v) $\mathbf{\Theta\eta}^+\varnothing \cap \Sigma^\times$ *is a hereditarily undecidable theory (in $\mathcal{L}^\times$).*

Obviously $\mathbf{\Theta\eta}^+\varnothing\cap\Sigma^\times$ is a theory in $\mathcal{L}^\times$ by 3.4(iii). Let $\mathcal{S}^\times$ be any one of those $\mathfrak{Q}$-systems in $\mathcal{L}^\times$ with axiom set $\mathbf{A}\xi^\times$ whose existence is claimed in (iii), i.e., for which $\mathbf{\Theta\eta}^\times\mathbf{A}\xi^\times$ is hereditarily (and essentially) undecidable. Since $\mathbf{\Theta\eta}^+\varnothing \cap \Sigma^\times$ is a subtheory of $\mathbf{\Theta\eta}^\times\mathbf{A}\xi^\times$ by the second part of 4.5(v)(γ3), our theorem follows at once.

An immediate consequence of (v) is:

(vi) $\mathbf{\Theta\eta}^\times\varnothing$ *is an undecidable theory in $\mathcal{L}^\times$.*[6]

(Concerning the relationship between the theories in (v) and (vi), cf. the penultimate paragraph of §3.4.)

More generally, (v) obviously expresses the fact that every theory in $\mathcal{L}^\times$ which is a subtheory of $\mathbf{\Theta\eta}^+\varnothing \cap \Sigma^\times$ is undecidable. This applies, in particular, to $\mathbf{\Theta\eta}_n^+\varnothing \cap \Sigma^\times$ for every integer $n \geq 3$, i.e., to the theory in $\mathcal{L}^\times$ consisting of just those sentences in $\Sigma^\times$ that are logically provable in $\mathcal{L}_n^+$.

[5] Theorem (v) was first announced, in a different formulation, in Tarski [**1941**], p. 88; see also Chin–Tarski [**1951**], p. 341, and p. 343, footnote 4.

[6] *In a private communication Maddux has informed us that (v) and (vi) remain valid when referred to the weaker formalism $\mathcal{L}w^\times$ defined in §3.10. On the other hand, for the still weaker formalism $\mathcal{L}o^\times$, it is proved in Németi [**1987**] that $\mathbf{\Theta\eta}\varnothing[\mathcal{L}o^\times]$ is a decidable theory.

For later use in Chapter 8 we formulate here (iii) in a more specific and indeed somewhat stronger form. Recall that a set $\Phi \subseteq \Sigma^{\times}$ is compatible with a set $\Psi \subseteq \Sigma^{\times}$ if $\Phi \cup \Psi$ is consistent.

(vii) *Let A, B be the predicates defined in* 4.6(ii) *and let* $\mathbf{\Omega}$ *be the set of the following three sentences*:

$$\boldsymbol{K}_{AB}(\boldsymbol{\exists}_x\boldsymbol{\forall}_y(\boldsymbol{\neg}\boldsymbol{y}\boldsymbol{E}\boldsymbol{x})),\ \boldsymbol{K}_{AB}(\boldsymbol{\forall}_{xy}\boldsymbol{\exists}_z\boldsymbol{\forall}_u(\boldsymbol{u}\boldsymbol{E}\boldsymbol{z} \leftrightarrow \boldsymbol{u}\boldsymbol{E}\boldsymbol{x} \vee \boldsymbol{u}\mathring{\mathbf{1}}\boldsymbol{y})),\ \boldsymbol{Q}_{AB}.$$

Then every theory $\Theta \subseteq \Sigma^{\times}$ that is compatible with $\mathbf{\Omega}$ *is undecidable. In particular,* $\mathbf{\Theta\eta}^{\times}\mathbf{\Omega}$ *is hereditarily and essentially undecidable.*

The essential (and hereditary) undecidability of $\mathbf{\Theta\eta}^{\times}\mathbf{\Omega}$ follows at once from the proof of (iii) and Definition 4.5(iii). To derive the undecidability of every theory $\Theta \subseteq \Sigma^{\times}$ compatible with $\mathbf{\Omega}$, we recall that $\mathcal{L}^{\times}$ is a formalism with a deduction theorem (cf. §3.3), and then, using the essential undecidability and finite axiomatizability of $\mathbf{\Theta\eta}^{\times}\mathbf{\Omega}$, we argue as in Tarski–Mostowski–Robinson [**1953**], pp. 17–18.

We shall now give some generalizations and improvements of 4.6(ii),(iv) which present some intrinsic interest and will be applied in later parts of our work.

We consider a system $\boldsymbol{U}$ of sentences in Σ^{+} indexed by arbitrary predicates C; it is determined by the formula

(viii) $U_C \;=\; \boldsymbol{\forall}_{xy}\boldsymbol{\exists}_z\boldsymbol{\forall}_u(\boldsymbol{u}C\boldsymbol{z} \leftrightarrow \boldsymbol{u}\mathring{\mathbf{1}}\boldsymbol{x} \vee \boldsymbol{u}\mathring{\mathbf{1}}\boldsymbol{y}).$

This formula obviously generalizes 4.6(i) in that we have $\boldsymbol{P} = \boldsymbol{U}_{\boldsymbol{E}}$.

It is convenient to use terminology borrowed from analytic geometry in order to describe the mathematical content of sentences $\boldsymbol{U}_C$. We treat binary relations R (included in a unit relation $U \times U$) as point-sets of an (abstract) two-dimensional Cartesian space $U \times U$. A relation R is said to be a *universal relation for two-element sets* if it satisfies the following condition: for every set S on the first axis consisting of two (not necessarily distinct) points there is a line L parallel to the first axis such that the projection of $L \cap R$ on the first axis coincides with S. The sentence $\boldsymbol{U}_C$, with $C \in \mathbf{\Pi}$, expresses the fact that the relation denoted by C (in any given realization $\mathfrak{U} = \langle U, E \rangle$) is just a universal relation for two-element sets.

By generalizing 4.6(ii), with practically no change in the proof, we obtain:

(ix) *For every $C \in \mathbf{\Pi}$ there are $A, B \in \mathbf{\Pi}$ such that*

$$\boldsymbol{U}_C \;\equiv\; \boldsymbol{Q}_{AB} \;\equiv\; (A^{\smile} \odot B = \mathbf{1}).$$

A and B can be defined just as in 4.6(ii), *replacing* "$\boldsymbol{E}$" *by* "C".

In view of 4.6(iii), the symbol $\equiv$ in its second occurrence in (ix) may be replaced by $\equiv^{\times}$.

We can also easily prove

(x) *For any $A, B \in \mathbf{\Pi}$ there is a $C \in \mathbf{\Pi}$ such that $\boldsymbol{Q}_{AB} \vdash \boldsymbol{U}_C$. Such a C can be obtained by setting $C = (A \boldsymbol{+} B)^{\smile}$.*

Using sentences $\boldsymbol{U}_C$ instead of $\boldsymbol{Q}_{AB}$, we obtain new characterizations of $\mathcal{Q}$-systems; they are analogous to, but somewhat simpler than, those given in 4.5(i),(ii).

(xi) *Let $\mathcal{S}$ be any system in $\mathcal{L}$ (or $\mathcal{L}^+$) with the axiom set $\mathbf{A}\boldsymbol{\xi}$.*

- (α) *$\mathcal{S}$ is a $\mathcal{Q}$-system iff there is a predicate C in $\mathcal{L}^+$ such that $\mathbf{A}\boldsymbol{\xi} \vdash^+ \boldsymbol{U}_C$ or, equivalently, $\mathbf{A}\boldsymbol{\xi} \vdash \boldsymbol{G}(\boldsymbol{U}_C)$.*
- (β) *$\mathcal{S}$ is a $\mathcal{Q}$-system iff there is an $F \in \mathbf{\Phi}$ satisfying the following conditions:*
  - (β1) $\boldsymbol{\Upsilon\phi} F = \{\boldsymbol{x}, \boldsymbol{y}\}$;
  - (β2) $\mathbf{A}\boldsymbol{\xi} \ \vdash \ \boldsymbol{\forall}_{zu}\boldsymbol{\exists}_y\boldsymbol{\forall}_x(F \leftrightarrow \boldsymbol{x}\mathring{\mathbf{1}}\boldsymbol{z} \vee \boldsymbol{x}\mathring{\mathbf{1}}\boldsymbol{u})$;
  - (β3) *F contains no variables different from $\boldsymbol{x}, \boldsymbol{y}, \boldsymbol{z}$.*

Part (α) follows directly from (ix) and (x), by 4.5(i). To derive (β) we argue as in the proof of 4.5(ii), using the definition of $\boldsymbol{U}_C$ in (viii).

4.8. The undecidability of first-order logics with finitely many variables, and the relative equipollence of $\mathcal{L}_3$ with $\mathcal{L}$

To conclude Chapter 4 we shall discuss various applications of our results to the formalism $\mathcal{L}_3$. In view of the equipollence of $\mathcal{L}_3$ and $\mathcal{L}^{\times}$ which was shown in §§3.8, 3.9, it will be easy to extend to $\mathcal{L}_3$, and indeed to $\mathcal{L}_n$ for $n \geq 3$, various results previously established for $\mathcal{L}^{\times}$. As an example we mention Theorem 4.7(v) which, when transferred to $\mathcal{L}_n$, runs as follows.

(i) *$\boldsymbol{\Theta\eta}\varnothing \cap \Sigma_n$ is a hereditarily undecidable theory (in $\mathcal{L}_n$) for every natural number $n \geq 3$.*

To prove this in case $n = 3$ we set $\Theta = \boldsymbol{\Theta\eta}^+\varnothing \cap \Sigma_3^+$, and with the help of 2.3(x), 3.8(viii), 3.9(i), we get

$$\Theta \cap \Sigma_3 = \boldsymbol{\Theta\eta}\varnothing \cap \Sigma_3 \quad \text{and} \quad \Theta \cap \Sigma^{\times} = \boldsymbol{\Theta\eta}^+\varnothing \cap \Sigma^{\times}.$$

In view of 4.7(v) and 3.9(xii)(ς), this immediately yields the desired conclusion.

In case $n > 3$, observe that if Δ is a subtheory of $\boldsymbol{\Theta\eta}\varnothing \cap \Sigma_n$, then $\Delta \cap \Sigma_3$ is a subtheory of $\boldsymbol{\Theta\eta}\varnothing \cap \Sigma_3$. Since our theorem has been shown to hold for $n = 3$, we infer that $\Delta \cap \Sigma_3$ is an undecidable theory (in $\mathcal{L}_3$). The set Σ_3 being recursive, we conclude immediately that Δ is an undecidable theory (in $\mathcal{L}_n$). Consequently, $\boldsymbol{\Theta\eta}\varnothing \cap \Sigma_n$ is hereditarily undecidable.

As an immediate corollary of (i) we obtain:

(ii) *$\boldsymbol{\Theta\eta}_n\varnothing$ is an undecidable theory in $\mathcal{L}_n$ for every natural number $n \geq 3$.*

We thus see that in every language $\mathcal{L}_n$ with $n \geq 3$ both the set of logically true sentences and that of logically provable sentences are not recursive. Since we do not concern ourselves here with formalisms $\mathcal{L}_1$ and $\mathcal{L}_2$, Theorem (ii) loses its meaning for values of n smaller than 3. On the other hand, the problem whether the set $\mathbf{\Theta}\boldsymbol{\eta}\varnothing \cap \Sigma_n$ from (i) is recursive continues to be meaningful for these values of n. In fact, it is known that the set $\mathbf{\Theta}\boldsymbol{\eta}\varnothing \cap \Sigma_n$ is recursive for $n = 2$, and hence also for $n = 1$; cf. Scott [**1962**].[7]

From now on we shall concentrate chiefly on systems in $\mathcal{L}_3$. In §4.5 we established the equipollence of any two correlated $\mathfrak{Q}$-systems in $\mathcal{L}$ and $\mathcal{L}^\times$. When comparing this result with the equipollence of $\mathcal{L}^\times$ and $\mathcal{L}_3$ we are naturally led to extending the notion of a $\mathfrak{Q}$-system to $\mathcal{L}_3$ and to proving the equipollence of any such system with the correlated system in $\mathcal{L}$ (cf. 2.4(ix) and its proof).

Let $\mathcal{S}$, $\mathcal{S}^+$, and $\mathcal{S}^\times$ be any three correlated $\mathfrak{Q}$-systems in $\mathcal{L}$, $\mathcal{L}^+$, and $\mathcal{L}^\times$ respectively. Thus $\mathcal{S}$ and $\mathcal{S}^+$ have the common axiom set $\mathbf{A}\boldsymbol{\xi}$, $\mathcal{S}^\times$ has the axiom set $\mathbf{A}\boldsymbol{\xi}^\times$, and there are predicates $A, B \in \mathbf{\Pi}$ such that

(iii) $\mathbf{A}\boldsymbol{\xi} \vdash^+ \boldsymbol{Q}_{AB}$ *and* $\mathbf{A}\boldsymbol{\xi}^\times = \boldsymbol{K}^*_{AB}\mathbf{A}\boldsymbol{\xi} \cup \{\boldsymbol{Q}_{AB}\}$.

A system $\mathcal{S}_3$ in $\mathcal{L}_3$ will be referred to as the *$\mathfrak{Q}$-system in $\mathcal{L}_3$ correlated with* $\mathcal{S}$, and hence also *with* $\mathcal{S}^+$ *and* $\mathcal{S}^\times$, if its axiom set $\mathbf{A}\boldsymbol{\xi}_3$ is determined by

(iv) $\mathbf{A}\boldsymbol{\xi}_3 = \boldsymbol{G}^*\mathbf{A}\boldsymbol{\xi}^\times$.

(We can characterize the correlated $\mathfrak{Q}$-system $\mathcal{S}_3^+$ in $\mathcal{L}_3^+$ in an analogous way, but this will not be involved in our observations.) Just as in the case of $\mathcal{S}^\times$ (cf. 4.5(iv) and the immediately following remarks), the system $\mathcal{S}_3$ so defined does not depend in any essential way on the choice of predicates A, B. Throughout the subsequent discussion we shall treat $\mathcal{S}$ as a fixed system in $\mathcal{L}$ with axiom set $\mathbf{A}\boldsymbol{\xi}$, and A, B as fixed predicates such that $\mathbf{A}\boldsymbol{\xi} \vdash^+ \boldsymbol{Q}_{AB}$. The remaining systems $\mathcal{S}^+$, $\mathcal{S}^\times$, $\mathcal{S}_3$, and their respective axiom sets $\mathbf{A}\boldsymbol{\xi}$, $\mathbf{A}\boldsymbol{\xi}^\times$, $\mathbf{A}\boldsymbol{\xi}_3$, are then uniquely determined by (iii) and (iv).

We first observe that $\mathcal{S}_3$ is a subsystem of $\mathcal{S}$ in the following sense.

(v) $\Sigma_3 \subseteq \Sigma$ (*and indeed* $\mathbf{\Phi}_3 \subseteq \mathbf{\Phi}$).

(vi) *Whenever* $\Psi \subseteq \Sigma_3$, $X \in \Sigma_3$, *and* $\Psi \vdash X$ $[\mathcal{S}_3]$, *then* $\Psi \vdash X$ $[\mathcal{S}]$.

Clearly, since $\mathcal{L}_3$ is a subformalism of $\mathcal{L}$, the proof of (vi) reduces to establishing the formula

$$\mathbf{A}\boldsymbol{\xi} \vdash \mathbf{A}\boldsymbol{\xi}_3,$$

which can be derived in a straightforward way from (iii) and (iv) by using 4.4(xxxiii)(δ) and 2.3(iv)(β),(v).

Next, we define a translation mapping $\boldsymbol{N}_{AB}$ by stipulating:

[7] *Andréka and Németi have called our attention to the fact that the proof in Scott [**1962**] is defective. It is based upon an assertion of Gödel [**1933**] that has recently been shown to be false, cf. Goldfarb [**1984**]. A correct proof is given in Mortimer [**1975**]; Scott's theorem can also be derived from Theorem 4.2.9 in Henkin–Monk–Tarski [**1985**]. Related but weaker results are given in von Wright [**1950**], [**1952**]; see also Ackermann [**1952**], [**1954**], pp. 88 ff.

(vii) $\boldsymbol{N}_{AB}$ *is the unique function* F *such that*

(α) $DoF = \Sigma$,

(β) $FX = X$ *in case* $X \in \Sigma_3$,

(γ) $FX = \boldsymbol{GK}_{AB}X$ *in case* $X \in \Sigma \sim \Sigma_3$.

The following properties of $\boldsymbol{N}_{AB}$ can easily be derived from the above definition.

(viii) (α) $\boldsymbol{N}_{AB}$ *is a recursive function.*

(β) $\boldsymbol{N}_{AB}$ *maps* Σ *onto* Σ_3.

(γ) $\boldsymbol{N}_{AB}X = X$ *iff* $X \in \Sigma_3$.

(δ) $\boldsymbol{N}_{AB}X \equiv_{\mathbf{A\xi}} X$ *for every* $X \in \Sigma$, *and hence* $\boldsymbol{N}^*_{AB}\Psi \equiv_{\mathbf{A\xi}} \Psi$ *for every* $\Psi \subseteq \Sigma$.

The proof of the next theorem is somewhat more difficult.

(ix) *For any given* $X \in \Sigma_3$ *we have*

$$\mathbf{A\xi}_3 \vdash_3 \boldsymbol{GK}_{AB}X \leftrightarrow X.$$

In fact, we obtain step-by-step:

(1)	$\vdash_3^+ \boldsymbol{H}X \leftrightarrow X$	by 3.9(iii)(ε);
(2)	$\boldsymbol{H}X \equiv^+ X$	by (1), $\mathcal{L}_3^+$ being a subformalism of $\mathcal{L}^+$;
(3)	$\boldsymbol{K}_{AB}\boldsymbol{H}X \equiv^{\times}_{\mathbf{A\xi}^{\times}} \boldsymbol{K}_{AB}X$	by (2), (iii), 4.4(xxxiv);
(4)	$\mathbf{A\xi}^{\times} \vdash_3^+ \boldsymbol{H}X \leftrightarrow \boldsymbol{K}_{AB}X$	by (3), 3.9(ii),(iii)(δ1), 4.4(xxxiii)(γ);
(5)	$\mathbf{A\xi}^{\times} \vdash_3^+ \boldsymbol{K}_{AB}X \leftrightarrow X$	by (1), (4).

By applying 3.8(ix),(xi) to (5) we get directly the conclusion.

We now arrive at the main mapping theorem for $\mathcal{S}$ and $\mathcal{S}_3$.

(x) *For every* $\Psi \subseteq \Sigma$ *and* $X \in \Sigma$, *we have* $\Psi \vdash X$ $[\mathcal{S}]$ *iff* $\boldsymbol{N}^*_{AB}\Psi \vdash \boldsymbol{N}_{AB}X$ $[\mathcal{S}_3]$.

We have indeed, by 4.5(v)(β4),

$$(1) \qquad \mathbf{A\xi} \cup \Psi \vdash X \quad \text{iff} \quad \mathbf{A\xi}^{\times} \cup \boldsymbol{K}^*_{AB}\Psi \vdash^{\times} \boldsymbol{K}_{AB}X.$$

From 3.8(xi) and 3.9(ix) we see that

$$(2) \qquad \mathbf{A\xi}^{\times} \cup \boldsymbol{K}^*_{AB}\Psi \vdash^{\times} \boldsymbol{K}_{AB}X \quad \text{iff} \quad \boldsymbol{G}^*\mathbf{A\xi}^{\times} \cup \boldsymbol{GK}^*_{AB}\Psi \vdash_3 \boldsymbol{GK}_{AB}X.$$

By the definition of $\boldsymbol{N}_{AB}$ and (iv), (ix) we get

$$(3) \qquad \boldsymbol{G}^*\mathbf{A\xi}^{\times} \cup \boldsymbol{GK}^*_{AB}\Psi \vdash_3 \boldsymbol{GK}_{AB}X \quad \text{iff} \quad \mathbf{A\xi}_3 \cup \boldsymbol{N}^*_{AB}\Psi \vdash_3 \boldsymbol{N}_{AB}X.$$

The conclusion follows directly from (1)–(3).

From (x), with the help of (v), (vi), and (viii)(γ), we easily obtain the proper equipollence theorems.

(xi) *For every $X \in \Sigma$ there is a $Y \in \Sigma_3$, and for every $Y \in \Sigma_3$ there is an $X \in \Sigma$, such that $X \equiv Y[\mathcal{S}]$.*

(xii) *For every $\Psi \subseteq \Sigma_3$ and $X \in \Sigma_3$, we have $\Psi \vdash X\ [\mathcal{S}]$ iff $\Psi \vdash X\ [\mathcal{S}_3]$.*

We thus see that *for every $\mathcal{Q}$-system $\mathcal{S}$ in $\mathcal{L}$ the correlated $\mathcal{Q}$-system $\mathcal{S}_3$ in $\mathcal{L}_3$ is equipollent with $\mathcal{S}$ in means of expression and proof.*[8] In consequence, by our observations in §4.6, *practically all set-theoretical systems familiar from the literature can be equipollently formalized in the language $\mathcal{L}_3$ of predicate logic with just three distinct variables.* Thus $\mathcal{L}_3$, just as $\mathcal{L}^\times$, provides an adequate framework for the development of set theory, and hence of all of mathematics.

From the equipollence results (x)–(xii) we can derive various conclusions concerning primarily $\mathcal{S}_3$. The proofs are easy and will, for the most part, be omitted.

Recall here that $\mathbf{\Theta\eta}\Psi[\mathcal{S}]$ coincides with $\mathbf{\Theta\eta}(\Psi \cup \mathbf{A\xi})[\mathcal{L}]$, and analogously for $\mathbf{\Theta\eta}\Psi[\mathcal{S}_3]$.

(xiii) *If $\Psi \subseteq \Sigma_3$, then $\mathbf{\Theta\eta}\Psi[\mathcal{S}_3] = \mathbf{\Theta\eta}\Psi[\mathcal{S}] \cap \Sigma_3$.*

(xiv) *If Φ is a theory in $\mathcal{S}$ (i.e., $\mathbf{A\xi} \subseteq \Phi$), then $\Phi \cap \Sigma_3$ is a theory in $\mathcal{S}_3$ and $\Phi = \mathbf{\Theta\eta}(\Phi \cap \Sigma_3)[\mathcal{S}]$. If, moreover, Φ or $\Phi \cap \Sigma_3$ satisfies one of the conditions (α)–(ε) in 2.3(xi), then both Φ and $\Phi \cap \Sigma_3$ satisfy this condition.*

As the last immediate consequence of the equipollence results we give a theorem which concerns arbitrary $\mathcal{Q}$-systems $\mathcal{T}$ formalized in $\mathcal{L}_3$, and not only systems $\mathcal{S}_3$ constructed as correlates of some $\mathcal{Q}$-systems $\mathcal{S}$ in $\mathcal{L}$. We obtain characterizations of $\mathcal{T}$ (or $\mathcal{T}^+$) as a $\mathcal{Q}$-system in $\mathcal{L}_3$ (or $\mathcal{L}_3^+$) by replacing $\mathcal{S}$ with $\mathcal{T}$ and $\vdash$ with $\vdash_3$ everywhere in 4.5(i),(ii). With obvious changes the proof of 4.5(ii) remains valid, so that the two characterizations of $\mathcal{Q}$-systems in $\mathcal{L}_3$ again turn out to be equivalent. Obviously, if $\mathcal{S}$ is a $\mathcal{Q}$-system in $\mathcal{L}$, then $\mathcal{S}_3$ is a $\mathcal{Q}$-system in $\mathcal{L}_3$ in the sense just defined.

---

[8]*Andréka and Németi have pointed out to us that Németi [**1985**] contains a strengthening of this result. Let $\mathcal{L}_3^*$ be the formalism discussed at the beginning of §3.7. Thus, neither (AIX′) nor (AX) occurs as an axiom schema of $\mathcal{L}_3^*$, only (AI)–(AVIII) and (AIX). In particular, the more complicated notion of substitution introduced in §3.7 is not involved in this axiomatization. Németi has shown that every $\mathcal{Q}$-system in $\mathcal{L}$ is equipollent to a $\mathcal{Q}$-system developed in a subformalism of $\mathcal{L}_3^*$. (This subformalism may, in particular, have fewer sentences than $\mathcal{L}_3^*$.) More precisely, for any predicates A, B in $\mathbf{\Pi}$, he defines a recursive mapping N'_{AB} from Σ into (but not necessarily onto) Σ_3 such that N'_{AB} has a recursive range and satisfies (viii)(δ). With every system $\mathcal{S}$ in $\mathcal{L}$ such that $\mathbf{A\xi}[\mathcal{S}] \vdash^+ Q_{AB}$ he then correlates a well-determined subsystem $\mathcal{S}'_3$ developed in a subformalism of $\mathcal{L}_3^*$ and defined in terms of N'_{AB}, much as $\mathcal{S}_3$ is defined in terms of $\boldsymbol{GK}_{AB}$, and he proves for N'_{AB} and $\mathcal{S}'_3$ the analogue of the main mapping theorem (x). Németi has constructed examples to show that in his theorem the development of $\mathcal{S}'_3$ in a subformalism of $\mathcal{L}_3^*$, instead of in $\mathcal{L}_3^*$ itself, is essential.

Thus, $\mathcal{L}_3^*$ provides an adequate basis for the formalization of most systems of set theory. Since $\mathcal{L}_3^*$ is a subformalism of the formalism $\mathcal{L}s_3$, Németi's results remain valid when $\mathcal{L}_3^*$ is replaced by $\mathcal{L}s_3$. Finally, since Maddux proved that $\mathcal{L}s_3$ is equipollent with $\mathcal{L}w^\times$ (see §3.10), we eventually conclude that $\mathcal{L}w^\times$ also provides an adequate basis for the formalization of set theory. (See the footnote, p. 90.)

(xv) *Every $\mathcal{Q}$-system $\mathcal{T}$ in $\mathcal{L}_3$ is semantically complete, i.e., for every $\Psi \subseteq \Sigma_3$ and $X \in \Sigma_3$, we have*

$$\Psi \models X\ [\mathcal{T}] \quad \textit{iff} \quad \Psi \vdash X\ [\mathcal{T}],$$

or, equivalently,

$$\Psi \vdash_{\mathbf{A}\xi} X\ [\mathcal{L}] \quad \textit{iff} \quad \Psi \vdash_{\mathbf{A}\xi} X\ [\mathcal{L}_3],$$

where $\mathbf{A}\xi = \mathbf{A}\xi[\mathcal{T}]$.

In fact, let $\mathcal{S}'$ be the system in $\mathcal{L}$ determined by $\mathbf{A}\xi[\mathcal{S}'] = \mathbf{A}\xi[\mathcal{T}]$; obviously $\mathcal{S}'$ is a $\mathcal{Q}$-system. It is easily seen that $\mathcal{S}'_3$ and $\mathcal{T}$ are identical since the relations $\vdash [\mathcal{T}]$ and $\vdash [\mathcal{S}'_3]$ coincide. From (viii)(γ), (x), and the semantical completeness of $\mathcal{L}$, we readily obtain the semantical completeness of $\mathcal{S}'_3$, and hence of $\mathcal{T}$.

We can of course extend the notion of a $\mathcal{Q}$-system to all of the formalisms $\mathcal{L}_{n+1}$ with $n \geq 3$. We have not studied such systems intensively. Many basic problems concerning them are open. In particular, it is not known whether, for any given $n \geq 3$, every $\mathcal{Q}$-system in $\mathcal{L}_{n+1}$ is semantically complete. However, the following weaker result, due to Givant, is known to hold.

(xvi) *Let $\mathcal{T}$ be a $\mathcal{Q}$-system in $\mathcal{L}_{n+1}$ ($n \geq 3$) with base $\mathbf{A}\xi$, and suppose $\mathbf{A}\xi \subseteq \Sigma_n$. Then for every $\Psi \subseteq \Sigma_n$ and $X \in \Sigma_n$, we have*

$$\Psi \models_{\mathbf{A}\xi} X\ [\mathcal{L}_{n+1}] \quad \textit{iff} \quad \Psi \vdash_{\mathbf{A}\xi} X\ [\mathcal{L}_{n+1}].$$

In establishing (xvi) it proves more convenient to work within the formalism $\mathcal{L}^{+}_{n+1}$ than within $\mathcal{L}_{n+1}$. We recall that, by the definition of a $\mathcal{Q}$-system, there are predicates $A, B \in \mathbf{\Pi}$ for which

(1) $\mathbf{A}\xi \vdash^{+}_{n+1} \boldsymbol{Q}_{AB}$;

we choose two such predicates and consider them as fixed. We then prove

(2) for every $X \in \mathbf{\Phi}_n$ with canonical sequence $\langle \boldsymbol{v}_{k_0}, \ldots, \boldsymbol{v}_{k_{m-1}} \rangle$ we have

$$\boldsymbol{Q}_{AB} \vdash^{+}_{n+1} [\boldsymbol{v}_n \boldsymbol{P}_{k_0} \boldsymbol{v}_{k_0} \wedge \cdots \wedge \boldsymbol{v}_n \boldsymbol{P}_{k_{m-1}} \boldsymbol{v}_{k_{m-1}} \rightarrow (X \leftrightarrow \boldsymbol{v}_n \boldsymbol{M}_{AB} X \boldsymbol{v}_n)]$$

(where the $\boldsymbol{P}_{k_i}$ are the predicates defined in 4.4(i) in terms of the fixed A, B, and $\boldsymbol{M}_{AB}$ is the mapping introduced in 4.4(vii)). (2) is established by induction on the structure of the formula X. In several places the argument can be simplified by using Theorem (xv), or rather its $\mathcal{L}^{+}_3$ version. The argument is not quite easy in the case when X is of the form $\forall_x Y$. From (2) we immediately derive

(3) for every $X \in \Sigma_n$ we have $\boldsymbol{Q}_{AB} \vdash^{+}_{n+1} X \leftrightarrow \forall_{\boldsymbol{v}_n} (\boldsymbol{v}_n \boldsymbol{M}_{AB} X \boldsymbol{v}_n)$.

We now turn to the conclusion of our theorem. Our task is to show that the conditions

(4) $\Psi \models_{\mathbf{A}\xi} X \; [\mathcal{L}^+_{n+1}]$

and

(5) $\Psi \vdash_{\mathbf{A}\xi} X \; [\mathcal{L}^+_{n+1}]$

are equivalent. That (5) implies (4) is obvious. To obtain the implication in the opposite direction, assume that (4) holds. By 4.4(xxxi) and the semantic completeness theorem for $\mathcal{L}^+$ we infer

(6) $\{\mathbf{M}_{AB}Y = \mathbf{1} \colon Y \in \Psi \cup \mathbf{A}\xi\} \vdash^{\times}_{\mathbf{Q}_{AB}} \mathbf{M}_{AB}X = \mathbf{1}.$

$\mathcal{L}^\times$ being a subformalism of $\mathcal{L}^+_3$ by 3.9(i),(ii), we may replace $\vdash^\times$ by $\vdash^+_3$ in (6); since, moreover, by (xv) in its $\mathcal{L}^+_3$ version and 4.4(x) we have

$$\mathbf{Q}_{AB} \vdash^+_3 \mathbf{M}_{AB}Y = \mathbf{1} \leftrightarrow \mathbf{\forall}_{v_0}(v_0 \mathbf{M}_{AB} Y v_0)$$

for every $Y \in \Sigma^+$, we conclude

(7) $\{\mathbf{\forall}_{v_0}(v_0\mathbf{M}_{AB}Yv_0) \colon Y \in \Psi \cup \mathbf{A}\xi\} \cup \{\mathbf{Q}_{AB}\} \vdash^+_3 \mathbf{\forall}_{v_0}(v_0\mathbf{M}_{AB}Xv_0).$

Combining (3) and (7) we see that

$$\Psi \cup \mathbf{A}\xi \cup \{\mathbf{Q}_{AB}\} \vdash^+_{n+1} X.$$

Hence, in view of (1), we obtain (5). Thus the conclusion of our theorem has been established for the formalism $\mathcal{L}^+_{n+1}$. Since this formalism is equipollent with $\mathcal{L}_{n+1}$, the proof is complete.

CHAPTER 5

Some Improvements of the Equipollence Results

In this chapter we shall consider some problems regarding possible improvements of the main results established in our earlier discussion. The problems can be divided into two classes. Those of the first class concern the possibility of improving the contents of some of the results by means of which the relative equipollence of the formalisms $\mathcal{L}$, $\mathcal{L}^+$, $\mathcal{L}^\times$—and hence of three correlated set-theoretical systems $\mathcal{S}$, $\mathcal{S}^+$, $\mathcal{S}^\times$ developed in these formalisms—has been expressed and established (without changing the structure of the systems and formalisms involved in these results). The problems of the second class concern the possibility of improving the formalism $\mathcal{L}^\times$ (and hence also the corresponding system $\mathcal{S}^\times$), i.e., of replacing $\mathcal{L}^\times$ by an equipollent formalism with a still simpler structure.

5.1. One-one translation mappings

In §4.5 we have summarized the main results concerning the equipollence of the correlated systems $\mathcal{S}$, $\mathcal{S}^+$, $\mathcal{S}^\times$ formalized respectively in $\mathcal{L}$, $\mathcal{L}^+$, $\mathcal{L}^\times$. In particular, 4.5(v)(β) contains the results formulated in terms of two recursive functions G and K (called translation mappings) which respectively map the set of sentences Σ^+ into, and actually onto, its subsets Σ and $\Sigma^\times$.

The problems of the first class with which we deal here concern exclusively 4.5(v)(β). We are interested in the question whether the function G can be replaced by a function H mapping Σ^+ onto Σ in a one-one way and otherwise preserving the essential properties of G; we are also interested whether K can be analogously replaced by a one-one function L.

The use of the new functions does not seem to enhance the intuitive value of the results stated in 4.5(v) by making the systems involved in some sense "more equipollent" than they appear to be in the original formulation of the results. It seems to the authors, however, that the use of one-one functions would at any rate enhance the mathematical elegance of the results. Also it seems plausible that the reformulated results may prove helpful in a study of further problems concerning relationships between the systems and formalisms involved.

The solution of the problem formulated above proves to be affirmative.

(i) *Let* $\mathcal{S}$ (*with the axiom set* $\mathbf{A\xi}$), $\mathcal{S}^+$ (*with the same axiom set*), *and* $\mathcal{S}^\times$ (*with the axiom set* $\mathbf{A\xi}^\times$) *be three correlated* $\mathfrak{Q}$*-systems formalized respectively in* $\mathcal{L}$, $\mathcal{L}^+$, *and* $\mathcal{L}^\times$. *Then there are recursive functions* H *and* L *with the following properties.*

(α) H *maps* Σ^+ *onto* Σ *in a one-one way; similarly for* L, Σ^+, *and* $\Sigma^\times$.

(β) $HX \equiv_{\mathbf{A\xi}} X$ *for every* $X \in \Sigma$; $LX \equiv^\times_{\mathbf{A\xi}^\times} X$ *for every* $X \in \Sigma^\times$.

(γ) $HX \equiv^+_{\mathbf{A\xi}} X \equiv^+_{\mathbf{A\xi}} LX$ *for every* $X \in \Sigma^+$.

(δ) *For every* $\Psi \subseteq \Sigma^+$ *and* $X \in \Sigma^+$ *the following three conditions are equivalent*:

$$\Psi \vdash^+_{\mathbf{A\xi}} X, \quad H^*\Psi \vdash_{\mathbf{A\xi}} HX, \quad \textit{and} \quad L^*\Psi \vdash^\times_{\mathbf{A\xi}^\times} LX.$$

The proofs for both functions, H and L, are entirely analogous. We shall outline the argument for L.

First, employing a device essentially due to Craig [**1953**], we introduce an auxiliary one-one function F which maps Σ^+ into $\Sigma^\times$. To this end notice that, the set Σ^+ being infinite and recursive, it is possible to arrange all sentences of Σ^+ in an infinite recursive sequence $\langle S_0, \ldots, S_n, \ldots\rangle$ without repeating terms. By 4.5(v)(β1), for any given $X \in \Sigma^+$ we have $KX \in \Sigma^\times$; thus KX is always an equation both sides of which, $(KX)^\ell$ and $(KX)^r$, are predicates. Keeping these observations in mind, we define

(1) $FX = [C_0 \mathbf{+} \cdots \mathbf{+} C_{n+1} \mathbf{=} (KX)^r]$, where n is such that $S_n = X$, and where $C_0 = \cdots = C_{n+1} = (KX)^\ell$.

From (1) and the recursiveness of K we easily derive the following properties of F:

(2) F is a recursive one-one function;

(3) F maps Σ^+ into $\Sigma^\times$;

(4) $FX \equiv^\times KX$ for every $X \in \Sigma^+$.

Using F, and following the lines of a well-known proof of the classical Cantor–Bernstein theorem, we now construct a one-one function L which maps the set Σ^+, not only into, but actually onto its subset $\Sigma^\times$ (compare, e.g., Sierpiński [**1958**], pp. 30–32). In fact, we set

(5) $\Phi = \bigcup_{n\in\omega}[(F^n)^*\Sigma^+ \sim (F^n)^*\Sigma^\times]$,

(6) $\Psi = \bigcup_{n\in\omega}[(F^{n+1})^*\Sigma^+ \sim (F^{n+1})^*\Sigma^\times]$,

(7) $\Theta = \bigcup_{n\in\omega}[(F^n)^*\Sigma^\times \sim (F^{n+1})^*\Sigma^+] \cup \bigcap_{n\in\omega}(F^n)^*\Sigma^\times$.

Hence we derive

(8) $\Phi \cup \Theta = \Sigma^+, \ \Psi \cup \Theta = \Sigma^\times,$

(9) $\Phi \cap \Theta = \varnothing = \Psi \cap \Theta,$

(10) $F^*\Phi = \Psi.$

From (2), (3), (8), (9), (10) it easily follows that there is a uniquely determined function L with the following properties:

(11) $LX = FX$ for every $X \in \Phi$; $LX = X$ for every $X \in \Theta$;

(12) $DoL = \Sigma^+$, $RnL = \Sigma^\times$, and L is one-one.

Thus L can be concisely defined by the formula

(13) $L = \Phi \rceil F \cup \Theta \rceil Id.$

Using (4), (8), (11), and 4.5(v)(β2), we obtain

(14) $LX \equiv^\times KX$ for every $X \in \Sigma^+$.

From (12) and (14) we conclude with the help of 4.5(v)(β2)–(β4) that L has all the properties stated in parts (α)–(δ) of our theorem.

It remains to be shown that L is recursive. From (1) it obviously follows that, for every $X \in \Sigma^+$, the sentence KX is shorter than FX. By 4.5(v)(β2) this implies that X itself is shorter than FX in case $X \in \Sigma^\times$. Hence, with the help of (2) and (3), we get

(15) $\bigcap_{n\in\omega}(F^n)^*\Sigma^\times = \varnothing;$

for if (15) did not hold, we could construct an infinite sequence of sentences in $\Sigma^\times$, $\langle X, F^{-1}X, \ldots, F^{-n}X, \ldots\rangle$, with strictly decreasing lengths. By (7) and (15) we have

(16) $\Theta = \bigcup_{n\in\omega}[(F^n)^*\Sigma^\times \sim (F^{n+1})^*\Sigma^+].$

An easy argument using (1)–(3), (5), (8), (9), (16), and based upon the intuitive notion of recursiveness, shows that the sets Φ and Θ are recursive. This argument can be loosely described as follows. By (1)–(3) there is a procedure which allows us, for any given sentence $X \in \Sigma^+$ and $n \in \omega$, to determine in a finite number of steps whether $X \in (F^n)^*\Sigma^+$ and whether $X \in (F^n)^*\Sigma^\times$; hence we can also determine whether X is in $(F^n)^*\Sigma^+ \sim (F^n)^*\Sigma^\times$ or in $(F^n)^*\Sigma^\times \sim (F^{n+1})^*\Sigma^+$. We now proceed stepwise and first decide whether $X \in (F^0)^*\Sigma^+ \sim (F^0)^*\Sigma^\times$. If so, then $X \in \Phi$ and $X \notin \Theta$ by (5) and (9), and the procedure stops. If not, we decide whether $X \in (F^0)^*\Sigma^\times \sim (F^1)^*\Sigma^+$. If so, then $X \in \Theta$ and $X \notin \Phi$ by (16) and (9), and, as before, the procedure stops. If not, we decide whether

$X \in (F^1)^*\Sigma^+ \sim (F^1)^*\Sigma^\times$, and so on. By (8), (5), and (16), X must be in just one of the sets $(F^n)^*\Sigma^+ \sim (F^n)^*\Sigma^\times$ or $(F^n)^*\Sigma^\times \sim (F^{n+1})^*\Sigma^+$ for some $n \in \omega$. Thus the procedure must stop in finitely many steps, and we arrive at the conclusion whether X is in Φ, but not in Θ, or *vice versa*.

Hence, by (2) and (13) we conclude that the function L is also recursive, and the proof is complete.

Notice that, for obvious reasons, 4.5(v)(β2) cannot be carried over to the one-one functions H and L. It has been replaced in our new result by a weaker condition, (β), which is still adequate for our purposes. For instance, (β) and (α) together make it possible to derive directly the proper equipollence theorems, 4.5(v)(γ1),(γ2), from the new main mapping theorem, (δ). (Compare 2.4(v).)

Just as in the case of 4.5(v)(β),(γ), those parts of (α)–(γ) above which concern the mapping H can be extended from the sets of sentences Σ and Σ^+ to the sets of formulas $\boldsymbol{\Phi}$ and $\boldsymbol{\Phi}^+$.

Using 4.5(v) we could express and assert some direct relations between the systems $\mathcal{S}$ and $\mathcal{S}^\times$ (disregarding the fact that they are subsystems of the same system $\mathcal{S}^+$). To this end, we would have to use both functions, G and K, with domains respectively restricted to $\Sigma^\times$ and Σ; moreover, some parts of 4.5(v) would have to be omitted, and in some other parts the equivalence relation $\equiv^+$ or $\equiv^+_{\mathbf{A}\xi}$ would have to be interpreted semantically. As we know, sentences X in Σ and Y in $\Sigma^\times$, when treated as sentences in Σ^+, are semantically equivalent (or semantically equivalent relative to $\mathbf{A}\xi$) provided $X \equiv^+ Y$ (or $X \equiv^+_{\mathbf{A}\xi} Y$). For $X \in \Sigma$ and $Y \in \Sigma^\times$ this definition can equivalently be replaced by one in which notions of $\mathcal{L}^+$ do not occur: X and Y are semantically equivalent (or semantically equivalent relative to $\mathbf{A}\xi$) iff every realization of $\mathcal{L}$ (or every model of $\mathbf{A}\xi$) that is a model of one of these two sentences is a model of the other. Similarly, if we replace X and Y by sets of sentences $\Psi \subseteq \Sigma$ and $\Delta \subseteq \Sigma^\times$ in this last statement, we arrive at a definition of the relation of semantical equivalence between sets of sentences in $\mathcal{L}$ and $\mathcal{L}^\times$ that does not involve $\mathcal{L}^+$. (In our earlier discussions we used the notion of semantical equivalence only as a relation between sentences, or sets of sentences, in one formalism.)

By means of (i) we can state the results mentioned in the previous paragraph in an improved form in which the two many-one functions are replaced by a single one-one function.

(ii) *Let $\mathcal{S}$ (with the axiom set $\mathbf{A}\xi$) and $\mathcal{S}^\times$ (with the axiom set $\mathbf{A}\xi^\times$) be two correlated $\mathcal{Q}$-systems respectively formalized in $\mathcal{L}$ and $\mathcal{L}^\times$.*

(α) *$\mathbf{A}\xi$ and $\mathbf{A}\xi^\times$ are semantically equivalent.*

(β) *There is a recursive function F with the following properties.*

(β1) *F maps Σ onto $\Sigma^\times$ in a one-one way.*

(β2) *For every $X \in \Sigma$, the sentences X and FX are semantically equivalent relative to $\mathbf{A}\xi$ (or $\mathbf{A}\xi^\times$).*

(β3) *For every* $\Psi \subseteq \Sigma$ *and* $X \in \Sigma$, *the conditions* $\Psi \vdash_{\mathbf{A}\xi} X$ *and* $F^*\Psi \vdash^\times_{\mathbf{A}\xi^\times} FX$ *are equivalent.*

This is a direct corollary of (i)(α),(γ),(δ), and 4.5(α); F is defined as the composition $L \circ H^{-1}$.

From an intuitive point of view, the relations between systems $\mathcal{S}$ and $\mathcal{S}^\times$ stated in (α) and (β) appear to be much weaker and less complete than those between $\mathcal{S}$ and $\mathcal{S}^+$, or $\mathcal{S}^\times$ and $\mathcal{S}^+$, described above and in §4.5. This applies especially to those portions of our results which are related to the problem of equipollence in means of proof. Conditions (β1) and (β2) in (ii) obviously imply the semantical equipollence of $\mathcal{S}$ and $\mathcal{S}^\times$ in means of expression, i.e., an analogue of 4.5(v)(γ1) (with $\mathbf{\Sigma}^+$ replaced by $\mathbf{\Sigma}^\times$, and with the relation $\equiv^+_{\mathbf{A}\xi}$ interpreted semantically). We see, however, no possibility of deriving from (α) and (β) any analogue of 4.5(v)(γ2), that is, of a theorem asserting directly the equipollence in means of proof; actually, we do not even know any intuitively plausible formulation of such an analogue.

Assume for a while that (β1) and (β3) in (ii) refer to systems $\mathcal{S}$ and $\mathcal{S}^\times$ in two arbitrary formalisms, $\mathcal{F}$ and $\mathcal{F}^\times$. To simplify the situation assume further that $\mathcal{S}$ and $\mathcal{S}^\times$ have empty axiom sets, so that the discussion concerns essentially the formalisms themselves; the relativization to $\mathbf{A}\xi$ and $\mathbf{A}\xi^\times$ in (β3) is then omitted. In view of the difficulties which were pointed out above, the idea may occur to use (β1) and (β3) as a definition of equipollence in means of proof. It turns out, however, that such a definition has some implications which are quite dubious from an intuitive point of view; in particular, it may happen that $\mathcal{F}$ is a subformalism of $\mathcal{F}^\times$ and is equipollent with $\mathcal{F}^\times$ in means of proof according to this definition, although $\mathcal{F}^\times$ has greater deductive power than $\mathcal{F}$ according to our stipulations in §2.4. This is one of the reasons why we have decided not to discuss in this work the equipollence of two formalisms in means of proof unless they are regarded as subformalisms of a third formalism given in advance (cf. §2.5). The same applies to the equipollence of systems.

5.2. Reducing the number of primitive notions of $\mathcal{L}^\times$: definitionally equivalent variants of $\mathcal{L}^\times$

In turn we take up the problem of simplifying the formalism $\mathcal{L}^\times$ and systems $\mathcal{S}^\times$ constructed in it. The simplifications we have in mind consist in reducing the number of logical constants.

Consider the binary operation O defined for arbitrary relations R and S by the formula

$$O(R,S) \ = \ {\sim} R \cup ({\sim} S)^{-1}.$$

We shall see that an operator denoting this operation, say $\dagger$, can replace the three operators $+$, $^-$, and $^\smile$.

Let $\overline{\mathcal{L}}^\times$ be the formalism obtained from $\mathcal{L}^\times$ by including the operator $\dagger$ in the vocabulary and by adding the following axiom schema (I) to the set of schemata

(BI)–(BX) in §3.1 defining the logical axioms of $\mathcal{L}^{\times}$:

$$A \dagger B = A^{-} + B^{-\smile}. \tag{I}$$

The necessary changes in the definitions of predicates, sentences, and derivability are obvious; the variables occurring in (BI)–(BX) and (I) are assumed to range over arbitrary predicates of $\overline{\mathcal{L}}^{\times}$.

Since (I) is a possible definition (or rather definition schema) of $\dagger$ in $\mathcal{L}^{\times}$, the formalism $\overline{\mathcal{L}}^{\times}$ is a definitional extension of $\mathcal{L}^{\times}$.

To establish the equipollence of $\mathcal{L}^{\times}$ and $\overline{\mathcal{L}}^{\times}$ in means of expression and proof, we follow strictly the lines of the argument in §2.3. The translation mapping G which we use in this context is defined by a simple recursion over all predicates and sentences in $\overline{\mathcal{L}}^{\times}$; when applied to any such predicate or sentence, G eliminates from it the operator $\dagger$ on the basis of (I).

We easily show that, for any given $A, B \in \Pi[\overline{\mathcal{L}}^{\times}]$, the following equations are logically provable in $\overline{\mathcal{L}}^{\times}$:

$$A + B = [A \dagger (A \dagger (A \dagger A))] \dagger [(B \dagger (B \dagger B)) \dagger B], \tag{II}$$

$$A^{-} = A \dagger (A \dagger (A \dagger A)), \tag{III}$$

$$A^{\smile} = (A \dagger A) \dagger (A \dagger (A \dagger A)). \tag{IV}$$

On the basis of (II)–(IV) we can transform in an obvious way the schemata (BI)–(BX) and (I) by eliminating from them the operators $+$, $^{-}$, and $^{\smile}$. Let (BI′)–(BX′) and (I′) be the schemata thus transformed. Assume that we replace the original set of logical axiom schemata for $\overline{\mathcal{L}}^{\times}$ by the set consisting of (BI′)–(BX′), (I′), and (II)–(IV); it is readily seen that this replacement does not affect the notion of derivability in $\overline{\mathcal{L}}^{\times}$, so that the formalism remains essentially unchanged.

Finally, we construct a subformalism $\mathcal{L}_{\mathrm{a}}^{\times}$ of $\overline{\mathcal{L}}^{\times}$ by deleting $+$, $^{-}$, $^{\smile}$ from the vocabulary and by defining the set of logical axioms as the set of all instances of (BI′)–(BX′) and (I′) (with variables ranging over arbitrary predicates in $\mathcal{L}_{\mathrm{a}}^{\times}$). Since (II)–(IV) are possible definitions of the operators $+$, $^{-}$, $^{\smile}$ in $\mathcal{L}_{\mathrm{a}}^{\times}$, the formalism $\overline{\mathcal{L}}^{\times}$ proves to be a definitional extension of $\mathcal{L}_{\mathrm{a}}^{\times}$ as well. Thus $\mathcal{L}^{\times}$ and $\mathcal{L}_{\mathrm{a}}^{\times}$ have a common definitional extension and are therefore definitionally equivalent.

As a consequence, all three formalisms, $\mathcal{L}^{\times}$, $\overline{\mathcal{L}}^{\times}$, $\mathcal{L}_{\mathrm{a}}^{\times}$, are equipollent in means of expression and proof. Hence, in particular, for every system $\mathcal{S}^{\times}$ in $\mathcal{L}^{\times}$ we can construct an equipollent system $\mathcal{S}_{\mathrm{a}}^{\times}$ in $\mathcal{L}_{\mathrm{a}}^{\times}$; no special assumptions concerning the axiom set of $\mathcal{S}^{\times}$ are needed. $\mathcal{L}_{\mathrm{a}}^{\times}$ is simpler than $\mathcal{L}^{\times}$ in that its vocabulary is smaller: $\mathcal{L}^{\times}$ has six and $\mathcal{L}_{\mathrm{a}}^{\times}$ only four logical constants.

Similar simplifications of $\mathcal{L}^{\times}$ can be achieved in several other ways. For instance, we can replace the operators $+$ and $^{-}$ by a binary operator $\|$ (the *exclusion symbol*, synonymous with Sheffer's stroke), and the operators $\odot$ and

$^{\smile}$ by another binary operator, $\otimes$. Instead of (I) we use two definition schemata,

(V) $$A \| B = A^{-} + B^{-},$$

(VI) $$A \otimes B = A \odot B^{\smile},$$

and, instead of (II)–(IV), four such schemata,

(VII) $$A + B = (A \| A) \| (B \| B),$$

(VIII) $$A^{-} = A \| A,$$

(IX) $$A \odot B = A \otimes (\mathring{1} \otimes B),$$

(X) $$A^{\smile} = \mathring{1} \otimes A.$$

The new binary operators in the resulting simplified formalism $\mathcal{L}_{\mathsf{b}}^{\times}$ have an intuitively clearer and more natural meaning than the (highly artificial) operator $\dagger$ in $\mathcal{L}_{\mathsf{a}}^{\times}$. Also, the set of logical axiom schemata for $\mathcal{L}_{\mathsf{b}}^{\times}$ obtained by the elimination method is simpler than the corresponding set for $\mathcal{L}_{\mathsf{a}}^{\times}$, and can be further simplified by means of special devices. A certain theoretical advantage of $\mathcal{L}_{\mathsf{a}}^{\times}$ consists in that, in opposition to $\mathcal{L}_{\mathsf{b}}^{\times}$, the symbol $\mathring{1}$ is not involved in any definition schemata used in the construction of $\mathcal{L}_{\mathsf{a}}^{\times}$. This detail will play some role in our further discussion (cf. §5.3).

The problem arises whether a formalism could be constructed which would be definitionally equivalent with $\mathcal{L}^{\times}$ in exactly the same sense as $\mathcal{L}_{\mathsf{a}}^{\times}$ and $\mathcal{L}_{\mathsf{b}}^{\times}$ are, and in which a single binary operator would replace all the logical operators of $\mathcal{L}^{\times}$, perhaps even including the predicate (operator of rank 0) $\mathring{1}$. The problem is open; an affirmative solution does not seem to be likely.[1]

5.3. Eliminating the symbol $\mathring{1}$ as a primitive notion from systems of set theory in $\mathcal{L}^{\times}$

We shall now show that, for the purpose of equipollent formalization of all the set-theoretical systems in which we are interested here, we can use formalisms obtained from $\mathcal{L}^{\times}$ and its variants such as $\mathcal{L}_{\mathsf{a}}^{\times}$ or $\mathcal{L}_{\mathsf{b}}^{\times}$ by deleting $\mathring{1}$ from the

[1] *Quite recently, a positive solution to this problem, presented in terms of relation algebras, has been given in Börner [**1986**]. To describe this solution in terms of the formalism $\mathcal{L}^{\times}$, let $\diamond$ be a new binary operation symbol. We define $\overline{\mathcal{L}}^{\times}$ to be the formalism obtained from $\mathcal{L}^{\times}$ by adding $\diamond$ as a new operator, and by taking as axioms all instances of the schemata (BI)–(BX), with $A, B, C \in \Pi[\overline{\mathcal{L}}^{\times}]$, and all instances of the following definition schema for $\diamond$:

$$A \diamond B = (A+B)^{-} + [A^{\smile} \cdot B^{\smile -} \cdot \mathring{0} \odot (A \cdot \mathring{1} + B \cdot \mathring{1})] \odot [A^{\smile} \cdot B^{\smile -} \cdot \mathring{0} \cdot [1 \odot (A \cdot B \cdot \mathring{1}) \odot 1] + (A^{\smile -} \cdot B^{\smile -} \cdot \mathring{0} \cdot [1 \odot (B \cdot \mathring{1}) \odot 1] + \mathring{1} \cdot [1 \odot (B \cdot \mathring{1}) \odot 1]^{-}) \cdot [1 \odot (A \cdot B \cdot \mathring{1}) \odot 1]^{-}],$$

where $A, B \in \Pi[\overline{\mathcal{L}}^{\times}]$. Clearly, $\overline{\mathcal{L}}^{\times}$ is a definitional extension of $\mathcal{L}^{\times}$. Moreover, there are possible definition schemata for $+$, $^{-}$, $\odot$, $^{\smile}$, and $\mathring{1}$ in terms of $\diamond$, all instances of which are provable in $\overline{\mathcal{L}}^{\times}$. For example, let Z be the predicate $[\boldsymbol{E} \diamond (\boldsymbol{E} \diamond \boldsymbol{E})] \diamond [(\boldsymbol{E} \diamond (\boldsymbol{E} \diamond \boldsymbol{E})) \diamond (\boldsymbol{E} \diamond (\boldsymbol{E} \diamond \boldsymbol{E}))]$. Then the following equations are provable in $\overline{\mathcal{L}}^{\times}$:

$$A^{-} = A \diamond A \qquad \text{for every } A \in \Pi[\overline{\mathcal{L}}^{\times}],$$
$$\mathring{1} = [(Z \diamond Z) \diamond Z] \diamond [(Z \diamond Z) \diamond Z].$$

vocabularies. This result, however, does not have a general metalogical character since it depends on a specific property of the systems involved, namely on the fact that in each of them the logical predicate $\mathring{\mathbf{1}}$ is definable in terms of the nonlogical predicate $\boldsymbol{E}$ and the remaining logical constants.

Indeed, in every set-theoretical system $\mathcal{S}$ formalized in $\mathcal{L}$ which is known from the literature, the sentence

$$\forall_{xy}[x\mathring{\mathbf{1}}y \leftrightarrow \forall_z((z\boldsymbol{E}x \leftrightarrow z\boldsymbol{E}y) \wedge (x\boldsymbol{E}z \leftrightarrow y\boldsymbol{E}z))]$$

proves to be valid. Hence the equation V defined by the formula

$$\text{(i)} \qquad V = [\mathring{\mathbf{1}} = (\boldsymbol{E}^{-\smile} \oplus \boldsymbol{E}) \cdot (\boldsymbol{E}^{\smile} \oplus \boldsymbol{E}^{-}) \cdot (\boldsymbol{E}^{-} \oplus \boldsymbol{E}^{\smile}) \cdot (\boldsymbol{E} \oplus \boldsymbol{E}^{-\smile})]$$

is provable in the correlated system $\mathcal{S}^{\times}$ formalized in $\mathcal{L}^{\times}$. Obviously, V is a possible definition of $\mathring{\mathbf{1}}$ in terms of $\boldsymbol{E}$ and the logical constants in $\mathcal{L}^{\times}$ that are different from $\mathring{\mathbf{1}}$.

Actually, it turns out that V can always be replaced by some simpler sentences. In fact, if $\mathcal{S}$ is a system which excludes proper classes and in which the pair axiom $\boldsymbol{P}$ (or merely the singleton axiom) is valid, then the simple equation

$$\text{(ii)} \qquad V' = (\mathring{\mathbf{1}} = \boldsymbol{E}^{-} \oplus \boldsymbol{E}^{\smile})$$

proves to be valid in the correlated system $\mathcal{S}^{\times}$; this is independent of whether $\mathcal{S}$ admits individuals. If, on the other hand, $\mathcal{S}$ excludes individuals (in the first conception of this notion), and if the extensionality axiom (T_6 in §4.6) is valid in it, then the equation

$$\text{(iii)} \qquad V'' = [\mathring{\mathbf{1}} = (\boldsymbol{E}^{-\smile} \oplus \boldsymbol{E}) \cdot (\boldsymbol{E}^{\smile} \oplus \boldsymbol{E}^{-})]$$

is provable in $\mathcal{S}^{\times}$. Finally, if $\mathcal{S}$ admits both proper classes and individuals, and if both the restricted singleton axiom (T_1 in §4.6) and the restricted extensionality axiom (T_6' in §4.6) are valid in it, then we cannot use V' or V'', but we can still simplify V by omitting the factor $\boldsymbol{E}^{-} \oplus \boldsymbol{E}^{\smile}$ (or $\boldsymbol{E} \oplus \boldsymbol{E}^{-\smile}$). It can readily be shown that V is derivable in $\mathcal{L}^{\times}$ from each of the simpler equations just mentioned.

As regards the possible definition schemata for $+$, $^{\smile}$, and $\odot$, although the intuitive idea behind them is rather clear, the schemata themselves, when written without abbreviations, are not simple. In fact, the equation defining $A \odot B$ is so involved that the total number of occurrences in it of the predicates A and B, and of the symbols $\Diamond$ and $\boldsymbol{E}$, seems to be in the trillions.

Imitating our construction of the formalism $\mathcal{L}_{\mathrm{a}}^{\times}$, one can now construct the formalism $\mathcal{L}_{\mathrm{d}}^{\times}$ with $\Diamond$ and $=$ as the only logical symbols (and $\boldsymbol{E}$ as the only nonlogical symbol) such that $\overline{\mathcal{L}}^{\times}$ is a definitional extension of $\mathcal{L}_{\mathrm{d}}^{\times}$. Therefore, we get that $\mathcal{L}^{\times}$ and $\mathcal{L}_{\mathrm{d}}^{\times}$ are definitionally equivalent.

It should be mentioned that the paper Andréka–Comer–Németi [**1985**] contains various improvements of the results in this section which predate Börner's work. For example, it follows at once from their results that in $\mathcal{L}^{\times}$ the operators $\odot$ and $^{\smile}$, together with the predicate $\mathring{\mathbf{1}}$, can be replaced by a single binary operator; moreover, all of the operators of $\mathcal{L}^{\times}$, together with $\mathring{\mathbf{1}}$, can be replaced by a single ternary operator.

Börner's ingenious definition of a binary operator that can replace all the operators of $\mathcal{L}^{\times}$ (including $\mathring{\mathbf{1}}$) has a rather *ad hoc* character. It would be quite interesting to find a binary operator with a more natural and intuitive meaning which would serve the same purpose.

Let $\mathcal{S}^\times$ be any of the familiar set-theoretical systems formalized in $\mathcal{L}^\times$, and let $\mathbf{A}\boldsymbol{\xi}^\times$ be the set of its nonlogical axioms. We assume that $\mathring{\mathbf{i}}$ is definable in terms of other constants occurring in $\mathcal{S}^\times$ and that, specifically,

$$\textbf{(iv)} \qquad \mathbf{A}\boldsymbol{\xi}^\times \vdash^\times V.$$

Under this assumption it is a routine matter to construct a system $\mathcal{S}^\frown$ which is equipollent with $\mathcal{S}^\times$ and is formalized in a language $\mathcal{L}^\frown$ obtained from $\mathcal{L}^\times$ by deleting $\mathring{\mathbf{i}}$ from the vocabulary (and changing the definitions of predicates and sentences accordingly). With any given sentence $X \in \Sigma^\times$ we correlate the sentence $FX \in \Sigma^\frown$ which we obtain by replacing each occurrence of $\mathring{\mathbf{i}}$ in X by V^τ. The set $\boldsymbol{\Lambda}^\frown$ of logical axioms of $\mathcal{L}^\frown$ is defined by the formula

$$\boldsymbol{\Lambda}^\frown = F^*\boldsymbol{\Lambda}^\times.$$

This amounts to saying that $\boldsymbol{\Lambda}^\frown$ is determined by the same schemata (BI)–(BX) as $\boldsymbol{\Lambda}^\times$, with the exception of (BVI), which is replaced by

$$\text{(I)} \qquad A \odot V^\tau = A,$$

and with the restriction of the range of variables occurring in these schemata to $\boldsymbol{\Pi}^\frown$. The definition of derivability remains virtually unchanged.

System $\mathcal{S}^\frown$ is determined by the stipulation

$$\textbf{(v)} \qquad \mathbf{A}\boldsymbol{\xi}^\frown = F^*\mathbf{A}\boldsymbol{\xi}^\times.$$

Formulas (iv) and (v) clearly imply that $\mathcal{S}^\frown$ is a subsystem of $\mathcal{S}^\times$. $\mathcal{S}^\frown$ can easily be shown to be equipollent with $\mathcal{S}^\times$ in means of expression and proof. The equipollence can be described and established with the help of the mapping F in exactly the same way in which it was described and established in 4.5(v)(β),(γ) for $\mathcal{S}^+$ and $\mathcal{S}$ with the help of G, or for $\mathcal{S}^+$ and $\mathcal{S}^\times$ with the help of K. However, 4.5(v)(α) has to be replaced by a weaker condition,

$$\mathbf{A}\boldsymbol{\xi}^\times \equiv^\times_V \mathbf{A}\boldsymbol{\xi}^\frown.$$

There are some dubious points in the construction just outlined, which may readily escape the attention of the reader. As was pointed out in §1.6, we wish to deal in this work with formalisms which are semantically sound. Hence the problem naturally arises whether the formalism $\mathcal{L}^\frown$ described above is indeed semantically sound. Since $\mathcal{L}^\times$ is known to be so, the answer to the question will prove to be affirmative if we manage to solve affirmatively the problem of whether $\mathcal{L}^\frown$ is a subformalism of $\mathcal{L}^\times$.

It is easy to realize that the latter problem (which presents some interest in its own right) reduces in turn to a much more special problem: is it true that all the logical axioms of $\mathcal{L}^\frown$ which are instances of Schema (I) are logically provable in $\mathcal{L}^\times$? In connection with this special problem notice first that Schema (I) can be equivalently replaced by two schemata,

$$\text{(I}'\text{)} \quad A \leq A \odot V^\tau \qquad \text{and} \qquad \text{(I}''\text{)} \quad A \odot V^\tau \leq A.$$

With regard to (I′) the solution of the special problem is affirmative. Indeed, we have

$$\vdash^{\times} \mathring{\mathbf{1}} \lesssim V^{r}$$

by (i), 3.2(xi),(i), and BA; hence, by applying 3.2(vii),(ix), we get

$$\vdash^{\times} A \lesssim A \odot V^{r}$$

for every $A \in \mathbf{\Pi}$.

As regards (I″) the situation is more involved. Assume first (as we implicitly do in the prevailing part of this discussion) that $\mathcal{L}^{\times}$ and hence also $\mathcal{L}^{\frown}$ contain $\boldsymbol{E}$ as the only nonlogical constant. In this case the solution of our problem is still affirmative. We prove that

$$\vdash^{\times} A \odot V^{r} \lesssim A$$

for every $A \in \mathbf{\Pi}^{\frown}$ by an easy induction on predicates, using various parts of §3.2. Thus, in the case discussed, $\mathcal{L}^{\frown}$ is a subformalism of $\mathcal{L}^{\times}$ and is therefore semantically sound.

The situation changes if the vocabularies of $\mathcal{L}^{\times}$ and $\mathcal{L}^{\frown}$ contain, in addition to $\boldsymbol{E}$, some other (binary) predicates as nonlogical constants. Assume, for example, that there is just one such predicate, say the predicate $\boldsymbol{I}$ (used in the formalization of set-theoretical systems with individuals). In this case there are many instances of (I″) which are not logically provable in $\mathcal{L}^{\times}$ and are not even logically true; the simplest example is $\boldsymbol{I} \odot V^{r} \lesssim \boldsymbol{I}$. Thus, formalism $\mathcal{L}^{\frown}$ now proves to be semantically unsound.

Two methods of correcting the situation by modifying the construction of $\mathcal{L}^{\frown}$ and $\mathcal{S}^{\frown}$ are available. The first consists in replacing (I) by (I′) on the list of logical axiom schemata and in securing in some other way that all the instances of (I) be, while not necessarily logically provable, at least provable in the system $\mathcal{S}^{\frown}$ (for otherwise we could not establish the equipollence of $\mathcal{S}^{\times}$ and $\mathcal{S}^{\frown}$). Fortunately, this can be done by including in $\mathbf{A}\boldsymbol{\xi}^{\frown}$ just two simple instances of (I″), namely

$$\boldsymbol{I} \odot V^{r} \lesssim \boldsymbol{I} \quad \text{and} \quad \boldsymbol{I}^{\smile} \odot V^{r} \lesssim \boldsymbol{I}^{\smile}.$$

By induction on predicates in $\mathcal{L}^{\frown}$ we can then show that all instances of (I″) are indeed derivable from the set $\mathbf{A}\boldsymbol{\xi}^{\frown}$ thus extended; this fully suffices for the proof of equipollence of $\mathcal{S}^{\times}$ and $\mathcal{S}^{\frown}$ outlined above. Notice that, if the axiom set of the original system $\mathcal{S}^{\frown}$ is finite, the same applies to the modified system.

The second method consists in modifying the equation V which served as a base for the construction of $\mathcal{L}^{\frown}$. In fact, we replace V by the equation

$$\textbf{(vi)} \qquad \overline{V} = [\mathring{\mathbf{1}} = V^{r} \cdot (\boldsymbol{I}^{-\smile} \oplus \boldsymbol{I}) \cdot (\boldsymbol{I}^{\smile} \oplus \boldsymbol{I}^{-}) \cdot (\boldsymbol{I}^{-} \oplus \boldsymbol{I}^{\smile}) \cdot (\boldsymbol{I} \oplus \boldsymbol{I}^{-\smile})].$$

Obviously, $\overline{V}$ (just as V) is a possible definition of $\mathring{\mathbf{1}}$ in the language $\mathcal{L}^{\frown}$. Using (vi), (i), 3.2(xi),(i), and BA we show that $V \vdash^{\times} \overline{V}$. Hence $\overline{V}$ is valid in every system in which V is valid; if, in particular, $\mathcal{S}$ is one of the known set-theoretical systems which are formalized in $\mathcal{L}$ and admit individuals, then $\overline{V}$ is provable in the correlated system $\mathcal{S}^{\times}$.

If now we follow strictly the lines of arguments in the earlier parts of this section, replacing V by $\overline{V}$, we readily convince ourselves that $\mathcal{S}^\times$ is equipollent with the modified system $\mathcal{S}^\frown$, and that the modified formalism $\mathcal{L}^\frown$ is a subformalism of $\mathcal{L}^\times$ and hence is semantically sound.

With obvious changes all the observations above extend to the case when the vocabulary of $\mathcal{L}^\times$ contains any finite number of nonlogical predicates.

We mention in passing some results concerning the relationship between the formalisms $\mathcal{L}^\times$ and $\mathcal{L}^\frown$ themselves (and not between systems formalized in them). It is easily seen that $\mathcal{L}^\times$ and $\mathcal{L}^\frown$ are not equipollent in means of expression, either syntactically or even semantically. On the other hand, it is not difficult to show that they are equipollent in means of proof.

Our discussion can be extended from the formalism $\mathcal{L}^\times$ to its definitionally equivalent, and hence equipollent, variants such as the formalisms $\mathcal{L}^\times_\mathsf{a}$ and $\mathcal{L}^\times_\mathsf{b}$ described in §5.2. In the case of $\mathcal{L}^\times_\mathsf{a}$ the procedure is straightforward. With the equation V in $\mathcal{L}^\times_\mathsf{a}$ we correlate the equation V_a in $\mathcal{L}^\times_\mathsf{a}$ obtained by eliminating the operators $+$, $^-$, $^\smile$ on the basis of Schemata 5.2(II)–(IV). Unfortunately, V_a is much too long to be formulated here explicitly. (Even a much simplified form of V_a contains almost 3600 occurrences of $\boldsymbol{E}$, $\dagger$, and $\odot$.) Obviously, V_a is provable in a system $\mathcal{S}^\times_\mathsf{a}$ (in $\mathcal{L}^\times_\mathsf{a}$) iff V is provable in the correlated system $\mathcal{S}^\times$ (in $\mathcal{L}^\times$). It is clear that V and V_a are logically equivalent (in the common definitional extension $\overline{\mathcal{L}}^\times$ of $\mathcal{L}^\times$ and $\mathcal{L}^\times_\mathsf{a}$); moreover, V_a is a possible definition of $\mathring{\mathbf{1}}$ in terms of other constants (namely $\dagger$, $\odot$, and $\boldsymbol{E}$) occurring in $\mathcal{L}^\times_\mathsf{a}$. From $\mathcal{L}^\times_\mathsf{a}$ we construct the formalism $\mathcal{L}^\frown_\mathsf{a}$ in the same way in which $\mathcal{L}^\frown$ was constructed from $\mathcal{L}^\times$; we use V_a instead of V as a base for this construction. If $\mathcal{S}^\times_\mathsf{a}$ is any system in $\mathcal{L}^\times_\mathsf{a}$ in which V_a is provable, then the correlated system $\mathcal{S}^\frown_\mathsf{a}$ in $\mathcal{L}^\frown_\mathsf{a}$ proves to be equipollent with $\mathcal{S}^\times_\mathsf{a}$ in means of expression and proof.

If we apply the same procedure to the formalism $\mathcal{L}^\times_\mathsf{b}$, we come across a difficulty: the equation V_b obtained from V by eliminating the operators $+$, $^-$, $\odot$, $^\smile$ on the basis of Schemata 5.2(VII)–(X) is not a possible definition of $\mathring{\mathbf{1}}$, since this symbol occurs on the right sides of (IX) and (X); we do not know any possible definition of $\mathring{\mathbf{1}}$ in terms of $\boldsymbol{E}$, $\|$, and $\otimes$ which is logically equivalent with V (in the common definitional extension of $\mathcal{L}^\times$ and $\mathcal{L}^\times_\mathsf{b}$). The difficulty vanishes if we concentrate entirely on those systems $\mathcal{S}^\times$ in $\mathcal{L}^\times$ in which the equation V' is provable, and on the correlated systems $\mathcal{S}^\times_\mathsf{b}$ in $\mathcal{L}^\times_\mathsf{b}$. We can then use V' instead of V for constructing the formalism $\mathcal{L}^\frown$ and the corresponding systems $\mathcal{S}^\frown$ (although in this case the remarks concerning the problem of soundness undergo some modifications). As is easily seen from 5.2(V),(VI), the equation

$$\mathring{\mathbf{1}} = [\boldsymbol{E} \otimes (\boldsymbol{E} \| \boldsymbol{E})] \| [\boldsymbol{E} \otimes (\boldsymbol{E} \| \boldsymbol{E})]$$

in $\mathcal{L}^\times_\mathsf{b}$, which is obviously a possible definition of $\mathring{\mathbf{1}}$, is logically equivalent with V'. Hence this equation can be used for constructing the formalism $\mathcal{L}^\frown_\mathsf{b}$ and for correlating equipollent systems $\mathcal{S}^\frown_\mathsf{b}$ with those systems $\mathcal{S}^\times_\mathsf{b}$ on which we are concentrating.

We wish still to mention a formalism $\mathcal{L}_c^\times$ which differs from $\mathcal{L}_b^\times$ only in that the operator $\otimes$ occurs in it with a somewhat different (though closely related) meaning; more specifically, in constructing $\mathcal{L}_c^\times$, Schema 5.2(VI) is replaced by

$$A \otimes B = A^{\smile} \odot B,$$

and Schemata 5.2(IX),(X) are modified correspondingly. $\mathcal{L}_c^\times$ proves useful if we concentrate on those systems $\mathcal{S}^\times$ in which V'' is provable. We can again formulate in $\mathcal{L}_c^\times$ a relatively simple equation which is a possible definition of $\mathring{\mathbf{1}}$ and is logically equivalent with V'', and we can use it for constructing the formalism $\mathcal{L}_c^\frown$ and systems $\mathcal{S}_c^\frown$.

It should perhaps be pointed out that in languages like $\mathcal{L}^\frown$ from which $\mathring{\mathbf{1}}$ has been eliminated, the predicates $\mathbf{1}$ and $\mathbf{0}$ defined as in 2.1(iii) are, of course, not present. Since, however, predicates with the same meanings may be needed in the discussion, we can redefine $\mathbf{1}$ and $\mathbf{0}$ by setting

$$\mathbf{1} = \boldsymbol{E} + \boldsymbol{E}^-, \qquad \mathbf{0} = (\boldsymbol{E} + \boldsymbol{E}^-)^-.$$

5.4. Eliminating the symbol = as a primitive notion from $\mathcal{L}^\times$: the reduced formalism $\mathcal{L}_r^\times$

The last of the possible simplifications of $\mathcal{L}^\times$ (and related formalisms) which will be discussed here differs from those previously considered in that it has a heavier impact on various aspects of the formalism involved.

We know that any equation $A = B$ in $\mathcal{L}^\times$ is logically equivalent to an equation whose right side is $\mathbf{1}$, namely to $(A = B)^t = \mathbf{1}$, that is, to $A \cdot B + A^- \cdot B^- = \mathbf{1}$. Thus it is possible to replace $\mathcal{L}^\times$ by an equipollent subformalism in which the only admitted sentences are equations of the form $C = \mathbf{1}$ (where C is an arbitrary predicate). We notice further that in this formalism there is a one-one correspondence between predicates C and sentences $C = \mathbf{1}$. Hence the idea occurs to dispense with equations altogether, to delete the equality symbol = from the vocabulary, and to use each predicate of $\mathcal{L}^\times$ with a double meaning: as a predicate, i.e., the designation of a binary relation, and as a sentence stating that the relation designated by the predicate is the universal relation.

We shall refer to the formalism embodying this idea as the *reduced version* of $\mathcal{L}^\times$ and shall denote it by "$\mathcal{L}_r^\times$". When referring to $\mathcal{L}_r^\times$, the predicates of $\mathcal{L}^\times$ can be called *predicate-sentences*, although we shall sometimes call them simply predicates or sentences; either "$\boldsymbol{\Pi}$" or "$\Sigma_r^\times$" can be used to denote the set of all predicate-sentences.

The problem of defining for $\mathcal{L}_r^\times$ the basic semantical notions—those of (possible) realization, denotation, and truth—presents no difficulties if we keep in mind the remarks in §3.1 concerning the analogous problem for $\mathcal{L}^\times$. $\mathcal{L}_r^\times$ is clearly assumed to have the same realizations as $\mathcal{L}^\times$ (and hence as $\mathcal{L}$ and $\mathcal{L}^+$), namely structures of the form $\mathfrak{U} = \langle U, E\rangle$, where U is any nonempty set and E any binary relation on U. Given any such structure $\mathfrak{U}$ and any predicate-sentence A in $\mathcal{L}_r^\times$, we define by recursion which binary relation on U is denoted by A.

We then stipulate that A is a true sentence of $\mathfrak{U}$ iff it denotes in $\mathfrak{U}$ the universal relation $U \times U$. (We see no way of defining directly the notion of truth for $\mathcal{L}_r^\times$, since the truth or falsity of, say, the compound predicate-sentence $B \odot C$ is not determined by the truth or falsity of B and C.)

More involved is the task of selecting an appropriate set of logical axioms for $\mathcal{L}_r^\times$ and defining with its help the relation of derivability, $\vdash_r^\times$. We can obtain adequate definitions of these notions by translating step-by-step those given for $\mathcal{L}^\times$ in §3.1. However, an algebraic-type definition of derivability may seem very unnatural for a formalism which, like $\mathcal{L}_r^\times$, contains no equations. On the other hand, the deduction theorem for $\mathcal{L}^\times$, 3.3(ii), suggests the possibility of applying in the present case a different method: we can define derivability for $\mathcal{L}_r^\times$ in exactly the same way as this was done for $\mathcal{L}$, i.e., using the appropriately formulated *modus ponens* as the only direct rule of inference (cf. §1.3). The same method, by the way, could be employed to formulate a definition of $\vdash^\times$ equivalent to the one which was given in 3.1(iii).

To simplify the formulations we define a new operator on and to predicate-sentences, which in view of 3.3(ii) adequately replaces in $\mathcal{L}_r^\times$ the operation $\rightarrow$ employed in the usual formulation of *modus ponens* in $\mathcal{L}$. Since the symbol " $\rightarrow$ " has not been used in connection with $\mathcal{L}^\times$ and $\mathcal{L}_r^\times$, we can use it to denote the new operation. Indeed, for any $A, B \in \boldsymbol{\Pi}$ we set

(i) $$(A \rightarrow B) \;=\; \mathbf{1} \odot A^{-} \odot \mathbf{1} + B$$

(where "$\mathbf{1}$" is of course the abbreviation introduced in 2.1(iii)). It turns out to be convenient to use a symbol of a similar shape, " $\Rightarrow$ ", to denote a related operation on predicate-sentences of a purely Boolean-algebraic character; we set

(ii) $$(A \Rightarrow B) \;=\; A^{-} + B.$$

To grasp the difference between the meanings of $\rightarrow$ and $\Rightarrow$, notice that the sentences $A \rightarrow B$ and $A \Rightarrow B$ in $\mathcal{L}_r^\times$ are respectively equivalent to the sentences

$$\forall_{xy}(xAy) \rightarrow \forall_{xy}(xBy) \quad \text{and} \quad \forall_{xy}(xAy \rightarrow xBy)$$

in $\mathcal{L}^+$. We can refer to $\Rightarrow$ as the *strong* (or *absolute*) *implication*, in opposition to $\rightarrow$, the *weak* (or *relative*) *implication*.

We shall also use the operation $\Leftrightarrow$ defined by

(iii) $$(A \Leftrightarrow B) \;=\; (A \Rightarrow B)\cdot(B \Rightarrow A),$$

i.e.,

$$(A \Leftrightarrow B) \;=\; [(A^{-} + B)^{-} + (B^{-} + A)^{-}]^{-}.$$

As is easily seen,

(iv) $$(A = B) \;\equiv^\times\; (A\cdot B + A^{-}\cdot B^{-} = \mathbf{1}) \;\equiv^\times\; [(A \Leftrightarrow B) = \mathbf{1}].$$

Notice that by (iii) and 3.1(i) we also have

$$\vdash^\times A \Leftrightarrow B = (A = B)^t.$$

We describe the logical axioms of $\mathcal{L}_r^\times$ by specifying an appropriate set of axiom schemata. The set consists of 16 schemata, (I)–(XVI). To obtain the first ten of them we recall that each of the schemata for $\mathcal{L}^\times$, (BI)–(BX) in 3.1(ii), has the form of an equation, say "$F = G$", and we replace it by "$F \Leftrightarrow G$".

The next five schemata are substitutes for the direct rules of inference that are stated for $\mathcal{L}^\times$ in 3.1(iii)(δ),(ε). (We disregard 3.1(iii)(γ) and the two parts of 3.1(iii)(ε) which are superfluous according to a remark following 3.1(iii).) The schemata run as follows:

(XI) $(A \Leftrightarrow B) \rightarrow [(A \Leftrightarrow C) \rightarrow (B \Leftrightarrow C)]$,

(XII) $(A \Leftrightarrow B) \rightarrow (A + C \Leftrightarrow B + C)$,

(XIII) $(A \Leftrightarrow B) \rightarrow (A^- \Leftrightarrow B^-)$,

(XIV) $(A \Leftrightarrow B) \rightarrow (A \odot C \Leftrightarrow B \odot C)$,

(XV) $(A \Leftrightarrow B) \rightarrow (A^\smile \Leftrightarrow B^\smile)$.

The last schema is

(XVI) $(A \Leftrightarrow \mathbf{1}) \rightarrow A$.

It seems that without the help of (XVI), given a logically provable sentence in $\mathcal{L}^\times$ of the form $A = \mathbf{1}$, we could not, in general, obtain the corresponding sentence in $\mathcal{L}_r^\times$ in the simple form A, but only in the more complicated form $A \Leftrightarrow \mathbf{1}$.

We now define $\vdash_r^\times$ by repeating almost literally the definition of $\vdash$ in 1.3(ii).

The set of logical axiom schemata just listed was especially designed to make the equipollence of $\mathcal{L}^\times$ and $\mathcal{L}_r^\times$ as obvious as possible. Treated, however, as a base for derivations in $\mathcal{L}_r^\times$, this set is unnecessarily complex; the involved structure of the schemata is concealed by the use of "$\Leftrightarrow$". The set can be simplified in various ways. We give here an improved version of the set of axiom schemata for $\mathcal{L}_r^\times$ in which the number of schemata is smaller and the structure of particular schemata is simpler. The definition of $\vdash_r^\times$ based upon the new set of schemata can be shown to be equivalent with the original definition; the proof is not quite trivial, but presents no appreciable difficulties and will not be given here.

The new set consists of thirteen schemata:

($\mathrm{I_r}$) $(A + A) \Rightarrow A$,

($\mathrm{II_r}$) $A \Rightarrow (A + B)$,

($\mathrm{III_r}$) $(B \Rightarrow C) \Rightarrow [(A + B) \Rightarrow (C + A)]$,

($\mathrm{IV_r}$) $[(A \odot B) \odot C] \Rightarrow [A \odot (B \odot C)]$,

($\mathrm{V_r}$) $[(A + B) \odot C] \Rightarrow (A \odot C + B \odot C)$,

($\mathrm{VI_r}$) $(A^\smile \odot B)^\smile \Rightarrow (B^\smile \odot A)$,

($\mathrm{VII_r}$) $[A^\smile \odot (A \odot B)^-] \Rightarrow B^-$,

($\mathrm{VIII_r}$) $(A \odot \mathring{\mathbf{1}}) \Rightarrow A$,

($\mathrm{IX_r}$) $A \Rightarrow (A \odot \mathring{\mathbf{1}})$,

$$(\mathrm{X_r}) \qquad (A \Rightarrow B) \to (A \to B),$$

$$(\mathrm{XI_r}) \qquad (A \Rightarrow B) \to [(A \odot C) \Rightarrow (B \odot C)],$$

$$(\mathrm{XII_r}) \qquad (A^{\smile} \Rightarrow B) \to (A \Rightarrow B^{\smile}),$$

$$(\mathrm{XIII_r}) \qquad (A \Rightarrow B^{\smile}) \to (A^{\smile} \Rightarrow B).$$

We wish to point out some easy consequences of the definition of $\vdash_r^\times$ based upon these schemata.

Using $(\mathrm{X_r})$ we conclude at once that, for every $A, B \in \mathbf{\Pi}$,

(v) $$A \Rightarrow B \vdash_r^\times A \to B,$$

and consequently

(vi) $$\{A, A \Rightarrow B\} \vdash_r^\times B.$$

It follows from (vi) that, when deriving a sentence in $\mathcal{L}_r^\times$ from a set of such sentences, we can use, in addition to the original rule of *modus ponens* formulated in terms of $\to$, also the derivative rule formulated in terms of $\Rightarrow$; we shall refer briefly to these two rules as *modus ponens for* $\to$ and *modus ponens for* $\Rightarrow$. However, this new rule could not adequately replace the original one in the definition of $\vdash_r^\times$. It can readily be seen that " $\to$ " cannot be replaced by " $\Rightarrow$ " in any of the schemata $(\mathrm{XI_r})$–$(\mathrm{XIII_r})$; in fact, many instances of $(\mathrm{XI_r})$–$(\mathrm{XIII_r})$ thus modified are not logically true. For the same reason we cannot interchange the expressions "$A \Rightarrow B$" and "$A \to B$" in $(\mathrm{X_r})$ or in (v). It seems plausible that there is no equivalent definition of $\vdash_r^\times$ which is based on a finite set of logical axiom schemata and in which *modus ponens* for $\Rightarrow$ is used as the only direct rule of inference.

We easily notice a close relationship between Schemata $(\mathrm{I_r})$–$(\mathrm{III_r})$ and a set of axioms for two-valued sentential logic which is stated in Church [**1956**], p. 137. This relationship is essentially the same as the one that was pointed out at the end of §3.2, concerning the set of schemata (BI)–(BIII) in 3.1(ii) and a set of postulates for Boolean algebras. As a consequence of this observation, all sentences in $\mathcal{L}_r^\times$ that correspond to true sentences of two-valued sentential logic (with the disjunction and negation symbols as the only connectives) not only are logically provable, but can be derived from $(\mathrm{I_r})$–$(\mathrm{III_r})$ using exclusively *modus ponens* for $\Rightarrow$.

Intuitively it seems obvious that the formalisms $\mathcal{L}^\times$ and $\mathcal{L}_r^\times$ are equipollent in means of expression and proof. However, neither of them is a subformalism of the other, if only for the reason that the sets $\mathbf{\Sigma}^\times$ and $\mathbf{\Sigma}_r^\times$ are disjoint. Hence, to establish the equipollence of the two formalisms, we have to treat them as subformalisms of a fixed third formalism.

A formalism $\mathcal{F}$ which is formally suitable for this purpose can be constructed from $\mathcal{L}^\times$ in a perfectly trivial way. We do not change either the vocabulary of $\mathcal{L}^\times$ or the definition of predicates, and we define the set of sentences in $\mathcal{F}$ simply as the union $\mathbf{\Sigma}^\times \cup \mathbf{\Pi}$ (i.e., $\mathbf{\Sigma}^\times \cup \mathbf{\Sigma}_r^\times$). The definition of derivability in

$\mathfrak{F}$ is obtained from that of $\vdash^{\times}$ by adding two clauses which secure, for every $A \in \mathbf{\Pi}$, the derivability of $A = \mathbf{1}$ from A and of A from $A = \mathbf{1}$. The proof that $\mathfrak{F}$ is equipollent with both $\mathcal{L}^{\times}$ and $\mathcal{L}_{\mathsf{r}}^{\times}$ is obvious.

The formalism $\mathfrak{F}$ thus described is an artificial structure which lacks intrinsic interest from either an intuitive or a formal point of view. We wish, therefore, to describe briefly another formalism, $\mathcal{L}_{\mathsf{s}}^{\times}$, which is also a common extension of the formalisms $\mathcal{L}^{\times}$ and $\mathcal{L}_{\mathsf{r}}^{\times}$ (and, for that matter, of $\mathfrak{F}$ as well) and which proves again to be equipollent with both of them. The construction of $\mathcal{L}_{\mathsf{s}}^{\times}$ is more interesting and natural than that of $\mathfrak{F}$; moreover, $\mathcal{L}_{\mathsf{s}}^{\times}$ will be involved in our later discussion.

Essentially, $\mathcal{L}_{\mathsf{s}}^{\times}$ is a formalism of the same type as $\mathcal{L}_{\mathsf{r}}^{\times}$; that is, every meaningful expression in $\mathcal{L}_{\mathsf{s}}^{\times}$ is treated as both a predicate and a sentence. When constructing $\mathcal{L}_{\mathsf{r}}^{\times}$ we eliminated $=$ from the vocabulary of $\mathcal{L}^{\times}$. To obtain $\mathcal{L}_{\mathsf{s}}^{\times}$ we reintroduce $=$ into the vocabulary of $\mathcal{L}_{\mathsf{r}}^{\times}$, treating it now as a binary operator which functions in exactly the same way as $+$ and $\odot$; thus, for any two predicate-sentences A, B in $\mathcal{L}_{\mathsf{s}}^{\times}$, $A = B$ is also a predicate-sentence. Hence, $\mathcal{L}^{\times}$ and $\mathcal{L}_{\mathsf{s}}^{\times}$ have the same vocabulary; however, $=$ can occur arbitrarily many times in a sentence of $\mathcal{L}_{\mathsf{s}}^{\times}$, whereas it occurs exactly once in each sentence of $\mathcal{L}^{\times}$. If A and B are regarded as predicates, i.e., designations of binary relations, then, semantically, $A = B$ is assumed to denote the same relation as $A \cdot B + A^{-} \cdot B^{-}$ or $A \Leftrightarrow B$ (cf. (iv)); thus, the binary operation on relations denoted by $=$ is the dual of the well-known operation of symmetric subtraction, and can be referred to as symmetric division. The notion of truth is defined for sentences of $\mathcal{L}_{\mathsf{s}}^{\times}$ exactly as it was defined for sentences of $\mathcal{L}_{\mathsf{r}}^{\times}$. Hence, it is easily seen that, in case A and B do not contain $=$, the expression $A = B$, when interpreted as a sentence, has the same meaning in $\mathcal{L}^{\times}$ and $\mathcal{L}_{\mathsf{s}}^{\times}$.

The definition of derivability for $\mathcal{L}_{\mathsf{s}}^{\times}$ differs from that for $\mathcal{L}_{\mathsf{r}}^{\times}$ in two points: the range of variables in the logical axiom schemata (thus, in (I)–(XVI) or else in ($\mathrm{I_r}$)–($\mathrm{XIII_r}$)) is extended to all predicate-sentences in $\mathcal{L}_{\mathsf{s}}^{\times}$, and the set of these schemata is supplemented by one new schema,

$$(\mathrm{I_s}) \qquad (A = B) \Leftrightarrow (A \Leftrightarrow B).$$

Schema ($\mathrm{I_s}$) can be regarded as a possible definition of $=$ in $\mathcal{L}_{\mathsf{r}}^{\times}$. If we wish, we can replace ($\mathrm{I_s}$) by the following three schemata:

$$(\mathrm{II_s}) \qquad (A = B) \Rightarrow (A \Rightarrow B),$$

$$(\mathrm{III_s}) \qquad (A = B) \Rightarrow (B \Rightarrow A),$$

$$(\mathrm{IV_s}) \qquad (A \Rightarrow B) \Rightarrow [(B \Rightarrow A) \Rightarrow (A = B)].$$

$\mathcal{L}^{\times}$ and $\mathcal{L}_{\mathsf{r}}^{\times}$ are easily shown to be subformalisms of $\mathcal{L}_{\mathsf{s}}^{\times}$. To prove the equipollence of $\mathcal{L}_{\mathsf{s}}^{\times}$ with $\mathcal{L}^{\times}$ and $\mathcal{L}_{\mathsf{r}}^{\times}$ we need, as always, two translation mappings: G from $\Sigma_{\mathsf{s}}^{\times}$ onto $\Sigma^{\times}$, and H from $\Sigma_{\mathsf{s}}^{\times}$ onto $\Sigma_{\mathsf{r}}^{\times}$. H is defined by recursion. Loosely speaking, to obtain HX from any given $X \in \Sigma_{\mathsf{s}}^{\times}$, we eliminate $=$ in X on the basis of ($\mathrm{I_s}$) (i.e., we replace step-by-step every equation $A = B$ which is a subsentence of X by $A \Longleftrightarrow B$). G can be defined by setting

$$GX = (HX = \mathbf{1}).$$

If, however, we wish G to be a translation mapping in the strict sense of 2.4(iii), we apply the above formula only in the case $X \in \Sigma_s^\times \sim \Sigma^\times$, and we set

$$GX = X \quad \text{for every} \quad X \in \Sigma^\times$$

(cf. 2.4(v)). With these mappings the equipollence proof is routine.

The following simple theorem exhibits a peculiar feature of $\mathcal{L}_s^\times$.

(vii) $\vdash_s^\times (A = \mathbf{1}) = A$ *for every* $A \in \Sigma_s^\times$.

The construction of a reduced formalism of the type $\mathcal{L}_r^\times$ can be based, of course, not only on the formalism $\mathcal{L}^\times$, but also (with some obvious changes) on various related formalisms described in §5.2 and §5.3. If, for instance, the construction is based on one of the formalisms $\mathcal{L}_a^\frown - \mathcal{L}_c^\frown$ discussed in §5.3, then the vocabulary of the resulting reduced formalisms $\mathcal{L}_{ar}^\frown - \mathcal{L}_{cr}^\frown$ will consist of two logical constants (namely two binary operators, $\dagger$ and $\odot$ in the case of $\mathcal{L}_{ar}^\frown$, and $\|$ and $\otimes$ in the case of $\mathcal{L}_{br}^\frown$, $\mathcal{L}_{cr}^\frown$) and in addition, for most purposes, of a single nonlogical constant, the predicate $\boldsymbol{E}$.

As can be seen from §5.3, the formalisms $\mathcal{L}_a^\frown$–$\mathcal{L}_c^\frown$ are adequate for the formalization either of almost all familiar systems of set theory (in the case of $\mathcal{L}_a^\frown$) or at least of comprehensive classes of such systems (in the case of $\mathcal{L}_b^\frown$, $\mathcal{L}_c^\frown$); the same obviously applies to the equipollent reduced formalisms $\mathcal{L}_{ar}^\frown$–$\mathcal{L}_{cr}^\frown$. It may seem surprising that a formal language with such a poor vocabulary proves adequate for this purpose. One should keep in mind, of course, that the number of symbols in the vocabulary is by no means the only yardstick by which the simplicity of a formalism can and ought to be measured.

The discussion of the reduced formalism $\mathcal{L}_r^\times$ (and its variants) will be completed by some observations concerning the peculiarity of this formalism and its relationship to similar features of certain formalisms known from the literature.

An essential feature of $\mathcal{L}_r^\times$ is the abolishment of any difference in the treatment of predicates (relation terms) and sentences (formulas). There is only one kind of meaningful compound expression in $\mathcal{L}_r^\times$, namely the predicate-sentences, and each such expression can be interpreted in a twofold way, as a predicate and as a sentence.

In this respect $\mathcal{L}_r^\times$ is similar to the formalism of what was referred to in §4.6 as the original, nonconventional version of Morse's system of set theory. In fact, the similarity is much closer than it might seem at first glance. Just as $\mathcal{L}^\times$ can serve as a formalism of very simple logical structure (without variables, with a finite vocabulary, etc.) in which the conventional version of Morse's system can be equipollently developed, $\mathcal{L}_r^\times$—and even more so $\mathcal{L}_s^\times$—can serve this purpose for the nonconventional version. We have given a brief outline of the proof that $\mathcal{L}^\times$ is equipollent (in means of expression and proof) with $\mathcal{L}_s^\times$ and $\mathcal{L}_r^\times$.

Following roughly the lines of this argument we can establish the equipollence of the two versions of Morse's system, and hence we can conclude that the nonconventional version is consistent if (and, trivially, only if) the conventional version is consistent. The relative consistency of these two versions of Morse's system was actually established in this way by Tarski and was announced without proof in Morse [**1965**], p. xxiii.

We should like to emphasize here that, in our belief, the lack of distinction between terms and formulas should not lead by itself to any confusion or errors. One could claim that the semantical ambiguity in $\mathcal{L}_r^\times$ or in Morse's system does not exceed similar ambiguities which are widely tolerated in other formal languages, and which come to light when these languages are translated into common English. To explain what we have in mind, we refer to the formalism $\mathcal{L}^+$. (If this formalism does not seem to be conventional enough, the remarks below can be referred instead to some conventional formalism of second-order logic.) Consider two formulas in this formalism, $\boldsymbol{xEy}$ and $\boldsymbol{E} = \boldsymbol{E}^{\smile}$. When translated into common English, the first formula reads "x is a member of y" (assuming that $\boldsymbol{x}, \boldsymbol{y}$ denote respectively the variables "x", "y"), while the second reads "the membership relation is identical with the converse of the membership relation". Thus, the symbol in $\mathcal{L}^+$ denoted by $\boldsymbol{E}$ is synonymous in the first case with the phrase "is a member of" and in the second with the phrase "the membership relation"; in the first case it functions as a verb, and in the second as a noun. We hope that at least some of the readers will agree with this trivial observation and that none of them will be bothered by it.

The idea of abolishing the distinctions between terms and formulas can be traced back, in an inceptive form, to some earlier work in mathematical logic. We recall a curious postulate which appears in Schröder [**1891**], p. 52 (see also Couturat [**1914**], p. 84, and Lewis [**1918**], p. 223):

(viii) $(A = 1) \;=\; A.$

Schröder seemed to believe that what is now called the two-valued logic could be treated as a specialized form of the theory of Boolean algebras, and that the specialization would simply consist in adding (viii) to an ordinary set of postulates for that theory.

For a contemporary logician the problem of clarifying the connection between Boolean algebras and sentential logic is not as simple as it seemed to Schröder. In Schröder's work the theory of Boolean algebras is presented as a nonformalized mathematical theory developed in the common language. In this presentation (viii) appears to be a meaningless expression which cannot influence in any way the development of the theory. The situation may change, however, if the theory of Boolean algebras is constructed in a formal language, with rigorous rules of inference and precise definitions of basic semantical notions. Imagine, for instance, that we are specifically interested in the equational theory of Boolean algebras. This theory can be clearly developed on the basis of equational logic, in a formalism closely related to $\mathcal{L}^\times$ (though differing from the latter in that its vocabulary is provided with variables); cf. Tarski [**1968**]. However, we can also use for this purpose an equipollent but structurally richer formalism which closely resembles the formalism $\mathcal{L}_s^\times$. In the new formalism the difference between terms and formulas vanishes. Expression (viii) becomes a meaningful sentence and actually a provable sentence in the theory of Boolean algebras; thus, adding it to the set of postulates does not specialize this theory, and actually does not affect it at all. (In this context it is instructive to recall (vii).) Finally, the new formal version of the theory of Boolean algebras admits an alternative interpretation in which it proves to coincide with the two-valued sentential logic. These facts vindicate to some extent Schröder's idea, but also show that the role assigned by him to (viii) is by no means justified.

5.5. Undecidable subsystems of sentential logic

The preceding remarks suggest investigating the structural relation between formal languages of type $\mathcal{L}_r^\times$ and those used for the development of sentential logic. To fix the ideas we concentrate on the formalism $\mathcal{L}_r^\times$ itself; we shall compare it with an appropriately designed formalism $\mathfrak{T}$ of sentential logic. Just as the vocabulary of $\mathcal{L}_r^\times$, that of $\mathfrak{T}$ has no variables. It contains two sentential constants (atomic sentences): the logical constant $\mathring{\mathbf{1}}$, functioning as a fixed true sentence (usually "$\boldsymbol{T}$" is used instead of "$\mathring{\mathbf{1}}$"), and the nonlogical constant $\boldsymbol{E}$, functioning as a fixed sentence whose meaning and truth value are not determined. In addition, the vocabulary of $\mathfrak{T}$ contains four sentential connectives, two binary—the disjunction symbol $\boldsymbol{+}$ and the conjunction symbol $\boldsymbol{\odot}$—and two unary—the negation symbol $^{-}$ and the affirmation symbol $^{\smile}$. While the connectives $\boldsymbol{+}$, $\boldsymbol{\odot}$, and $^{-}$ are interpreted in $\mathfrak{T}$ as synonyms of the usual connectives $\boldsymbol{\vee}$, $\boldsymbol{\wedge}$, and $\boldsymbol{\neg}$, the affirmation symbol $^{\smile}$ can be translated into common language by the phrase "it is the case that". (The affirmation symbol occurs rather infrequently in formalizations of sentential logic and is essentially superfluous.) All the meaningful compound expressions of $\mathfrak{T}$ are sentences; they are constructed in a familiar way from the atomic sentences $\mathring{\mathbf{1}}$ and $\boldsymbol{E}$ by means of sentential connectives. The metalogical abbreviations adopted for the discussion of $\mathcal{L}^\times$ and $\mathcal{L}_r^\times$ extend to that of $\mathfrak{T}$; this applies, in particular, to 5.4(i)–(iii).

Once the meaning of the sentential constant $\boldsymbol{E}$ has been fixed in some definite way, the intuitive meaning of any sentence in $\mathfrak{T}$ appears to be clear. It is, however, not quite obvious how the underlying semantical notions for $\mathfrak{T}$—those of realization and truth—are to be precisely defined. To achieve this we can follow, for instance, the method of Frege [**1879**] and assume that every sentence designates one of two special objects, truth T and falsehood F, jointly referred to as *truth values*. Thus $\mathring{\mathbf{1}}$ designates T, while the truth value designated by $\boldsymbol{E}$ is, in general, not determined. A possible realization of the formalism $\mathfrak{T}$ is any structure $\langle U, E\rangle$, where $U = \{T, F\}$ and E is either T or F; hence $\mathfrak{T}$ has just two realizations. The sentential connectives $\boldsymbol{+}$, $\boldsymbol{\odot}$, $^{-}$, and $^{\smile}$ are assumed to denote the (binary and unary) operations on and to truth values that are described in the familiar way by means of truth tables. On the basis of these descriptions we define recursively which sentences are true and which are false in a given realization $\langle U, E\rangle$. A sentence in $\mathcal{T}$ is logically true iff it holds in both realizations, i.e., if and only if it is true, independent of the truth value denoted by $\boldsymbol{E}$.

The notion of derivability for $\mathcal{T}$ is defined as usual in terms of logical axioms, using *modus ponens* for $\Rightarrow$ (not for $\rightarrow$) as the only direct rule of inference. The set of logical axioms is selected in such a way that the logically provable sentences coincide with the sentential tautologies (i.e., the logically true sentences in the sense of sentential logic). This set is described as the set of all instances of certain axiom schemata in which the variables range over arbitrary sentences of $\mathcal{T}$. In particular, following Church [**1956**], p. 137, we use ($\mathrm{I_r}$)–($\mathrm{III_r}$) in §5.4 as the schemata involving exclusively $\boldsymbol{+}$ and $^{-}$. In addition, some logical axioms are needed that involve essentially $\mathring{\mathbf{1}}$, $\boldsymbol{\odot}$, and $^{\smile}$. To this end we can choose, e.g., one axiom for $\mathring{\mathbf{1}}$,

$$\mathring{\mathbf{1}} \Leftrightarrow \boldsymbol{E}\boldsymbol{+}\boldsymbol{E}^{-},$$

and one axiom schema for each of $\boldsymbol{\odot}$ and $^{\smile}$, respectively

$$A\boldsymbol{\odot}B \Leftrightarrow A\cdot B \text{ (i.e., } A\boldsymbol{\odot}B \Leftrightarrow (A^{-}\boldsymbol{+}B^{-})^{-} \text{) and } A^{\smile} \Leftrightarrow A.$$

In view of their formal structure, all three can be regarded as possible definitions or definition schemata of $\mathring{\mathbf{1}}$, $\boldsymbol{\odot}$, and $^{\smile}$ in terms of the remaining constants of $\mathcal{T}$.

From the definitions of derivability and realization for $\mathcal{T}$ it can be concluded that the formalism $\mathcal{T}$ is semantically complete.

It is clear from the above discussion that the formalisms $\mathcal{L}_r^{\times}$ and $\mathcal{T}$ differ essentially in their semantical aspects. On the other hand, they are very closely related from a syntactical point of view. In fact the predicate-sentences of $\mathcal{L}_r^{\times}$ obviously coincide with the sentences of $\mathcal{T}$. As can easily be checked, all the logical axioms of $\mathcal{L}_r^{\times}$ specified in §5.4 are logically provable in $\mathcal{T}$. Furthermore, whenever sentences $A, A \rightarrow B \in \Sigma[\mathcal{T}]$ are derivable from a set $\Psi \subseteq \Sigma[\mathcal{T}]$ in $\mathcal{T}$, B is also derivable from Ψ in $\mathcal{T}$; in other words, *modus ponens* for $\rightarrow$ can be used in the derivations in $\mathcal{T}$ as a derivative rule of inference. As a consequence, $\mathcal{L}_r^{\times}$ proves to be, in its syntactical part, a subformalism of $\mathcal{T}$ and in fact a subformalism with the same vocabulary and the same set of sentences. Therefore every theory in $\mathcal{T}$ can be treated as a theory in $\mathcal{L}_r^{\times}$. This applies, in particular, to the logic of $\mathcal{T}$, $\boldsymbol{\Theta\eta}\varnothing\,[\mathcal{T}]$, i.e., to the ordinary (two-valued) sentential logic. By treating $\boldsymbol{\Theta\eta}\varnothing\,[\mathcal{T}]$ as a theory in $\mathcal{L}_r^{\times}$ we shall be able to speak of proper subtheories of it, and, more generally, of theories (in the same language) which are not extensions of it. (This is obviously impossible when $\boldsymbol{\Theta\eta}\varnothing\,[\mathcal{T}]$ is treated as a theory in $\mathcal{T}$.)

Since $\mathcal{L}_r^{\times}$ is a subformalism of $\mathcal{T}$, all sentences logically provable in $\mathcal{L}_r^{\times}$ are also logically provable in $\mathcal{T}$. The converse is, in general, not true, except for sentences constructed exclusively from $\boldsymbol{E}$, $\boldsymbol{+}$, and $^{-}$. For example, in contrast to their definability in $\mathcal{T}$, none of the symbols $\mathring{\mathbf{1}}$, $\odot$, and $\smile$ is definable in $\mathcal{L}_{\mathsf{r}}^{\times}$ in terms of the remaining symbols. Thus the logic of $\mathcal{L}_{\mathsf{r}}^{\times}$, $\boldsymbol{\Theta\eta}_{\mathsf{r}}^{\times}\varnothing$, is a proper subtheory of the ordinary sentential logic, $\boldsymbol{\Theta\eta}\varnothing\,[\mathcal{T}]$.

It turns out (but it is not obvious) that we can make $\mathcal{L}_{\mathsf{r}}^{\times}$ into a formalism identical with $\mathcal{T}$ in its syntactical part if we add just one rather simple axiom to the set $\boldsymbol{\Lambda}_{\mathsf{r}}^{\times}$ (while continuing to use *modus ponens* for $\rightarrow$ as the direct rule of inference). In fact, we can take

$$\mathring{\mathbf{1}}$$

as this additional axiom. (We assure the reader that there is no misprint or omission in the preceding line.) In other words, $\boldsymbol{\Theta\eta}\varnothing\,[\mathcal{T}] = \boldsymbol{\Theta\eta}_{\mathsf{r}}^{\times}\mathring{\mathbf{1}}$.

There are many other simple sentences and schemata which can be used in place of $\mathring{\mathbf{1}}$ to achieve the same effect. For instance, we can enrich the set of axiom schemata $(\mathrm{I_r})$–$(\mathrm{XIII_r})$ for $\mathcal{L}_{\mathsf{r}}^{\times}$ listed in §5.4 by the following schema, which in a sense is dual to $(\mathrm{X_r})$:

$$(A \rightarrow B) \Rightarrow (A \Rightarrow B).$$

In the presence of the set of axiom schemata thus enriched we can obviously use either of the two forms of *modus ponens* as the only direct rule of inference without affecting the notion of derivability. If, on the other hand, we decide to use *modus ponens* for $\Rightarrow$, then $(\mathrm{X_r})$ can be deleted. Notice, by the way, that, for any $A, B \in \boldsymbol{\Pi}$, we have

$$(A \rightarrow B) \;\equiv\; (A \Rightarrow B)\; [\mathcal{T}].$$

Thus, on the basis of the logic of $\mathcal{T}$, both forms of *modus ponens* are essentially equivalent.

The above observations may seem somewhat paradoxical. The formalism $\mathcal{T}$ of two-valued sentential logic is usually regarded as the simplest and most trivial logical formalism, with an almost empty mathematical content. Nevertheless, the formalism $\mathcal{L}_{\mathsf{r}}^{\times}$, so closely related to $\mathcal{T}$ in its syntactical part, presents an adequate basis for the development of set theory, which is, in a sense, the richest mathematical discipline; and even in its logical part $\mathcal{L}_{\mathsf{r}}^{\times}$ embodies an interesting and far from trivial mathematical theory, namely the equational theory of relation algebras (cf. the remarks in §3.2).

One conclusion emerges from our discussion: the connection between the formal structure of a language and its intended semantical interpretation is much looser than we might be inclined to believe.

It may be interesting to observe that the logic of $\mathcal{T}$ is well known to be decidable, while the logic of $\mathcal{L}_{\mathsf{r}}^{\times}$ is undecidable, in view of 4.7(vi) and the equipollence of $\mathcal{L}_{\mathsf{r}}^{\times}$ with $\mathcal{L}^{\times}$. Thus, we have obtained an example of an undecidable subtheory of the two-valued sentential logic (in fact a subtheory based upon a finite set of axiom schemata). Also, by 4.7(vii), this subtheory can be supplemented by means of finitely many axioms to form an essentially undecidable theory Θ; this will be a theory in the same formalism (i.e., $\mathcal{L}_r^\times$) as sentential logic, but clearly not a subtheory of that logic.[2,3]

As a base for these undecidability results related to sentential logic we could take, instead of $\mathcal{L}_r^\times$, some definitionally equivalent variant of it, $\mathcal{L}_r'$, with a different set of operation symbols (e.g., $\mathcal{L}_{ar}^\times$ or $\mathcal{L}_{br}^\times$). The corresponding formalism $\mathfrak{T}'$ of sentential logic could then assume a more conventional and natural form; for instance, the (essentially superfluous) affirmation symbol $\smile$ could be eliminated. The results discussed here can also be easily extended to formalisms of sentential logic whose vocabularies include arbitrarily many sentential constants.

[2]The result concerning the undecidable subtheory of sentential logic (but not concerning the essentially undecidable theory Θ) was announced in Tarski [**1953b**]. For other examples of undecidable subtheories of the two-valued sentential logic, constructed by different methods, see for instance Linial–Post [**1949**], Yntema [**1964**], and Singletary [**1964**]. However, no other examples of essentially undecidable theories developed in a formalism suitable for sentential logic (with finitely many axiom schemata or axioms) seem to be known in the literature.

[3]*The following quote is from a letter written by Tarski to W. V. O. Quine on March 27, 1942 while at the Institute for Advanced Study in Princeton. It sheds light on the question of when the results in this section, and, more generally, the results in Chapter 4, were obtained. (Givant learned of the letter's existence from Quine in 1985. The quote is published here with Quine's permission.)

"I have some new results in logic and am sorry that there is no opportunity to discuss them with you. Perhaps you will find some interest in one of these results. I have formulated an axiom system S for a partial system of the two-valued sentential calculus. It consists of a few relatively simple and mostly well-known formulas such as the law of syllogism, De Morgan's laws and so on. I have shown that there is no decision procedure for this system (the class of sentences derivable from S is not generally recursive). And, moreover, with practically every mathematical problem (Fermat's theorem, continuum hypothesis, and so on) a [well-] determined formula F of the sentential calculus can be correlated such that the problem whether F is derivable from S is equivalent to the mathematical problem in question. (All formulas of S, as well as the formula F, are true formulas of the ordinary two-valued logic.) Isn't [it] a nice thing 'pour épater les logiciens-bourgeois'."

CHAPTER 6

Implications of the Main Results for Semantic and Axiomatic Foundations of Set Theory

The content of this chapter is clearly indicated by the title. In the first part of the chapter we concern ourselves with semantic foundations, and in the much longer second part with axiomatic foundations, of set theory. The connections between the two parts are not close. Nevertheless, some results stated in the first part simplify the discussion in the second.

6.1. Denotation and truth in $\mathcal{L}^{\times}$

At the end of §2.2 we briefly discussed semantical notions for the language $\mathcal{L}^{+}$. Two of these notions have a basic character, the notion of denotation, which applies to predicates (relation terms), and that of satisfaction, which applies to formulas. The two notions are defined by recursions on predicates and formulas respectively. The definition of denotation is both intuitively and formally quite simple. The definition of satisfaction, based upon that of denotation, appears to be more involved in both respects; the understanding of the definition seems to present intuitive difficulties to people who are interested in the semantics of formal languages, but are not sufficiently well acquainted with the conceptual apparatus of modern mathematics.

If we now turn to the language $\mathcal{L}^{\times}$, the problem of introducing semantical notions becomes considerably simpler; this is due to the structural poverty of $\mathcal{L}^{\times}$, and in particular to the lack of variables in its vocabulary. As was pointed out in §3.1, the simple definition of denotation is carried over without change from $\mathcal{L}^{+}$ to $\mathcal{L}^{\times}$; the notion of satisfaction becomes trivial and need not be introduced at all. Other semantical notions, in particular the important notions of truth and model, are defined for $\mathcal{L}^{\times}$ directly (without recursion) in terms of denotation alone.

Since the discussion of $\mathcal{L}^{\times}$ is the focus of this monograph, we wish to state here the basic semantical definitions for $\mathcal{L}^{\times}$ in a formal way. We consider any realization $\mathfrak{U} = \langle U, E \rangle$ of $\mathcal{L}^{\times}$, with the universe U and the fundamental relation

E, and we assume it to be fixed for the purposes of the subsequent discussion. (If the vocabulary of $\mathcal{L}^\times$ contained more than one nonlogical predicate, then $\mathfrak{U}$ would have to be provided with a correspondingly larger number of fundamental relations.) The definition of denotation in reference to $\mathcal{L}^\times$ runs as follows.

(i) *By the denotation function in $\mathfrak{U}$, symbolically $\mathbf{De}_\mathfrak{U}$, we understand the (uniquely determined) function F whose domain is the set $\mathbf{\Pi}$ (of all predicates of $\mathcal{L}^\times$) and which satisfies the following conditions (where $A, B \in \mathbf{\Pi}$ are arbitrary):*

(α) $F\boldsymbol{E} = E$,

(β) $F\mathring{\mathbf{1}} = U\upharpoonright Id$,

(γ) $F(A^{-}) = {\sim}FA$ (*i.e.*, $= U \times U \sim FA$),

(δ) $F(A^{\smile}) = (FA)^{-1}$,

(ε) $F(A \boldsymbol{+} B) = FA \cup FB$,

(ς) $F(A \boldsymbol{\odot} B) = FA|FB$.

The relation R such that $\mathbf{De}_\mathfrak{U}\, A = R$ is said to be denoted in $\mathfrak{U}$ by the predicate A.

Strictly speaking, instead of "$\mathbf{De}_\mathfrak{U}\, A$" we should write "$\mathbf{De}^\times_\mathfrak{U}\, A$", thus exhibiting the relativization to both $\mathfrak{U}$ and $\mathcal{L}^\times$.

The definition just stated obviously becomes simpler if it is referred not to $\mathcal{L}^\times$, but to some variants of $\mathcal{L}^\times$ with fewer logical constants, e.g., the formalism $\mathcal{L}^{\frown}_{\mathsf{a}}$ mentioned in §5.3. In this case condition (β) is eliminated while (γ)–(ε) are replaced by one condition:

(η) $F(A \dagger B) = {\sim}FA \cup ({\sim}FB)^{-1}$.

Of course in this case the symbol "$\mathbf{\Pi}$" in (i) must be interpreted as referring to the set of predicates of $\mathcal{L}^{\frown}_{\mathsf{a}}$.

The notions of truth and model in $\mathcal{L}^\times$ can now be defined simultaneously in the following way.

(ii) *For any sentence $X \in \Sigma^\times$ we say that X is true of $\mathfrak{U}$, or holds in $\mathfrak{U}$, and also that $\mathfrak{U}$ is a model of X, if* $\mathbf{De}_\mathfrak{U}\, X^{\ell} = \mathbf{De}_\mathfrak{U}\, X^{r}$.

If, instead of $\mathcal{L}^\times$, we use a restricted formalism of the type of $\mathcal{L}^\times_{\mathsf{r}}$ (see §5.4), this definition assumes the following form.

(iii) *A predicate-sentence A is said to be true of $\mathfrak{U}$, and $\mathfrak{U}$ is called a model of A, if $\mathbf{De}_\mathfrak{U}\, A = U \times U$.*

The definitions stated in this section provide an adequate foundation for the whole semantics of $\mathcal{L}^\times$. In particular, they supply a base for rigorous proofs of those results involving semantical notions for $\mathcal{L}^\times$ which were mentioned in our earlier discussion (such as the semantical soundness and semantical incompleteness of $\mathcal{L}^\times$).

6.2. The denotability of first-order definable relations in $\mathfrak{Q}$-structures

The results in Chapters 3 and 4 lead to some interesting conclusions concerning the notion of definability in systems of set theory. As we shall see, the conclusions apply to systems formalized in conventional predicate logic and do not depend on the peculiarities of the formalism $\mathcal{L}^{\times}$.

We concentrate on the notion of definability applied to binary relations. As before, we consider a structure $\mathfrak{U} = \langle U, E\rangle$ (assumed to be fixed) and relations R on the universe U of this structure; we shall often refer to such relations as *relations in* $\mathfrak{U}$.

We find it convenient to refer the discussion at first to the formalism $\mathcal{L}^{+}$. Two semantical notions of definability are available in this context, one of which is more restricted and hence stronger than the other. We shall use the term "denotable" for relations definable in the stronger sense and reserve the term "definable" for those definable in the weaker sense.

(i) *A relation R in $\mathfrak{U}$ is said to be denotable (in $\mathcal{L}^{+}$) if there is a predicate A such that* $\mathbf{De}_{\mathfrak{U}} A = R$.

We shall not refer (i) to the language $\mathcal{L}$, since this would lead to a trivial conclusion: E and $U1Id$ are the only relations in $\mathfrak{U}$ which are denotable in $\mathcal{L}$. On the other hand, (i) can obviously be referred to $\mathcal{L}^{\times}$ instead of $\mathcal{L}^{+}$, without affecting the extent of the term "denotable".

What is more important, instead of referring (i) to $\mathcal{L}^{+}$ or $\mathcal{L}^{\times}$, we can characterize the notion of denotability in purely mathematical terms, independently of any formal language. In fact, let us agree to call a family F of relations a *relation ring on the set* U if F satisfies the following conditions: $R \subseteq U \times U$ for every $R \in F$; $U1Id \in F$; $\sim R$ (i.e., $U \times U \sim R$) and R^{-1} belong to F whenever $R \in F$; $R \cup S$, $R|S \in F$ whenever $R, S \in F$. Notice that the set U is uniquely determined by the relation ring F; for this reason we shall sometimes speak of a relation ring without referring to a set U. The relation ring F is said to be *generated by* a family G of relations if $G \subseteq F$ and there is no relation ring F' (on the same set U) such that $G \subseteq F' \subset F$. In case G is a singleton, $G = \{R\}$, we simply say that F is *generated by* R.

Using this terminology, we easily derive the following conclusion from (i) and the definition of $\mathbf{De}_{\mathfrak{U}}$ in 6.1(i).

(ii) *A relation R in $\mathfrak{U} = \langle U, E\rangle$ is denotable iff it belongs to the relation ring on U generated by E.*

As regards the notion of definability (in the weaker sense) we stipulate:

(iii) (α) *A formula $X \in \mathbf{\Phi}^{+}$ is said to define (in $\mathcal{L}^{+}$) a relation R in $\mathfrak{U} = \langle U, E\rangle$ if $\Upsilon\phi X = \{\boldsymbol{x}, \boldsymbol{y}\}$ and R consists of all ordered pairs $\langle x, y\rangle \in U \times U$ such that every infinite sequence $\langle u_0, \ldots, u_n, \ldots\rangle$ of elements of U with $u_0 = x$ and $u_1 = y$ satisfies X (in $\mathfrak{U}$).*

(β) *A relation R in $\mathfrak{U}$ is said to be definable (in $\mathcal{L}^{+}$) if there is a formula $X \in \mathbf{\Phi}^{+}$ which defines R (in $\mathcal{L}^{+}$).*

Just as (i), Definition (iii)(β) can be reformulated in purely mathematical terms without referring to any formal language, but the formulation is considerably more involved than in the case of denotability; compare Tarski [**1956**], Article VI.

Definition (iii) is obviously not applicable to the formalism $\mathcal{L}^{\times}$. On the other hand, it extends automatically to $\mathcal{L}$ (and to many other formalisms of a related structure, e.g., to the formalisms $\mathcal{L}_n$ and $\mathcal{L}_n^+$ discussed in Chapter 3). From 2.3(i),(iv)(γ)–(ε) we easily conclude that a relation R in $\mathfrak{U}$ is definable in $\mathcal{L}$ iff it is definable in $\mathcal{L}^+$.

Obviously, a relation R which is denoted by a predicate A in the sense of 6.1(i) is defined by the formula $\boldsymbol{x}A\boldsymbol{y}$ in the sense of (iii)(α). Hence, we obtain:

(iv) *Every relation R in $\mathfrak{U}$ which is denotable is also definable (in $\mathcal{L}^+$).*

The converse of (iv) does not hold:

(v) *Let U be a set with at least 7 different elements. Then there are relations E and R on U such that R is definable in $\mathfrak{U} = \langle U, E\rangle$ but not denotable.*

Indeed, given any subset V of U with cardinality 3, set

$$E = V \times V \cup (U \sim V) \times (U \sim V), \qquad R = V \times V.$$

As is easily seen, R consists of all ordered pairs $\langle x, y\rangle$ belonging to $U \times U$ and satisfying the following condition: there are exactly two elements $u, v \in U$ different from y such that, for every $z \in U$, xEz iff $z = y$ or $z = u$ or $z = v$. Hence R is definable in $\mathfrak{U}$. On the other hand, the relation ring on U generated by E contains just eight relations, and R is not among them.

In connection with (iv) and (v) we state the following consequence of the results in §§3.8 and 3.9, in particular of 3.8(ix) and 3.9(iii)(γ_2),(ε).

(vi) *A relation R in $\mathfrak{U}$ is denotable (in $\mathcal{L}^+$) iff it is definable in $\mathcal{L}_3^+$ (or, equivalently, in $\mathcal{L}_3$).*

This result can be improved using 3.10(ii).

In the present work we are not so much interested in arbitrary realizations of $\mathcal{L}$, $\mathcal{L}^+$, and $\mathcal{L}^{\times}$ as we are in those realizations that are models of set-theoretical systems discussed in the literature. We also know that the only property common to all these systems which is relevant for our discussion is that of being a $\mathfrak{Q}$-system. This property is by no means characteristic for set-theoretical systems; it applies to many other systems as well.

We agree to call a relational structure $\mathfrak{U} = \langle U, E\rangle$ a *$\mathfrak{Q}$-structure* if it is a model of a $\mathfrak{Q}$-system. Hence, by 4.5(i), 4.7(xi)(α), we obtain:

(vii) *Each of the following two conditions is necessary and sufficient for $\mathfrak{U} = \langle U, E\rangle$ to be a $\mathfrak{Q}$-structure:*

(α) *there are $A, B \in \mathbf{\Pi}$ such that $\mathfrak{U}$ is a model of $\boldsymbol{Q}_{AB}$ or, equivalently, $\mathbf{De}_{\mathfrak{U}}\, A$ and $\mathbf{De}_{\mathfrak{U}}\, B$ are conjugated quasiprojections;*

(β) *there is a $C \in \mathbf{\Pi}$ such that $\mathfrak{U}$ is a model of $\boldsymbol{U}_C$ or, equivalently, $\mathbf{De}_{\mathfrak{U}}\, C$ is a universal relation for two-element sets.*

Notice also that, by (i), (ii), (vii), the notion of a $\mathfrak{Q}$-structure itself can be characterized without any reference to formal languages.

(viii) *Each of the following two conditions is necessary and sufficient for* $\mathfrak{U} = \langle U, E\rangle$ *to be a* $\mathfrak{Q}$*-structure*:

(α) *the relation ring on* U *generated by* E *contains two conjugated quasiprojections*;

(β) *the relation ring on* U *generated by* E *contains a universal relation for two-element sets.*

The relationship between denotability and definability undergoes an essential change if we refer these notions to $\mathfrak{Q}$-structures. In fact, with the help of 4.4(xiii) we easily show that a relation in a $\mathfrak{Q}$-structure $\mathfrak{U}$ which is definable in $\mathcal{L}^+$ (or $\mathcal{L}$) is also denotable. Combining this result with those in (ii), (iv), and (vi), we arrive at the following theorem, which provides a characterization of definability for $\mathfrak{Q}$-structures.

(ix) *For a binary relation* R *in a* $\mathfrak{Q}$*-structure* $\mathfrak{U} = \langle U, E\rangle$ *the following conditions are equivalent*:

(α) R *is definable in* $\mathcal{L}^+$ (*or* $\mathcal{L}$);

(β) R *is definable in* $\mathcal{L}_3^+$ (*or* $\mathcal{L}_3$);

(γ) R *is denotable*;

(δ) R *is in the relation ring on* U *generated by* E.

The notion of definability extends in an obvious way from binary to n-ary relations for every positive integer n. Especially interesting is the case of $n = 1$, that is, of unary relations in a structure $\mathfrak{U}$; they are identified here with arbitrary sets in $\mathfrak{U}$, i.e., subsets of the universe U. The definability of unary relations easily reduces to that of binary relations.

(x) *A set* S *in* $\mathfrak{U}$ *is definable* (*in* $\mathcal{L}^+$) *iff it is the domain of a definable binary relation* (*or, equivalently, iff* $S{\upharpoonright}Id$ *is a definable binary relation*).

In fact, if a set S is definable, then the binary relation $S \times U$ (or $(S \times S) \cap Id$) is also definable, and $S = Do(S \times U)$.

The notion of denotability (in $\mathcal{L}^+$ or $\mathcal{L}^\times$) does not extend in a direct way to n-ary relations with $n \neq 2$. We obtain, however, a rather natural extension of this notion to the case $n = 1$ by stipulating that a set S in $\mathfrak{U}$ is denotable iff it is the domain of a denotable binary relation in $\mathfrak{U}$ (or, alternately, iff $S{\upharpoonright}Id$ is denotable). Under this stipulation all the results stated in (ii), (iv)–(vi), (ix) can be carried over to definable and denotable sets in $\mathfrak{U}$. (This is not quite obvious in the case of (v).) We state here explicitly the result obtained in this way from Theorem (ix).

(xi) *For a set* S *in a* $\mathfrak{Q}$*-structure* $\mathfrak{U} = \langle U, E\rangle$ *the following conditions are equivalent*:

(α) S *is definable in* $\mathcal{L}^+$ (*or* $\mathcal{L}$);

(β) *S is definable in $\mathcal{L}_3^+$ (or $\mathcal{L}_3$);*

(γ) *S is denotable;*

(δ) *S is the domain of a relation which belongs to the relation ring on U generated by E.*

In connection with this result recall that all the set-theoretical structures $\mathfrak{U} = \langle U, E \rangle$ which are models of known systems of set theory are $\mathfrak{Q}$-structures. These structures have also the following property: for any members $u_0, \ldots, u_{n-1}$ of U the ordered n-tuple $\langle u_0, \ldots, u_{n-1} \rangle$ exists and belongs to U. (If, however, the system of set theory under consideration admits proper classes, we cannot use the classical construction of ordered n-tuples in this context; cf. remarks in §4.6 referring to the case $n = 2$.) Since an n-ary relation among elements of U is construed in set theory as a set of ordered n-tuples, such a relation is always a subset of U, i.e., a set in $\mathfrak{U}$.

Theorem (xi) provides a simple and uniform characterization of definable sets (and thus also of definable finitary relations) for all known systems of set theory (formalized in first-order logic). It may be hoped that, because of their simplicity and generality, both (ix) and (xi) will find some interesting applications in future research.

With this we have completed the semantical part of the discussion in the present chapter.

6.3. The $\mathcal{L}^\times$-expressibility of certain relativized sentences

Before turning to problems concerning the axiomatic foundations of set theory, we introduce some definitions and establish some lemmas that will be needed in our consideration of those problems. We first define precisely what is meant by the formula obtained from a given formula X by restricting the range of bound variables to sets; the formula thus obtained will be denoted by "X^{s}". Clearly, we shall deal here with an operation on and to formulas which assigns to any given formula X a new formula X^{s}. Since the subsequent discussion will be carried on in the extended formalism $\mathcal{L}^+$, we define this operation for all formulas of $\mathcal{L}^+$. The following abbreviation will be used:

(i) $\boldsymbol{S}x = \boldsymbol{\exists}_u(x\boldsymbol{E}u)$, *where* $x, u \in \boldsymbol{\Upsilon}$ *and* $\boldsymbol{in}\, u = \boldsymbol{in}\, x + 1$.

Obviously $\boldsymbol{S}x$ expresses the fact that in a set-theoretical realization of $\mathcal{L}^+$ the object represented by x is a set.

The formal definition of the operation is based upon a double recursion on predicates and formulas, and runs as follows.

(ii) *We set* $X^{\mathsf{s}} = GX$ *for every* $X \in \boldsymbol{\Phi}^+$, *where G is the unique function satisfying the conditions* (α)–(θ) *for all* $A, B \in \boldsymbol{\Pi}$, *all* $x, y, z \in \boldsymbol{\Upsilon}$ *such that* $\boldsymbol{in}\, z = \max(\boldsymbol{in}\, x, \boldsymbol{in}\, y) + 1$, *and all* $X, Y \in \boldsymbol{\Phi}^+$:

(α) $DoG = \boldsymbol{\Phi}^+$,

(β) $G(x\mathring{\boldsymbol{1}}y) = x\mathring{\boldsymbol{1}}y$ *and* $G(x\boldsymbol{E}y) = x\boldsymbol{E}y$,

(γ) $G(xA \boldsymbol{+} By) = G(xAy) \boldsymbol{\vee} G(xBy)$,

(δ) $G(xA^{-}y) = \neg G(xAy)$ *and* $G(xA^{\smile}y) = G(yAx)$,
(ε) $G(xA \odot By) = \exists_z[Sz \wedge G(xAz) \wedge G(zBy)]$,
(ς) $G(A = B) = \forall_{xy}(Sx \wedge Sy \to [G(xAy) \leftrightarrow G(xBy)])$,
(η) $G(X \to Y) = GX \to GY$ *and* $G(\neg X) = \neg GX$,
(θ) $G(\forall_y X) = \forall_y(Sy \to GX)$.

The definition becomes much simpler, of course, if it is applied exclusively to formulas $X \in \mathbf{\Phi}$, since in this case conditions (β)–(ς) are disregarded. The method of constructing the formula X^{s} from a given formula X is a particular case of a general method known as the *relativization of quantifiers*; X^{s} can be referred to as the *formula obtained by relativizing the quantifiers to formulas* Sx.

We state here some consequences of (ii). The first five are obvious.

(iii) (α) *For any* $X, Y \in \mathbf{\Phi}^+$ *and* $x \in \mathbf{\Upsilon}$ *we have*
$(X \vee Y)^{\mathsf{s}} \equiv X^{\mathsf{s}} \vee Y^{\mathsf{s}}$, $(X \wedge Y)^{\mathsf{s}} \equiv X^{\mathsf{s}} \wedge Y^{\mathsf{s}}$,
$(X \leftrightarrow Y)^{\mathsf{s}} \equiv X^{\mathsf{s}} \leftrightarrow Y^{\mathsf{s}}$, *and* $(\exists_x Y)^{\mathsf{s}} \equiv \exists_x(Sx \wedge Y^{\mathsf{s}})$.
(β) *If* $X \in \mathbf{\Phi}^+$, *then* $X^{\mathsf{s}} \in \mathbf{\Phi}$ *and* $\mathbf{\Upsilon}\phi X = \mathbf{\Upsilon}\phi X^{\mathsf{s}}$.
(γ) *If* $X \in \mathbf{\Phi}^+_{(3)}$, *then* $X^{\mathsf{s}} \in \mathbf{\Phi}_{(3)}$ *and* $\mathbf{\Upsilon}\phi X = \mathbf{\Upsilon}\phi X^{\mathsf{s}}$.
(δ) *If* $X \in \Sigma^+$, *then* $X^{\mathsf{s}} \in \Sigma$.
(ε) *If* $X \in \Sigma^+_{(3)}$, *then* $X^{\mathsf{s}} \in \Sigma_{(3)}$.

Statements (γ) and (ε) make use of the notation introduced in §3.10.

The following theorem is important for our purposes.

(iv) *For every* $\Psi \subseteq \Sigma^+$ *and* $X \in \Sigma^+$, *if* $\Psi \vdash X$, *then* $\{Y^{\mathsf{s}} : Y \in \Psi\} \cup \{\exists_x Sx\} \vdash X^{\mathsf{s}}$.

The proof, by induction on sentences derivable from Ψ, is straightforward. We consider the set Θ of all sentences $Z \in \Sigma^+$ satisfying the condition

$$\{Y^{\mathsf{s}} : Y \in \Psi\} \cup \{\exists_x Sx\} \vdash Z^{\mathsf{s}}.$$

To prove that $X \in \Theta$ whenever $\Psi \vdash X$ it suffices to show that Θ satisfies three conditions: (1) $\Psi \subseteq \Theta$, (2) $\mathbf{\Lambda}^+ \subseteq \Theta$, (3) Θ is closed under *modus ponens*. (1) and (3) are obvious, and (2) can be checked without difficulty, using Definition (ii), for all the different types of axioms which constitute $\mathbf{\Lambda}^+$.

Theorem (iv) is a particular case of a general result on the relativization of quantifiers stated (implicitly) in Tarski–Mostowski–Robinson [**1953**], p. 25f. It is also known from general metalogical considerations that the presence of the sentence $\exists_x Sx$ in the conclusion of (iv) is essential. If this sentence is eliminated, then condition (2) in the proof of (iv) cannot be verified, but only in the case when Z is an instance of Schema (AVI). In fact, if we take $\varnothing$ for Ψ and, for instance, $\exists_x(x\mathring{\mathbf{1}}x)$ for X in (AVI), then the resulting instance Z of (AVI) is of course a logical axiom, but its relativization Z^{s} is not even logically provable.

In this connection notice the following corollary of (iv).

(v) *For every* $\Psi \subseteq \Sigma^+$ *and every sentence* $X \in \Sigma^+$ *which either begins with the quantifier* $\forall$ *or else is an equation, if* $\Psi \vdash X$ *then* $\{Y^{\mathsf{s}} : Y \in \Psi\} \vdash X^{\mathsf{s}}$.

Indeed, by hypothesis and (ii)(ς),(θ) we see that X^{s} must have either the form $\forall_x(Sx \to Z)$ or $\forall_{xy}(Sx \wedge Sy \to Z)$. Hence, obviously, $\neg\exists_x Sx \vdash X^{\mathrm{s}}$; together with the conclusion of (iv) this yields the conclusion of (v).

With the help of (v) we now establish a result of a less general character which depends in an essential way on the discussion in Chapter 4. It is the principal result from this section which will be used in §6.4. Recall that $\boldsymbol{P}$ is the pair axiom formulated in 4.6(i). In $\boldsymbol{P}^{\mathrm{s}}$ we easily recognize a stronger form of the restricted pair axiom S_2 given in 3.6(ii).

(vi) *For any $x, y \in \Upsilon$ and any $X \in \boldsymbol{\Phi}^+$ with $\Upsilon\phi X \subseteq \{x, y\}$, there is an $A \in \Pi$ such that $\boldsymbol{P}^{\mathrm{s}} \vdash \forall_{xy}[Sx \wedge Sy \to (X^{\mathrm{s}} \leftrightarrow (xAy)^{\mathrm{s}})]$.*

In fact, by 4.4(xiii) and 4.6(ii), for any given formula X with $\Upsilon\phi X \subseteq \{x, y\}$ there is a predicate A such that $\boldsymbol{P} \vdash \forall_{xy}(X \leftrightarrow xAy)$. Hence the conclusion follows from (v) with the help of (iii)(α) and (ii)(θ).

To conclude this section we state two results which are of some intrinsic interest, but which are not relevant for the subsequent discussion.

(vii) *If X is $\mathcal{L}^{\times}$-expressible, then $X^{\mathrm{s}} \wedge \exists_x Sx$ is $\mathcal{L}^{\times}$-expressible in the general case, and X^{s} is $\mathcal{L}^{\times}$-expressible in the case when X begins with the quantifier $\forall$.*

In fact, by hypothesis and 3.10(iv) there is a $Y \in \Sigma_{(3)}$ such that $\vdash X \leftrightarrow Y$. Then $\exists_x Sx \vdash X^{\mathrm{s}} \leftrightarrow Y^{\mathrm{s}}$ by (iv), (iii)(α), whence

$$(1) \qquad X^{\mathrm{s}} \wedge \exists_x Sx \;\equiv\; Y^{\mathrm{s}} \wedge \exists_x Sx.$$

Since $\exists_x Sx \in \Sigma_{(3)}$ and $Y \in \Sigma_{(3)}$, we get $(Y^{\mathrm{s}} \wedge \exists_x Sx) \in \Sigma_{(3)}$ by (iii)(ε). From this and (1) we conclude, using 3.10(iv), that $X^{\mathrm{s}} \wedge \exists_x Sx$ is $\mathcal{L}^{\times}$-expressible.

In case X begins with $\forall$, it follows from (ii)(θ) that

$$(2) \qquad X^{\mathrm{s}} \wedge \neg\exists_x Sx \equiv \neg\exists_x Sx.$$

Now (1) and (2) imply $X^{\mathrm{s}} \equiv (\exists_x Sx \to Y^{\mathrm{s}})$; since $(\exists_x Sx \to Y^{\mathrm{s}}) \in \Sigma_{(3)}$ by (iii)(ε), we get that X^{s} is $\mathcal{L}^{\times}$-expressible.

Finally, we know from 4.6(ii) that $\boldsymbol{P}$ is $\mathcal{L}^{\times}$-expressible. Hence we conclude by (vii):

(viii) *$\boldsymbol{P}^{\mathrm{s}}$ is $\mathcal{L}^{\times}$-expressible.*

It is interesting to compare (viii) with the result stated in 3.6(ii), by which the sentence S_2, so closely related to $\boldsymbol{P}^{\mathrm{s}}$, is not $\mathcal{L}^{\times}$-expressible.

It may be observed that (vii) and (viii) continue to hold if we consider sentences relativized, not necessarily to formulas Sx, but to arbitrary formulas $F \in \boldsymbol{\Phi}_{(3)}$ with one free variable. This enables us to prove the $\mathcal{L}^{\times}$-expressibility of many sentences to which the method discussed in the first part of §4.7 does not seem to be directly applicable.

6.4. The finite axiomatizability of predicative systems of set theory admitting proper classes

We now turn to some results and problems concerning axiomatic foundations of set theory.

So far we have not had occasion to discuss a schema which, in one form or another, is present in every axiomatic system of set theory—in the sense that all instances of this schema either are included among the nonlogical axioms or at least are derivable from them. The presence of this schema is probably the most characteristic structural property of set-theoretical systems. (We speak here of systems designed as frameworks for the actual development of set theory, and not of various fragmentary systems used, e.g., as illustrations of, or counterexamples to, some general metamathematical statements.)

The schema in question is known as the comprehension schema or the schema of class construction. Its simplest and strongest form is

$$[\exists_s \forall_x (xEs \leftrightarrow F)] \quad \textit{with} \quad F \in \mathbf{\Phi} \quad \textit{and} \quad s \notin \Upsilon\phi F,$$

and in this form it will be referred to as the *unrestricted comprehension schema.*

In the simplest case when $\Upsilon\phi F = \{x\}$, the instances of this schema express in their totality the following fact: if $\mathfrak{U} = \langle U, E\rangle$ is a set-theoretical model of the system discussed and F is any formula (in the language of this system) expressing a condition imposed on members of U, then there is a member of U which is a class consisting of just those members of U that satisfy this condition. Thus, one can claim that the comprehension schema embodies the most basic intuition relating to the notion of a class.

It is well known that the unrestricted comprehension schema cannot actually be used in constructing consistent set-theoretical systems, since some of its particular instances are self-contradictory. Therefore, beginning with the classical work of Zermelo [**1908**], this schema was always replaced by a weaker and plausibly consistent variant. We shall discuss here some of these variants, namely those which occur in systems descending in a direct line from Zermelo's system. Thus we shall not concern ourselves with the variant of the comprehension schema which is based upon the idea of stratification and which occurs as a characteristic and essential component in the systems presented in Quine [**1937**] and Quine [**1951**].

We shall only consider systems excluding individuals (in the sense of §4.6); the extension to systems with individuals is obvious. We first concentrate on systems of set theory which admit proper classes. The variant of the unrestricted comprehension schema which naturally suggests itself for constructing such systems is

$$\text{(C)} \qquad [\exists_s \forall_x (xEs \leftrightarrow F \wedge Sx)] \quad \textit{with} \quad F \in \mathbf{\Phi} \quad \textit{and} \quad s \notin \Upsilon\phi F.$$

We shall denote the set of all particular instances of (C) by "$\mathbf{\Omega}$"; i.e., we set

(i) $\mathbf{\Omega} = \{X : X = [\exists_s \forall_x(xEs \leftrightarrow F \wedge Sx)] \text{ for some } F \in \mathbf{\Phi} \text{ with } s \notin \Upsilon\phi F\}$.

(This set should not be confused with the set $\mathbf{\Omega}$ defined in 4.7(vii).)

One could claim that (C) is the strongest possible variant of the comprehension schema which can be used for the construction of systems admitting proper classes. Indeed, once we reconcile ourselves to the possibility that some objects in the universe of a model of our set-theoretical system are not elements of any members of this universe, we must exclude those objects as possible elements of any class whose existence is asserted. Hence, we must exclude from the comprehension schema those formulas F which are satisfied by some proper classes—and this is what we do when we replace arbitrary formulas F by formulas of the type $F \wedge Sx$, i.e., when we pass from the unrestricted comprehension schema to the new schema (C).

Schema (C) has indeed been adopted as an axiom schema in some set-theoretical systems. Thus, it occurs as one of two comprehension schemata in Quine [**1951**], p. 162; the other comprehension schema, which plays a more substantial part in the development of that system, is also a weaker form of the unrestricted comprehension schema, but its construction is based on the idea of stratification. In Morse's system (see §4.6) Schema (C) is the most essential component of the axiom set.

On the other hand, in Bernays' system and Gödel's variant of it (see again §4.6), (C) is replaced by a weaker schema. This weaker schema has essentially the form of (C), but F is no longer an arbitrary formula since, loosely speaking, the range of every variable which occurs bound in it is restricted to sets. It can be formulated as follows:

$$(\mathrm{C^s}) \qquad [\exists_s \forall_x(xEs \leftrightarrow F^{\mathrm{s}} \wedge Sx)] \quad \text{with} \quad F \in \mathbf{\Phi} \quad \text{and} \quad s \notin \Upsilon\phi F.$$

Notice that replacing $\mathbf{\Phi}$ by $\mathbf{\Phi}^+$ in ($\mathrm{C^s}$) (or in (C)) would not increase the deductive power of the schema, but would somewhat complicate the discussion. In analogy to (i) we set

(ii) $\mathbf{\Omega}^{\mathrm{s}} = \{X : X = [\exists_s \forall_x(xEs \leftrightarrow F^{\mathrm{s}} \wedge Sx)] \text{ for some } F \in \mathbf{\Phi} \text{ with } s \notin \Upsilon\phi F\}$.

Let $\mathcal{S}$ be any system of set theory formalized in $\mathcal{L}^\times$ which admits proper classes (and hence does not contain $\forall_x Sx$ among its provable sentences) and whose axiom set $\mathbf{A}\xi[\mathcal{S}]$ includes $\mathbf{\Omega}$. Let $\mathcal{S}^{\mathrm{s}}$ be the system in $\mathcal{L}^+$ which we obtain from $\mathcal{S}$ if we replace $\mathbf{\Omega}$ by $\mathbf{\Omega}^{\mathrm{s}}$ in $\mathbf{A}\xi$, i.e., whose axiom set is $\mathbf{A}\xi \sim \mathbf{\Omega} \cup \mathbf{\Omega}^{\mathrm{s}}$.

Schema ($\mathrm{C^s}$) is often referred to as the *predicative* version (or variant) of Schema (C), and (C) is referred to as the *impredicative* version of ($\mathrm{C^s}$). This terminology extends also to the sets $\mathbf{\Omega}$ and $\mathbf{\Omega}^{\mathrm{s}}$ as well. We shall not discuss the question of to what extent the use of the terms "predicative" and "impredicative" in these contexts is consistent with the intuitive meanings of the two terms.

Since $\mathbf{\Omega}^{\mathrm{s}} \subseteq \mathbf{\Omega}$, the system $\mathcal{S}^{\mathrm{s}}$ is always a subsystem of $\mathcal{S}$. It is known that, under certain general assumptions, $\mathcal{S}^{\mathrm{s}}$ is essentially weaker than $\mathcal{S}$. This can easily be concluded, for instance, by comparing the results in Section VII of Chuaqui

[**1980**] with 6.4(vi) below. At any rate, this is true in case the two systems involved are the system of Morse referred to in §4.6, which we temporarily denote by $\mathcal{M}$, and its predicative version $\mathcal{M}^{\mathrm{s}}$. The fact that $\mathcal{M}^{\mathrm{s}}$ is weaker than $\mathcal{M}$ is a consequence of two results: $\mathcal{M}^{\mathrm{s}}$ is finitely axiomatizable and $\mathcal{M}$ is not. The second of these results is due to Chuaqui [**1980**]. The first result follows from certain metamathematical arguments which can be found in Bernays [**1937**], Part I, pp. 72–76, and Gödel [**1940**], pp. 8–14. While the axiom set of $\mathcal{M}^{\mathrm{s}}$ is obviously infinite (since it includes $\boldsymbol{\Omega}^{\mathrm{s}}$), Bernays' system and Gödel's variant of it (presented in the papers cited above) are provided with finite axiom sets; however, from the arguments just mentioned it is seen directly that the three axiom sets are essentially equivalent. For this reason, when referring to $\mathcal{M}^{\mathrm{s}}$ in our further discussion we shall have in mind Bernays' system, or more precisely, Gödel's variant of it, for brevity $\mathcal{BG}$.

The results mentioned in the preceding paragraph lead naturally to the problem of determining the classes of set-theoretical systems $\mathcal{S}$ and $\mathcal{S}^{\mathrm{s}}$ to which the results can be extended. We shall be interested here in this problem only with reference to systems $\mathcal{S}^{\mathrm{s}}$ and their finite axiomatizability. Indeed, we shall establish a theorem which implies the finite axiomatizability of a large class of these systems. It is assumed that the sentence $\boldsymbol{P}^{\mathrm{s}}$ (discussed in §6.3) holds in each of these systems. Under this assumption the theorem supplies a set Σ of a few relatively simple sentences by which the set $\boldsymbol{\Omega}^{\mathrm{s}}$ can be equivalently replaced. The class of systems to which this theorem applies comprehends, in particular, $\mathcal{BG}$; thus, our discussion provides a new proof of the finite axiomatizability of that system.

An essential feature of the theorem which will be established here is its close relationship (in both the statement of the theorem and the method of proof) to some notions and results discussed in earlier portions of the present work. This relationship will be detected, we hope, by all who read the proof with some attention; they will notice, in particular, the crucial role played in the proof by Theorem 6.3(vi), which is a specialized consequence of some of the main results of Chapter 4. While the proof we are going to outline is essentially of a syntactical character, the relationship referred to would become even more obvious if we were to give here a model-theoretical version of the proof, using instead of 6.3(vi) some semantical consequences of results of Chapter 4, namely 6.2(ix),(xi).

We should like to mention that the idea of applying some results and methods of the present work to the discussion of the finite axiomatizability of $\mathcal{BG}$ emerged from conversations which Tarski had with Gödel sometime in the early forties.

In formulating the proposed theorem we use a new abbreviation:

(iii) $$C(x,y,z) = \exists_{uv}\forall_w[(w\boldsymbol{E}x \leftrightarrow w\mathring{\mathbf{1}}u \vee w\mathring{\mathbf{1}}v) \wedge (w\boldsymbol{E}u \leftrightarrow w\mathring{\mathbf{1}}y) \wedge (w\boldsymbol{E}v \leftrightarrow w\mathring{\mathbf{1}}y \vee w\mathring{\mathbf{1}}z)],$$

where $x,y,z,u,v,w \in \Upsilon$, *with* $\mathit{in}\,u = \max(\mathit{in}\,x, \mathit{in}\,y, \mathit{in}\,z) + 1$, $\mathit{in}\,v = \mathit{in}\,u + 1$, *and* $\mathit{in}\,w = \mathit{in}\,v + 1$.

The formula $\boldsymbol{C}(x, y, z)$ obviously expresses the fact that the object represented by x is an ordered couple whose first and second terms are respectively the objects represented by y and z.

In the subsequent discussion we shall use extensively the definitions of $\boldsymbol{S}x$ and $\boldsymbol{C}(x, y, z)$ without referring to them explicitly. The same applies to some obvious consequences of these definitions, for example:

$$\forall_{xyz}[C(x,y,z) \to Sy \wedge Sz],$$

$$\forall_{xyzuw}[C(x,y,z) \wedge C(x,u,w) \to y\mathring{\mathbf{I}}u \wedge z\mathring{\mathbf{I}}w].$$

Also, the following equivalence is easily seen to hold:

$$P^{\mathsf{S}} \equiv \forall_{yz}[Sy \wedge Sz \to \exists_x(Sx \wedge C(x,y,z))].$$

The formulation of the theorem follows.

(iv) *Let* Σ *be the set of the following five sentences:*

$$\begin{aligned}
S_1 &= \exists_s\forall_x[xEs \leftrightarrow \exists_{yz}(C(x,y,z) \wedge yEz) \wedge Sx],\\
S_2 &= \forall_{pq}\exists_s\forall_x(xEs \leftrightarrow xEp \wedge \neg xEq),\\
S_3 &= \forall_{pq}\exists_s\forall_x[xEs \leftrightarrow \exists_{yzrtw}(yEp \wedge zEq \wedge\\
&\qquad\qquad C(x,r,t) \wedge C(y,r,w) \wedge C(z,t,w)) \wedge Sx],\\
S_4 &= \forall_p\exists_s\forall_x[xEs \leftrightarrow \exists_y(xEy \wedge yEp)],\\
S_5 &= \forall_p\exists_s\forall_x[xEs \leftrightarrow \exists_y(yEp \wedge \forall_w(wEx \leftrightarrow w\mathring{\mathbf{I}}y)) \wedge Sx].
\end{aligned}$$

In addition, let

$$T = \forall_{pq}[Sp \wedge Sq \to \exists_s(Ss \wedge \forall_x(xEs \leftrightarrow x\mathring{\mathbf{I}}p \vee x\mathring{\mathbf{I}}q))].$$

Under these assumptions we have

$$\mathbf{\Omega}^{\mathsf{S}} \vdash \Sigma \quad \textit{and} \quad \mathbf{\Omega}^{\mathsf{S}} \equiv_T \Sigma.$$

The meaning of the sentences $S_1 - S_5$ above can be easily decoded. For instance, in reference to set-theoretical models, S_3 expresses the fact that for any two binary relations P and Q there exists a relation S which is the relative product of P and Q^{-1}, while S_5 states that for any class P there exists a class S consisting of all singletons of elements of P. Obviously the sentence T is logically equivalent to (but formally a little simpler than) the sentence $\boldsymbol{P}^{\mathsf{S}}$.

It is readily seen that each of the sentences $S_1 - S_5$ (but not T) is logically equivalent to some instance of the schema $(\mathrm{C}^{\mathsf{S}})$, and hence $\mathbf{\Omega}^{\mathsf{S}} \vdash \Sigma$. In consequence, to prove (iv) we have only to show that $\Sigma \vdash_T \mathbf{\Omega}^{\mathsf{S}}$.

By (ii) each sentence in $\mathbf{\Omega}^{\mathsf{S}}$ has the form

(1) $\quad [\exists_s\forall_x(xEs \leftrightarrow F^{\mathsf{S}} \wedge Sx)]$ for some $F \in \mathbf{\Phi}$ with $s \notin \Upsilon\phi F$.

We first deal with the case when $\Upsilon\phi F \subseteq \{x\}$, so that we have to prove

(2) $\quad \Sigma \vdash_T \exists_s\forall_x(xEs \leftrightarrow F^{\mathsf{S}} \wedge Sx)$.

Now, by 6.3(vi) (with x and y replaced by $\boldsymbol{x}$), we can find for each F a predicate A such that

$$(3) \qquad \boldsymbol{P}^{\mathsf{s}} \vdash \forall_{\boldsymbol{x}}(F^{\mathsf{s}} \wedge S\boldsymbol{x} \leftrightarrow (\boldsymbol{x}A\boldsymbol{x})^{\mathsf{s}} \wedge S\boldsymbol{x}).$$

Hence,

$$(4) \qquad T \vdash \forall_{\boldsymbol{x}}(F^{\mathsf{s}} \wedge S\boldsymbol{x} \leftrightarrow (\boldsymbol{x}A\boldsymbol{x})^{\mathsf{s}} \wedge S\boldsymbol{x}).$$

By (2) and (4) our task reduces to showing that, for every $A \in \boldsymbol{\Pi}$,

$$(5) \qquad \Sigma \vdash_T \exists_{\boldsymbol{s}}\forall_{\boldsymbol{x}}(\boldsymbol{x}\boldsymbol{E}\boldsymbol{s} \leftrightarrow (\boldsymbol{x}A\boldsymbol{x})^{\mathsf{s}} \wedge S\boldsymbol{x}).$$

To this end consider the set of predicates defined by the following formula:

$$(6) \qquad \Gamma = \{B : B \in \boldsymbol{\Pi} \text{ and } \Sigma \vdash_T \exists_{\boldsymbol{s}}\forall_{\boldsymbol{x}}[\boldsymbol{x}\boldsymbol{E}\boldsymbol{s} \leftrightarrow \exists_{\boldsymbol{y}\boldsymbol{z}}(C(\boldsymbol{x},\boldsymbol{y},\boldsymbol{z}) \wedge (\boldsymbol{y}B\boldsymbol{z})^{\mathsf{s}}) \wedge S\boldsymbol{x}]\}.$$

Predicates in Γ could be referred to as predicates representable by classes. We obtain successively:

(7) $\boldsymbol{E} \in \Gamma$ — by (6), using S_1;

(8) $B \cdot C^{-}, B \odot C^{\smile} \in \Gamma$ whenever $B, C \in \Gamma$ — by (6) and 6.3(ii)(γ)–(ε), using S_2, S_3;

(9) $\boldsymbol{E} \odot \boldsymbol{E}^{\smile} \in \Gamma$ and $T \vdash (\boldsymbol{E} \odot \boldsymbol{E}^{\smile} = \mathbf{1})^{\mathsf{s}}$ — by (7), (8), 6.3(ii)(β)–(ς);

(10) If $B \in \Gamma$ and $\Sigma \vdash_T (B = C)^{\mathsf{s}}$, then $C \in \Gamma$ — by (6), 6.3(ii)(ς);

(11) $\mathbf{1} \in \Gamma$ — by (9), (10);

(12) $B^{-}, B + C \in \Gamma$ whenever $B, C \in \Gamma$ — by (8), (10), (11), 6.3(v);

(13) $(\boldsymbol{E}^{-} \odot \boldsymbol{E}^{\smile})^{-} \in \Gamma$ and $T \vdash ((\boldsymbol{E}^{-} \odot \boldsymbol{E}^{\smile})^{-} = \mathring{\mathbf{i}})^{\mathsf{s}}$ — by (7), (8), (12), 6.3(ii)(β),(δ)–(ς);

(14) $\mathring{\mathbf{i}} \in \Gamma$, and $B^{\smile}, B \odot C \in \Gamma$ whenever $B, C \in \Gamma$ — by (10), (13), (8), 6.3(v);

(15) $\boldsymbol{\Pi} \subseteq \Gamma$ — by (7), (12), (14), and induction on predicates;

(16) $\Sigma \vdash_T \exists_{\boldsymbol{s}}\forall_{\boldsymbol{x}}[\boldsymbol{x}\boldsymbol{E}\boldsymbol{s} \leftrightarrow \exists_{\boldsymbol{y}}(C(\boldsymbol{x},\boldsymbol{y},\boldsymbol{y}) \wedge (\boldsymbol{y}A\boldsymbol{y})^{\mathsf{s}}) \wedge S\boldsymbol{x}]$ — by (15), (6) (taking $A \cdot \mathring{\mathbf{i}}$ for B).

From (16), using S_4 twice, as well as T, we obtain (5) and hence (2). Observe that S_5 has so far not been used.

Now consider the case when in F, from (1), there is just one variable different from $\boldsymbol{x}$, say $\boldsymbol{u}$, which occurs free, so that $\Upsilon\phi F \subseteq \{\boldsymbol{x}, \boldsymbol{u}\}$. Thus we have to prove

$$(17) \qquad \Sigma \vdash_T \forall_{\boldsymbol{u}} \exists_{\boldsymbol{s}} \forall_{\boldsymbol{x}} (\boldsymbol{x E s} \leftrightarrow F^{\mathrm{s}} \wedge S\boldsymbol{x}).$$

Without loss of generality we can, of course, stipulate that $\boldsymbol{u}$ does not occur bound in F. Thus, let $\boldsymbol{\Phi}^*$ be the set of all $F \in \boldsymbol{\Phi}$ for which $\Upsilon\phi F \subseteq \{\boldsymbol{x}, \boldsymbol{u}\}$ and in which $\boldsymbol{u}$ does not occur bound.

Using S_5 twice, we immediately obtain

$$(18) \qquad \Sigma \vdash_T \forall_{\boldsymbol{u}} \exists_{\boldsymbol{s}} \forall_{\boldsymbol{x}} [\boldsymbol{x E s} \leftrightarrow \exists_{\boldsymbol{y}} (\boldsymbol{y E u} \wedge C(\boldsymbol{x}, \boldsymbol{y}, \boldsymbol{y})) \wedge S\boldsymbol{x}].$$

This is trivially equivalent to a particular instance of (17).

Next we establish (17) for those special formulas F in which every atomic subformula containing $\boldsymbol{u}$ is of the form $y\boldsymbol{E u}$ with $y \neq \boldsymbol{u}$. To this end we consider (exclusively for the purposes of the present argument) two auxiliary formalisms, $\overline{\mathcal{L}}$ and $\overline{\mathcal{L}}^+$. They are obtained from $\mathcal{L}$ and $\mathcal{L}^+$, respectively, by adjoining an additional atomic binary predicate, say $\boldsymbol{D}$, to the vocabularies, without changing the description of the formalisms otherwise. When referring to the new formalisms we shall still use the ordinary derivability symbol "$\vdash$" without any superscript.

As was pointed out earlier, the results in Chapter 4 extend almost automatically to formalisms thus constructed. To extend the results of §6.3, we supplement the definition of X^{s} by adding the formula $G(x\boldsymbol{D}y) = x\boldsymbol{D}y$ to those in 6.3(ii)(β), so that $(x\boldsymbol{D}y)^{\mathrm{s}} = x\boldsymbol{D}y$ for any $x, y \in \Upsilon$. We then obtain, in particular, 6.3(vi) in its application to $\overline{\mathcal{L}}^+$.

When trying to extend (2) to $\overline{\mathcal{L}}^+$, we meet with an obstacle: it is not in general true that $\boldsymbol{D} \in \Gamma$, and this formula is needed to complete the induction on predicates in proving (15). Instead of $\boldsymbol{D} \in \Gamma$ we can obtain a weaker conclusion (which, however, suffices for our purposes). In fact, set

$$(19) \qquad X = \forall_{\boldsymbol{xy}} (\boldsymbol{x D y} \leftrightarrow \boldsymbol{x} \mathring{\mathbf{I}} \boldsymbol{y} \wedge \boldsymbol{y E u});$$

obviously $\boldsymbol{u}$ is the only variable occurring free in the formula X. Our conclusion runs then as follows:

$$\Sigma \cup \{\exists_{\boldsymbol{u}} X\} \vdash_T \exists_{\boldsymbol{s}} \forall_{\boldsymbol{x}} [\boldsymbol{x E s} \leftrightarrow \exists_{\boldsymbol{yz}} (C(\boldsymbol{x}, \boldsymbol{y}, \boldsymbol{z}) \wedge (\boldsymbol{y D z})^{\mathrm{s}}) \wedge S\boldsymbol{x}];$$

it is easily derived with the help of (18). Hence, by following strictly the lines of argument in the proof of (2), and, in particular, by imitating (6)–(16), we eventually obtain

$$(20) \qquad \Sigma \cup \{\exists_{\boldsymbol{u}} X\} \vdash_T \exists_{\boldsymbol{s}} \forall_{\boldsymbol{x}} (\boldsymbol{x E s} \leftrightarrow G^{\mathrm{s}} \wedge S\boldsymbol{x}) \text{ for every } G \in \overline{\boldsymbol{\Phi}} \text{ with } \Upsilon\phi G \subseteq \{\boldsymbol{x}\}$$

(where $\overline{\boldsymbol{\Phi}}$ is of course the set of all formulas of $\overline{\mathcal{L}}$).

Let Ψ be the set of all $F \in \mathbf{\Phi}^*$ such that every atomic subformula of F in which $\boldsymbol{u}$ occurs is of the form $y\boldsymbol{E}\boldsymbol{u}$ with $y \in \Upsilon \sim \{\boldsymbol{u}\}$. For any $F \in \Psi$ let $\overline{F}$ be the formula in $\overline{\mathbf{\Phi}}$ which is obtained from F when every subformula of F of the form $y\boldsymbol{E}\boldsymbol{u}$ is replaced by $y\boldsymbol{D}y$. Using (19), we get by an easy induction on formulas in Ψ that

$$(21) \qquad \vdash [X \rightarrow (F^{\mathrm{s}} \leftrightarrow \overline{F}^{\mathrm{s}})] \quad \text{for every} \quad F \in \Psi.$$

Let F be any formula in Ψ. Clearly $\overline{F} \in \overline{\mathbf{\Phi}}$ and $\Upsilon\phi\overline{F} \subseteq \{\boldsymbol{x}\}$. Therefore (20) holds for $G = \overline{F}$, and together with (21) this yields

$$\Sigma \vdash_T \forall_{\boldsymbol{u}}[X \rightarrow \exists_{\boldsymbol{s}}\forall_{\boldsymbol{x}}(\boldsymbol{x}\boldsymbol{E}\boldsymbol{s} \leftrightarrow F^{\mathrm{s}} \wedge S\boldsymbol{x})].$$

Hence, we infer that (17) holds for all $F \in \Psi$. We can prove this easily, making use of the semantical completeness of $\mathcal{L}$ and $\overline{\mathcal{L}}$; the argument, by extending models of $\Sigma \cup \{T\}$ in $\mathcal{L}$ to those in $\overline{\mathcal{L}}$, is routine.

Our task will be completed if we extend (17) from formulas in Ψ to arbitrary formulas in $\mathbf{\Phi}^*$. Let F be a fixed formula in $\mathbf{\Phi}^*$. First consider the statement

$$(22) \qquad \Sigma \vdash_T \forall_{\boldsymbol{u}}[S\boldsymbol{u} \rightarrow \exists_{\boldsymbol{s}}\forall_{\boldsymbol{x}}(\boldsymbol{x}\boldsymbol{E}\boldsymbol{s} \leftrightarrow F^{\mathrm{s}} \wedge S\boldsymbol{x})].$$

To prove (22) we pick any two distinct variables, y and v, which do not occur in F at all and are different from $\boldsymbol{x}$ and $\boldsymbol{s}$. Next, we replace in F every subformula $\boldsymbol{u}\boldsymbol{E}z$ by $\forall_y(y\boldsymbol{E}v \rightarrow y\boldsymbol{E}z)$, $z\boldsymbol{E}\boldsymbol{u}$ by $\forall_y(y\boldsymbol{E}v \rightarrow z\boldsymbol{E}y)$, $\boldsymbol{u}\boldsymbol{E}\boldsymbol{u}$ by $\forall_y(y\boldsymbol{E}v \rightarrow y\boldsymbol{E}y)$, $z\mathring{\mathbf{I}}\boldsymbol{u}$ and $\boldsymbol{u}\mathring{\mathbf{I}}z$ by $z\boldsymbol{E}v$, and $\boldsymbol{u}\mathring{\mathbf{I}}\boldsymbol{u}$ by $\forall_y(y\mathring{\mathbf{I}}y)$; in these subformulas z is assumed to be different from $\boldsymbol{u}$. Let G be the resulting formula. As is easily seen, by repeated applications of an appropriate variant of the well-known schema of equivalent replacement (R) in 3.7(i), we have

$$(23) \qquad \vdash \forall_{\boldsymbol{u}v}[\forall_{\boldsymbol{x}}(\boldsymbol{x}\boldsymbol{E}v \leftrightarrow \boldsymbol{x}\mathring{\mathbf{I}}\boldsymbol{u}) \rightarrow \forall_{\boldsymbol{x}}(F^{\mathrm{s}} \leftrightarrow G^{\mathrm{s}})].$$

Notice also that $\boldsymbol{u}$ does not occur in G and that the formula H obtained from G by changing v everywhere to $\boldsymbol{u}$ belongs to Ψ. Since (17) holds for all formulas in Ψ, we conclude, in particular, that the sentence

$$\forall_{\boldsymbol{u}}\exists_{\boldsymbol{s}}\forall_{\boldsymbol{x}}(\boldsymbol{x}\boldsymbol{E}\boldsymbol{s} \leftrightarrow H^{\mathrm{s}} \wedge S\boldsymbol{x})$$

is derivable from Σ on the basis of T. If in this sentence we rename variables, replacing $\boldsymbol{u}$ by v everywhere, then H goes back into G and we arrive at

$$(24) \qquad \Sigma \vdash_T \forall_v\exists_{\boldsymbol{s}}\forall_{\boldsymbol{x}}(\boldsymbol{x}\boldsymbol{E}\boldsymbol{s} \leftrightarrow G^{\mathrm{s}} \wedge S\boldsymbol{x}).$$

In view of the obvious fact that

$$T \vdash \forall_{\boldsymbol{u}}[S\boldsymbol{u} \rightarrow \exists_v\forall_{\boldsymbol{x}}(\boldsymbol{x}\boldsymbol{E}v \leftrightarrow \boldsymbol{x}\mathring{\mathbf{I}}\boldsymbol{u})],$$

from (23) and (24) we derive (22) directly.

Now consider the statement

(25) $\Sigma \vdash_T \forall_u[\neg Su \rightarrow \exists_s \forall_x (xEs \leftrightarrow F^s \wedge Sx)]$.

We replace in F the atomic formula $u\mathring{\imath}u$ (if it occurs as a subformula of F at all) by $x\mathring{\imath}x$, and any other atomic subformula which contains u, but is not of the form yEu with $y \neq u$, by $\neg x\mathring{\imath}x$. For the resulting $G \in \Psi$ we have

(26) $\Sigma \vdash_T \forall_u \exists_s \forall_x (xEs \leftrightarrow G^s \wedge Sx)$

and

(27) $\vdash \forall_u[\neg Su \rightarrow \forall_x (F^s \wedge Sx \leftrightarrow G^s \wedge Sx)]$.

Statement (26) is a direct consequence of the fact that (17) has been established for all formulas in Ψ. To prove (27) we again apply repeatedly the schema of equivalent replacement. (25) follows at once from (26) and (27).

Statements (22) and (25) obviously imply (17) for the fixed formula F, and hence for all formulas of Φ in which at most one variable different from x occurs free. Essentially the same argument, with some minor complications, can be used to show that all sentences in Ω^s (i.e., all sentences of the form (1) with arbitrarily many variables occurring free in F) are derivable from Σ on the basis of T, and thus to complete the proof of (iv).

Some modifications of Theorem (iv) are known, none of which is substantial. We give here a variant of (iv) which is formally simpler than the original statement.

(v) *Under the assumptions of* (iv), *let* $\Sigma' = \{S_1', S_2, S_3, S_4\}$ *where*

$$S_1' = \forall_p \exists_s \forall_x [xEs \leftrightarrow \exists_{yz}(yEp \wedge C(x,y,z) \wedge yEz) \wedge Sx].$$

Then $\Omega^s \vdash \Sigma'$ *and* $\Omega^s \equiv_T \Sigma'$.

Since S_1' is obviously equivalent with a sentence in Ω^s, we have to show that

(1) $\Sigma' \vdash_T S_1$

and

(2) $\Sigma' \vdash_T S_5$.

In fact, using S_1' and T, and applying S_4 three times in succession, we conclude that

$$\Sigma' \vdash_T \exists_u \forall_y (Sy \rightarrow yEu).$$

Hence, using S_1' again (this time with p replaced by u), we obtain (1).

Recall now that in the proof of (iv) we have derived statement (2) without using S_5; thus, in view of (1) we have shown that

(3) $\Sigma' \vdash_T \exists_s \forall_x [xEs \leftrightarrow \exists_y C(x,y,y) \wedge Sx]$.

Also, using S_1', S_3, and T we easily see that

(4) $\Sigma' \vdash_T \forall_p \exists_s \forall_x [xEs \leftrightarrow \exists_{yz}(C(x,y,z) \wedge yEp \wedge zEp) \wedge Sx]$.

From (3) and (4), applying S_2 twice and then S_4, we derive (2) above, which completes the proof.

An obvious corollary of (iv) and (v) is:

(vi) *Assume that* T, Σ, *and* Σ' *are defined as in* (iv) *and* (v). *Let* $\mathcal{S}$ *be any system formalized in* $\mathcal{L}$ *whose axiom set* $\mathbf{A}\xi$ *satisfies the following conditions:* $\mathbf{\Omega}^s \subseteq \mathbf{A}\xi$, *the set* $\mathbf{A}\xi \sim \mathbf{\Omega}^s$ *is finite, and* $\mathbf{A}\xi \vdash T$. *Then* $\mathcal{S}$ *is finitely axiomatizable, i.e., there is a finite set* Ξ *such that* $\mathbf{A}\xi \equiv \Xi$; *in fact, the set* $\mathbf{A}\xi \sim \mathbf{\Omega}^s \cup \Sigma \cup \{T\}$ *or else* $\mathbf{A}\xi \sim \mathbf{\Omega}^s \cup \Sigma' \cup \{T\}$ *can be taken for* Ξ.

Results very similar to those stated above can be obtained by direct inspection of the proofs of Bernays and Gödel referred to earlier in this section. Indeed, the results we have in mind establish the finite axiomatizability of a comprehensive class of systems which are formalized in $\mathcal{L}$ and whose axiom sets are finite extensions of the set $\mathbf{\Omega}^s$, and they also supply for each of these systems a specific finite set of sentences which can be used as the axiom set. It turns out, however, that the results stated in (iv)–(vi) apply to a wider class of systems, and the finite axiomatizations provided by them appear to be formally simpler.

To be more specific, consider the result derived from the proof of Gödel. It differs from (iv), (vi) only in that the sentence T and the set Σ are replaced by two sets of sentences, Δ and Θ. Here $\Delta \cup \Theta$ is the set of all the sentences of Gödel's axiom set which are used in his proof; Θ consists of those sentences which are derivable from $\mathbf{\Omega}^s$, and Δ consists of the remaining ones.

From a remark in Gödel [**1940**], p. 18, ℓ. 12f, it is seen that the set of Gödel's axioms A1–A4 (*op. cit.*, p. 3) is intended to be taken for Δ. Since, however, we consider here a version of $\mathcal{BG}$ whose only nonlogical constant is $\boldsymbol{E}$ (cf. §4.6), A1 and A2 fall away; thus Δ consists of two sentences, the extensionality axiom A3 and the pair axiom A4, i.e., our sentences T_6 from §4.6 and T from (iv).[1] At any rate, the premise $\mathbf{A}\xi \vdash T$ in (vi) is now replaced by a stronger premise, $\mathbf{A}\xi \vdash \Delta$; this of course weakens the result by narrowing its range of applications. (It is possible, however, that this replacement may be shown to be unnecessary.)

The set Θ consists of all Gödel's axioms of class construction, B1–B8 (*op. cit.*, p. 5). When Θ, Σ, and Σ' are compared with respect to the number and the formal structure of their constituents, Θ appears to be more complex than Σ and *a fortiori* more complex than Σ'. In this connection it should perhaps be pointed

[1] Gödel's argument uses in one place (*op. cit.*, p. 10) his Axiom $\boldsymbol{D}$, the well-foundedness axiom, which should therefore also be included in Δ. Actually, however, this use of Axiom $\boldsymbol{D}$ can easily be eliminated (cf. Mendelson [**1964**], p. 164).

out that to compare the structural simplicity of sentences in a formal language one should formulate them without using any abbreviations or defined terms such as $\boldsymbol{C}(x,y,z)$, $\langle x,y\rangle$, or $\langle x,y,z\rangle$. It then becomes clear that the formulations of the sentences B7 and B8 in Θ are quite involved, and that of the sentence S_3 in Σ and Σ' is not much simpler; the presence of some of these sentences may be regarded as an "aesthetic defect" of any axiom set in which they are included. It would therefore be interesting to know whether, for instance, S_3 could not be replaced in (iv) and (vi) by a substantially simpler sentence or a couple of such sentences. If no way of achieving this is found, we may be inclined to think that the complicated formal structure of some set-theoretical axioms is an unavoidable consequence of the "unnatural" tendency to adopt membership as the only primitive notion and of the resulting necessity of defining in its terms such intuitively distant concepts as ordered couple and relation.

Theorems (iv), (vi), and the corresponding result derived from Gödel's proof jointly imply that the sets Σ, Σ', and Θ are equivalent under Δ. In consequence, our results enable us to simplify every axiom set which includes both Δ and Θ and, in fact, to replace in it Θ by Σ or Σ'. In particular, the finite axiom set for $\mathcal{BG}$ given in Gödel [**1940**], pp. 3–6, can be simplified in that way. This observation has already been used in the literature to obtain a concise and elegant formulation of the constructibility axiom; see Scott [**1961**], p. 521.[2]

With minor changes, all the above remarks apply also to results derived from the proof of Bernays [**1937**], pp. 73–76. The set replacing T is in this case even stronger than Δ.

In connection with (iv)–(vi), certain problems arise which present perhaps some intrinsic interest, although they do not seem to be significant for the study of axiomatic foundations of set theory. From a general metamathematical point of view the interesting part of (vi) is the fact that the set $\mathbf{\Omega}^{\mathrm{s}} \cup \{T\}$ is finitely based, i.e., that there is a finite set of sentences logically equivalent with it; the actual composition of that finite set is rather irrelevant. The problem naturally arises whether the set $\mathbf{\Omega}^{\mathrm{s}}$ itself is finitely based. If the solution is negative (and this seems to be plausible), then one can ask further questions concerning the set Γ of all sentences Y such that $\mathbf{\Omega}^{\mathrm{s}} \cup \{Y\}$ is finitely based. In particular, we may inquire whether various specific sentences possessing some intrinsic interest belong to Γ. In view of (iv) a sufficient condition for a sentence Y to belong to Γ is that Y imply T, either logically or at least relative to $\mathbf{\Omega}^{\mathrm{s}}$, i.e., that $\mathbf{\Omega}^{\mathrm{s}} \cup \{Y\} \vdash T$. As an example of a sentence satisfying this condition we mention

$$\forall_{pqs}[\forall_x(xEs \leftrightarrow x\mathring{\imath}p \vee x\mathring{\imath}q) \rightarrow Ss],$$

which is somewhat simpler (though logically neither stronger nor weaker) than T. On the other hand, the problem seems to be open whether Γ contains, e.g., the restricted pair axiom given as S_2 in 3.6(ii) or the restricted singleton axiom in the form

$$\forall_p[Sp \rightarrow \exists_s(Ss \wedge \forall_x(xEs \leftrightarrow x\mathring{\imath}p))].$$

[2] There is a misprint in that formulation: "$x|x$" should be replaced by "$x|y$".

6.5. The finite axiomatizability of predicative systems of set theory excluding proper classes

We now turn to comprehension schemata in systems of set theory which exclude proper classes, and in which, therefore, the sentence $\mathbf{V}_x Sx$ holds. We have in mind primarily the system of Zermelo, together with its variants and extensions.

It is clear that neither the impredicative schema (C) in §6.4 nor its predicative variant ($\mathrm{C^s}$) can be used in constructing axiom sets for systems of this kind. Indeed, as is easily seen, Schema ($\mathrm{C^s}$) not only admits models with proper classes, but actually implies the existence of such classes, i.e., the negation of $\mathbf{V}_x Sx$, and the same applies *a fortiori* to Schema (C).

On the other hand, the following axiom schema, which occurs in the modern version $\mathcal{Z}$ of the Zermelo system (and corresponds to the "Aussonderungsaxiom" in the original version), is well adapted to the construction of systems excluding proper classes:

$$\text{(Z)} \qquad [\mathbf{V}_u \mathbf{\exists}_s \mathbf{V}_x (xEs \leftrightarrow F \wedge xEu)] \quad \textit{with} \quad F \in \mathbf{\Phi} \quad \textit{and} \quad s \notin \Upsilon\phi F.$$

We can also consider a predicative variant of (Z) which is obtained from (Z) by restricting the range of variables occurring bound in F to elements of the class represented by u. A formula obtained in this way from a given formula X can be denoted by "X^{u}". A formal definition of X^{u} differs from the definition of X^{s} only in that expressions of the form Sx in 6.3(ii) are replaced by the corresponding expressions xEu. For obvious reasons, we avoid applying the formal definition of X^{u} to formulas X in which u occurs bound. The predicative variant of (Z) now assumes the form

$$(\mathrm{Z^u}) \qquad [\mathbf{V}_u \mathbf{\exists}_s \mathbf{V}_x (xEs \leftrightarrow F^{\mathrm{u}} \wedge xEu)], \quad \textit{where} \quad F \in \mathbf{\Phi},\ s \notin \Upsilon\phi F, \quad \textit{and } u \textit{ does not occur bound in } F.$$

As in the case of $\mathbf{\Omega}$ and $\mathbf{\Omega}^{\mathrm{s}}$ we set

$$\text{(i)} \quad \mathbf{\Psi} = \{X: X = [\mathbf{V}_u \mathbf{\exists}_s \mathbf{V}_x (xEs \leftrightarrow F \wedge xEu)] \textit{ for some } F \in \mathbf{\Phi} \textit{ with } s \notin \Upsilon\phi F\},$$

$$\text{(ii)} \quad \mathbf{\Psi}^{\mathrm{u}} = \{X: X = [\mathbf{V}_u \mathbf{\exists}_s \mathbf{V}_x (xEs \leftrightarrow F^{\mathrm{u}} \wedge xEu)] \textit{ for some } F \in \mathbf{\Phi} \textit{ with } s \notin \Upsilon\phi F \textit{ and } u \textit{ not occurring bound in } F\}.$$

Let $\mathcal{S}$ be any system of set theory formalized in $\mathcal{L}^+$ whose axiom set $\mathbf{A}\xi$ includes $\mathbf{\Psi}$, and let $\mathcal{S}^{\mathrm{u}}$ be the system whose axiom set is $\mathbf{A}\xi \sim \mathbf{\Psi} \cup \mathbf{\Psi}^{\mathrm{u}}$. It is known from the literature that some important such systems $\mathcal{S}$ are not finitely axiomatizable; for $\mathcal{Z}$ and all its consistent extensions this has been shown in Montague [**1961**]. Our next theorem implies that the predicative variants $\mathcal{S}^{\mathrm{u}}$ of many such systems $\mathcal{S}$ are finitely axiomatizable.

(iii) *Let* $\Sigma = \{S_1, S_2, S_3\}$, *where*

$$S_1 = \forall_u \exists_s \forall_x [xEs \leftrightarrow \exists_{yz}(C(x,y,z) \wedge yEz) \wedge xEu],$$

$$S_2 = \forall_{up} \exists_s \forall_x (xEs \leftrightarrow \neg xEp \wedge xEu),$$

$$S_3 = \forall_{upq} \exists_s \forall_x [xEs \leftrightarrow \exists_{yzrtw}(yEp \wedge zEq \wedge C(x,r,t) \wedge C(y,r,w) \wedge C(z,t,w)) \wedge xEu].$$

Furthermore, let $\Theta = \{T_1, T_2, T_3\}$, *where*

$$T_1 = \forall_p \exists_s \forall_x [xEs \leftrightarrow \exists_y (xEy \wedge yEp)],$$

$$T_2 = \forall_{pq} \exists_s \forall_x [xEs \leftrightarrow \exists_{yz}(yEp \wedge zEq \wedge \forall_w (wEx \leftrightarrow w\mathring{I}y \vee w\mathring{I}z)) \wedge Sx],$$

$$T_3 = \forall_{pq}[Sp \wedge Sq \rightarrow \exists_s (Ss \wedge \forall_x (xEs \leftrightarrow x\mathring{I}p \vee x\mathring{I}q))].$$

Then we have $\mathbf{\Psi}^{\mathsf{u}} \equiv_\Theta \Sigma$.

We shall not outline here a proof of (iii). Obviously, Theorem (iii) is closely related to 6.4(iv); just as the latter, it has been strongly influenced by the discussion in Chapter 4. Each of the special sentences in (iii) corresponds in a natural way to one in 6.4(iv), with which it either coincides or differs only in details. (The pair of corresponding sentences T_2 and S_5 could be regarded as an exception to this statement; they are related in content, but S_5 has a much more special character than T_2.) The ideas underlying the proofs of the two theorems are also similar; technically, however, the proofs are different. We see no way of applying the results of Chapter 4 to the proof of (iii). What was obtained in the proof of 6.4(iv) by means of those results has now to be achieved in a direct way; this lengthens and complicates the argument.

We may mention that the results of Chapter 4 could be applied to the proof of (iii) if we included in the set Θ the following sentence:

$$T_4 = \forall_u \exists_s [\forall_x (xEu \rightarrow xEs) \wedge \forall_{xyz}(yEs \wedge zEs \wedge Sx \wedge \forall_w (wEx \leftrightarrow w\mathring{I}y \vee w\mathring{I}z) \rightarrow xEs)].$$

This would, however, considerably weaken the applicability of our result; for instance it could not be applied to the system $\mathfrak{Z}^{\mathsf{u}}$, the predicative version of Zermelo's system.

Notice a difference between Theorems (iii) above and 6.4(iv): in (iii) it is not claimed that all sentences of Σ are derivable from $\mathbf{\Psi}^{\mathsf{u}}$. In fact, it seems unlikely that either $\mathbf{\Psi}^{\mathsf{u}} \vdash S_1$ or $\mathbf{\Psi}^{\mathsf{u}} \vdash S_3$ holds. However, we can replace S_1 and S_3 by two related (though more complicated) sentences, $\overline{S}_1$ and $\overline{S}_3$, which (like S_2) are members of $\mathbf{\Psi}^{\mathsf{u}}$, without affecting the validity of the conclusion of (iii).

We proceed as follows. S_1 and S_3 are obviously in $\mathbf{\Psi}$ by (i). Let X be in $\mathbf{\Psi}$. We pick a variable v which does not occur in X, and let F' be the formula obtained from F by replacing $\boldsymbol{u}$ with v everywhere (where F is the subformula of X involved in (i)). Set

$$\overline{X} = [\forall_u \exists_s \forall_x (xEs \leftrightarrow (F' \wedge xEv)^{\mathsf{u}} \wedge xEu)].$$

By (ii) we obviously have $\overline{X} \in \mathbf{\Psi}^{\mathsf{u}}$; hence, if we set $\overline{\Sigma} = \{\overline{S}_1, S_2, \overline{S}_3\}$, we obtain $\mathbf{\Psi}^{\mathsf{u}} \vdash \overline{\Sigma}$. On the other hand, it is not difficult to show that $\overline{X} \vdash_\Theta X$ for $X = S_1$

and $X = S_3$, and therefore $\mathbf{\Psi}^{\mathsf{u}} \equiv_{\Theta} \bar{\Sigma}$. Actually, the argument just outlined is a part of the proof of (iii).

In this connection we should like to mention two schemata closely related to $(\mathrm{C}^{\mathsf{s}})$ and $(\mathrm{Z}^{\mathsf{u}})$. The first of them is

$(\mathrm{C}^{\mathsf{s}'})$ $[\exists_s \forall_x (x\boldsymbol{E}s \leftrightarrow F \wedge \boldsymbol{S}x)]$, *where* $F \in \mathbf{\Phi}$, $s \notin \Upsilon\phi F$, *and every subformula of* F *which begins with* $\forall$ *has the form* $\forall_y (y\boldsymbol{E}z \rightarrow G)$ *with* $y, z \in \Upsilon$, $y \neq z$, *and* $G \in \mathbf{\Phi}$.

The second schema, $(\mathrm{Z}^{\mathsf{u}'})$, differs from $(\mathrm{C}^{\mathsf{s}'})$ only in that $\boldsymbol{S}x$ is replaced by $x\boldsymbol{E}u$ at the end of the schema.

It is readily seen that $(\mathrm{C}^{\mathsf{s}})$ and $(\mathrm{C}^{\mathsf{s}'})$ are equivalent; we use the fact that $\mathbf{\Omega}^{\mathsf{s}} \vdash \exists_u \forall_x (x\boldsymbol{E}u \leftrightarrow \boldsymbol{S}x)$. However, $(\mathrm{Z}^{\mathsf{u}'})$ appears to be stronger than $(\mathrm{Z}^{\mathsf{u}})$. If $\mathbf{\Psi}^{\mathsf{u}'}$ is the set of all instances of $(\mathrm{Z}^{\mathsf{u}'})$, then obviously $\mathbf{\Psi}^{\mathsf{u}'} \supseteq \mathbf{\Psi}^{\mathsf{u}}$. Furthermore, S_1 and S_3 are logically equivalent with some sentences in $\mathbf{\Psi}^{\mathsf{u}'}$, so that $\mathbf{\Psi}^{\mathsf{u}'} \vdash \Sigma$, while from what was stated above we are inclined to believe that $\mathbf{\Psi}^{\mathsf{u}} \vdash \Sigma$ does not hold. On the other hand, if $X \in \mathbf{\Psi}^{\mathsf{u}'}$ and $\overline{X}$ is constructed in the way indicated above, we can show that $\overline{X} \vdash_{\Theta} X$. Consequently $\mathbf{\Psi}^{\mathsf{u}'} \equiv_{\Theta} \mathbf{\Psi}^{\mathsf{u}}$ and hence $\mathbf{\Psi}^{\mathsf{u}'} \equiv_{\Theta} \Sigma$, so that $\mathbf{\Psi}^{\mathsf{u}'}$ can be used instead of $\mathbf{\Psi}^{\mathsf{u}}$ to obtain the conclusion of (iii) in the desired form, without changing Σ.

It may be noticed that Schemata $(\mathrm{C}^{\mathsf{s}'})$ and $(\mathrm{Z}^{\mathsf{u}'})$ seem to be better adapted to the actual development of set theory than $(\mathrm{C}^{\mathsf{s}})$ and $(\mathrm{Z}^{\mathsf{u}})$, but, at the same time, they seem more distant from the intuitive notion of predicativity.

Theorem (iii) obviously implies the following corollary (an analogue of 6.4(vi)).

(iv) *Under the assumptions of* (iii) *let* $\mathcal{S}$ *be any system formalized in* $\mathcal{L}$ *with an axiom set* $\mathbf{A}\xi$ *such that* $\mathbf{A}\xi \vdash \Theta$, $\mathbf{\Psi}^{\mathsf{u}} \subseteq \mathbf{A}\xi$, *and the set* $\mathbf{A}\xi \sim \mathbf{\Psi}^{\mathsf{u}}$ *is finite. Then there is a finite set* Ξ *such that* $\mathbf{A}\xi \equiv \Xi$; *e.g.*, $\Xi = \mathbf{A}\xi \sim \mathbf{\Psi}^{\mathsf{u}} \cup \Sigma \cup \Theta$.

If we are interested specifically in systems which exclude proper classes, we can obviously employ the pair axiom $\boldsymbol{P}$ in place of the restricted pair axiom T_3 in formulating both (iii) and (iv) above.

With this modification we can use (iii) to obtain a finite axiom set for the predicative version $\mathcal{Z}^{\mathsf{u}}$ of the system $\mathcal{Z}$ of Zermelo. The axiom set will consist of those axioms of $\mathcal{Z}$ which are not instances of (Z), as well as of the three sentences S_1, S_2, S_3 of (iii). Notice that $\boldsymbol{P}$ and T_1 are actually axioms of $\mathcal{Z}$, while T_2 can easily be derived from the axiom set for $\mathcal{Z}^{\mathsf{u}}$ just described. (A detailed derivation of T_2 has been worked out by Maddux.)

The system $\mathcal{Z}^{\mathsf{u}}$ deserves perhaps more attention than it has been paid so far. It seems to provide a sufficient basis for the reconstruction of a large part of mathematics and, in particular, practically all of classical analysis. It is finitely axiomatizable and, as we have just seen, it can be based upon a rather simple finite axiom set. As a consequence of finite axiomatizability, its semantics can be adequately developed within $\mathcal{Z}$; this is seen from the results in Levy [**1965**].

The set of sentences $\mathbf{\Psi}$ and *a fortiori* its predicative version $\mathbf{\Psi}^{\mathsf{u}}$ do not imply $\forall_x \boldsymbol{S}x$, i.e., do not exclude proper classes, and this is true even in case these sets are supplemented by sentences $T_1 - T_3$ of (iii). Hence, Schemata (Z) and (Z^u) can be used to construct set-theoretical systems admitting proper classes.

$\mathbf{\Psi}$ is weaker than the set $\mathbf{\Omega}$, and similarly $\mathbf{\Psi}^u$ is weaker than $\mathbf{\Omega}^s$. However, the difference in deductive powers is not large. The sentence $\mathbf{\exists}_u\mathbf{\forall}_x(xEu \leftrightarrow Sx)$, stating the existence of the universal class, is derivable from $\mathbf{\Omega}$ and $\mathbf{\Omega}^s$, but not from $\mathbf{\Psi}$ or $\mathbf{\Psi}^u$. It is easily seen, however, that

(v) $\mathbf{\Omega} \equiv \mathbf{\Psi} \cup \{\mathbf{\exists}_u\mathbf{\forall}_x(xEu \leftrightarrow Sx)\}$

and, similarly,

(vi) $\mathbf{\Omega}^s \equiv \mathbf{\Psi}^u \cup \{\mathbf{\exists}_u\mathbf{\forall}_x(xEu \leftrightarrow Sx)\}$.

A consequence of these simple remarks is that the discussion of the schemata (Z) and (Z^u) has indeed a wide range of applications. In particular, in view of (vi), Theorem (iv) secures the finite axiomatizability of $\mathcal{Z}^u$, $\mathcal{BG}$, and all their finite extensions. It may also be noticed that 6.4(vi) could be derived rather easily as a corollary of (iv) (and similarly 6.4(iv) as a corollary of (iii)).[3]

The Zermelo-Fraenkel system has not yet been involved in the discussion of this chapter. It is based on the so-called replacement schema, which is quite different from the comprehension schema. The results of the present work do not seem to lead to any interesting conclusions concerning the axiomatic foundations of the Zermelo-Fraenkel system. It is known that the system is not finitely axiomatizable and cannot even be provided with an axiom set which is a finite extension of the set $\mathbf{\Psi}$; cf. Montague [**1962**]. On the other hand, some systems are known which can be regarded as predicative variants of that system and can be proved to be finitely axiomatizable. Certain results in this direction obtained by Tarski (but not published) have been superseded by results in Levy [**1965**]; see here also Thiele [**1968**].

To conclude, we may mention that there are some interesting open problems which concern the set $\mathbf{\Psi}^u$ and are analogous to those discussed at the end of §6.4. In particular, while 6.4(iv) implies the finite axiomatizability of $\mathbf{\Omega}^s \cup \{T\}$, (iii) implies the same for the set $\mathbf{\Psi}^u \cup \Theta$. Θ is, however, much stronger than T since, in addition to $T_3 = T$, it contains two other sentences. Hence, the problem arises whether the presence of all these sentences in Θ is essential for the conclusion. It would be interesting to know, in particular, whether $\mathbf{\Psi}^u$ by itself is finitely axiomatizable and, if not, whether this holds for $\mathbf{\Psi}^u \cup \{T\}$.

[3] The main results in §§6.4 and 6.5 were found by Tarski in the late 1940's. To our knowledge the result in 6.5(iv), as applied to the system $\mathcal{Z}^u$ (or, more exactly, $\mathcal{Z}^{u'}$) was first stated in print in Mostowski [**1954**], p. 24, as an unpublished result of Tarski.

CHAPTER 7

Extension of Results to Arbitrary Formalisms of Predicate Logic, and Applications to the Formalization of the Arithmetics of Natural and Real Numbers

We shall concern ourselves in this chapter with systems formalized in arbitrary languages of predicate logic, and not necessarily in the particular language $\mathcal{L}$ underlying the discussion in the earlier chapters of our work. The notion of a $\mathcal{Q}$-system can readily be extended to such systems in an adequate way. The main conclusion of a general character which will be established here, in 7.2(iv), is that for practically every $\mathcal{Q}$-system in a language of predicate logic with finitely many nonlogical constants an equipollent system can be constructed in the simple language $\mathcal{L}^{\times}$ described in Chapter 3. The best known and mathematically most important examples of systems to which this conclusion applies are—besides the theory of sets—the arithmetic of natural numbers (elementary number theory) and the arithmetic of real numbers, as well as some of their axiomatic subsystems, in particular, Peano arithmetic.[1]

7.1. Extension of equipollence results to $\mathcal{Q}$-systems in first-order formalisms with just binary relation symbols

We first consider formalisms $\mathcal{M}^{(n)}$, for an arbitrary natural number n, which differ from $\mathcal{L}$ only in that their vocabularies are provided with $n+1$ nonlogical constants. (Because of the triviality of formalisms without nonlogical constants, we do not care to subsume them under our discussion; formally, the results and proofs in this chapter apply, with minimal changes, to such formalisms as well.)

All the nonlogical constants of $\mathcal{M}^{(n)}$ are (atomic) binary predicates; they are assumed to be arranged in a finite sequence without repeating terms, $\langle \boldsymbol{F}_0, \ldots, \boldsymbol{F}_n \rangle$. Just as in the case of $\mathcal{L}$, we correlate with each $\mathcal{M}^{(n)}$ an extended formalism $\mathcal{M}^{(n)+}$ and a simple subformalism of the latter (without variables, quantifiers, and sentential connectives), $\mathcal{M}^{(n)\times}$. Thus $\mathcal{M}^{(n)+}$ has the same logical constants

[1] The main results of this chapter were announced in Tarski [**1954**] and [**1954a**].

as $\mathcal{L}^+$ and the same nonlogical constants as $\mathcal{M}^{(n)}$. The syntactical and semantical notation established for $\mathcal{L}, \mathcal{L}^+, \mathcal{L}^\times$ extends with obvious changes to the new formalisms; this applies, in particular, to the notions of a $\mathcal{Q}$-system and a $\mathfrak{Q}$-structure. Realizations of $\mathcal{M}^{(n)}, \mathcal{M}^{(n)+}$, and $\mathcal{M}^{(n)\times}$ are, of course, relational structures $\mathfrak{U} = \langle U, F_0, \ldots, F_n \rangle$, where $F_0, \ldots, F_n$ are arbitrary binary relations on the set U.

In our previous discussion we referred occasionally to formalisms $\mathcal{M}^{(n)}, \mathcal{M}^{(n)+}$, $\mathcal{M}^{(n)\times}$ (without introducing special designations for them); specifically, the formalism $\mathcal{M}^{(1)}$, with two nonlogical predicates, was involved in the discussion of set-theoretical systems with individuals in §4.6. It was pointed out several times that the validity of our results and their proofs, referring to formalisms $\mathcal{L}, \mathcal{L}^+, \mathcal{L}^\times$, is not essentially affected when we pass to formalisms with virtually the same structure but with a vocabulary containing a larger number of nonlogical binary predicates. This applies, in particular, to the main equipollence results in Chapter 4. Thus, every $\mathcal{Q}$-system $\mathcal{T}$ in $\mathcal{M}^{(n)}$ (or $\mathcal{T}^+$ in $\mathcal{M}^{(n)+}$) is equipollent in means of expression and proof with the correlated system $\mathcal{T}^\times$ in $\mathcal{M}^{(n)\times}$. In Theorems (ii)–(iv) below we shall go a step further and show that, under certain mild restrictions on $\mathcal{T}$, a system $\mathcal{S}^\times$ equipollent with $\mathcal{T}$ can also be constructed in the original formalism $\mathcal{L}^\times$.

It may be recalled that a general policy in this work is not to discuss the equipollence of two systems in means of expression and proof unless they can be presented as subsystems of a given third system (cf. §2.5). This is the reason why, in formulating various equipollence results in this chapter, we shall always point out specific common equipollent extensions of the systems involved, instead of simply stating the equipollence of these systems.

We should also remember that, in agreement with our observations in 2.4(iii), a translation mapping constructed to establish the equipollence of a system with one of its subsystems must always be recursive (assuming that the notion of recursiveness has been appropriately extended to sets of expressions of the formalisms concerned, and to relations between and functions on such sets—and this assumption is satisfied in all cases discussed below). The recursiveness of the translation mappings involved in the subsequent discussion will always follow easily from their (explicit or implicit) constructions, and we shall usually not even bother to point it out.

We begin with a simple lemma which will play an important role in a subsequent argument (but which, because of its content, could have been placed somewhere in the earliest sections of Chapter 4).

(i) *Let U be a set with $|U| > 1$. If R and S are conjugated quasiprojections on U, then so are $(R|S) \cap Di$ and $S \cap Di$.*

In fact, since R and S are functions, by the definition of quasiprojections in §4.1, it is obvious that $(R|S) \cap Di$ and $S \cap Di$ are also functions. Thus, the proof reduces to showing that, for any $y, z \in U$, there is an $x \in U$ different from y and z, and such that $xR|Sy$ and xSz. Notice that by 4.1(v) and the hypothesis of (i) the set U must be infinite. Hence, *a fortiori* there are three distinct elements $u_1, u_2, u_3 \in U$. Since R and S are quasiprojections, there exist elements $v_i \in U$ with v_iRu_i and v_iSy for $i = 1, 2, 3$. Therefore, R being a function, the elements v_1, v_2, v_3 must also be distinct. Analogously, there must be three distinct elements $w_1, w_2, w_3 \in U$ with w_iRv_i and w_iSz for $i = 1, 2, 3$. Consequently we have $w_iR|Sy$ and w_iSz for $i = 1, 2, 3$, and at least one of the three elements w_i must be different from y and z. Taking this element for x we complete the proof.

(ii) *Let $\mathcal{M}^{(n)}$, for a given $n \in \omega$, be a formalism with $n+1$ nonlogical (atomic) predicates $\boldsymbol{F}_0, \ldots, \boldsymbol{F}_n$ different from the predicate $\boldsymbol{E}$ of $\mathcal{L}$, and let $\mathcal{T}$ be a $\mathcal{Q}$-system in $\mathcal{M}^{(n)}$ such that*

(∗) *for each m with $0 \le m \le n$ we have either*

$$\mathsf{A}\xi[\mathcal{T}] \vdash \boldsymbol{\exists}_{\boldsymbol{xy}}(\boldsymbol{x}\boldsymbol{F}_m\boldsymbol{y}) \quad or \quad \mathsf{A}\xi[\mathcal{T}] \vdash \boldsymbol{\forall}_{\boldsymbol{xy}}(\boldsymbol{\neg x}\boldsymbol{F}_m\boldsymbol{y}).$$

Under the above assumptions there is a $\mathcal{Q}$-system $\mathcal{S}$ in the formalism $\mathcal{L}$ and a $\mathcal{Q}$-system $\overline{\mathcal{T}}$ in the formalism $\mathcal{M}^{(n+1)}$ with $n+2$ nonlogical predicates, $\boldsymbol{F}_0, \ldots, \boldsymbol{F}_n$, $\boldsymbol{E}$, satisfying the following conditions.

(α) *$\overline{\mathcal{T}}$ is a common definitional extension of the systems $\mathcal{S}$ and $\mathcal{T}$; consequently, $\mathcal{S}$ and $\mathcal{T}$ are definitionally equivalent, and hence equipollent with each other in means of expression and proof.*

(β) *The system $\overline{\mathcal{T}}^+$ in the extended formalism $\mathcal{M}^{(n+1)+}$ is a common equipollent extension, not only of $\mathcal{T}$ and $\mathcal{S}$, but also of the system $\mathcal{S}^\times$ in $\mathcal{L}^\times$; hence $\mathcal{T}$ and $\mathcal{S}^\times$ (treated as subsystems of $\overline{\mathcal{T}}^+$) are equipollent with each other as well.*

In outlining the proof of Theorem (ii) we find it more convenient to consider first, instead of systems $\mathcal{T}, \mathcal{S}$, and $\overline{\mathcal{T}}$, the correlated systems $\mathcal{T}^+, \mathcal{S}^+$, and $\overline{\mathcal{T}}^+$ in the extended formalisms $\mathcal{M}^{(n)+}$, $\mathcal{L}^+$, and $\mathcal{M}^{(n+1)+}$.

Thus $\mathcal{T}^+$ is assumed to be a $\mathcal{Q}$-system in $\mathcal{M}^{(n)+}$ satisfying (∗) (with $\mathcal{T}$ replaced, of course, by $\mathcal{T}^+$). Hence, by 4.5(i) (applied to $\mathcal{M}^{(n)+}$) there are $A', B' \in \mathbf{\Pi}[\mathcal{M}^{(n)+}]$ such that

(1) $\quad \mathsf{A}\xi[\mathcal{T}^+] \vdash \boldsymbol{Q}_{A'B'}.$

Keeping in mind the semantical completeness of the formalism $\mathcal{M}^{(n)+}$, we obtain, as a metalogical translation of Lemma (i),

(2) $\quad \boldsymbol{Q}_{A'B'} \vdash \boldsymbol{\neg}\mathring{\mathbf{0}} = \mathbf{0} \rightarrow \boldsymbol{Q}_{A' \odot B' \cdot \mathring{0},\ B' \cdot \mathring{0}}.$

Notice also that by 4.1(ii) we obviously have

(3) $\quad \vdash \mathring{\mathbf{0}} = \mathbf{0} \rightarrow \boldsymbol{Q}_{\mathbf{1},\mathbf{1}}.$

Set

(4) $\quad A'' = \mathbf{0} \oplus \mathring{\mathbf{1}} \oplus \mathbf{0} + A' \odot B' \cdot \mathring{\mathbf{0}}, \qquad B'' = \mathbf{0} \oplus \mathring{\mathbf{1}} \oplus \mathbf{0} + B' \cdot \mathring{\mathbf{0}}.$

Since by 2.2(iii) and 3.2(iii),(xiii),

(5) $\quad \vdash \neg \mathring{\mathbf{0}} = \mathbf{0} \rightarrow \mathbf{0} \oplus \mathring{\mathbf{1}} \oplus \mathbf{0} = \mathbf{0}$ and $\vdash \mathring{\mathbf{0}} = \mathbf{0} \rightarrow \mathbf{0} \oplus \mathring{\mathbf{1}} \oplus \mathbf{0} = \mathbf{1},$

we get from (4),

(6) $\quad \vdash \neg \mathring{\mathbf{0}} = \mathbf{0} \rightarrow (A'' = A' \odot B' \cdot \mathring{\mathbf{0}} \wedge B'' = B' \cdot \mathring{\mathbf{0}})$ and
$\quad \vdash \mathring{\mathbf{0}} = \mathbf{0} \rightarrow (A'' = \mathbf{1} \wedge B'' = \mathbf{1}).$

Steps (2), (3), and (6) yield directly that $\boldsymbol{Q}_{A'B'} \vdash \boldsymbol{Q}_{A''B''}$, and hence we get by (1),

(7) $\quad \mathbf{A}\xi[\mathfrak{T}^+] \vdash \boldsymbol{Q}_{A''B''}.$

Therefore, if we set $C = (A'' + B'')^{\smile}$, we conclude by 4.7(x) that

(8) $\quad \mathbf{A}\xi[\mathfrak{T}^+] \vdash \boldsymbol{U}_C.$

Moreover, the definition of C, together with (4), (5), imply an additional property of C, namely

(9) $\quad \mathbf{A}\xi[\mathfrak{T}^+] \vdash C \cdot \mathring{\mathbf{1}} = \mathbf{0} \oplus \mathring{\mathbf{1}} \oplus \mathbf{0}.$

As we shall see, this additional property plays an essential part in our argument. This explains why we have constructed the predicate C satisfying (8) in a roundabout way, instead of directly deriving its existence from 4.7(x) and the hypothesis of our theorem.

Setting

(10) $\quad D = C^{\smile} \odot [C^{\smile} \cdot (C^{\smile -} \oplus \mathring{\mathbf{1}})], \qquad F = C^{\smile} \odot C^{\smile},$

(11) $\quad A = D \cdot (D^{-} \oplus \mathring{\mathbf{1}}), \qquad B = F \cdot (F^{-} + A \oplus \mathring{\mathbf{1}}),$

we obtain by (8) and 4.7(ix) that

(12) $\quad \mathbf{A}\xi[\mathfrak{T}^+] \vdash \boldsymbol{Q}_{AB}.$

In the subsequent discussion we shall disregard entirely the original predicates A' and B', as well as the predicates A'' and B'' constructed in terms of them, and we shall use instead the new predicates, A and B. The reason why A and B are better suited for our purposes is that they have been constructed, with the help of logical symbols, from a single nonlogical predicate C.

Our next task is to construct a system $\bar{\mathcal{T}}^+$ which is a common extension of both the given system $\mathcal{T}^+$ and a system $\mathcal{S}^+$ (in $\mathcal{L}^+$) to be subsequently described. In agreement with the statement of our theorem, $\bar{\mathcal{T}}^+$ is to be developed in the formalism $\mathcal{M}^{(n+1)+}$ with $n+2$ nonlogical atomic predicates, $\boldsymbol{F}_0, \ldots, \boldsymbol{F}_n, \boldsymbol{F}_{n+1}$, where $\boldsymbol{E}$, the only nonlogical atomic predicate of $\mathcal{L}^+$, is taken for $\boldsymbol{F}_{n+1}$.

To describe $\bar{\mathcal{T}}^+$ we notice that, by the condition $(*)$ of our theorem, every number $k \in \omega$ with $0 \leq k \leq n$ can be put into one of two mutually exclusive sets, depending on whether $\mathbf{A}\xi[\mathcal{T}^+] \vdash \neg \boldsymbol{F}_k = \mathbf{0}$ or $\mathbf{A}\xi[\mathcal{T}^+] \vdash \boldsymbol{F}_k = \mathbf{0}$. Clearly, without loss of generality we can assume that for some $m = 0, \ldots, n+1$,

$$\text{(13)} \qquad \mathbf{A}\xi[\mathcal{T}^+] \vdash \neg \boldsymbol{F}_k = \mathbf{0} \quad \text{iff} \quad 0 \leq k < m,$$

$$\text{(14)} \qquad \mathbf{A}\xi[\mathcal{T}^+] \vdash \boldsymbol{F}_k = \mathbf{0} \quad \text{iff} \quad m \leq k \leq n;$$

we include 0 and $n+1$ in the range of m so as not to exclude the possibility that the set of numbers k involved in (13) or (14) is empty.

We now define an auxiliary predicate G in $\mathcal{M}^{(n)+}$ by setting

$$\text{(15)} \qquad G = (\boldsymbol{P}_{AB}^{(0)} \odot \boldsymbol{F}_0 \odot \boldsymbol{P}_{AB}^{(0)\smile}) \cdot \cdots \cdot (\boldsymbol{P}_{AB}^{(m-1)} \odot \boldsymbol{F}_{m-1} \odot \boldsymbol{P}_{AB}^{(m-1)\smile}),$$

where $\boldsymbol{P}_{AB}^{(0)}, \ldots, \boldsymbol{P}_{AB}^{(m-1)}$ are the predicates constructed in 4.2(i). $\bar{\mathcal{T}}^+$ is uniquely described by adjoining the equation I determined by

$$\text{(16)} \qquad I = (\boldsymbol{E} = C \cdot \mathring{\mathbf{0}} + A \odot G \odot B^{\smile} \cdot \mathring{\mathbf{1}})$$

to the axiom set of $\mathcal{T}^+$, so that

$$\text{(17)} \qquad \mathbf{A}\xi[\bar{\mathcal{T}}^+] = \mathbf{A}\xi[\mathcal{T}^+] \cup \{I\} = \mathbf{A}\xi[\mathcal{T}^+] \cup \{\boldsymbol{E} = C \cdot \mathring{\mathbf{0}} + A \odot G \odot B^{\smile} \cdot \mathring{\mathbf{1}}\}.$$

Notice that, by (16), the equation I is a possible definition of $\boldsymbol{E}$ in $\mathcal{M}^{(n)+}$. Hence, by (17), $\bar{\mathcal{T}}^+$ is not only an extension, but actually a definitional extension of $\mathcal{T}^+$. Since $\mathcal{T}^+$ is a $\mathcal{Q}$-system by hypothesis, the same is obviously true for its extension $\bar{\mathcal{T}}^+$.

We proceed to the description of $\mathcal{S}^+$, and we begin by listing several consequences of the assumptions and stipulations made above.

$$\text{(18)} \qquad \mathbf{A}\xi[\bar{\mathcal{T}}^+] \vdash C = \mathbf{0} \oplus \mathring{\mathbf{1}} \oplus \mathbf{0} + \boldsymbol{E} \cdot \mathring{\mathbf{0}} \qquad \text{by (9), (16);}$$

$$\text{(19)} \qquad \mathbf{A}\xi[\bar{\mathcal{T}}^+] \vdash A \odot G \odot B^{\smile} \cdot \mathring{\mathbf{1}} = \boldsymbol{E} \cdot \mathring{\mathbf{1}} \qquad \text{by (16);}$$

$$\text{(20)} \qquad \mathbf{A}\xi[\bar{\mathcal{T}}^+] \vdash G = A^{\smile} \odot (A \odot G \odot B^{\smile} \cdot \mathring{\mathbf{1}}) \odot B \qquad \text{by (12), (17), 4.1(ix)}(\beta);$$

$$\text{(21)} \qquad \mathbf{A}\xi[\bar{\mathcal{T}}^+] \vdash G = A^{\smile} \odot (\boldsymbol{E} \cdot \mathring{\mathbf{1}}) \odot B \qquad \text{by (19), (20);}$$

(22) $\mathbf{A\xi}[\bar{\mathcal{T}}^+] \vdash \boldsymbol{F}_k = \boldsymbol{P}_{AB}^{(k)\smile} \odot G \odot \boldsymbol{P}_{AB}^{(k)}$ for $0 \le k < m$ by (12), (13), (15), 4.2(viii)(γ).

We can now establish the following conclusion: for every nonlogical predicate P in $\mathcal{M}^{(n+1)+}$ involved so far in our argument, a well-determined predicate P^∇ in $\mathcal{L}^+$ can be constructed so that the equation $P = P^\nabla$ is provable in $\bar{\mathcal{T}}^+$. Indeed, we establish this conclusion successively, first for C by (18), next for A and B by (10) and (11), then for G by (21), and finally for $\boldsymbol{F}_k$ by (22) and 4.2(i) when $0 \le k < m$ and by (14) when $m \le k \le n$. We thus obtain

(23) $$\mathbf{A\xi}[\bar{\mathcal{T}}^+] \vdash \{\boldsymbol{F}_0 = \boldsymbol{F}_0^\nabla, \ldots, \boldsymbol{F}_n = \boldsymbol{F}_n^\nabla\}.$$

The equations in (23) are clearly possible definitions in the language $\mathcal{L}^+$ of $\boldsymbol{F}_0, \ldots, \boldsymbol{F}_n$, the nonlogical constants of $\mathcal{M}^{(n)+}$. We now define a function L on the set $\Sigma[\mathcal{M}^{(n+1)+}]$ as follows: given any sentence X in $\mathcal{M}^{(n+1)+}$, LX is the sentence obtained from X when the predicates $\boldsymbol{F}_0, \ldots, \boldsymbol{F}_n$, wherever they occur in X, are respectively replaced by $\boldsymbol{F}_0^\nabla, \ldots, \boldsymbol{F}_n^\nabla$. We clearly have

(24) $\{\boldsymbol{F}_0 = \boldsymbol{F}_0^\nabla, \ldots, \boldsymbol{F}_n = \boldsymbol{F}_n^\nabla\} \vdash X \leftrightarrow LX$ and $LX \in \Sigma[\mathcal{L}^+]$ for every $X \in \Sigma[\mathcal{M}^{(n+1)+}]$.

System $\mathcal{S}^+$ can now be determined by setting

(25) $$\mathbf{A\xi}[\mathcal{S}^+] = \{LX : X \in \mathbf{A\xi}[\bar{\mathcal{T}}^+]\}.$$

Consider the set

(26) $$\Theta = \mathbf{A\xi}[\mathcal{S}^+] \cup \{\boldsymbol{F}_0 = \boldsymbol{F}_0^\nabla, \ldots, \boldsymbol{F}_n = \boldsymbol{F}_n^\nabla\}.$$

From (23)–(26) it clearly follows that Θ is a subset of $\Sigma[\mathcal{M}^{(n+1)+}]$ which is logically equivalent with $\mathbf{A\xi}[\bar{\mathcal{T}}^+]$. Θ is therefore a base of $\bar{\mathcal{T}}^+$ and could be used to replace the original axiom set of this system. Hence, in view of (26), $\bar{\mathcal{T}}^+$ proves to be an extension of $\mathcal{S}^+$, and actually, because of the definitional character of equations $\boldsymbol{F}_0 = \boldsymbol{F}_0^\nabla, \ldots, \boldsymbol{F}_n = \boldsymbol{F}_n^\nabla$, it is a definitional extension of $\mathcal{S}^+$.

We have thus shown that systems $\mathcal{T}^+$ and $\mathcal{S}^+$ have system $\bar{\mathcal{T}}^+$ as a common definitional extension. Hence we conclude, by means of a routine argument, that $\bar{\mathcal{T}}^+$ is equipollent with both $\mathcal{T}^+$ and $\mathcal{S}^+$; see 2.4(xiv). As a consequence, $\mathcal{T}^+$ and $\mathcal{S}^+$ are equipollent with each other. We know that $\mathcal{T}^+$ and $\bar{\mathcal{T}}^+$ are $\mathcal{Q}$-systems; hence, from the equipollence of $\bar{\mathcal{T}}^+$ with $\mathcal{S}^+$ we easily conclude that $\mathcal{S}^+$ is a $\mathcal{Q}$-system as well.

The conclusions obtained can readily be transferred from systems in the extended formalisms $\mathcal{L}^+$, $\mathcal{M}^{(n)+}$, and $\mathcal{M}^{(n+1)+}$ to the correlated systems in $\mathcal{L}$, $\mathcal{M}^{(n)}$, and $\mathcal{M}^{(n+1)}$. This requires, however, some minor modifications in the preceding discussion. As we know, any two correlated systems such as $\mathcal{T}$ and

$\mathcal{T}^+$ are assumed to have the same sets of nonlogical axioms (cf. §2.3). By this assumption all members of $\mathbf{A}\boldsymbol{\xi}[\mathcal{T}^+]$ are sentences, not only in $\mathcal{M}^{(n)+}$, but actually in $\mathcal{M}^{(n)}$. Similarly, the axiom sets of the constructed systems $\bar{\mathcal{T}}^+$ and $\mathcal{S}^+$ must consist exclusively of sentences in $\mathcal{M}^{(n+1)}$ and $\mathcal{L}$ respectively, if these two systems are to be correlated with appropriate systems $\bar{\mathcal{T}}$ and $\mathcal{S}$. This requires some changes in formulas (17) and (25) defining $\mathbf{A}\boldsymbol{\xi}[\bar{\mathcal{T}}^+]$ and $\mathbf{A}\boldsymbol{\xi}[\mathcal{S}^+]$. We use here the function $\boldsymbol{G}$, which was originally constructed in §2.3 as a mapping from $\mathcal{L}^+$ into $\mathcal{L}$, but can be extended in an obvious way to a mapping from $\mathcal{M}^{(n+1)+}$ into $\mathcal{M}^{(n+1)}$. In (17) we replace I by the (logically equivalent) sentence $\boldsymbol{G}I$, and we take the set thus obtained for the common axiom set of $\bar{\mathcal{T}}$ and $\bar{\mathcal{T}}^+$. Similarly, we replace in (25) all the sentences LX by $\boldsymbol{G}LX$. A detailed argument presents no difficulty. This completes the proof of (α).

Turning now to (β) we notice that $\mathcal{T}$ and $\mathcal{S}^\times$ are respectively subsystems of $\mathcal{T}^+$ and $\mathcal{S}^+$; hence $\bar{\mathcal{T}}^+$, which is an extension of $\mathcal{T}^+$ and $\mathcal{S}^+$, is also a common extension of $\mathcal{T}$ and $\mathcal{S}^\times$. Actually, as is easily seen, $\bar{\mathcal{T}}^+$ is a definitional extension of $\mathcal{T}$ (though not of $\mathcal{S}^\times$). On the other hand, the function L previously used in this proof is a translation mapping from $\bar{\mathcal{T}}^+$ onto $\mathcal{S}^+$, while the function $\boldsymbol{K}_{AB}$, which was defined and studied in §4.4, is a translation mapping from $\mathcal{S}^+$ onto $\mathcal{S}^\times$; hence the composition $\boldsymbol{K}_{AB} \circ L$ proves to be a translation mapping from $\bar{\mathcal{T}}^+$ onto $\mathcal{S}^\times$ by means of which the equipollence of $\bar{\mathcal{T}}^+$ and $\mathcal{S}^\times$ can be established. In consequence, $\mathcal{T}$ and $\mathcal{S}^\times$, treated as subsystems of $\bar{\mathcal{T}}^+$, turn out to be equipollent with each other.

It goes without saying that everything in this work which is established for given formalisms remains true for any formalisms which differ from the given ones only in the shape of the symbols (constants or variables) occurring in them. In particular, in formulating (ii) we can omit the restriction that $\boldsymbol{E}$ is not a nonlogical constant in $\mathcal{M}^{(n)}$ provided that at the same time we replace our original formalism $\mathcal{L}$ by any formalism of type $\mathcal{M}^{(0)}$ with a nonlogical predicate $\boldsymbol{E}'$ different from $\boldsymbol{F}_0, \ldots, \boldsymbol{F}_n$. Analogous remarks apply of course to all later theorems in this chapter, and, in particular, to (iii) and 7.2(iv) below.

The restriction $(*)$ in the hypotheses of (ii) may seem unimportant from the point of view of applications. However, this is not correct since the theorem as formulated above does not apply to various interesting and natural $\mathcal{Q}$-systems. Examples of such systems that are of special interest to us can be found among the systems of set theory discussed in §4.6; these are systems that admit individuals but may not imply the existence of individuals. We know from §4.6 that such systems may be developed in a language with two binary predicates, $\boldsymbol{E}$ and $\boldsymbol{I}$. If in some such system $\mathcal{T}$ the pair axiom $\boldsymbol{P}$ formulated in 4.6(i) is provable, then $\mathcal{T}$ is certainly a $\mathcal{Q}$-system and, moreover, the sentence $\boldsymbol{\exists}_{\boldsymbol{xy}}(\boldsymbol{xEy})$ is also provable. In addition $\mathcal{T}$ may have many other axioms containing $\boldsymbol{E}$ alone, or even $\boldsymbol{E}$ and $\boldsymbol{I}$, provided they do not imply $\boldsymbol{\exists}_{\boldsymbol{xy}}(\boldsymbol{xIy})$. Using only (ii) we cannot prove that $\mathcal{T}$ can be equipollently formalized in a language with one binary predicate.

In view of these remarks the following improvement of Theorem (ii), due to Givant, deserves attention, since it enables us to show that the conclusions of that theorem hold, in particular, for the set-theoretical systems with individuals discussed above.

(iii) *Theorem* (ii) *remains true if we replace the condition* $(*)$ *by the following condition*:

$(**)$ *there is at most one* m *with* $0 \le m \le n$ *for which we have neither*

$$\mathbf{A}\xi[\mathfrak{T}] \vdash \exists_{xy}(x\mathring{0}y) \lor \exists_{xy}(xF_m y)$$

nor

$$\mathbf{A}\xi[\mathfrak{T}] \vdash \exists_{xy}(x\mathring{0}y) \lor \forall_{xy}(\neg xF_m y).$$

The proof of (iii) is a modification of that of (ii). Steps (1)–(12) remain the same. Because of $(**)$, steps (13) and (14) require some alteration. Instead we now introduce two subsets, V and W, of $\{0,\ldots,n\} = n+1$ by stipulating that

$$(13') \quad k \in V \ \text{ iff } \ \mathbf{A}\xi[\mathfrak{T}^+] \vdash \mathring{0} = 0 \rightarrow \neg F_k = 0,$$

$$(14') \quad k \in W \ \text{ iff } \ \mathbf{A}\xi[\mathfrak{T}^+] \vdash \mathring{0} = 0 \rightarrow F_k = 0.$$

By $(**)$ there is at most one number in $(n+1) \sim V \cup W$; we take p to be this unique number if it exists, and otherwise we take p to be 0.

Next we define, instead of G and I, the corresponding predicate G' and equation I' by the following conditions, which replace (15) and (16):

$$(15') \quad G' = [P^{(0)}_{AB} \odot (\mathring{0} + A \odot F_0 \odot B^{\smile}) \odot P^{(0)\smile}_{AB}] \cdot \cdots \cdot [P^{(n)}_{AB} \odot (\mathring{0} + A \odot F_n \odot B^{\smile}) \odot P^{(n)\smile}_{AB}],$$

$$(16') \quad I' = (E = C \cdot \mathring{0} + A \odot [1 \odot \mathring{0} \odot 1 \cdot G' + (0 \oplus \mathring{1} \oplus 0) \cdot F_p] \odot B^{\smile} \cdot \mathring{1}).$$

As in the proof of (ii) the formalism $\overline{\mathfrak{T}}^+$ is uniquely described by the statement

$$(17') \quad \mathbf{A}\xi[\overline{\mathfrak{T}}^+] = \mathbf{A}\xi[\mathfrak{T}^+] \cup \{I'\}.$$

Statement (18′) coincides with (18), while (19′)–(21′) are obtained from (19)–(21) by replacing G with $1 \odot \mathring{0} \odot 1 \cdot G' + (0 \oplus \mathring{1} \oplus 0) \cdot F_p$. The successive derivations of (18′)–(21′) are fully analogous to those of (18)–(21) and are left to the reader.

In the next portion of our proof we shall establish certain facts which are closely related to (22) and which play an entirely analogous role in the subsequent discussion. However, the argument here is considerably more involved than in the case of (ii). We shall need three statements, $(22_1{}')$, $(22_2{}')$, and $(22_3{}')$, to adequately replace (22), and in addition they will be preceded by an auxiliary statement, $(22_0{}')$:

$$(22_0{}') \quad \mathbf{A}\xi[\overline{\mathfrak{T}}^+] \vdash \neg\mathring{0} = 0 \rightarrow F_k = A^{\smile} \odot (P^{(k)\smile}_{AB} \odot [1 \odot \mathring{0} \odot 1 \cdot G' + (0 \oplus \mathring{1} \oplus 0) \cdot F_p] \odot P^{(k)}_{AB} \cdot \mathring{1}) \odot B \ \text{ for } \ 0 \le k \le n.$$

To establish ($22_0'$) we first use (5) and 3.2(iii), and then we reason as in (22), applying 4.2(xi)(γ) instead of 4.2(viii)(γ).

$$(22_1')\quad \mathbf{A\xi}[\bar{\mathfrak{T}}^+] \vdash \boldsymbol{F}_k = A^{\smile} \odot \big(\boldsymbol{P}_{AB}^{(k)\smile} \odot [\mathbf{1} \odot \mathring{\mathbf{0}} \odot \mathbf{1} \cdot G' + (\mathbf{0} \oplus \mathring{\mathbf{1}} \oplus \mathbf{0}) \cdot \boldsymbol{F}_p] \odot \boldsymbol{P}_{AB}^{(k)} \cdot \mathring{\mathbf{1}}\big) \odot B + (\mathbf{0} \oplus \mathring{\mathbf{1}} \oplus \mathbf{0}) \quad \text{for } k \in V$$

by (5), (13′), and ($22_0'$);

$$(22_2')\quad \mathbf{A\xi}[\bar{\mathfrak{T}}^+] \vdash \boldsymbol{F}_k = A^{\smile} \odot \big(\boldsymbol{P}_{AB}^{(k)\smile} \odot [\mathbf{1} \odot \mathring{\mathbf{0}} \odot \mathbf{1} \cdot G' + (\mathbf{0} \oplus \mathring{\mathbf{1}} \oplus \mathbf{0}) \cdot \boldsymbol{F}_p] \odot \boldsymbol{P}_{AB}^{(k)} \cdot \mathring{\mathbf{1}}\big) \odot B \cdot (\mathbf{1} \odot \mathring{\mathbf{0}} \odot \mathbf{1}) \quad \text{for } k \in W$$

by (5), 3.6(iii), (14′), and ($22_0'$);

$$(22_3')\quad \mathbf{A\xi}[\bar{\mathfrak{T}}^+] \vdash \boldsymbol{F}_k = A^{\smile} \odot \big(\boldsymbol{P}_{AB}^{(k)\smile} \odot [\mathbf{1} \odot \mathring{\mathbf{0}} \odot \mathbf{1} \cdot G' + (\mathbf{0} \oplus \mathring{\mathbf{1}} \oplus \mathbf{0}) \cdot \boldsymbol{F}_p] \odot \boldsymbol{P}_{AB}^{(k)} \cdot \mathring{\mathbf{1}}\big) \odot B \quad \text{for } k = p$$

by (5), 3.6(iii),(xiii), 4.1(vii), 4.2(v), and ($22_0'$).

The remaining part of the argument presents no difficulties. As in the proof of (ii), we correlate successively with every predicate P in $\mathcal{M}^{(n+1)+}$ a predicate $P^{\nabla\prime}$ in $\mathcal{L}^+$ such that the equation $P = P^{\nabla\prime}$ is provable in $\bar{\mathfrak{T}}^+$. (We may notice that in determining $P^{\nabla\prime}$ for $P = \boldsymbol{F}_p$ we disregard ($22_3'$) in case $p = 0$ and $p \in V \cup W$, and use instead either ($22_1'$) or ($22_2'$).) We then define the mapping L' from $\Sigma[\mathcal{M}^{(n+1)+}]$ onto $\Sigma[\mathcal{L}^+]$ in analogy with the definition of L in the proof of (ii), and we complete the argument following strictly the lines of that proof.

Theorem (iii) implies the following corollary, which may be more convenient for applications.

(iv) *Theorem* (ii) *remains true if we replace the condition* (*) *with*

$$(***)\qquad \mathbf{A\xi}[\mathfrak{T}^+] \vdash \exists_{xy}(x\mathring{\mathbf{0}}y).$$

As was noticed by Givant, Theorem (ii) ceases to hold if we omit condition (*) entirely. This can be shown using an example which, in view of (iii), is the simplest possible. Indeed, let $\mathfrak{T}$ be the system in $\mathcal{M}^{(1)}$ which has $\forall_{xy}(x\mathring{\mathbf{1}}y)$ as its only nonlogical axiom and which, therefore, is a $\mathfrak{Q}$-system (cf. step (3) in the proof of (ii)). If (ii) (with condition (*) omitted) applied to $\mathfrak{T}$, we could conclude that there is a system $\mathcal{S}$ in $\mathcal{L}$ equipollent with $\mathfrak{T}$; therefore the sentence $\forall_{xy}(x\mathring{\mathbf{1}}y)$ would also hold in $\mathcal{S}$ as well, and $\mathcal{S}$ would also be a $\mathfrak{Q}$-system. This is, however, impossible. Indeed, in the language of $\mathfrak{T}$ there are two different nonlogical constants, and consequently at least (in fact, exactly) 16 sentences no two of which are equivalent on the basis of $\mathbf{A\xi}[\mathfrak{T}]$, while in the language of $\mathcal{S}$ there is only one nonlogical constant, and hence at most 4 such nonequivalent sentences. (A detailed proof of these last properties of $\mathcal{T}$ and $\mathcal{S}$ is facilitated if we exploit fully the fact that $\mathcal{T}$ and $\mathcal{S}$ are $\mathcal{Q}$-systems and apply, for instance, the results of 4.5(v)(β1)–(β3).)

Using the fact that the $\mathcal{Q}$-systems $\mathcal{T}$ and $\mathcal{S}$ involved in (ii) are not only equipollent, but actually definitionally equivalent, we obtain, with the help of 6.2(ix), the following rather interesting conclusion.

(v) *For any given $\mathfrak{Q}$-structure $\mathfrak{U} = \langle U, F_0, \ldots, F_n\rangle$, where $F_0, \ldots, F_n$ are binary relations, there is a binary relation E on U such that the relation rings on U generated respectively by $\{F_0, \ldots, F_n\}$ and by E are identical.*

This conclusion can also be established directly, by a purely set-theoretical argument (which, however, follows closely the lines of the proof of (ii) outlined above). It then becomes clear that the proofs of both (ii) and (v) depend essentially on the set-theoretical statement given in 4.2(x).

7.2. Extension of equipollence results to weak $\mathcal{Q}$-systems in arbitrary first-order formalisms

We turn now to formalisms $\mathcal{P}$ of predicate logic with finitely many nonlogical constants of an arbitrary character. In principle, we could simplify the metalogical discussion of these formalisms by restricting our attention to languages of predicate logic in which the only nonlogical constants are predicates of various ranks. This is because of the well-known fact that with every operation Q of rank k on a set U we can correlate in a one-one way a relation $\overline{Q}$ of rank $k+1$, which is defined for every $k+1$-termed sequence $\langle x_0, \ldots, x_k\rangle$ of elements of U by the stipulation

$$\overline{Q}(x_0, \ldots, x_k) \text{ iff } Q(x_0, \ldots, x_{k-1}) = x_k.$$

However, languages which admit operation symbols in addition to predicates usually prove to be more convenient for formalizing various special mathematical theories and are frequently used for this purpose. The results which are formally stated in the subsequent discussion, such as (iii), (iv), apply to formalisms of predicate logic admitting both predicates and operation symbols. On the other hand, in the outlines of the proofs of such results, and in the informal parts of the text, we restrict ourselves as a rule to formalisms without operation symbols (and without predicates of rank 0); in application to formalisms with operation symbols our remarks may require some modifications and elaborations.

We shall consider various systems $\mathcal{U}$ developed in the formalism $\mathcal{P}$, and we shall be particularly interested in $\mathcal{Q}$-systems. Recall that in 4.5(i)(β),(ii) two different, but logically equivalent, characterizations of $\mathcal{Q}$-systems in $\mathcal{L}$ are stated. Using the semantical terminology introduced in 6.2(i),(iii), we can restate these characterizations in the following way: a system $\mathcal{S}$ in $\mathcal{L}$ is a $\mathcal{Q}$-system by 4.5(i)(β) iff there are two predicates in the extended formalism $\mathcal{L}^+$ which denote two conjugated quasiprojections in every realization of $\mathcal{S}$; $\mathcal{S}$ is a $\mathcal{Q}$-system by 4.5(ii) iff there are two formulas in $\mathcal{L}$ which define two conjugated quasiprojections in every realization of $\mathcal{S}$ and which contain at most three different variables.

A direct extension of the first characterization of $\mathcal{Q}$-systems to systems formalized in $\mathcal{P}$ would require the preliminary construction of an extended formalism $\mathcal{P}^{\oplus}$. The problem of adequately constructing such a formalism $\mathcal{P}^{\oplus}$ (and the correlated formalism $\mathcal{P}^{\otimes}$ analogous to $\mathcal{L}^{\times}$) is not quite simple, and we do not wish to be involved here in the discussion of this problem. We may mention only that $\mathcal{P}^{\oplus}$ would be provided with the possibility of constructing compound predicates of various ranks from atomic predicates of the same, or even different, ranks—and, in opposition to $\mathcal{L}^{+}$, this possibility would not be restricted to predicates of rank 2. (A formal framework for such a construction could be found, not in abstract relation algebras as in the case of $\mathcal{L}^{+}$ and $\mathcal{L}^{\times}$, but in cylindric algebras studied in Henkin–Monk–Tarski [**1971**], [**1985**], or in related algebraic structures discussed in Halmos [**1962**] and Craig [**1974**].)

On the other hand, the characterization given in 4.5(ii) can be literally extended to the formalism $\mathcal{P}$ and adopted as the definition of $\mathcal{Q}$-systems in this formalism. It enables us, in particular, to formulate and establish a result which embodies the main purpose of this portion of our discussion, namely an extension of 7.1(ii),(iii) to arbitrary formalisms of predicate logic. It turns out that the result thus obtained can be further improved by using, instead of the notion of a $\mathcal{Q}$-system, a simpler and wider notion, which appears perhaps more natural in the general context of predicate logic—the notion of a *weak $\mathcal{Q}$-system* (or a *$\mathcal{Q}$-system in the wider sense*). The definition of the new notion differs from 4.5(ii) only in that condition (δ), restricting the number of variables which occur in formulas D and E, is deleted. For later reference we state this definition explicitly.

(i) *A system $\mathfrak{U}$ in the formalism $\mathcal{P}$ is a weak $\mathcal{Q}$-system iff there are formulas $D, E \in \mathbf{\Phi}[\mathcal{P}]$ satisfying the following conditions:*

(α) $\mathbf{\Upsilon}\phi D = \mathbf{\Upsilon}\phi E = \{\mathbf{x}, \mathbf{y}\}$;

(β) $\mathbf{A}\xi[\mathfrak{U}] \vdash \{\forall_{xyz}(D[\mathbf{x},\mathbf{y}] \wedge D[\mathbf{x},\mathbf{z}] \rightarrow \mathbf{y}\mathring{\mathbf{1}}\mathbf{z}), \forall_{xyz}(E[\mathbf{x},\mathbf{y}] \wedge E[\mathbf{x},\mathbf{z}] \rightarrow \mathbf{y}\mathring{\mathbf{1}}\mathbf{z})\}$;

(γ) $\mathbf{A}\xi[\mathfrak{U}] \vdash \forall_{xy}\exists_{z}(D[\mathbf{z},\mathbf{x}] \wedge E[\mathbf{z},\mathbf{y}])$.

There is also a somewhat simpler characterization of weak $\mathcal{Q}$-systems which is obtained from 4.7(xi)(β) in the same way in which (i) is obtained from 4.5(ii).

(ii) *For a system $\mathfrak{U}$ in $\mathcal{P}$ to be a weak $\mathcal{Q}$-system, it is necessary and sufficient that there be an $F \in \mathbf{\Phi}[\mathcal{P}]$ satisfying the following conditions:*

(α) $\mathbf{\Upsilon}\phi F = \{\mathbf{x}, \mathbf{y}\}$,

(β) $\mathbf{A}\xi[\mathfrak{U}] \vdash \forall_{zu}\exists_{y}\forall_{x}(F \leftrightarrow \mathbf{x}\mathring{\mathbf{1}}\mathbf{z} \vee \mathbf{x}\mathring{\mathbf{1}}\mathbf{u})$.

In other words, a system $\mathfrak{U}$ in $\mathcal{P}$ is a weak $\mathcal{Q}$-system iff there is a formula F in $\mathcal{P}$ that defines in any given model of $\mathfrak{U}$ a binary relation universal for two-element sets.

To construct D, E in (i) from F in (ii), and conversely, we imitate the construction of corresponding predicates A, B, and C in 4.7(ix),(x) (keeping in mind the logical axiom schemata (DI)–(DIV) in §2.2 and the symbolic conventions 2.1(ii),(iii) accepted for $\mathcal{L}^+$). In particular, F is obtained from D, E simply by setting $F = D[\boldsymbol{y}, \boldsymbol{x}] \vee E[\boldsymbol{y}, \boldsymbol{x}]$.

In opposition to weak $\mathcal{Q}$-systems, $\mathcal{Q}$-systems in the original sense can be referred to as *strong $\mathcal{Q}$-systems* (or *$\mathcal{Q}$-systems in the narrower sense*). More often, however, we shall continue to refer to such systems simply as $\mathcal{Q}$-systems, without any qualification.

Obviously, every strong $\mathcal{Q}$-system is also a weak $\mathcal{Q}$-system. It seems likely that the converse does not hold, i.e., that there are weak $\mathcal{Q}$-systems which are not strong. To our knowledge, however, no suitable example has yet been constructed, even in the simplest case when $\mathcal{P}$ coincides with our original formalism $\mathcal{L}$.

We now state a lemma, which will be followed by the main result of the present chapter, Theorem (iv).

(iii) *Let $\mathcal{P}$ be a formalism of predicate logic with $n+1$ distinct nonlogical constants $\boldsymbol{C}_0, \ldots, \boldsymbol{C}_n$; let $\mathcal{U}$ be a weak $\mathcal{Q}$-system formalized in $\mathcal{P}$. Then there is a $\mathcal{Q}$-system $\mathcal{T}$ in a formalism $\mathcal{M}^{(n+2)}$ with $n+3$ distinct nonlogical binary predicates $\boldsymbol{F}_0, \ldots, \boldsymbol{F}_{n+2}$, not occurring in the vocabulary of $\mathcal{P}$, as well as a $\mathcal{Q}$-system $\overline{\mathcal{U}}$ in the formalism $\overline{\mathcal{P}}$ with nonlogical constants $\boldsymbol{C}_0, \ldots, \boldsymbol{C}_n$, $\boldsymbol{F}_0, \ldots, \boldsymbol{F}_{n+2}$, such that $\overline{\mathcal{U}}$ is a common definitional extension of the two systems $\mathcal{T}$ and $\mathcal{U}$. Consequently, $\mathcal{T}$ and $\mathcal{U}$ (treated as subsystems of $\overline{\mathcal{U}}$) are definitionally equivalent and hence are equipollent with each other in means of expression and proof.*

In outlining the proof of (iii) we assume that all the constants $\boldsymbol{C}_0, \ldots, \boldsymbol{C}_n$ are predicates with positive ranks $\rho 0, \ldots, \rho n$. For the binary predicates $\boldsymbol{F}_0, \ldots, \boldsymbol{F}_{n+2}$ we choose any $n+3$ distinct symbols which do not occur in the vocabulary of $\mathcal{P}$.

Our main task is to construct the systems $\mathcal{T}$ and $\overline{\mathcal{U}}$ by describing their sets of nonlogical axioms, $\mathbf{A}\boldsymbol{\xi}[\mathcal{T}]$ and $\mathbf{A}\boldsymbol{\xi}[\overline{\mathcal{U}}]$. We start with $\overline{\mathcal{U}}$. To this end some preliminary remarks are needed.

Since $\mathcal{U}$ is by hypothesis a weak $\mathcal{Q}$-system, there exist formulas D and E in $\mathcal{P}$ which satisfy conditions (i)(α)–(γ) or, in other words, which define two conjugated quasiprojections in every model of $\mathbf{A}\boldsymbol{\xi}[\mathcal{U}]$. From 4.2(i)–(iii) we know that if the relations denoted by two given predicates $A, B \in \mathbf{\Pi}[\mathcal{L}^+]$ in a possible realization $\mathfrak{U}$ of $\mathcal{L}^+$ form a pair of conjugated quasiprojections, then the relations denoted by the predicates $\boldsymbol{P}_{AB}^{(0)}, \ldots, \boldsymbol{P}_{AB}^{(m)}$ ($m \in \omega$) in $\mathfrak{U}$ form an $(m+1)$-termed sequence of conjugated quasiprojections; in other words, using the notations introduced in 4.1(i) and 4.2(ix) we have $\boldsymbol{Q}_{AB} \vdash^+ \boldsymbol{Q}(\boldsymbol{P}_{AB}^{(0)}, \ldots, \boldsymbol{P}_{AB}^{(m)})$. From 4.2(i) we also easily see that $\vdash^+ \boldsymbol{P}_{AB}^{(0)} = B$ and $\vdash^+ \boldsymbol{P}_{AB}^{(m+1)} = A \odot \boldsymbol{P}_{AB}^{(m)}$ for every $m \in \omega$, which indicates that Definition 4.2(i) could equivalently be replaced by a recursive construction. We now imitate this recursive construction in the formalism $\mathcal{P}$; in fact, using the formulas D, E instead of the predicates A, B, we define by recursion the sequence of formulas $\langle Q_0, \ldots, Q_m, \ldots \rangle$, setting

(1) $Q_0 = E$ and $Q_{m+1} = \exists_z(D[x,z] \wedge Q_m[z,y])$ for every $m \in \omega$.

As in 4.2(ii),(iii), we conclude from (1) and (i)(α)–(γ), by induction on m, that for each $m \in \omega$ the formulas $Q_0, \ldots, Q_m$ satisfy the following conditions:

(2) $Q_m \in \Phi[\mathcal{P}]$ and $\Upsilon\phi Q_m = \{x, y\}$,

(3) $A\xi[\mathcal{U}] \vdash \forall_{xyz}(Q_m[x,y] \wedge Q_m[x,z] \rightarrow y \mathring{1} z)$,

(4) $A\xi[\mathcal{U}] \vdash \forall_{v_1 \ldots v_{m+1}} \exists_x (Q_0[x, v_1] \wedge \cdots \wedge Q_m[x, v_{m+1}])$.

With every atomic formula $C_k(v_1, \ldots, v_{\rho k})$ in $\mathcal{P}$ we correlate the formula H_k defined for each $k = 0, \ldots, n$ by the stipulation

(5) $H_k = x\mathring{1}y \wedge \exists_{v_1 \ldots v_{\rho k}}(C_k(v_1, \ldots, v_{\rho k}) \wedge Q_0[x, v_1] \wedge \cdots \wedge Q_{\rho k - 1}[x, v_{\rho k}])$.

Steps (2) and (5) obviously imply

(6) $H_k \in \Phi[\mathcal{P}]$ and $\Upsilon\phi H_k = \{x, y\}$.

Finally, we determine the axiom set $A\xi[\overline{\mathcal{U}}]$ by setting

(7) $A\xi[\overline{\mathcal{U}}] = A\xi[\mathcal{U}] \cup \{R_0, \ldots, R_{n+2}\}$,

where $R_0, \ldots, R_{n+2}$ are sentences in $\overline{\mathcal{P}}$ defined as follows:

(8) $R_k = \forall_{xy}(xF_k y \leftrightarrow H_k)$ for $k = 0, \ldots, n$,

(9) $R_{n+1} = \forall_{xy}(xF_{n+1}y \leftrightarrow D)$, $R_{n+2} = \forall_{xy}(xF_{n+2}y \leftrightarrow E)$.

In view of (5), (7), and (8) we can say that, on the basis of $A\xi[\overline{\mathcal{U}}]$, each binary predicate F_k with $k = 0, \ldots, n$ represents in $\mathcal{M}^{(n+2)}$ the correlated predicate C_k with rank ρk, or the atomic formula $C_k(v_1, \ldots, v_{\rho k})$, from $\mathcal{P}$. (This notion of a predicate representing a formula is closely related to the one introduced after Theorem 4.4(xii) to express the intuitive content of that theorem. The relationship becomes even clearer when we compare not only condition 4.4(xii)(β') with conditions (5), (7), and (8), but also condition 4.4(xii)(α') with (10) and (11), or with (14) and (15) below.)

We see from (6), (8), (9), and (i)(α) that each sentence R_k for $k = 0, \ldots, n+2$ is a possible definition in $\mathcal{P}$ of the predicate F_k from $\mathcal{M}^{(n+2)}$. (In fact, the reason for adding the formula $x\mathring{1}y$ as a conjunct of H_k in (5) is to make the second part of (6) true, since this condition must be satisfied if R_k is to be a possible definition of F_k; see 2.4(xii).) Thus, in view of (7), system $\overline{\mathcal{U}}$ is a definitional extension of the given system $\mathcal{U}$. (It follows that we can define a recursive translation mapping H from $\overline{\mathcal{P}}$ into $\mathcal{P}$ which correlates with every formula $X \in \Phi[\overline{\mathcal{P}}]$ a formula $HX \in \Phi[\mathcal{P}]$; HX is obtained from X by eliminating the nonlogical constants $\boldsymbol{F}_0, \ldots, \boldsymbol{F}_{n+2}$ on the basis of the sentences $R_0, \ldots, R_{n+2}$, and X and HX are equivalent relative to the set of these sentences. This mapping will not be involved in the remainder of the present proof, but it will play a role in the proofs of 7.2(iv) and 7.3(i),(ii).)

We turn now to the construction of system $\mathcal{T}$. Consider first the following sentences S_k with $k = 0, \ldots, n$:

(10) $S_k = [\boldsymbol{C}_k(\boldsymbol{v}_1, \ldots, \boldsymbol{v}_{\rho k}) \leftrightarrow \exists_{\boldsymbol{x}}(\boldsymbol{x}\boldsymbol{F}_k\boldsymbol{x} \wedge Q_0[\boldsymbol{x}, \boldsymbol{v}_1] \wedge \cdots \wedge Q_{\rho k-1}[\boldsymbol{x}, \boldsymbol{v}_{\rho k}])]$.

With the help of (3)–(5), (7), and (8) it is easy to show that

(11) $\mathbf{A}\boldsymbol{\xi}[\overline{\mathcal{U}}] \vdash \{S_0, \ldots, S_n\}$.

We define by recursion the formulas P_m by setting

(12) $P_0 = \boldsymbol{x}\boldsymbol{F}_{n+2}\boldsymbol{y}$, and $P_{m+1} = \exists_{\boldsymbol{z}}(\boldsymbol{x}\boldsymbol{F}_{n+1}\boldsymbol{z} \wedge P_m[\boldsymbol{z}, \boldsymbol{y}])$ for every $m \in \omega$.

Hence, obviously,

(13) $P_m \in \boldsymbol{\Phi}[\mathcal{M}^{(n+2)}]$ and $\Upsilon\phi P_m = \{\boldsymbol{x}, \boldsymbol{y}\}$ for every $m \in \omega$.

For each $k = 0, \ldots, n$ let

(14) $T_k = [\boldsymbol{C}_k(\boldsymbol{v}_1, \ldots, \boldsymbol{v}_{\rho k}) \leftrightarrow \exists_{\boldsymbol{x}}(\boldsymbol{x}\boldsymbol{F}_k\boldsymbol{x} \wedge P_0[\boldsymbol{x}, \boldsymbol{v}_1] \wedge \cdots \wedge P_{\rho k-1}[\boldsymbol{x}, \boldsymbol{v}_{\rho k}])]$.

Comparing (12) with (1), and (14) with (10), we easily conclude with the help of (7) and (9) that the formulas Q_m and P_m for each $m \in \omega$, and hence also the sentences S_k and T_k for each $k = 0, \ldots, n$, are equivalent under $\mathbf{A}\boldsymbol{\xi}[\overline{\mathcal{U}}]$. Therefore, by (11),

(15) $\mathbf{A}\boldsymbol{\xi}[\overline{\mathcal{U}}] \vdash \{T_0, \ldots, T_n\}$.

Notice also that, in view of (13) and (14), each sentence T_k is a possible definition in $\mathcal{M}^{(n+2)}$ of the nonlogical constant $\boldsymbol{C}_k$ from $\mathcal{P}$. Thus, we have constructed $n+1$ sentences which are possible definitions in $\mathcal{M}^{(n+2)}$ of all nonlogical constants of $\mathcal{P}$ and are provable in the system $\overline{\mathcal{U}}$.

Once this has been achieved, the remaining part of the proof is rather routine. We follow closely the pattern of the last portion of the proof of 7.1(ii) (i.e., the portion which succeeds step (23)). Thus, we define a (recursive) translation mapping K from $\overline{\mathcal{P}}$ into $\mathcal{M}^{(n+2)}$ which correlates with every formula $X \in \boldsymbol{\Phi}[\overline{\mathcal{P}}]$ a formula $KX \in \boldsymbol{\Phi}[\mathcal{M}^{(n+2)}]$ obtained from X by eliminating the nonlogical constants $\boldsymbol{C}_0, \ldots, \boldsymbol{C}_n$ on the basis of (14). We obviously have

(16) $\{T_0, \ldots, T_n\} \vdash X \leftrightarrow KX$ and $KX \in \Sigma[\mathcal{M}^{(n+2)}]$
for every $X \in \Sigma[\overline{\mathcal{P}}]$.

We now determine the system $\mathcal{T}$ by stipulating that

$$(17) \qquad \mathbf{A\xi}[\mathcal{T}] = \{KX : X \in \mathbf{A\xi}[\overline{\mathcal{U}}]\}.$$

Consider the set

$$(18) \qquad \Theta = \mathbf{A\xi}[\mathcal{T}] \cup \{T_0, \ldots, T_n\}.$$

From (15)–(18) it clearly follows that Θ is a subset of $\Sigma[\overline{\mathcal{P}}]$ which is logically equivalent with $\mathbf{A\xi}[\overline{\mathcal{U}}]$ and hence is a base for $\overline{\mathcal{U}}$. Therefore, in view of the definitional character of sentences $T_0, \ldots, T_n$, the system $\overline{\mathcal{U}}$ is a definitional extension of system $\mathcal{T}$. Thus, $\mathcal{U}$ and $\mathcal{T}$ have a common definitional extension $\overline{\mathcal{U}}$, and in consequence these three systems are equipollent in means of expression and proof.

Finally, recall that formulas D and E satisfy conditions (α)–(γ) of (i). Hence, from (7), (9) it is seen that conditions (α)–(δ) of 4.5(ii) are satisfied if $\boldsymbol{x}\boldsymbol{F}_{n+1}\boldsymbol{y}$, $\boldsymbol{x}\boldsymbol{F}_{n+2}\boldsymbol{y}$, and $\mathbf{A\xi}[\overline{\mathcal{U}}]$ are respectively taken for D, E, and $\mathbf{A\xi}$. Thus $\overline{\mathcal{U}}$ is a (strong) $\mathcal{Q}$-system. Since $\boldsymbol{x}\boldsymbol{F}_{n+1}\boldsymbol{y}$ and $\boldsymbol{x}\boldsymbol{F}_{n+2}\boldsymbol{y}$ are formulas in $\mathcal{M}^{(n+2)}$, and $\overline{\mathcal{U}}$ is an equipollent extension of $\mathcal{T}$, we conclude that $\mathcal{T}$ is a $\mathcal{Q}$-system as well, and the proof of Lemma (iii) has been completed.

To formulate our main result we need the notion of a formalism $\mathcal{P}^+$ constructed from any given formalism $\mathcal{P}$ of predicate logic in *exactly* the same manner in which $\mathcal{L}^+$ has been constructed from $\mathcal{L}$. Thus, the vocabulary of $\mathcal{P}^+$ differs from the vocabulary of $\mathcal{P}$ only in that it contains, in addition to the constants of $\mathcal{P}$, the operators $\boldsymbol{+}, \boldsymbol{^-}, \boldsymbol{\odot}, \boldsymbol{^\smile}$, and the second identity symbol, $\boldsymbol{=}$. In consequence, $\mathcal{P}^+$ is provided with the possibility of constructing compound predicates from atomic ones, but this possibility is restricted entirely to binary predicates. (Thus, if no binary predicates occur among the nonlogical constants of $\mathcal{P}$, then $\mathring{\mathbf{1}}$ is the only atomic predicate from which compound predicates can be constructed.) In this respect $\mathcal{P}^+$ differs essentially from the formalism $\mathcal{P}^\oplus$ alluded to at the beginning of this section. With every system $\mathcal{U}$ in $\mathcal{P}$ we correlate a system $\mathcal{U}^+$ in $\mathcal{P}^+$ by taking $\mathbf{A\xi}[\mathcal{U}]$ for the set of axioms of $\mathcal{U}^+$. By arguing as in §2.3 we show that $\mathcal{P}^+$ and $\mathcal{U}^+$ are respectively definitional extensions of, and hence equipollent with, $\mathcal{P}$ and $\mathcal{U}$.

(iv) *Let $\mathcal{P}$ be any formalism of predicate logic with $n+1$ distinct nonlogical constants $\boldsymbol{C}_0, \ldots, \boldsymbol{C}_n$ different from $\boldsymbol{E}$, and let $\mathcal{U}$ be a weak $\mathcal{Q}$-system formalized in $\mathcal{P}$ for which*

(∗) *there is at most one m with $0 \le m \le n$ such that $\boldsymbol{C}_m$ is a relation (and not an operation) symbol and we have neither*

$$\mathbf{A\xi}[\mathcal{U}] \vdash \boldsymbol{\exists}_{\boldsymbol{xy}}(\boldsymbol{x}\mathring{\mathbf{0}}\boldsymbol{y}) \boldsymbol{\vee} \boldsymbol{\exists}_{\boldsymbol{v}_0\ldots\boldsymbol{v}_{\rho m-1}} \boldsymbol{C}_m(\boldsymbol{v}_0, \ldots, \boldsymbol{v}_{\rho m-1})$$

nor

$$\mathbf{A\xi}[\mathcal{U}] \vdash \boldsymbol{\exists}_{\boldsymbol{xy}}(\boldsymbol{x}\mathring{\mathbf{0}}\boldsymbol{y}) \boldsymbol{\vee} \boldsymbol{\forall}_{\boldsymbol{v}_0\ldots\boldsymbol{v}_{\rho m-1}} \boldsymbol{\neg} \boldsymbol{C}_m(\boldsymbol{v}_0, \ldots, \boldsymbol{v}_{\rho m-1}).$$

Then there exists a $\mathcal{Q}$-system $\mathcal{S}$ in the original formalism $\mathcal{L}$, as well as a $\mathcal{Q}$-system $\mathcal{U}^*$ in the formalism $\mathcal{P}^*$ of predicate logic with $n+2$ nonlogical constants $C_0, \ldots, C_n, E$, satisfying the following conditions.

(α) *$\mathcal{U}^*$ is a common definitional extension of the systems $\mathcal{U}$ and $\mathcal{S}$; consequently, $\mathcal{U}$ and $\mathcal{S}$ are definitionally equivalent, and hence equipollent with each other in means of expression and proof.*

(β) *The system $\mathcal{U}^{*+}$ in the extended formalism $\mathcal{P}^{*+}$ is a common equipollent extension, not only of $\mathcal{U}$ and $\mathcal{S}$, but also of the system $\mathcal{S}^\times$ in $\mathcal{L}^\times$; hence $\mathcal{U}$ and $\mathcal{S}^\times$ (treated as subsystems of $\mathcal{U}^{*+}$) are equipollent with each other as well.*[2]

Essentially, to establish this result, we first apply Lemma (iii) above to the given system $\mathcal{U}$ in $\mathcal{P}$ and obtain an auxiliary system $\mathcal{T}$ in $\mathcal{M}^{(n+2)}$; we then apply Theorem 7.1(iii) to $\mathcal{T}$ and arrive at a system $\mathcal{S}$ in $\mathcal{L}$ which has the desired properties. (Whenever we refer, here or below, to the formulation or the proof of 7.1(iii), we assume that n has been replaced everywhere in 7.1(iii) by $n+2$.)

Actually, our task is somewhat more involved, since in addition to $\mathcal{S}$ we have to construct, in an appropriate formalism $\mathcal{P}^*$, a system $\mathcal{U}^*$ which is a common equipollent extension of $\mathcal{U}$ and $\mathcal{S}$. For this purpose, however, the formulations of (iii) and 7.1(iii) do not suffice by themselves, and we have to make use of the constructions described in the proofs of these two results.

As usual, we stipulate that in the formalism $\mathcal{P}$ all the nonlogical constants are atomic predicates $C_0, \ldots, C_n$ with positive ranks $\rho 0, \ldots, \rho n$. In addition to $C_0, \ldots, C_n$ and E, we shall use the $n+3$ nonlogical binary predicates $F_0, \ldots, F_{n+2}$ which occur in $\mathcal{M}^{(n+2)}$ and $\mathcal{M}^{(n+2)+}$; again, we assume that none of them occurs in $\mathcal{P}^*$.

We shall use the translation mapping G constructed in §2.3 for $\mathcal{L}^+$ and appropriately extended to the formalism $\mathcal{M}^{(n+3)+}$ (with nonlogical constants $F_0, \ldots, F_{n+2}, E$), as well as the equation I' in $\mathcal{M}^{(n+3)+}$ defined in the proof of 7.1(iii), formula (16′). From the definition of I' we see that $I'^{\ell} = E$, while $I'^{r} \in \Pi[\mathcal{M}^{(n+2)+}]$ (see 3.1(i)). Therefore

$$GI' = \forall_{xy}[xEy \leftrightarrow G(xI'^{r}y)], \quad \text{where} \quad G(xI'^{r}y) \in \Phi[\mathcal{M}^{(n+2)}].$$

The sentences $R_0, \ldots, R_{n+2}$ (in the formalism $\overline{\mathcal{P}}$ with the nonlogical constants $C_0, \ldots, C_n, F_0, \ldots, F_{n+2}$, and hence also in the formalism $\overline{\mathcal{P}}^*$ with nonlogical constants $C_0, \ldots, C_n, E, F_0, \ldots, F_{n+2}$) defined in the proof of Lemma (iii), formulas (8) and (9), are respectively possible definitions of $F_0, \ldots, F_{n+2}$ in $\mathcal{P}$ and hence also in $\mathcal{P}^*$. On the basis of these sentences we can define in the usual way a translation mapping H from $\overline{\mathcal{P}}^*$ into $\mathcal{P}^*$ which also maps $\overline{\mathcal{P}}$ into $\mathcal{P}$; in particular, H correlates with each $X \in \Phi[\mathcal{M}^{(n+3)}]$ and each $Y \in \Phi[\mathcal{M}^{(n+2)}]$ well-determined formulas $HX \in \Phi[\mathcal{P}^*]$ and $HY \in \Phi[\mathcal{P}]$ which are respectively

[2]Theorem (ii) without condition (∗) was first announced in Tarski [**1954**]. It was pointed out by Givant that this announced form of (ii) is incorrect, and the present corrected form is due to him; cf. Theorem 7.1(iii) and the remarks following 7.1(iv).

equivalent with X and Y relative to $\{R_0, \ldots, R_{n+2}\}$. We obviously have

$$HGI' = \forall_{xy}[xEy \leftrightarrow HG(xI'^{\tau}y)], \quad \text{where} \quad HG(xI'^{\tau}y) \in \Phi[\mathcal{P}].$$

Consequently, HGI' is a possible definition of E in $\mathcal{P}$. We now define the system $\mathcal{U}^*$ by setting

$$\mathbf{A\xi}[\mathcal{U}^*] = \mathbf{A\xi}[\mathcal{U}] \cup \{HGI'\}.$$

We see at once that $\mathcal{U}^*$ is a definitional extension of $\mathcal{U}$.

Recall now the mapping K from $\mathcal{P}$ (actually from $\overline{\mathcal{P}}$) into $\mathcal{M}^{(n+2)}$ defined in the proof of (iii), and the mapping L' from $\mathcal{M}^{(n+2)}$ (actually from $\mathcal{M}^{(n+3)+}$) into $\mathcal{L}^+$ introduced in the proof of 7.1(iii). If X is any sentence in $\Sigma[\mathcal{P}]$, then $KX \in \Sigma[\mathcal{M}^{(n+2)}]$ and $GL'K(X) \in \Sigma[\mathcal{L}]$. We determine the system $\mathcal{S}$ by setting

$$\mathbf{A\xi}[\mathcal{S}] = \{GL'K(X) \colon X \in \mathbf{A\xi}[\mathcal{U}^*]\}.$$

Recall also the sentences T_k, with $k = 0, \ldots, n$, defined in the proof of (iii), formula (14). Each of them has the form

$$T_k = [C_k(v_1, \ldots, v_{\rho k}) \leftrightarrow X_k], \quad \text{where} \quad X_k \in \Phi[\mathcal{M}^{(n+2)}].$$

Set

$$T'_k = [C_k(v_1, \ldots, v_{\rho k}) \leftrightarrow GL'X_k]$$

for $k = 0, \ldots, n$. Since $GL'X_k \in \Phi[\mathcal{L}]$, the sentences $T'_0, \ldots, T'_n$ are possible definitions of $C_0, \ldots, C_n$ in $\mathcal{L}$. Consider the subset Ω of $\Sigma[\mathcal{P}^*]$ determined by

$$\Omega = \mathbf{A\xi}[\mathcal{S}] \cup \{T'_0, \ldots, T'_n\}.$$

The sets $\mathbf{A\xi}[\mathcal{U}^*]$ and Ω turn out to be logically equivalent. The proof is rather straightforward, but somewhat complicated in details; it can be considerably simplified by making use of the semantical completeness of the formalisms involved. It appears to be convenient to use the auxiliary system $\overline{\mathcal{U}}^*$ which is constructed (in $\overline{\mathcal{P}}^*$) from $\mathcal{U}^*$ in the same way as $\overline{\mathcal{U}}$ was constructed from $\mathcal{U}$ in (iii). Thus

$$\mathbf{A\xi}[\overline{\mathcal{U}}^*] = \mathbf{A\xi}[\mathcal{U}^*] \cup \{R_0, \ldots, R_{n+2}\}.$$

In addition, we make use of the fact that the set

$$\Omega' = \Omega \cup \{G(F_0 = F_0^{\nabla'}), \ldots, G(F_{n+2} = F_{n+2}^{\nabla'})\}$$

is also a base for $\overline{\mathcal{U}}^*$; here the equations $F_k = F_k^{\nabla'}$, for $k = 0, \ldots, n+2$, are as defined in the proof of 7.1(iii). (The condition (*) in the hypothesis of our theorem is needed to show that (**) in 7.1(iii) holds.) In particular, we arrive at the formulas

$$\mathbf{A\xi}[\overline{\mathcal{U}}^*] \vdash \Omega \quad \text{and} \quad \Omega' \vdash \mathbf{A\xi}[\mathcal{U}^*].$$

From the first formula, using the fact that $\overline{\mathcal{U}}^*$ is a definitional extension of $\mathcal{U}^*$ and hence equipollent with it in means of proof, we see that $\mathbf{A\xi}[\mathcal{U}^*] \vdash \Omega$. In a completely analogous fashion we obtain from the second formula that $\Omega \vdash \mathbf{A\xi}[\mathcal{U}^*]$. Thus Ω is a base for $\mathcal{U}^*$, and therefore $\mathcal{U}^*$ is a definitional extension of $\mathcal{S}$ as well as of $\mathcal{U}$. We conclude that the three systems $\mathcal{U}, \mathcal{S}$, and $\mathcal{U}^*$ are equipollent in means of proof.

To extend the conclusion to the correlated system $\mathcal{S}^\times$ in $\mathcal{L}^\times$, we pass from the system $\mathcal{U}^*$ in $\mathcal{P}^*$ to the correlated system $\mathcal{U}^{*+}$ in the extended formalism $\mathcal{P}^{*+}$, and we proceed as in the last part of the proof of 7.1(ii). This completes the proof.

From the statement of (iv) it follows that *condition* $(*)$ *is superfluous in case all (or all but one) of the nonlogical constants are operation symbols.* It may further be remarked that $(*)$ is automatically satisfied if the sentence $\exists_{xy}(x\mathring{0}y)$ is provable in $\mathcal{U}$. If we wanted to prove (iv) under the stronger assumptions corresponding to those in 7.1(ii)$(*)$, then instead of the mapping L' from 7.1(iii) we could use the simpler mapping L from 7.1(ii). In particular, this applies to the case when all nonlogical constants are operation symbols.

To conclude this section we wish to give an easy consequence of (iv) of a semantical nature, established with the help of 6.2(ix). Recall that by a (strong) $\mathfrak{Q}$-structure we understand any model of a $\mathcal{Q}$-system (in a given formalism). The notion of a weak $\mathfrak{Q}$-structure is defined analogously. For the notion of definability see 6.2(iii).

(v) *Let $\mathcal{P}$ be a formalism of predicate logic with a finite number of nonlogical constants. For any weak $\mathfrak{Q}$-structure $\mathfrak{U}$, with universe U, which is a realization of $\mathcal{P}$, there is a binary relation E definable in $\mathfrak{U}$ such that any given binary relation is definable in $\mathfrak{U}$ iff it belongs to the relation ring on U generated by E.*

Indeed, from (iv)(α) we can, without essential difficulty, infer the existence of a binary relation E on U such that $\langle U, E\rangle$ is a (strong) $\mathfrak{Q}$-structure, and the set of relations definable in $\mathfrak{U}$ coincides with the set of relations definable in $\langle U, E\rangle$. On the other hand, by 6.2(ix) the set of binary relations definable in $\langle U, E\rangle$ is identical with the relation ring on U generated by E. Hence the conclusion follows at once.

7.3. The equipollence of weak $\mathcal{Q}$-systems with finite variable subsystems

In §§4.5 and 4.8 we have established two results of a related character concerning $\mathcal{Q}$-systems $\mathcal{S}$ in the formalism $\mathcal{L}$. First, we have shown in §4.5 that $\mathcal{S}$ is equipollent with a $\mathcal{Q}$-system $\mathcal{S}^\times$ developed in the formalism $\mathcal{L}^\times$. Secondly, using this result we have proved in §4.8 that $\mathcal{S}$ is equipollent to a $\mathcal{Q}$-system $\mathcal{S}_3$ in $\mathcal{L}_3$, which actually (in contrast to $\mathcal{S}^\times$) turns out to be a subsystem of $\mathcal{S}$.

In 7.2(iv) we have extended the first result, under mild restrictions, to all weak $\mathcal{Q}$-systems $\mathcal{U}$ formalized in predicate logic $\mathcal{P}$. We wish in this section to establish a theorem that in a sense can be regarded as an extension of the second result. In fact, assuming $\mathcal{P}$ has no operation symbols, we shall show that $\mathcal{U}$ is equipollent with a subsystem of itself formalized in a predicate logic with a finite number of variables; in general, however, this number is greater than 3.

Let m be a natural number ≥ 3. By $\mathcal{P}_m$ we understand the subformalism of $\mathcal{P}$ obtained in the same way as $\mathcal{L}_m$ was obtained from $\mathcal{L}$; see §§3.8 and 3.10. Thus, the set of formulas of $\mathcal{P}_m, \mathbf{\Phi}[\mathcal{P}_m]$, consists of just those formulas of $\mathcal{P}$ in which only variables from the set $\mathbf{\Upsilon}_m = \{\mathbf{v}_0, \ldots, \mathbf{v}_{m-1}\}$ occur; similarly for sentences of $\mathcal{P}_m$. The set $\mathbf{\Lambda}[\mathcal{P}_m]$ of logical axioms consists of all those sentences in $\Sigma[\mathcal{P}]$ that are instances of Schemata (AI)–(AVIII) in §1.3, or Schema (AIX′) in §3.7; however, in case $m = 3$ we also adjoin all instances of the associativity schema (AX) in §3.7. (For $m > 3$ such an adjunction is superfluous, just as in the case of $\mathcal{L}_m$; cf. §3.10.)

For any given $m \geq 3$ the formalism $\mathcal{P}_m$ just described is, in a sense, a natural m-variable restriction of $\mathcal{P}$. In particular, various familiar metalogical laws which prove to hold for the formalism $\mathcal{L}_m$ (cf. §§3.7, 3.8, and 3.10) can be extended to $\mathcal{P}_m$.

Unfortunately, the formalism $\mathcal{P}_m$ does not seem to be adequate for our purposes. It turns out that the difficulties which appear to arise can be overcome if we replace $\mathcal{P}_m$ by a stronger formalism $\mathcal{P}_{m+}$ which is a kind of hybrid between $\mathcal{P}_m$ and $\mathcal{P}_{m+1}$. The construction of $\mathcal{P}_{m+}$ may seem somewhat artificial: by $\mathcal{P}_{m+}$ we understand the formalism which has the same variables as $\mathcal{P}_m$, but in which the relation of derivability is that of $\mathcal{P}_{m+1}$. More precisely,

$$\mathbf{\Phi}[\mathcal{P}_{m+}] = \mathbf{\Phi}[\mathcal{P}_m] \quad \text{and hence} \quad \Sigma[\mathcal{P}_{m+}] = \Sigma[\mathcal{P}_m];$$

$$\Psi \vdash_{m+} X \text{ iff } \Psi \vdash_{m+1} X, \text{ provided } \Psi \subseteq \Sigma[\mathcal{P}_m] \text{ and } X \in \Sigma[\mathcal{P}_m].$$

(Formalisms closely related to $\mathcal{P}_{m+}$ are involved in the discussion in Henkin–Tarski [**1961**], p. 109.)

It is clear that, for each $m \geq 3, \mathcal{P}_m$ is a subformalism of $\mathcal{P}_{m+}$, and this in turn is a subformalism of $\mathcal{P}_{m+1}$. It may be noted that, in the particular case of the formalism $\mathcal{L}_{3+}$, the derivability relations $\vdash [\mathcal{L}_{3+}]$ and $\vdash [\mathcal{L}_3]$ are identical by a result of Maddux [**1978a**], p. 210, mentioned in §3.10, and hence the formalisms $\mathcal{L}_3$ and $\mathcal{L}_{3+}$ coincide. On the other hand, it is known that $\mathcal{L}_4$ and $\mathcal{L}_{4+}$ do not coincide. In fact, if T is the sentence given in 3.4(vi) and $\boldsymbol{G}$ is the translation mapping of §2.3, then $\boldsymbol{G}T \in \Sigma_4$ and we have $\vdash \boldsymbol{G}T[\mathcal{L}_{4+}]$, but not $\vdash \boldsymbol{G}T[\mathcal{L}_4]$; cf. §3.10. The problem whether $\mathcal{L}_m$ and $\mathcal{L}_{m+}$ coincide, and hence are equipollent, is open for every $m \geq 5$.[3] As regards the relationship between $\mathcal{L}_{m+}$ and $\mathcal{L}_{m+1}$ for $m \geq 3$, it follows from the results of Kwatinetz mentioned after 3.10(vi) that these two formalisms are never equipollent, and in fact that $\mathcal{L}_{m+1}$ is stronger than $\mathcal{L}_{m+}$ in means of expression. All the above observations extend to the formalisms $\mathcal{M}_m^{(n)}$, and $\mathcal{M}_{m+}^{(n)}$ with $n+1$ distinct binary predicates.

Consider now an arbitrary formalism $\mathcal{P}$ of predicate logic. In case $\mathcal{P}$ is provided with at least one predicate of rank ≥ 3, the problem whether $\mathcal{P}_m$ is equipollent with the corresponding $\mathcal{P}_{m+}$ is still open for all values of $m \geq 3$ (and thus, in particular, for $m = 3$ and $m = 4$).[4] On the other hand, Givant has shown that

[3] *See the footnote, p. 93.

[4] *Andréka and Németi have communicated to us the solution to a special case of this problem. Namely, if $\mathcal{P}$ contains a m-ary relation symbol, then $\mathcal{P}_m$ is not equipollent with

the results of Kwatinetz just mentioned can be extended to every formalism $\mathcal{P}$, without any special assumptions about its vocabulary; in consequence, for every $m \geq 3$ the formalism $\mathcal{P}_{m+}$ is poorer than $\mathcal{P}_{m+1}$ in means of expression.

The notions of a $\mathcal{Q}$-system and a weak $\mathcal{Q}$-system can clearly be extended to the formalisms $\mathcal{P}_m$ and $\mathcal{P}_{m+}$; e.g., 7.2(i), with obvious changes, provides a characterization of a weak $\mathcal{Q}$-system in $\mathcal{P}_m$ if we interpret the derivability symbol in (β) and (γ) as referring to the formalism $\mathcal{P}_m$ (and not $\mathcal{P}$).

In 4.8(xvi) we have essentially established the semantical completeness theorem for $\mathcal{Q}$-systems in $\mathcal{L}_{m+}$. This can easily be extended to formalisms of the type $\mathcal{M}^{(n)}_{m+}$. Using this observation we now establish the semantical completeness theorem for weak $\mathcal{Q}$-systems $\mathcal{U}$ formalized in $\mathcal{P}_{m+}$ by showing that the semantical relation of consequence, $\models [\mathcal{U}]$, coincides with the syntactical relation of derivability, $\vdash [\mathcal{U}]$. (This result is due to Givant.) In its application to the formalism $\mathcal{L}_{m+}$ this will provide an improved version of 4.8(xvi).

(i) *Let $m \geq 3$ and let $\mathcal{P}$ be a formalism of predicate logic with no operation symbols and with nonlogical atomic predicates $C_0, \ldots, C_n$, each of rank $< m$. If $\mathcal{U}$ is a weak $\mathcal{Q}$-system in $\mathcal{P}_{m+}$, then for every $\Psi \subseteq \Sigma[\mathcal{P}_m]$ and $X \in \Sigma[\mathcal{P}_m]$ we have*

$$\Psi \models X\,[\mathcal{U}] \quad \textit{iff} \quad \Psi \vdash X\,[\mathcal{U}]$$

(or, equivalently,

$$\Psi \vdash_{\mathbf{A}\mathfrak{E}[\mathcal{U}]} X\,[\mathcal{P}] \quad \textit{iff} \quad \Psi \vdash_{\mathbf{A}\mathfrak{E}[\mathcal{U}]} X\,[\mathcal{P}_{m+1}]).$$

We shall give here a rough outline of the proof, leaving many of the details (not all of them of a trivial character) to the reader. We use some of the methods applied in the proof of 7.2(iii). Thus, we correlate with the given weak $\mathcal{Q}$-system $\mathcal{U}$ in $\mathcal{P}_{m+}$ the (strong) $\mathcal{Q}$-systems $\overline{\mathcal{U}}$ and $\mathcal{T}$ formalized respectively in $\overline{\mathcal{P}}_{m+}$ and $\mathcal{M}^{(n+2)}_{m+}$, as well as the mappings H from $\overline{\mathcal{P}}_{m+}$ to $\mathcal{P}_{m+}$ and K from $\overline{\mathcal{P}}_{m+}$ to $\mathcal{M}^{(n+2)}_{m+}$. As opposed to K, the mapping H is only mentioned parenthetically in 7.2(iii), but it is introduced explicitly in the proof of 7.2(iv). (Strictly speaking, the mappings H and K to be used here are not exactly the mappings involved in 7.2(iii),(iv), but rather their restrictions to the set $\Sigma[\overline{\mathcal{P}}_{m+}]$.) We can then establish the following three statements:

(1) $\Psi \vdash X\,[\overline{\mathcal{U}}]$ implies $H^*\Psi \vdash HX\,[\mathcal{U}]$ whenever $\Psi \subseteq \Sigma[\overline{\mathcal{P}}_m]$ and $X \in \Sigma[\overline{\mathcal{P}}_m]$;

(2) $\vdash HKY \leftrightarrow Y\,[\overline{\mathcal{U}}]$ and $\boldsymbol{\Upsilon}\phi HKY = \boldsymbol{\Upsilon}\phi Y$ for every $Y \in \boldsymbol{\Phi}[\overline{\mathcal{P}}_m]$;

(3) $\Theta \vdash Z\,[\mathcal{T}]$ implies $H^*\Theta \vdash HZ\,[\overline{\mathcal{U}}]$ whenever $\Theta \subseteq \Sigma[\mathcal{M}^{(n+2)}_m]$ and $Z \in \Sigma[\mathcal{M}^{(n+2)}_m]$.

$\mathcal{P}_{m+}$, and in fact there is a weak $\mathcal{Q}$-system in $\mathcal{P}_m$ that cannot be equipollently formalized in $\mathcal{P}_{m+}$.

Statement (1) is proved by induction on sentences X derivable in $\overline{\mathfrak{U}}$ from Ψ, using formulas (8) and (9) in the proof of 7.2(iii) (upon which a precise definition of the mapping H is based). The only portion of the proof that is not straightforward is the one in which X is assumed to be an instance of (AIX′) or (AX); the argument in this portion depends essentially on certain specific properties of the notion of substitution applicable to logics with finitely many variables, which was discussed in §§3.7 and 3.8.

The proof of (2) is by induction on formulas in $\overline{\mathcal{P}}_m$. Only the beginning of the inductive procedure, i.e., the proof that (2) holds for atomic formulas, presents some difficulties. We apply here statements (14), (8), and (9) from the proof of 7.2(iii). Also, we make essential use of the formulas $Q_0, \ldots, Q_{m-2}$ that were introduced in 7.2(iii) and are involved in the definition of H; we have to show that various properties of these formulas which have been established in 7.2(iii) when $\mathfrak{U}$ is a weak $\mathcal{Q}$-system in $\mathcal{P}$ continue to hold when $\mathfrak{U}$ is a weak $\mathcal{Q}$-system in $\mathcal{P}_{m+}$.

The proof of (3) is analogous to the proof of (1). Since, however, $\mathbf{A}\xi[\mathfrak{T}]$ coincides with $K^*\mathbf{A}\xi[\overline{\mathfrak{U}}]$ by definition, we also use (2) in the argument.

We now take up the equivalence in the conclusion of our theorem. We consider only the implication from left to right, since the implication in the opposite direction is trivial. Thus, assume $\Psi \models X\,[\mathfrak{U}]$. Since KY is semantically equivalent with Y in $\overline{\mathfrak{U}}$ for each $Y \in \Sigma[\mathcal{P}_m]$, we get $K^*\Psi \models KX\,[\mathfrak{T}]$ and therefore $K^*\Psi \vdash KX\,[\mathfrak{T}]$, by 4.8(xvi) extended to $\mathcal{M}_{m+}^{(n+2)}$. Hence, we derive successively

$$HK^*\Psi \vdash HKX\,[\overline{\mathfrak{U}}],$$
$$\Psi \vdash X\,[\overline{\mathfrak{U}}],$$
$$\Psi \vdash X\,[\mathfrak{U}],$$

by (3), (2), (1). In the case of the last formula, which is just the one we wish to obtain, we also use the fact that $HY = Y$ for each $Y \in \Sigma[\mathcal{P}_m]$. The proof is thus complete.

We now formulate the main result of this section (a joint result of Givant and Tarski).

(ii) *If $\mathcal{P}$ is a formalism of predicate logic with no operation symbols and $\mathfrak{U}$ is a weak $\mathcal{Q}$-system formalized in $\mathcal{P}$, then there is a natural number $m \geq 3$ and a weak $\mathcal{Q}$-system $\mathcal{V}$ formalized in $\mathcal{P}_{m+}$ such that $\mathfrak{U}$ and $\mathcal{V}$ are equipollent in means of expression and proof. In fact, if $\mathbf{C}_0, \ldots, \mathbf{C}_n$ are all the distinct nonlogical constants of $\mathcal{P}$, and D, E are two formulas in $\mathcal{P}$ satisfying conditions (α)–(γ) in 7.2(i) (the definition of a weak $\mathcal{Q}$-system), then we can take m to be the maximum of the numbers 3, $\rho 0+1, \ldots, \rho n+1, d, e$, where $\rho 0, \ldots, \rho n$ are the respective ranks of $\mathbf{C}_0, \ldots, \mathbf{C}_n$, and d, e the respective numbers of distinct variables occurring in D, E.*

We shall again give only an outline of the proof, which is based on (i) and 7.2(iii). We shall use certain facts which either appear explicitly in the formulation of Lemma 7.2(iii), or can be easily obtained by analyzing its proof. Thus, with the given weak $\mathcal{Q}$-system $\mathcal{U}$ in $\mathcal{P}$ we correlate the (strong) $\mathcal{Q}$-systems $\overline{\mathcal{U}}$ in $\overline{\mathcal{P}}$ and $\mathcal{T}$ in $\mathcal{M}^{(n+2)}$, as well as the mappings H and K from $\overline{\mathcal{P}}$ to $\mathcal{P}$ and $\mathcal{M}^{(n+2)}$ respectively.

We now use some of the results of §4.8, primarily 4.8(x); as stated several times, these results, which are explicitly formulated for $\mathcal{L}$, extend to arbitrary formalisms of the type $\mathcal{M}^{(k)}$, and, in particular, to $\mathcal{M}^{(n+2)}$. Thus, following the lines of 4.8(iv),(vii), we can construct a subformalism $\mathcal{T}_3$ of $\mathcal{T}$, as well as a translation mapping N from $\mathcal{M}^{(n+2)}$ to $\mathcal{M}_3^{(n+2)}$, so that $\mathcal{T}$ and $\mathcal{T}_3$ turn out to be equipollent. It is easily seen that $\mathcal{T}_3$ is also an equipollent subformalism of $\overline{\mathcal{U}}$; in fact $N \circ K$ is a mapping from $\overline{\mathcal{P}}$ to $\mathcal{M}_3^{(n+2)}$ that proves to have all the desired properties of a translation mapping with respect to the $\mathcal{Q}$-systems $\overline{\mathcal{U}}$ and $\mathcal{T}_3$.

Let m be the number specified in the second part of our theorem (with D, E fixed). With the help of formulas (5), (8), and (9) in the proof of 7.2(iii) we readily show that

(1) $\quad HX \in \Sigma[\mathcal{P}_m]$ whenever $X \in \Sigma[\mathcal{M}_m^{(n+2)}]$.

Similarly, with the help of (14) in the proof of 7.2(iii) we get

(2) $\quad KX \in \Sigma[\mathcal{M}_m^{(n+2)}]$ whenever $X \in \Sigma[\mathcal{P}_m]$.

Steps (1) and (2) imply

(3) $\quad HKX \in \Sigma[\mathcal{P}_m]$ whenever $X \in \Sigma[\mathcal{P}_m]$.

Since $\Sigma[\mathcal{M}_3^{(n+2)}] \subseteq \Sigma[\mathcal{M}_m^{(n+2)}]$, we see from (1) that $H^*\mathbf{A\xi}[\mathcal{T}_3] \subseteq \Sigma[\mathcal{P}_m]$. Therefore we can determine a system $\mathcal{V}$ in $\mathcal{P}_{m+}$ by stipulating that

(4) $\quad \mathbf{A\xi}[\mathcal{V}] = H^*\mathbf{A\xi}[\mathcal{T}_3]$.

The construction of the system $\mathcal{V}$ has thus been completed.

Our task now is to establish the equipollence of $\mathcal{V}$ with $\mathcal{U}$; HNK will be used as the underlying mapping from $\mathcal{U}$ to $\mathcal{V}$. The result will be obtained as an immediate consequence of the next four statements, (5)–(8).

(5) $\quad \vdash HNK(X) \leftrightarrow X\,[\overline{\mathcal{U}}]$ for all $X \in \Sigma[\overline{\mathcal{P}}]$, and
$\quad\quad \vdash HNK(X) \leftrightarrow X\,[\mathcal{U}]$ for all $X \in \Sigma[\mathcal{P}]$.

This follows directly from the properties of the translation mappings H, N, and K, and from the fact that $\overline{\mathcal{U}}$ is a definitional extension of $\mathcal{U}$.

(6) $\quad$ System $\mathcal{V}$ in $\mathcal{P}_{m+}$ is a subsystem of $\mathcal{U}$.

This is readily obtained from (4), (5) and the definition of $\mathbf{A}\xi[\mathcal{T}_3]$ (see 4.8(iv) and formula (17) in the proof of 7.2(iii)).

(7) For every $\Psi \subseteq \Sigma[\mathcal{P}]$ and $X \in \Sigma[\mathcal{P}]$, we have

$$\Psi \vdash X\,[\mathcal{U}] \quad \text{iff} \quad HNK^*\Psi \vdash HNK(X)\,[\mathcal{V}].$$

In fact, the implication from right to left follows at once from (5) and (6). To obtain the implication in the opposite direction we first observe that, by the main mapping theorems for N and K,

$$\Psi \vdash X\ [\mathcal{U}] \quad \text{implies} \quad NK^*\Psi \vdash NK(X)\ [\mathcal{T}_3].$$

In view of this it suffices to show that, for every $\Delta \subseteq \Sigma[\mathcal{T}_3]$ and $Y \in \Sigma[\mathcal{T}_3]$,

$$\Delta \vdash Y\ [\mathcal{T}_3] \quad \text{implies} \quad H^*\Delta \vdash HY\ [\mathcal{V}].$$

This last implication is established in the same way as was statement (1) in the proof of (i) above.

(8) $\vdash HNK(X) \leftrightarrow X\ [\mathcal{V}]$ for every $X \in \Sigma[\mathcal{P}_m]$.

To show (8) we first observe that $\mathcal{V}$ is a weak $\mathfrak{Q}$-system, since $\mathcal{T}_3$ is a strong one. Also $HNK(X) \leftrightarrow X$ is a sentence in $\mathcal{P}_m$ by (1), (2). Thus (8) follows from Theorem (i) above, with the help of (5).

In view of (5)–(7), *HNK* is "almost" a translation mapping from $\mathcal{U}$ to $\mathcal{V}$. It isn't a translation mapping in the strict sense since it does not coincide with the identity on $\Sigma[\mathcal{V}] = \Sigma[\mathcal{P}_m]$. Instead, a weaker property of this mapping is stated in (8). From certain observations in our earlier discussions we know, however, that in such a situation we can construct a new mapping, closely related to the original one, which is a translation mapping in the strict sense and yields the equipollence of the formalisms or systems involved; see, e.g., 2.4(v). In the present case, this new mapping F is determined by

$$FX = X \quad \text{for} \quad X \in \Sigma[\mathcal{P}_m] \quad \text{and} \quad FX = HNK(X) \quad \text{for} \quad X \in \Sigma[\mathcal{P}] \sim \Sigma[\mathcal{P}_m],$$

and it yields the equipollence of the systems $\mathcal{U}$ and $\mathcal{V}$, thus completing the proof of our theorem.

As regards a formalism $\mathcal{P}$ with operation symbols, the problem of extending (i) and (ii) to such formalisms is not yet entirely settled; nor have the formalisms $\mathcal{P}_m$ been thoroughly investigated by us. Nevertheless, we can extend that part of (ii) which refers to equipollence in means of expression, without referring to formalisms $\mathcal{P}_m$ (or $\mathcal{P}_{m+}$).

(iii) *If $\mathcal{P}$ is a formalism of predicate logic (possibly with operation symbols) and $\mathcal{U}$ is a weak $\mathfrak{Q}$-system formalized in $\mathcal{P}$, then there is a natural number $m \geq 3$ such that for every $X \in \Sigma[\mathcal{P}]$ there is a $Y \in \Sigma[\mathcal{P}_m]$ satisfying $X \equiv Y\ [\mathcal{U}]$. In* *fact, m can be determined just as in* (ii) *except that, whenever $\boldsymbol{C}_k (0 \leq k \leq n)$ is an operation symbol, we replace $\rho k + 1$ with $\rho k + 2$.*

The proof of (iii) is just like a portion of the proof of (ii)—in fact it is essentially that portion in which steps (1) and (5) are carried through.

7.4. Comparison of equipollence results for strong and weak $\mathfrak{Q}$-systems

We wish to make here some observations contrasting the results obtained for strong $\mathfrak{Q}$-systems with those for weak $\mathfrak{Q}$-systems. This contrast will be brought more sharply into focus if we restrict our attention to the original formalism $\mathcal{L}$ and to related formalisms with a binary predicate as the only nonlogical constant; in particular, we shall deal with a formalism $\mathcal{L}'$ which differs from $\mathcal{L}$ only in that its unique nonlogical binary predicate $\boldsymbol{E}'$ is distinct from $\boldsymbol{E}$.

Consider first the results concerning equipollence with systems in formalisms of the type $\mathcal{L}^{\times}$. If $\mathcal{S}$ is a strong $\mathfrak{Q}$-system in $\mathcal{L}$, then by §4.5 it is equipollent with a $\mathfrak{Q}$-system $\mathcal{U}$ formalized in $\mathcal{L}^{\times}$, and in fact they are both equipollent subsystems of a certain system $\mathcal{T}$ of $\mathcal{L}^{+}$. (Indeed, we can take $\mathcal{U}$ and $\mathcal{T}$ to be $\mathcal{S}^{\times}$ and $\mathcal{S}^{+}$ respectively.) If, however, $\mathcal{S}$ is only assumed to be a weak $\mathfrak{Q}$-system, then on the basis of 7.2(iv) we can merely claim that it is equipollent with a $\mathfrak{Q}$-system $\mathcal{U}$ formalized in $\mathcal{L}'^{\times}$, $\mathcal{S}$ and $\mathcal{U}$ being treated as equipollent subsystems of a system $\mathcal{T}$ in a formalism $\mathcal{M}^{(1)+}$ with two nonlogical predicates, $\boldsymbol{E}$ and $\boldsymbol{E}'$. Actually, if $\mathcal{S}$ is a weak $\mathfrak{Q}$-system that is not strong, then, as is readily seen, it cannot be equipollent with any $\mathfrak{Q}$-system $\mathcal{U}$ in $\mathcal{L}^{\times}$, where $\mathcal{S}$ and $\mathcal{U}$ are to be treated as equipollent subsystems of some system in $\mathcal{L}^{+}$.

We turn now to the results concerning equipollence with systems in formalisms of type $\mathcal{L}_3$. If $\mathcal{S}$ is again a strong $\mathfrak{Q}$-system, then by §4.8 there exists a $\mathfrak{Q}$-system $\mathcal{U}$ in $\mathcal{L}_3$ which is an equipollent subsystem of $\mathcal{S}$. If $\mathcal{S}$ is only assumed to be a weak $\mathfrak{Q}$-system, then from §4.8 and 7.2(iv) we easily derive the existence of a $\mathfrak{Q}$-system $\mathcal{U}$ in $\mathcal{L}'_3$ which is equipollent with $\mathcal{S}$; trivially, however, $\mathcal{U}$ is not a subsystem of $\mathcal{S}$. If, in particular, $\mathcal{S}$ is a weak $\mathfrak{Q}$-system which is not strong, then, obviously it cannot be equipollent with any subsystem in $\mathcal{L}_3$ that is a $\mathfrak{Q}$-system. Nevertheless, in this case we can construct by 7.3(ii) an equipollent weak $\mathfrak{Q}$-system which is a subsystem of $\mathcal{S}$ developed in $\mathcal{L}_{m+}$ for the appropriately chosen m.

The difference between strong and weak $\mathfrak{Q}$-systems is brought even more sharply into focus if we look at some semantical properties of these systems. Given a structure $\mathfrak{U} = \langle U, E \rangle$, where E is a binary relation on U, consider the condition

(i) *Every binary relation definable in $\mathfrak{U} = \langle U, E \rangle$ belongs to the relation ring on U generated by E.*

By 6.2(ix) every strong $\mathfrak{Q}$-structure $\mathfrak{U}$ satisfies (i). However, this result cannot be extended to arbitrary weak $\mathfrak{Q}$-structures since, as is easily seen, a weak $\mathfrak{Q}$-structure that satisfies (i) must in fact be a strong one. On the other hand, as a direct consequence of 7.2(v), we conclude that in any given weak $\mathfrak{Q}$-structure $\mathfrak{U} = \langle U, E\rangle$ there is a definable binary relation E' such that every binary relation definable in $\mathfrak{U}$ belongs to the relation ring on U generated by E'.

It may be mentioned that, in case we consider not the formalism $\mathcal{L}$, but an arbitrary formalism $\mathcal{P}$ of predicate logic with at least one predicate of rank greater than 2 (or at least one operation symbol of rank greater than 1), we see no way of improving the results in §§7.2 and 7.3 if we restrict ourselves to strong $\mathfrak{Q}$-systems in $\mathcal{P}$ and their models.

7.5. The formalizability of the arithmetic of natural numbers in $\mathcal{L}^{\times}$

Among special mathematical systems to which the conclusions of Theorem 7.2(iv) apply, the best known is the system $\mathcal{N}$ of *elementary number theory*, also referred to as the system of the *arithmetic of natural numbers.*

Elementary number theory can be loosely characterized as that part of the general theory of natural numbers which can be formalized within (first-order) predicate logic. In a formal setting this theory can be defined semantically as the (first-order) theory $\mathbf{\Theta\rho}\mathfrak{N}$ of a definite algebraic structure $\mathfrak{N}$, i.e., the set of all sentences, in an appropriate formalism $\mathcal{P}$ of predicate logic, that are true of $\mathfrak{N}$. The universe of $\mathfrak{N}$ is the set N of natural numbers. (Thus, in this and the next two sections we shall use "N" instead of "ω".) The choice of fundamental notions for this theory, i.e., of operations and relations on N (possibly including some operations of rank 0, i.e., some particular natural numbers) which together with N constitute $\mathfrak{N}$, is to a large extent arbitrary. Following tradition, we select for this purpose the natural number 0, the unary successor operation S, and the binary operations $+$ and $\cdot$, so that $\mathfrak{N} = \langle N, 0, S, +, \cdot\rangle$.

System $\mathcal{N}$ is the system whose theory (the set of all provable sentences) coincides with $\mathbf{\Theta\rho}\mathfrak{N}$; thus, under our description of $\mathfrak{N}$, it is formalized in the language $\mathcal{P}^{\mathsf{N}}$ provided with four nonlogical constants: the individual constant $\mathbf{0}$, the unary operation symbol $\boldsymbol{S}$, and the binary operation symbols $\boldsymbol{+}$ and $\boldsymbol{\cdot}$. (The symbols "$\mathbf{0}$", "$\boldsymbol{+}$", "$\boldsymbol{\cdot}$" just introduced for the purpose of our discussion in this and the following two sections should not be confused with the ones introduced in §2.1 and frequently employed throughout this work; the same applies to the symbol "$\boldsymbol{=}$" which will be used below, instead of "$\mathring{\mathbf{1}}$", to denote the ordinary identity symbol of $\mathcal{P}^{\mathsf{N}}$, and which should not be confused with the symbol of the same shape that appears, e.g., in 2.2(DV). Finally, a similar remark applies to the symbol "$\boldsymbol{S}$", which is used in §6.3 and the subsequent sections of Chapter 6 with a quite different meaning.) As is well known, $\mathbf{\Theta\rho}\mathfrak{N}$ is complete, but not recursive, and hence does not have a recursive base (compare, e.g., Monk [**1976**], pp. 263, 280). In consequence, $\mathcal{N}$ cannot be presented as an axiomatic system, with a recursive axiom set. However, for our purposes we can treat $\mathcal{N}$ as a system with a nonrecursive axiom set, letting $\mathbf{A\xi}[\mathcal{N}] = \mathbf{\Theta\rho}\mathfrak{N}$. $\mathfrak{N}$ may be regarded as the standard model of $\mathcal{N}$.

By means of an elementary argument we can establish the following theorem.

(i) *$\mathcal{N}$ is a (strong) $\mathfrak{Q}$-system. In fact the formulas*

$$\exists_z[x = (y+z)\cdot(y+z)+y] \quad \textit{and} \quad \exists_z[x = (z+y)\cdot(z+y)+z]$$

satisfy conditions (α)–(δ) *in* 4.5(ii).

There are many other pairs of formulas that we could use in (i), but none of them is simpler than the pair we have actually used.

From (i) we see that the conclusions of 7.2(iv) do indeed apply to $\mathcal{N}$. In other words:

(ii) *$\mathcal{N}$ is definitionally equivalent with a $\mathfrak{Q}$-system $\mathcal{N}'$ formalized in $\mathcal{L}$, and hence is equipollent in means of expression and proof with the correlated $\mathfrak{Q}$-system $\mathcal{N}'^{\times}$ in $\mathcal{L}^{\times}$.*

It may be interesting to state here an equivalent model-theoretical formulation of (ii).

(iii) *There exists a binary relation E between natural numbers such that the structure $\mathfrak{N}' = \langle N, E\rangle$ is a strong $\mathfrak{Q}$-structure, and $\mathfrak{N}$ and $\mathfrak{N}'$ are first-order definitionally equivalent.*

Recall that two structures are said to be *first-order definitionally equivalent* if the fundamental operations and relations of each are first-order definable (cf. 6.2(iii)(α)) in the other (see, e.g., Henkin–Monk–Tarski [**1971**], pp. 56–57). The derivation of (iii) from (ii) is straightforward (as is the derivation in the opposite direction).

The relation E in (iii), when constructed by means of the general method, i.e., by analyzing the proof of 7.2(iv), is quite involved and has no clear mathematical content. It seems therefore interesting to outline here another way of deriving (iii) which does not make use of the general method, but rather depends on specific properties of the structure $\mathfrak{N}$. The relation underlying this proof is relatively simple and has a clear mathematical content.

Indeed, recall that any given natural number y can be uniquely represented in the form

$$y = 2^{z_0} + \cdots + 2^{z_{n-1}},$$

where $n, z_0, \ldots, z_{n-1} \in N$ and $z_0 < z_1 < \cdots < z_{n-1}$. (Of course, in case $y = 0$ we take $n = 0$, so that the sequence $\langle z_0, \ldots, z_{n-1}\rangle$ becomes the empty sequence.) For any $x, y \in N$ we stipulate that xEy if x coincides with one of the exponents $z_0, \ldots, z_{n-1}$ occurring in this representation of y. This can also be expressed by saying that in the dyadic expansion of y the $x + 1^{\text{st}}$ digit from the right is 1. An equivalent way of defining E is to say that y can be represented in the form $y = 2^x \cdot q + r$ where $r < 2^x$ and q is odd.

Clearly E is a universal relation for two-element sets. In fact E has a much stronger property: it is a *universal relation for all finite sets.* This means that for every finite set $Z \subseteq N$ there is a $y \in N$ such that, for each natural number

x, we have xEy iff $x \in Z$. Since in the present case each number y uniquely determines, and is uniquely determined by, the set Z, the relation E obviously induces a one-one correspondence F between natural numbers and finite sets of natural numbers; F is determined by the condition

$$Fy = \{x \colon xEy\}.$$

Thus, for example, $F7 = \{0, 1, 2\}$ since $7 = 2^0 + 2^1 + 2^2$.

Consider now a different kind of finite sets, namely all the sets that can be obtained from $\varnothing$ by applying any finite number of times the operations of forming singletons and of forming binary unions; as is easily seen, these two operations can equivalently by replaced by the single operator $\triangleright$ of adjunction of a given element y to a given set x:

$$x \triangleright y = x \cup \{y\}.$$

Let H be the class of all sets thus obtained. It is not hard to show that H coincides with the class of what are called in set theory the *hereditarily finite sets*, or *sets of finite rank*. By looking more closely at the correspondence F above, we readily see that in terms of E we can establish a one-one correspondence G between natural numbers and hereditarily finite sets. In fact, since $F0 = \varnothing$ and $F1 = \{0\}$, we set $G0 = \varnothing$ and $G1 = \{\varnothing\} = \varnothing \triangleright \varnothing$. Since $F3 = \{0, 1\}$ we put $G3 = \{\varnothing, \{\varnothing\}\} = (\varnothing \triangleright \varnothing) \triangleright (\varnothing \triangleright \varnothing)$, etc. In general, we define G recursively by stipulating for every $y \in N$ that

$$Gy = \{Gx \colon xEy\} \quad \text{or, equivalently,} \quad Gy = G^{*}Fy.$$

From this definition of G we arrive without difficulty at the following statement.

(iv) *G is an isomorphic transformation of the structure $\langle N, E\rangle$ onto the structure $\langle H, \in\rangle$.*

(Here $\in$ is the usual membership relation with its field restricted to H.) In addition to its intrinsic interest, this result may help the reader in grasping some of the intuitions behind the subsequent discussion.

We now state the improved version of (iii) at which we were aiming.

(v) *The relation E which is defined to hold between any two natural numbers x and y iff*

(α) *there are $z, u, v \in N$ such that $y = S(z + z) \cdot 2^x + u$ and $2^x = Su + v$*

satisfies all the conditions of (iii).

To prove (v) we have first to show that the structures $\mathfrak{N} = \langle N, 0, S, +, \cdot\rangle$ and $\mathfrak{N}' = \langle N, E\rangle$ are definitionally equivalent. In fact to show that E is definable in $\mathfrak{N}$ it suffices, in view of (α), to show that the function determined by $y = 2^x$ is (first-order) definable in $\mathfrak{N}$, and this is a well-known result due to Gödel (compare, e.g., Monk [**1976**], p. 251).

The proof in the opposite direction is considerably more involved. Our task is to show that all the fundamental notions of $\mathfrak{N}$ are definable in $\mathfrak{N}'$. We use some constructions in $\mathfrak{N}'$ of a set-theoretical nature, and we avail ourselves of some properties of these constructions which are obviously implied by (iv) (but which, because of their elementary character, can be established directly without difficulty). Thus, we say that a number x is *properly included* in a number y, in symbols $x \subset y$, provided $x \neq y$ and, for every number u, uEy whenever uEx. (Strictly speaking, we should not use here the set-theoretical term "properly included", but rather some similar but different term, say "properly pseudo-included". We are confident, however, that this abuse of set-theoretical terminology and notation will not lead to any confusion.) For any $x \in N$ we shall understand by the *singleton of* x, in symbols $\{x\}$, the unique $z \in N$ such that uEz iff $u = x$, for every $u \in N$. It may be noted that, by the definition of E, $\{x\}$ is just the number 2^x. By the pair of x and y (in N), in symbols $\{x, y\}$, we understand the unique z such that uEz iff $u = x$ or $u = y$, for every u. The *ordered pair* $\langle x, y\rangle$ is determined as usual by the condition

$$\langle x, y\rangle = \{\{x\}, \{x, y\}\}.$$

A number s is *transitive*, in symbols $T(s)$, if, for any x, y, the conditions xEy and yEs imply xEs. It is easily seen that, for every number x, the number $s = 2^0 + 2^1 + \cdots + 2^x$ is transitive; as a consequence, for any two numbers x, y, there is a transitive number s such that xEs and yEs.

It is convenient to establish the definability of the notions $0, S, +$, and $\cdot$ successively in the indicated order. The number 0 obviously can be defined in terms of E as the only natural number not in the range of E. To define S we find it convenient to introduce first an auxiliary notion, the usual natural order relation $<$. The way in which we define this relation is closely related to the way in which we establish the lexicographic ordering in the set of all finite, strictly decreasing sequences of natural numbers. Consider the following statement:

> $x < y$ iff either $x \subset y$ or else there are u, v satisfying the conditions: $uEx, vEy, u < v$, and, for all z, if zEx and $u < z$, then zEy and $v < z$.

It can be shown that this statement holds (for all $x, y \in N$) in the structure $\langle N, E, <\rangle$. Because of its form, the statement can be characterized as a recursive definition of $<$ in terms of E. We allude here to recursive definitions of the type often used in set theory, in which, for example, we show how to determine the value of a set-function F at a certain set X under the assumption that the values of F have been determined for all members of X. To formulate a regular (i.e., an explicit) definition of $<$ in terms of E we make use of the set-theoretical constructions available in $\mathfrak{N}'$. (In connection with the last remarks compare Monk [**1969**], pp. 86–89.) Since the regular definition of $<$ which we have in mind is rather involved, we give it in a symbolic transcription, using symbols "$\exists$", "$\forall$", "$\vee$", "$\wedge$", "$\rightarrow$", and "$\leftrightarrow$" simply as abbreviations for synonymous expressions of the common language.

$$x < y \leftrightarrow \exists_{rs}\Big(T(s) \wedge \langle x, y\rangle Er \wedge \forall_{x'y'}\big[\langle x', y'\rangle Er \leftrightarrow x'Es \wedge y'Es \wedge \big(x' \subset y' \vee \exists_{uv}[uEx' \wedge vEy' \wedge \langle u, v\rangle Er \wedge \forall_z(zEx' \wedge \langle u, z\rangle Er \rightarrow zEy' \wedge \langle v, z\rangle Er)]\big)\big]\Big).$$

It is now an easy matter to (explicitly) define S in terms of $<$:

$Sx = y$ iff $x < y$ and for no z do we have both $x < z$ and $z < y$.

As is well known, we can recursively define $+$ in terms of S and 0, and $\cdot$ in terms of $+$, S, and 0:

$x + y = z$ iff either $y = 0$ and $z = x$ or else there is a w for which $y = Sw$ and $z = S(x + w)$,

$x \cdot y = z$ iff either $y = 0$ and $z = 0$ or else there is a w for which $y = Sw$ and $z = x \cdot w + x$.

Just as in the case of $<$, these two recursive definitions can be replaced by equivalent explicit ones. We then conclude that $S, +$, and $\cdot$ are definable in terms of E alone. Therefore $\mathfrak{N}$ and $\mathfrak{N}'$ are first-order definitionally equivalent. Since, in addition, E is a universal relation for two-element sets, $\mathfrak{N}'$ is a strong $\mathfrak{Q}$-structure by 6.2(viii), and our outline of the proof of (v) has thus been completed.

With the help of (v) we can derive (ii) directly from the results of Chapter 4, without using the general methods developed in the present chapter. We should like to point out that the definability of the relation E in $\mathfrak{N}$ (but not the definability of $0, S, +$, and $\cdot$ in $\mathfrak{N}'$) was first established in Ackermann [**1937**].

An interesting corollary of (v) and 6.2(ix) is that *the relation ring on N generated by the relation E in* (v) *coincides with the set of all binary relations first-order definable in* $\mathfrak{N}$. In this connection we should like to call the attention of the reader to another structure definitionally equivalent to $\mathfrak{N}$ which, as opposed to $\mathfrak{N}'$, has two simple binary relations as its fundamental notions. It is the structure $\mathfrak{N}'' = \langle N, S', D\rangle$, where S' is the successor relation and D the divisibility relation. Obviously S' and D are definable in $\mathfrak{N}$ since we have

$$xS'y \text{ iff } Sx = y,$$
$$xDy \text{ iff } x \cdot z = y \text{ for some } z.$$

It is much less obvious that the fundamental notions of $\mathfrak{N}$ are definable in $\mathfrak{N}''$. For appropriate definitions of these notions see Robinson [**1949**], where the definitional equivalence of $\mathfrak{N}$ and $\mathfrak{N}''$ was first established. An obvious consequence of this equivalence is that the relations defined in $\mathfrak{N}$ by the displayed formulas in (i) are also definable in $\mathfrak{N}''$, and hence $\mathfrak{N}''$ is a weak $\mathfrak{Q}$-structure. It is, however, an open problem whether $\mathfrak{N}''$ is a strong $\mathfrak{Q}$-structure or, in an equivalent formulation (cf. the remarks in §7.4), whether all binary relations definable in $\mathfrak{N}$ belong to the relation ring on N generated by S' and D. The problem remains open in case the relation S' is replaced in the construction of $\mathfrak{N}''$ by $\leq$, the usual ordering relation on N.

The existence of a relatively simple binary relation E' between natural numbers such that the structure $\langle N, E'\rangle$ and $\mathfrak{N}$ are (first-order) definitionally equivalent was first established in Myhill [**1950**]; E' is determined by the stipulation

$$xE'y \quad \text{iff} \quad x = S(y \cdot z) \quad \text{for some} \quad z \in N.$$

Since, as Myhill points out, the binary relations of successor and divisiblity are definable in terms of E', the definitional equivalence of $\langle N, E'\rangle$ and $\mathfrak{N}$ follows from the results of Robinson [**1949**]. However, it is again an open problem whether $\langle N, E'\rangle$ is a strong $\mathfrak{Q}$-structure. Thus the result obtained in Myhill [**1950**] is apparently weaker than Theorem (iii).

We turn now to the problem of formalizing elementary number theory in predicate logic with finitely many variables. In view of the completeness of the theory of $\mathcal{N}$, every sentence of $\mathcal{P}^{\mathsf{N}}$ is obviously equivalent in $\mathcal{N}$ with a variable-free sentence of $\mathcal{P}^{\mathsf{N}}$, namely with $\mathbf{0} = \mathbf{0}$ or $\mathbf{0} = \boldsymbol{S}\mathbf{0}$. However, this correlation between sentences of $\mathcal{P}^{\mathsf{N}}$ and equivalent variable-free sentences is not recursive. On the other hand, as a consequence of 7.3(iii), with every sentence X of $\mathcal{P}^{\mathsf{N}}$ we can correlate in a recursive way a sentence $Y \in \Sigma_4[\mathcal{P}^{\mathsf{N}}]$ that is equivalent with X in $\mathcal{N}$. We now establish an improvement of this latter result, using some specific properties of $\mathcal{N}$.

(vi) *With every $X \in \Sigma[\mathcal{P}^{\mathsf{N}}]$ we can correlate in a recursive way a $Y \in \Sigma_3[\mathcal{P}^{\mathsf{N}}]$ such that $X \equiv Y$ $[\mathcal{N}]$.*

In the first part of the proof of (vi) we follow strictly the lines of the argument used to establish 7.2(iii). For our purposes we must introduce in this argument certain modifications that we now describe.

For $\mathcal{P}$ in 7.2(iii) we take $\mathcal{P}^{\mathsf{N}}$. Hence $n = 3$, the constants $\boldsymbol{C}_0, \boldsymbol{C}_1, \boldsymbol{C}_2, \boldsymbol{C}_3$, are respectively $\mathbf{0}, \boldsymbol{S}, \boldsymbol{+}, \boldsymbol{\cdot}$, and for $\mathfrak{U}$ we take $\mathcal{N}$.

With any two terms t and s in $\mathcal{P}^{\mathsf{N}}$ we correlate a new term, $t\triangle s$, by stipulating that

$$t\triangle s \;=\; (t \boldsymbol{+} s) \boldsymbol{\cdot} (t \boldsymbol{+} s) \boldsymbol{+} t.$$

(As regards the definition of terms in $\mathcal{P}^{\mathsf{N}}$, see §1.5.) We recall (i) above and set

$$D \;=\; \boldsymbol{\exists}_{\boldsymbol{z}}(\boldsymbol{x} = \boldsymbol{y}\triangle \boldsymbol{z}), \quad E \;=\; \boldsymbol{\exists}_{\boldsymbol{z}}(\boldsymbol{x} = \boldsymbol{z}\triangle \boldsymbol{y}).$$

Instead of the infinite sequence $\langle Q_0, \ldots, Q_m, \ldots\rangle$ of formulas introduced in the proof of 7.2(iii), step (1), we need only the first three terms of this sequence, and we modify their definitions as follows:

$$(1) \qquad Q_0 = D, \quad Q_1 = \boldsymbol{\exists}_{\boldsymbol{z}}(E[\boldsymbol{x}, \boldsymbol{z}] \wedge D[\boldsymbol{z}, \boldsymbol{y}]), \quad Q_2 = \boldsymbol{\exists}_{\boldsymbol{z}}(E[\boldsymbol{x}, \boldsymbol{z}] \wedge E[\boldsymbol{z}, \boldsymbol{y}]).$$

One readily shows that

$$(2) \qquad \boldsymbol{x} = \boldsymbol{y}\triangle(\boldsymbol{z}\triangle \boldsymbol{u}) \equiv Q_0[\boldsymbol{x}, \boldsymbol{y}] \wedge Q_1[\boldsymbol{x}, \boldsymbol{z}] \wedge Q_2[\boldsymbol{x}, \boldsymbol{u}] \;\; [\mathcal{N}].$$

On the basis of (2) we can equivalently replace the formulas $H_0, \ldots, H_3$ in the proof of 7.2(iii), step (5), with simpler formulas, which—and this is the crucial point—contain no variables different from $\boldsymbol{x}, \boldsymbol{y}$, and $\boldsymbol{z}$:

(3)
$$\begin{aligned} H_0 &= (\boldsymbol{x} = \boldsymbol{y} \wedge \boldsymbol{x} = \boldsymbol{0}), \\ H_1 &= [\boldsymbol{x} = \boldsymbol{y} \wedge \exists_{\boldsymbol{y}}(\boldsymbol{x} = \boldsymbol{y} \triangle S\boldsymbol{y})], \\ H_2 &= \big[\boldsymbol{x} = \boldsymbol{y} \wedge \exists_{\boldsymbol{yz}}\big(\boldsymbol{x} = \boldsymbol{y} \triangle [\boldsymbol{z} \triangle (\boldsymbol{y} + \boldsymbol{z})]\big)\big], \\ H_3 &= \big[\boldsymbol{x} = \boldsymbol{y} \wedge \exists_{\boldsymbol{yz}}\big(\boldsymbol{x} = \boldsymbol{y} \triangle [\boldsymbol{z} \triangle (\boldsymbol{y} \cdot \boldsymbol{z})]\big)\big]. \end{aligned}$$

All the subsequent steps in the proof of 7.2(iii) remain unchanged with just one exception. Namely, instead of the infinite sequence $\langle P_0, \ldots, P_m, \ldots \rangle$ of formulas defined in step (12) of that proof, we only need the first three terms of this sequence and, in analogy with (1) above, we modify their definitions as follows:

$$P_0 = \boldsymbol{x} F_4 \boldsymbol{y}, \quad P_1 = \exists_{\boldsymbol{z}}(\boldsymbol{x} F_5 \boldsymbol{z} \wedge \boldsymbol{z} F_4 \boldsymbol{y}), \quad P_2 = \exists_{\boldsymbol{z}}(\boldsymbol{x} F_5 \boldsymbol{z} \wedge \boldsymbol{z} F_5 \boldsymbol{y}).$$

Just as the original proof of 7.2(iii), our modified version of this proof leads to strong $\mathfrak{Q}$-systems $\overline{\mathcal{N}}$ in $\overline{\mathcal{P}}^{\mathsf{N}}$ and $\mathcal{T}$ in $\mathcal{M}^{(5)}$, as well as to translation mappings H' and K' from $\overline{\mathcal{P}}^{\mathsf{N}}$ to $\mathcal{P}^{\mathsf{N}}$ and $\mathcal{M}^{(5)}$ respectively. However, using (3) above we readily show that the mapping H' has the following important additional property:

(4) $\qquad H'Y \in \Sigma_3[\mathcal{P}^{\mathsf{N}}]$ whenever $Y \in \Sigma_3[\mathcal{M}^{(5)}]$.

As in the proof of 7.3(ii) we now construct a subsystem $\mathcal{T}_3$ of $\mathcal{T}$ in the formalism $\mathcal{M}_3^{(5)}$, and a translation mapping N' from $\mathcal{M}^{(5)}$ to $\mathcal{M}_3^{(5)}$, so that $\mathcal{T}$ and $\mathcal{T}_3$ turn out to be equipollent. Using the properties of the translation mappings H', K', and N', we easily establish

(5) $\qquad X \equiv H'N'K'(X)\,[\mathcal{N}]$ for all $X \in \Sigma[\mathcal{P}^{\mathsf{N}}]$.

Since $N'K'(X) \in \Sigma_3[\mathcal{M}^{(5)}]$ for each $X \in \Sigma[\mathcal{P}^{\mathsf{N}}]$, we see from (4) that

(6) $\qquad H'N'K'(X) \in \Sigma_3[\mathcal{P}^{\mathsf{N}}]$ for every $X \in \Sigma[\mathcal{P}^{\mathsf{N}}]$.

Steps (5) and (6) together yield directly the desired conclusion.

The mapping $H'N'K'$ used in the above argument is rather complicated, and seems unnatural because it transforms sentences of the formalism $\mathcal{P}^{\mathsf{N}}$ into sentences of the same formalism by passing through formalisms very distant from $\mathcal{P}$. It would be interesting to have a simple and natural mapping, constructed without reference to other formalisms, which would serve the same purpose, i.e., would transform any given sentence of the language of elementary number theory into an equivalent sentence of the same language, containing only three variables.

As mentioned above, we abstain in general from considering the equipollence of systems in predicate logic with operation symbols (cf. remarks preceding 7.3(iii)). In the case of $\mathcal{N}$, however, we can easily construct a special formalism $\mathcal{P}_3^{\mathsf{N}}$ and a system $\mathcal{V}$ in $\mathcal{P}_3^{\mathsf{N}}$ such that $\mathcal{V}$ and $\mathcal{N}$ are equipollent in means of expression and proof. The sentences in $\mathcal{P}_3^{\mathsf{N}}$ are just the sentences in $\Sigma_3[\mathcal{P}^{\mathsf{N}}]$. For the notion of derivability in $\mathcal{P}_3^{\mathsf{N}}$ we take, for instance, the notion of derivability in $\mathcal{P}^{\mathsf{N}}$ restricted to sentences in $\mathcal{P}_3^{\mathsf{N}}$. By analogy with the definition of $\mathcal{N}$ we define $\mathcal{V}$ to be the system whose theory is the set of sentences in $\mathcal{P}_3^{\mathsf{N}}$ true of $\mathfrak{N}$. The equipollence of $\mathcal{N}$ and $\mathcal{V}$ in means of expression is just the content of Theorem (vi) above; the equipollence in means of proof is deprived of much interest since it is a trivial consequence of the fact that the theories of these systems are complete.

By analyzing the proof of (vi) we observe that it can be generalized as follows. Let $\mathcal{P}$ be a formalism of predicate logic whose vocabulary contains no relation symbols (or, somewhat more generally, no relation symbols with rank > 2) and no operation symbols with rank > 2 among its nonlogical constants. Suppose $\mathfrak{U}$ is a system in $\mathcal{P}$ and there exists a term t with just two variables, $\boldsymbol{x}$ and $\boldsymbol{y}$, such that the sentence

$$\forall_{\boldsymbol{xyzu}}(t[\boldsymbol{x},\boldsymbol{y}] = t[\boldsymbol{z},\boldsymbol{u}] \rightarrow \boldsymbol{x} = \boldsymbol{z} \wedge \boldsymbol{y} = \boldsymbol{u})$$

holds in $\mathfrak{U}$. Under these assumptions, for every $X \in \Sigma[\mathcal{P}]$ there is a $Y \in \Sigma_3[\mathcal{P}]$ such that $X \equiv Y[\mathfrak{U}]$.

7.6. The formalizability of Peano arithmetic in $\mathcal{L}^{\times}$, and the definitional equivalence of Peano arithmetic with a system of set theory

A well-known and frequently discussed axiomatic subsystem of $\mathcal{N}$ is the system $\mathcal{N}_0$ of *Peano arithmetic*. It is developed in the same formalism $\mathcal{P}^{\mathsf{N}}$ as $\mathcal{N}$. Its axiom set $\mathbf{A}\mathfrak{x}[\mathcal{N}_0]$ consists of the three sentences

(PI) $$\forall_{\boldsymbol{xy}}[\neg S\boldsymbol{x} = \mathbf{0} \wedge (S\boldsymbol{x} = S\boldsymbol{y} \rightarrow \boldsymbol{x} = \boldsymbol{y})],$$

(PII) $$\forall_{\boldsymbol{xy}}[\boldsymbol{x} + \mathbf{0} = \boldsymbol{x} \wedge \boldsymbol{x} + S\boldsymbol{y} = S(\boldsymbol{x} + \boldsymbol{y})],$$

(PIII) $$\forall_{\boldsymbol{xy}}[\boldsymbol{x} \cdot \mathbf{0} = \mathbf{0} \wedge \boldsymbol{x} \cdot S\boldsymbol{y} = \boldsymbol{x} \cdot \boldsymbol{y} + \boldsymbol{x}],$$

as well as of all instances of the following axiom schema:

(In) $$[F[\mathbf{0}] \wedge \forall_{\boldsymbol{x}}(F[\boldsymbol{x}] \rightarrow F[S\boldsymbol{x}]) \rightarrow \forall_{\boldsymbol{x}} F[\boldsymbol{x}]], \quad \textit{where} \quad F \in \Phi[\mathcal{P}^{\mathsf{N}}].$$

(Since $\mathcal{P}^{\mathsf{N}}$ contains operation symbols, we assume in formulating (In) that the symbolic conventions in §1.2 concerning substitutions have been appropriately extended: the variable "y" in expressions like "$F[x/y]$", "$F[y]$", ... is assumed to range over, not only the variables, but also all the terms of $\mathcal{P}^{\mathsf{N}}$.) (In) is of course the familiar *induction schema.* As is well known, Schema (In) can be equivalently replaced by a weaker axiom schema obtained by adding the requirement that $\boldsymbol{x}$ is the only free variable of F (so that the symbols "**[**" and "**]**" become superfluous; see, e.g., Hilbert–Bernays [**1968**], pp. 348ff.).

Most observations in §7.5 concerning the system $\mathcal{N}$ can be carried over with minimal changes to its subsystem $\mathcal{N}_0$. In particular, 7.5(i),(ii),(vi) continue to hold if we replace $\mathcal{N}$ by $\mathcal{N}_0$. Thus, $\mathcal{N}_0$ is a $\mathcal{Q}$-system. It is definitionally equivalent with a strong $\mathcal{Q}$-system $\mathcal{N}_0'$ formalized in $\mathcal{L}$, and hence is equipollent in means of expression and proof with the correlated $\mathcal{Q}$-system $\mathcal{N}_0'^{\times}$ in $\mathcal{L}^{\times}$; $\mathcal{N}_0'$ and $\mathcal{N}_0'^{\times}$

are respectively subsystems of $\mathcal{N}'$ and $\mathcal{N}'^\times$. As a consequence, *Peano arithmetic, just like elementary number theory, admits a simple equipollent formalization of the same kind as all the familiar set-theoretical systems discussed in Chapter* 4. Finally, every sentence in the language of Peano arithmetic is equivalent to a sentence containing at most three different variables.

Since we see no interesting adaptation of 7.5(iii) to $\mathcal{N}_0$, we now turn to the remaining result of §7.5, namely 7.5(v). We briefly sketch how this result can be extended to $\mathcal{N}_0$. We consider the formalism $\overline{\mathcal{P}}^{\mathsf{N}}$ obtained from $\mathcal{P}^{\mathsf{N}}$ by including the predicate $\boldsymbol{E}$ in its vocabulary, and an axiomatic system $\mathcal{T}_0$ in $\overline{\mathcal{P}}^{\mathsf{N}}$ whose axiom set consists of the axioms of $\mathcal{N}_0$ together with a possible definition of $\boldsymbol{E}$. In fact, for this definition we take the sentence

$$\forall_{xy}(x\boldsymbol{E}y \leftrightarrow G),$$

where "G" stands for the formula that expresses in $\mathcal{P}^{\mathsf{N}}$ the content of (α) in 7.5(v) without the help of exponentiation (cf. the remark after 7.5(v)). The possible definitions of the constants $\mathbf{0}$, $\boldsymbol{S}$, $\boldsymbol{+}$, and $\cdot$ in terms of $\boldsymbol{E}$ which are referred to in the observations following the statement of 7.5(v) can be shown to be provable in $\mathcal{T}_0$; the formal proof of this assertion is lengthy and tedious, but presents no essential difficulties. Hence, by means of a routine argument which was already used several times in the earlier part of this work (cf., for instance, the latter part of the proof of 7.1(ii)) we construct an axiomatic system $\mathcal{S}_0$ in our original formalism $\mathcal{L}$, stipulating that $\mathbf{A}\boldsymbol{\xi}[\mathcal{S}_0]$ is obtained from $\mathbf{A}\boldsymbol{\xi}[\mathcal{T}_0]$ by eliminating all occurrences of the constants $\mathbf{0}, \boldsymbol{S}, \boldsymbol{+}, \cdot$ on the basis of the definitions just mentioned. Thus we arrive at the following theorem, which can be viewed as an analogue of 7.5(v).

(i) *The systems $\mathcal{N}_0$ in $\mathcal{P}^{\mathsf{N}}$, $\mathcal{T}_0$ in $\overline{\mathcal{P}}^{\mathsf{N}}$, and $\mathcal{S}_0$ in $\mathcal{L}$ are strong $\mathcal{Q}$-systems; moreover, $\mathcal{T}_0$ is a common definitional extension of $\mathcal{N}_0$ and $\mathcal{S}_0$, so that all three systems are equipollent in means of expression and proof.*

A defect of the system $\mathcal{S}_0$ is that its axioms have a complicated structure and lack a clear mathematical content. However, we know other sets of sentences of $\mathcal{L}$ that do not exhibit this defect and that are logically equivalent with $\mathbf{A}\boldsymbol{\xi}[\mathcal{S}_0]$, so that they can be used to replace the latter set. We shall deal here with two such sets, $\mathbf{A}\boldsymbol{\xi}'[\mathcal{S}_0]$ and $\mathbf{A}\boldsymbol{\xi}''[\mathcal{S}_0]$. The first set, $\mathbf{A}\boldsymbol{\xi}'[\mathcal{S}_0]$, is the simplest set known to us which can serve as a base for $\mathcal{S}_0$. It consists of the three sentences

(QI) $$\forall_{xy}(\forall_u(u\boldsymbol{E}x \leftrightarrow u\boldsymbol{E}y) \rightarrow x = y),$$

(QII) $$\exists_z\forall_u(\neg u\boldsymbol{E}z),$$

(QIII) $$\forall_{xy}\exists_z\forall_u(u\boldsymbol{E}z \leftrightarrow u\boldsymbol{E}x \lor u = y),$$

as well as all instances of the following axiom schema:

(Is) $$[\forall_x(\forall_u(\neg u\boldsymbol{E}x) \rightarrow F[x]) \land \forall_{xyz}([\forall_u(u\boldsymbol{E}z \leftrightarrow u\boldsymbol{E}x \lor u = y) \land F[x] \land F[y]] \rightarrow F[z]) \rightarrow \forall_x F[x]],$$

where $F \in \mathbf{\Phi}[\mathcal{L}]$.

Axiom (QII) asserts that the universe of every model of $\mathcal{S}_0$ contains an element which is not in the range of the relation E denoted by $\boldsymbol{E}$. Since the unicity of such an element is secured by (QI), we can introduce an individual constant, say "0", to denote this element. Axiom (QIII) asserts that, for any elements x, y in the universe of a model of $\mathcal{S}_0$, there is an element z in the universe which is obtained by adjoining y to x, cf. remarks preceding 7.5(iv). Again, (QI) secures the unicity of z for a given x and y, which justifies the introduction of a special symbol, "$\triangleright$", denoting this binary operation of adjunction. In view of (QI)–(QIII), the universe of every model contains the element 0 and is closed under $\triangleright$. By Axiom Schema (Is), every property (expressible in $\mathcal{S}_0$) which is true of 0 and is true of $x \triangleright y$ whenever it is true of x and y, must be true of all elements. Thus, (Is) can be called the *induction schema* (*for* $\mathcal{S}_0$). It can be shown that (Is) (just as the induction schema (In) for $\mathcal{N}_0$) can be equivalently replaced with a weaker axiom schema obtained by restricting the range of "F" to formulas with a single free variable.

It may also be noticed that we can delete (QI), provided we strengthen both (QII) and (QIII) by requiring that the element represented by $\boldsymbol{z}$ be unique. Speaking precisely, let H be the formula $\boldsymbol{\forall}_{\boldsymbol{u}}(\neg \boldsymbol{uEz})$ in the case of (QII) and the formula $\boldsymbol{\forall}_{\boldsymbol{u}}(\boldsymbol{uEz} \leftrightarrow \boldsymbol{uEx} \vee \boldsymbol{u} = \boldsymbol{y})$ in the case of (QIII). We strengthen the two axioms by replacing the subformula H in each of them with $\boldsymbol{\forall}_{\boldsymbol{w}}(H[\boldsymbol{z}/\boldsymbol{w}] \leftrightarrow \boldsymbol{z} = \boldsymbol{w})$. It is not difficult to show that from Axioms (QII) and (QIII) thus strengthened, with the help of an appropriate instance of (Is), we can derive (QI).

(If, instead of $\mathcal{L}$, we wish to use an enriched formalism obtained by adding the constants $\mathbf{0}$ and $\blacktriangleright$ to the vocabulary, then we can replace the axiom set $\mathbf{A}\boldsymbol{\xi}'[\mathcal{S}_0]$ by the set consisting of the two possible definitions of $\mathbf{0}$ and $\blacktriangleright$ in terms of $\boldsymbol{E}$,

$$\boldsymbol{\forall}_{\boldsymbol{z}}[\boldsymbol{z} = \mathbf{0} \leftrightarrow \boldsymbol{\forall}_{\boldsymbol{u}}(\neg \boldsymbol{uEz})],$$

$$\boldsymbol{\forall}_{\boldsymbol{xyz}}[\boldsymbol{x} \blacktriangleright \boldsymbol{y} = \boldsymbol{z} \leftrightarrow \boldsymbol{\forall}_{\boldsymbol{u}}(\boldsymbol{uEz} \leftrightarrow \boldsymbol{uEx} \vee \boldsymbol{u} = \boldsymbol{y})],$$

as well as of all instances of the induction schema

$$[(F[\mathbf{0}] \wedge \boldsymbol{\forall}_{\boldsymbol{xy}}(F[\boldsymbol{x}] \wedge F[\boldsymbol{y}] \rightarrow F[\boldsymbol{x} \blacktriangleright \boldsymbol{y}])) \rightarrow \boldsymbol{\forall}_{\boldsymbol{x}} F[\boldsymbol{x}]].$$

It may be noticed that from this axiom set we can easily derive the sentence

$$\boldsymbol{\forall}_{\boldsymbol{xy}}(\boldsymbol{xEy} \leftrightarrow \boldsymbol{y} \blacktriangleright \boldsymbol{x} = \boldsymbol{y}).$$

Thus $\boldsymbol{E}$, and hence also $\mathbf{0}$, prove to be definable in terms of $\blacktriangleright$, so that we can construct a system definitionally equivalent to Peano arithmetic, using $\blacktriangleright$ as the only nonlogical constant.)

The reader has probably noticed that Axioms (QI)–(QIII) become familiar set-theoretical laws if $\boldsymbol{E}$ is interpreted as denoting the membership relation. In fact, under this interpretation (QI) is the extensionality law, (QII) the law of the empty set, and (QIII) the law of adjoining an element to a set. Essentially, however, $\mathbf{A}\boldsymbol{\xi}'[\mathcal{S}_0]$ is in spirit very closely related to the axiom set $\mathbf{A}\boldsymbol{\xi}[\mathcal{N}_0]$ of Peano arithmetic, a fact which is clearly brought out in the parenthetical remark above. On the other hand, the axiom set $\mathbf{A}\boldsymbol{\xi}''[\mathcal{S}_0]$ that we now wish to discuss is based entirely on set-theoretical intuitions. Indeed, somewhat unexpectedly, this axiom set coincides with a certain variant of the axiom set of $\mathcal{Z}$, the modern version of Zermelo's system; cf. §4.6. It consists of seven particular sentences, together with all instances of a schema. Three of these particular sentences were formulated in earlier parts of our discussion: the extensionality axiom (the axiom (QI) in $\mathbf{A}\xi'[\mathcal{S}_0]$), the pair axiom (the sentence $\boldsymbol{P}$ in 4.6(i)), and the generalized union axiom (the sentence S_4 in 6.4(iv)). Also the schema was discussed before: it is the "Aussonderungsaxiom" from $\mathcal{Z}$, formulated as (Z) in §6.5. The remaining four axioms are the power set axiom, the well-foundedness axiom, the axiom of transitive embedding, and the finiteness axiom; we formulate them explicitly in that order:

$$\forall_x\exists_z\forall_y[yEz \leftrightarrow \forall_u(uEy \rightarrow uEx)],$$

$$\forall_x\big(\exists_y(yEx) \rightarrow \exists_z[zEx \wedge \forall_u(uEz \rightarrow \neg uEx)]\big),$$

$$\forall_x\exists_z[xEz \wedge \forall_{yu}(yEu \wedge uEz \rightarrow yEz)],$$

$$\forall_x\big[\exists_y(yEx) \rightarrow \exists_z\big(zEx \wedge \forall_u[uEx \wedge \neg u = z \rightarrow \exists_w(wEz \wedge \neg wEu)]\big)\big].$$

The first two of these four sentences are familiar axioms used in many formalizations of set theory. The third sentence is a weaker form of what is known as the law of transitive closure, and, together with the other axioms, can be used to derive this law. The last sentence expresses the fact that, in every set-theoretical model of our axioms, all sets are finite; cf. Tarski [**1924**], p. 49, footnote 2. Neither the well-foundedness axiom nor the axiom of transitive embedding is provable in the system $\mathcal{Z}$. Therefore, the set-theoretical system based upon $\mathbf{A}\xi''[\mathcal{S}_0]$ can be referred to as the *extended Zermelo-like theory of (hereditarily) finite sets.*

The fact that $\mathbf{A}\xi'[\mathcal{S}_0]$ and $\mathbf{A}\xi''[\mathcal{S}_0]$ are logically equivalent may seem somewhat surprising since the most essential components of these sets, the induction schema (Is) in $\mathbf{A}\xi'[\mathcal{S}_0]$ and the comprehension schema 6.5(Z) in $\mathbf{A}\xi''[\mathcal{S}_0]$ do not seem to exhibit any affinity in their contents. Nevertheless, the proof of this equivalence is straightforward and will hardly present any difficulties to a person with some skill in set-theoretical arguments. The proof of the equivalence of $\mathbf{A}\xi[\mathcal{S}_0]$ and $\mathbf{A}\xi'[\mathcal{S}_0]$ is less direct, and considerably more involved. Details will not be presented here but will be postponed to a separate publication.

We sum up the claimed results in the following theorem.

(ii) *The three sets* $\mathbf{A}\xi[\mathcal{S}_0], \mathbf{A}\xi'[\mathcal{S}_0]$, *and* $\mathbf{A}\xi''[\mathcal{S}_0]$ *are logically equivalent. Hence Peano arithmetic and the extended Zermelo-like theory of finite sets are definitionally equivalent, and therefore equipollent in means of expression and proof.*

The question naturally arises whether and how the situation would change if the extended Zermelo-like theory of finite sets were replaced with the proper Zermelo-like theory of finite sets, i.e., the system $\mathcal{Z}f$ in $\mathcal{L}$ whose axiom set is obtained from $\mathbf{A}\xi''[\mathcal{S}_0]$ by deleting the well-foundedness axiom and the axiom of transitive embeddings. It turns out that $\mathbf{A}\xi[\mathcal{Z}f]$ is in fact weaker than $\mathbf{A}\xi''[\mathcal{S}_0]$. Actually, the set of sentences obtained from $\mathbf{A}\xi''[\mathcal{S}_0]$ by deleting either one of these axioms is weaker than $\mathbf{A\xi}''[\mathcal{S}_0]$ and stronger than $\mathbf{A\xi}[\mathcal{Z}f]$. We can now ask the question whether $\mathcal{Z}f$ has an axiomatization similar in character to $\mathbf{A\xi}'[\mathcal{S}_0]$. The answer is affirmative; indeed, such an axiom set can be obtained from $\mathbf{A\xi}'[\mathcal{S}_0]$ by very slightly modifying (and actually simplifying) the induction schema: we just replace "$F[\boldsymbol{x}] \wedge F[\boldsymbol{y}]$" by "$F[\boldsymbol{x}]$" in this schema. We have not looked into the question of how to describe the subsystem of $\mathcal{N}_0$ (if there is any) which is definitionally equivalent to $\mathcal{Z}f$, under the same mutual definitions which we used for $\mathcal{N}_0$ and $\mathcal{S}_0$. The elaboration of all these observations will be found in the separate publication alluded to above. Both Theorem (ii) and the subsequent observations were announced in Givant-Tarski [**1977**].

7.7. The formalizability of the arithmetic of real numbers in $\mathcal{L}^\times$

Another system to which Theorem 7.2(iv) applies is the system $\mathcal{R}$ of the *elementary theory*, or *arithmetic*, *of real numbers*. It can be formalized in the language $\mathcal{P}^{\mathsf{R}}$ of predicate logic with 6 nonlogical constants: individual constants $\mathbf{0}$ and $\mathbf{1}$, a binary predicate $\leq$, binary operation symbols $+$ and $\cdot$, and a unary predicate $\boldsymbol{N}$. A possible realization of $\mathcal{P}^{\mathsf{R}}$ is, e.g., the algebraic structure $\mathfrak{R} = \langle R, 0, 1, \leq, +, \cdot, N\rangle$, where R is the set of reals, $0, 1, \leq, +,$ and $\cdot$ are the familiar notions of the arithmetic of reals, and N is the set of natural numbers. The system $\mathcal{R}$ is defined semantically by assuming that $\mathbf{A\xi}[\mathcal{R}]$ consists of just those sentences in $\mathcal{P}^{\mathsf{R}}$ which are true of $\mathfrak{R}$; $\mathfrak{R}$ is treated as the standard model of $\mathcal{R}$.

Of interest is an axiomatic subsystem $\mathcal{R}_0$ of $\mathcal{R}$. The axiom set $\mathbf{A\xi}[\mathcal{R}_0]$ consists of finitely many familiar sentences which characterize the structure $\langle R, 0, 1, \leq, +, \cdot\rangle$ as an ordered field with at least two different elements (see, for instance, Tarski [**1965a**], pp. 217–220); furthermore, of two sentences expressing some elementary properties of natural numbers:

$$\forall_x(Nx \rightarrow 0 \leq x), \qquad \forall_x[Nx \leftrightarrow (x = 0 \vee \exists_y(Ny \wedge x = y + 1))];$$

and, finally, of all instances of the following *continuity schema*:

$$\text{(Cn)}\quad [\forall_{xy}(F[x] \wedge G[y] \rightarrow x \leq y) \rightarrow \exists_z(\forall_{xy}(F[x] \wedge G[y] \rightarrow x \leq z \wedge z \leq y))],$$
$$\textit{where } F, G \in \mathbf{\Phi}[\mathcal{P}^{\mathsf{R}}] \textit{ and } y, z \notin \Upsilon\phi F, \textit{ while } x, z \notin \Upsilon\phi G.$$

System $\mathcal{R}$ is in a sense an extension of the system $\mathcal{N}$ of elementary number theory (though not in the precise sense in which the term "extension of a system" is used throughout this work). To describe the precise relationship, consider the following possible definition of the unary operation symbol $\boldsymbol{S}$ in the language $\mathcal{P}^{\mathsf{R}}$:

$$\forall_x(Sx = x + 1).$$

Given any sentence X in the language $\mathcal{P}^{\mathsf{N}}$ of $\mathcal{N}$, we first eliminate in X all the occurrences of $\boldsymbol{S}$ on the basis of the above definition, and then we relativize all the quantifiers to the unary predicate $\boldsymbol{N}$ (or rather, in agreement with our terminology in §6.3, to formulas $\boldsymbol{N}x$ with $x \in \Upsilon$). It is then easily seen that the resulting sentence X^* holds in $\mathcal{R}$ iff X holds in $\mathcal{N}$.

The relationship between $\mathcal{N}_0$ and $\mathcal{R}_0$ is less close. It is still true, and not difficult to prove, that X^* holds in R_0 whenever X holds in $\mathcal{N}_0$; this applies, in particular, to all the instances X of the induction schema (In) from §7.6. It is known, however, that the converse fails: a sentence X can be constructed which does not hold in $\mathcal{N}_0$, although X^* holds in $\mathcal{R}_0$.

The construction of such a sentence X depends essentially on the following facts: it is possible to define adequately in $\mathcal{R}_0$ the notion of a sentence in $\mathcal{P}^{\mathsf{N}}$ being true of $\mathfrak{N}$; on the other hand, it can be shown that an adequate definition of this notion cannot be constructed in $\mathcal{N}_0$, and not even in $\mathcal{N}$. To be fully understood, the observations just stated obviously require further elaborations and meticulous verification. (We have not checked these observations in all details, especially with regard to the specific axiomatization of $\mathcal{R}_0$ that we have adopted.) In this context we refer the reader to Tarski [**1956**], Article VIII, in particular, pp. 268–278 (where a reference to the underlying work of Gödel can also be found).

Both $\mathcal{R}$ and $\mathcal{R}_0$ are (weak) $\mathcal{Q}$-systems. However, all the known constructions of two conjugated quasiprojections elementarily definable in $\mathfrak{R}$ are quite involved, and no way of essentially simplifying them is seen. An idea of such a construction can be obtained by analyzing the proof of Theorem 1 in Sierpiński [**1958**], p. 61, and of some of the earlier results in that work. We shall outline here a variant of Sierpiński's construction. As is well known, any given real number r with $0 < r < 1$ has a unique representation as an infinite dyadic fraction

$$r = \sum_{i=0}^{\infty} 2^{-m_i} \quad ,$$

where the m_i are positive integers with $m_i < m_{i+1}$. Let s be any other real number with $0 < s < 1$, and let

$$s = \sum_{i=0}^{\infty} 2^{-n_i}$$

be the corresponding representation. We set

$$r * s = \sum_{i=0}^{\infty} 2^{-[(m_i+n_i)\cdot(m_i+n_i)+m_i]}$$

It is easily seen that the operation $*$ maps the Cartesian square of the open interval $(0,1)$ into $(0,1)$ in a one-one way. On the other hand, we readily convince ourselves that the conditions

$$x\cdot z\cdot(z-1) \;=\; 2\cdot z - 1 \quad \text{and} \quad 0 < z < 1$$

determine a function H (with argument x and value z) mapping the set of reals into the interval $(0,1)$ in a one-one way. For any $x, y \in R$, set

$$x \,\square\, y \;=\; Hx * Hy.$$

Obviously the operation $\square$ is a one-one mapping from $R \times R$ into R, and we can use it to define two conjugated quasiprojections F and G by stipulating that

$$Fz = x \quad \text{iff there is a } y \text{ for which} \quad x \,\square\, y = z,$$

$$Gz = y \quad \text{iff there is an } x \text{ for which} \quad x \,\square\, y = z.$$

The proof that the construction just outlined can be carried through within the framework of the systems $\mathcal{R}$ and $\mathcal{R}_0$ should not present the reader with any essential difficulties. Once two conjugated quasiprojections have been defined, we can apply the general method from the proof of 7.2(iv) to obtain an equipollent formalization of $\mathcal{R}$ and $\mathcal{R}_0$ in our original language $\mathcal{L}$, and hence also in $\mathcal{L}^{\times}$.

However, just as in the case of $\mathcal{N}$ and $\mathcal{N}_0$, an application of the general method in the present case can be avoided. The procedure leading to this conclusion is still not simple; we shall sketch it here only in an informal way.

From the discussion of the systems in §§7.5 and 7.6 it is seen that, formally, natural numbers can be associated with sets of finite ranks; as a consequence, the system $\mathcal{S}_0$ from §7.6, which is an equipollent form of Peano arithmetic, proves to be essentially identical with an axiomatic system of the theory of sets of finite ranks that is closely related to Zermelo's system of set theory. On the other hand, it is pretty well known that real numbers can be treated as sets of natural numbers. By combining these two observations, we arrive at systems $\mathcal{R}'$ and $\mathcal{R}'_0$ which are respectively equipollent, and in fact definitionally equivalent, with $\mathcal{R}$ and $\mathcal{R}_0$, but which can be treated as set-theoretical systems; when so treated, they appear to be systems admitting proper classes (but excluding individuals). They are both formalized in our original language $\mathcal{L}$. $\mathcal{R}'$ is defined as the system whose theory coincides with that of a structure $\mathfrak{R}' = \langle R, E \rangle$; in the set-theoretical conception of real numbers, R consists of all sets of sets of finite rank, while E is the membership relation with the field restricted to R (we use here the language of "naive set-theory"). A set-theoretical peculiarity of $\mathfrak{R}'$ is that all those and only those members of R are proper classes (relative to $\mathfrak{R}'$) which are infinite. $\mathcal{R}'_0$ is an axiomatic subsystem of $\mathcal{R}'$. As $\mathbf{A}\mathfrak{x}[\mathcal{R}'_0]$ we can select a set of sentences which exhibits a close relation to the axiom set of Morse's system (see §4.6). However, in opposition to the case of $\mathcal{S}_0$, we abstain from giving here a detailed description of $\mathbf{A}\mathfrak{x}[\mathcal{R}'_0]$.

The definition of the relation E from $\mathfrak{R}'$ in terms of fundamental notions of $\mathfrak{R}$ is much less simple than the corresponding definition in the case of $\mathfrak{N}'$ and $\mathfrak{N}$ which was given in 7.5(v); in constructing the definition for $\mathfrak{R}'$ an essential use is again made of the representability of real numbers x with $0 < x \leq 1$ as infinite dyadic fractions. As is seen from the outlined proof of 7.5(v), the available definitions of the fundamental notions of $\mathfrak{N}$ in terms of the relation E from $\mathfrak{N}'$ are quite involved. The corresponding definitions for $\mathfrak{R}$ and $\mathfrak{R}'$ appear to be incomparably more involved. The reader can get a partial idea of the complicated structure of these definitions from various mathematical textbooks in which the set-theoretical construction of the arithmetic of reals from the arithmetic of naturals (and through the arithmetic of rationals) is carried out with some details.

Even a rough estimate of the lengths of definitions discussed (written explicitly, without any abbreviations) would require considerable effort.

In any event we can conclude from the above remarks that *systems* $\mathcal{R}$ *and* $\mathcal{R}_0$ *indeed admit an equipollent formalization in the language* $\mathcal{L}$ (*and hence also in* $\mathcal{L}^{\times}$). Fortunately, however, our interest in such a formalization is purely theoretical, with no practical applications in view.

7.8. Remarks on first-order formalisms with limited vocabularies

In this chapter we have extended the equipollence results of Chapter 4 to much wider classes of formal systems. We wish to point out here that the applications of these results to systems developed in "poor" languages of predicate logic are without interest, since all these systems prove to be of a very trivial nature.

Primarily we have here in mind languages $\mathcal{P}^{(0)}$ in which all nonlogical constants are either predicates of rank at most 1, or else individual constants. If, in particular, $\mathcal{P}^{(0)}$ has at most finitely many nonlogical constants (as is usual in our discussions), then as is well known, every system in $\mathcal{P}^{(0)}$ which has a model with an infinite universe also has a model with a finite universe whose cardinality exceeds any natural number given in advance; cf. for instance, Hilbert–Bernays [**1968**], pp. 198 ff. On the other hand, by 4.1(v) no $\mathcal{Q}$-system (whether weak or strong) in any language of predicate logic has a model with a finite universe containing more than one element. *Consequently, the only* $\mathcal{Q}$*-systems in* $\mathcal{P}^{(0)}$ *are those which possess exclusively one-element models, i.e., in which the sentence* $\forall_{xy}(x\mathring{1}y)$ *holds.* The case when $\mathcal{P}^{(0)}$ has infinitely many nonlogical constants easily reduces to the preceding case.

The only languages of predicate logic to which the result just stated can be extended are the languages $\mathcal{P}^{(1)}$ which, in addition to individual constants and unary predicates, contain just one unary operation symbol. The proof of this result, which is due to A. Ehrenfeucht, is more difficult than in the case of languages $\mathcal{P}^{(0)}$. It has not yet been published.[5]

In contrast to the situation for $\mathcal{P}^{(0)}$ and $\mathcal{P}^{(1)}$, we can construct interesting examples of $\mathcal{Q}$-systems in any other language of predicate logic, thus in every language which contains among its nonlogical constants at least one predicate or operation symbol of rank two or more, or else at least two unary operation symbols.

The problem arises whether there exist systems of $\mathcal{L}$ that are not $\mathcal{Q}$-systems, but which are still equipollent with some system formalized in $\mathcal{L}^{\times}$ (or $\mathcal{L}_3$). Such systems do in fact exist. We can find them among those systems whose set of provable sentences is recursive and for which some kind of elimination of quantifiers has been established. As a simple example we may mention the system $\mathcal{S}$ in $\mathcal{L}^+$ with the axioms

$$\exists_{xyz}(\neg x\mathring{1}y \wedge \neg x\mathring{1}z \wedge \neg y\mathring{1}z \wedge \forall_u(u\mathring{1}x \vee u\mathring{1}y \vee u\mathring{1}z)), \quad \forall_{xy}(xEy),$$

[5]We are obliged to J. Mycielski for sending us an outline of Ehrenfeucht's proof.

and its subsystem $\mathcal{T}$ in $\mathcal{L}^{\times}$ with the axioms

$$\mathring{\mathbf{0}} \odot \mathring{\mathbf{0}} = \mathbf{1}, \quad \boldsymbol{E} = \mathbf{1}.$$

Since both of these systems are complete, their equipollence is a rather simple matter. Much less trivial examples are also known, and in fact among systems that are not complete and have, indeed, infinitely many complete extensions. However, we do not know any results of a rather general character which would extend the results in §§4.5, 4.8, and 7.2 to comprehensive classes of systems that are not $\mathfrak{Q}$-systems.

CHAPTER 8

Applications to Relation Algebras and to Varieties of Algebras

In the present chapter we shall apply some of the main theorems of this work, primarily those of Chapter 4, to obtain a number of results, both of a mathematical and metamathematical nature, in the theory of relation algebras. From these results we shall draw some consequences of a more general character concerning arbitrary algebras. In the first section we introduce some notions and notations from the domain of general algebra and equational logic. For the most part these notions are well known (although that of dual decidability and some closely related notions seem to be new). In §§8.2–8.4 the notions of equational logic will be used only in a casual way; in §§8.5 and 8.6, however, they will be essentially involved.

8.1. Equational formalisms

By an *algebra* we understand here any structure $\mathfrak{A} = \langle A, O \rangle$ formed by a nonempty set A, the *universe* of $\mathfrak{A}$, and a system $O = \langle O_i : i \in I \rangle$ of *fundamental operations* on A, each of a definite finite rank ρO_i. (The set $\{O_i : i \in I\}$ may be of arbitrary cardinality.) As mentioned in §1.5, we shall sometimes write "$\langle A, O_i \rangle_{i \in I}$" instead of "$\langle A, O \rangle$". Distinguished elements of the algebra are identified with nullary operations, i.e., operations of rank 0. (In Henkin–Monk–Tarski [**1971**], where algebras with nullary operations are not admitted, distinguished elements are identified with constant unary operations.) The system $\tau\mathfrak{A} = \langle \rho O_i : i \in I \rangle$ of ranks of the operations of $\mathfrak{A}$ is called the (*similarity*) *type* of $\mathfrak{A}$.

In metamathematical discussions we shall concern ourselves exclusively with the first-order (or elementary) theories of algebras. In such discussions the theory of algebras of the type of $\mathfrak{A} = \langle A, O_i \rangle_{i \in I}$ is assumed to be developed within an appropriate formalism $\mathcal{P}$ of predicate logic in which all nonlogical constants are operation symbols $\boldsymbol{O}_i$, $i \in I$; here each symbol $\boldsymbol{O}_i$ is assumed to have the same rank as the operation O_i. We find it more convenient, however, to assume that the identity symbol in our formalism is the usual symbol "=", and not "$\mathring{1}$", and to introduce the corresponding symbol "$=$" as its metamathematical designation. (Fortunately we shall not have to deal with the language $\mathcal{P}^+$, for otherwise the symbol "$=$" would be ambiguous.)

Among sentences in such a formalism $\mathcal{P}$ special attention has been paid to those which contain only universal quantifiers and the equality symbol $=$ (but no sentential connectives) as logical constants, and therefore have the form of universally quantified equations

(i) $\forall_{xyz\ldots}(s = t)$,

where s, t are terms of $\mathcal{P}$. A class K of (similar) algebras that can be characterized as the class of models of a set Γ of universally quantified equations is called an *equational class* or a *variety.* The set of all universally quantified equations true of a given algebra $\mathfrak{A}$, or of a given class K of such algebras, is referred to as the *equational theory* of $\mathfrak{A}$, or of K.

It proves convenient to develop the equational theories of algebras within autonomous formalisms with a very simple structure, the equational formalisms $\mathcal{E}$. To enhance the simplicity of equational formalisms, we do not introduce in them any universal quantifiers. Thus, instead of the quantified equation (i), we consider the equation

(ii) $s = t$,

i.e., the equational part of (i); instead of speaking of the sentence (i) being true in an algebra $\mathfrak{A}$, we speak of the formula (ii) (with any number of free variables) being *identically satisfied* in $\mathfrak{A}$, i.e., satisfied by every appropriate sequence of elements in the universe of $\mathfrak{A}$.

The vocabulary of an equational formalism $\mathcal{E}$ appropriate for the discussion of an algebra $\mathfrak{A} = \langle A, O_i \rangle_{i \in I}$, or of a class of algebras similar to $\mathfrak{A}$, consists of denumerably many variables arranged in an infinite sequence $\boldsymbol{v} = \langle \boldsymbol{v}_0, \boldsymbol{v}_1, \ldots \rangle$ without repeating terms, the equality predicate $=$ as the only logical constant, and a system $\boldsymbol{O} = \langle \boldsymbol{O}_i : i \in I \rangle$ of operation symbols. Nullary operation symbols are referred to as *individual constants.* Terms are constructed from variables and individual constants, using operation symbols of positive rank, in the manner described in §1.5. The set of all terms of the formalism $\mathcal{E}$ is denoted by "$\mathbf{T}\boldsymbol{\mu}[\mathcal{E}]$", or simply "$\mathbf{T}\boldsymbol{\mu}$". The only formulas are equations between terms; we use "$\boldsymbol{\Phi}[\mathcal{E}]$", or simply "$\boldsymbol{\Phi}$", to denote the set of all equations of $\mathcal{E}$. Since there are no quantifiers, no variables in an equation occur bound. As a consequence, the definition of the notion of substitution (of arbitrary terms for arbitrary variables) in an equation presents no difficulties. (Contrast this with the remarks in §1.2.)

The notion of derivability in an equational formalism $\mathcal{E}$ is defined as follows.

(iii) *Given an equation X and a set of equations Ψ in $\mathcal{E}$, we say that X is derivable from Ψ, in symbols $\Psi \vdash X$ $[\mathcal{E}]$ or simply $\Psi \vdash X$, if X belongs to every set Ω of equations satisfying the following conditions for any terms s, t, p, q in $\mathcal{E}$:*

(α) *Ψ is included in Ω;*

(β) *every tautology $s = s$ is in Ω;*

(γ) *if $s = t$ is in Ω, then every equation obtained from $s = t$ by substitution is in Ω;*

(δ) *if $s = t$ and $p = q$ are in Ω and if p occurs as a part in s or t, then every equation obtained from $s = t$ by replacing some occurrence of p with q also belongs to Ω.*

The rules of inference embodied in (γ) and (δ) are respectively called the *rule of substitution* and the *rule of replacement.* It is not difficult to see that condition (γ) may be deleted provided we strengthen (α) to read: *every substitution instance of an equation of Ψ is in Ω.*

It may be noticed that, in case the vocabulary of $\mathcal{E}$ contains some individual constants, there are formulas in $\mathcal{E}$ that are sentences in the sense of predicate logic. These are, of course, equations in which no free variables, and thus no variables at all, occur. Let $\Sigma[\mathcal{E}]$, or simply Σ, be the set of all such equations. This set plays a role in the metalogic of equational formalisms, but, as is seen from (iii), it is by no means the basic notion in terms of which the notion of derivability is defined.

The notions and notation defined for predicate logic in terms of derivability (see §1.3) extend with obvious changes to equational formalisms. For instance, the theory generated in $\mathcal{E}$ by a set Ψ of equations, symbolically $\mathbf{\Theta\eta}\Psi[\mathcal{E}]$ or simply $\mathbf{\Theta\eta}\Psi$, is determined by

$$\mathbf{\Theta\eta}\Psi = \{X\colon X \in \mathbf{\Phi} \text{ and } \Psi \vdash X\}.$$

Notice that $\mathbf{\Theta\eta}\varnothing$, i.e., what could be called the logic of $\mathcal{E}$, has a trivial character, since it consists exclusively of tautologies $s = s$.

Similarly, semantical notions and notations defined for predicate logic extend to equational formalisms. In particular, an algebra $\mathfrak{A}$ is said to be a model of an equation if this equation is identically satisfied in $\mathfrak{A}$. Modifying our earlier definition, we define the *equational theory* of an algebra $\mathfrak{A}$, or of a class K of algebras, to be the set of all equations X such that $\mathfrak{A}$, or every algebra in K, is a model of X (in $\mathcal{E}$). Notice that a variety K is just the class of all models of an equational theory Θ; K is said to be *finitely based* if there is a finite set Ψ of equations such that K is the class of models of Ψ.

By the well-known results in Birkhoff [**1935**], every equational formalism $\mathcal{E}$ is *semantically complete*, i.e., the notions $\vdash [\mathcal{E}]$ and $\models [\mathcal{E}]$ coincide (cf. §1.6). Hence there is no need of distinguishing between syntactical and semantical notions in discussing such formalisms.

There are certain differences between formalisms $\mathcal{P}$ of predicate logic and equational formalisms $\mathcal{E}$. In particular, a set of sentences of $\mathcal{P}$ is inconsistent iff it has no models. This characterization of inconsistent sets of sentences does not carry over to $\mathcal{E}$ since every set of equations has as models all one-element algebras of the appropriate similarity type. On the other hand, a set Δ of equations is inconsistent in $\mathcal{E}$ iff it has no model with more than one element, or, equivalently, iff $\Delta \vdash \boldsymbol{x} = \boldsymbol{y}$.

Another, somewhat related peculiarity of equational formalisms is the lack of operations on formulas which function like $\rightarrow$ and $\neg$ in formalisms of predicate logic; in particular, the formalisms $\mathcal{E}$ are not provided with deduction theorems. This strongly affects the discussion of decision problems where, as we recall, the deduction theorem plays an essential role. For example, the logic of $\mathcal{E}$, $\mathbf{\Theta}\boldsymbol{\eta}\varnothing[\mathcal{E}]$, consisting just of tautologies, is recursive; hence no theory in $\mathcal{E}$ can be hereditarily undecidable. The theorem by which every finite extension of a decidable theory is decidable fails in general for $\mathcal{E}$, and presumably this also applies to the theorem by which every consistent decidable theory has a complete and decidable extension. In connection with these remarks compare §§1.3 and 3.3.

On the other hand, as a consequence of the peculiarities of equational formalisms just discussed, some new notions appear. Loosely speaking, when studying a class K of structures within an appropriate formalism $\mathcal{F}$, we are interested in two kinds of problems: those of affirming and those of rejecting a sentence X expressed in this formalism, i.e., of showing that X is true of all structures in K, or that it is false for all such structures. To affirm X we can show that it is derivable from a set of accepted axioms. To reject X we can show that it is incompatible with this set of axioms. If, say, $\mathcal{F}$ is a formalism of predicate logic, then each of the two kinds of problems is reducible to the other, since, e.g., rejecting X is equivalent to affirming $\neg X$. However, in case $\mathcal{F}$ is an equational formalism, the two kinds of problems seem in general not to be mutually reducible. In particular, the existence of a "mechanical" method for affirming equations does not imply the existence of such a method for rejecting equations. Consequently, it seems worthwhile to study in equational logic, in addition to the usual notion of a decidable theory, that of a *dually decidable* theory.

(iv) (α) *A theory* Θ *in* $\mathcal{E}$ *is called dually decidable* (*dually undecidable*) *if the set of all equations that are incompatible with* Θ *is recursive* (*nonrecursive*).

(β) Θ *is called essentially dually undecidable if* Θ, *as well as every consistent theory extending* Θ *in* $\mathcal{E}$, *is dually undecidable.*

As an example, consider the equational formalism $\mathcal{E}$ with one binary operation symbol as its only nonlogical constant, and let Θ be the logic of $\mathcal{E}$, so that $\Theta = \mathbf{\Theta}\boldsymbol{\eta}\varnothing[\mathcal{E}]$. As was pointed out above, Θ is decidable. On the other hand, Θ is dually undecidable. In fact, it is shown in McNulty [**1976a**] that the set of equations X in $\mathcal{E}$ for which $\{X\}$ is consistent is not recursive. Hence, the complement of this set is also not recursive, and this is just the set of equations incompatible with Θ.

It may be noticed that the notions of decidability and dual decidability always coincide for complete theories.

In a sense every equational formalism $\mathcal{E}$ is a subformalism of the corresponding formalism $\mathcal{P}$ of predicate logic. To bring this statement into agreement with our use of the term "subformalism" (cf. §1.6), we would have to introduce explicitly the universal quantifier into the vocabulary of $\mathcal{E}$, to replace equations (ii) by the corresponding sentences (i), and to modify appropriately the definitions of the basic syntactical and semantical notions. (This would make some of these definitions more involved.)

We have restricted ourselves here to equational formalisms provided with a simple infinite sequence of distinct variables. We could, however, extend our discussion by considering equational formalisms provided with any finite or transfinite sequence of distinct variables. Such formalisms will not be discussed in this work. They can be employed, for instance, for a metamathematical construction of free algebras, cf. Henkin–Monk–Tarski [**1971**], pp. 143ff. On the other hand, a very special case of such formalisms, namely formalisms with the empty sequence of variables, i.e., with no variables, could be used for an alternative presentation of the material in §8.5; such a presentation would be somewhat more concise, but conceptually more complicated than the one we shall actually adopt here. We may mention that $\mathcal{L}^\times$ is, in some sense, an example of an equational formalism without variables.

8.2. Relation algebras

A familiar example of a variety is the class BA of all Boolean algebras. In this and the next sections we shall be concerned with another variety, the class RA, of much richer structures called (abstract) *relation algebras.* (We shall also use "RA" simply as an abbreviation for "relation algebra", and similarly for other analogous notations.)

(i) *A relation algebra is a structure*

$$\mathfrak{A} = \langle A, +, ^-, \odot, ^\smile, \mathring{1} \rangle$$

of type $\langle 2, 1, 2, 1, 0 \rangle$ *(so that* $+$, $\odot$ *are binary, and* $^-$, $^\smile$ *are unary, operations on* A, *while* $\mathring{1}$ *is a distinguished element of* A*) satisfying the following conditions for any* $x, y, z \in A$:

(Ra I)	$x + y = y + x,$
(Ra II)	$x + (y + z) = (x + y) + z,$
(Ra III)	$(x^- + y)^- + (x^- + y^-)^- = x,$
(Ra IV)	$x \odot (y \odot z) = (x \odot y) \odot z,$
(Ra V)	$(x + y) \odot z = x \odot z + y \odot z,$
(Ra VI)	$x \odot \mathring{1} = x,$
(Ra VII)	$x^{\smile\smile} = x,$
(Ra VIII)	$(x + y)^\smile = x^\smile + y^\smile,$
(Ra IX)	$(x \odot y)^\smile = y^\smile \odot x^\smile,$
(Ra X)	$x^\smile \odot (x \odot y)^- + y^- = y^-.$

(ii) RA *is the class of all relation algebras.* (Ra I)–(Ra X) *are referred to as the axioms or postulates for* RA.

The reader should observe the close relationship between (Ra I)–(Ra X) and the logical axiom schemata (BI)–(BX) of $\mathcal{L}^\times$ given in §3.1.

In common language we refer to the fundamental notions of relation algebras as follows: to $+$ as *absolute* or *Boolean addition*, to $^-$ as *complementation*, to $\odot$ as *relative* or *Peircean multiplication*, to $^\smile$ as *conversion* (or *formation of converses*), to $\mathring{1}$ as the *identity element* (or *Peircean unit*). In terms of these fundamental operations we define some further notions of a related character.

(iii) *Given an* RA $\mathfrak{A} = \langle A, +, {}^-, \odot, {}^\smile, \mathring{1}\rangle$ *we set, for any* $x, y \in A$:

$$x \cdot y = (x^- + y^-)^- \qquad (\textit{absolute or Boolean multiplication}),$$
$$x \oplus y = (x^- \odot y^-)^- \qquad (\textit{relative or Peircean addition}),$$
$$x \le y \ \textit{ iff } \ x + y = y \qquad (\textit{relation of inclusion}),$$
$$1 = \mathring{1} + \mathring{1}^- \qquad (\textit{Boolean unit}),$$
$$0 = (\mathring{1} + \mathring{1}^-)^- \qquad (\textit{Boolean zero}),$$
$$\mathring{0} = \mathring{1}^- \qquad (\textit{diversity element or Peircean zero}).$$

All four operations $+$, $\cdot$, $\oplus$, and $\odot$ prove to be associative; they are extended by recursion to operations Σ, Π, $\mathring{\Sigma}$, and $\mathring{\Pi}$ on finite sequences of elements of A. (Compare here, in part, §2.1.) Regarding the omission of parentheses in expressions that involve the operation symbols of a relation algebra, we shall follow the same convention as stated on p. 24 for the operators of $\mathcal{L}^+$.

The formalism of predicate logic appropriate for our discussion is the formalism $\mathcal{P}^{\mathsf{A}}$, with five nonlogical constants, $\mathbf{+}$, $\mathbf{^-}$, $\boldsymbol{\odot}$, $\mathbf{^\smile}$, and $\mathring{\mathbf{1}}$, which are operation symbols of ranks $2, 1, 2, 1$, and 0 respectively. (We are purposely choosing the nonlogical constants of $\mathcal{P}^{\mathsf{A}}$ to coincide with the logical constants of $\mathcal{L}^\times$. This choice will prove to be very convenient for the purposes of our subsequent discussion. We do not deny that, in principle, some confusion could arise from this choice, namely when we consider relation algebras, or similar algebraic structures, constructed from expressions of the language $\mathcal{L}^\times$. We feel, however, that in the present text, the reader will readily be able to distinguish between the different uses of the symbols, and in one place we shall point out which usage we intend.) Although "$\mathbf{+}$", "$\mathbf{^-}$", "$\boldsymbol{\odot}$", "$\mathbf{^\smile}$", and "$\mathring{\mathbf{1}}$" have been introduced as metalogical names of the corresponding operation symbols in the language of RA's, they can also be treated as names of appropriate metalogical operations on terms in $\mathcal{P}^{\mathsf{A}}$ (cf. §1.2). With the help of these operations on terms we can define various other notions of a related character corresponding to those defined in (iii) above. In fact, for any two terms s and t of $\mathcal{P}^{\mathsf{A}}$ we set

$$s \boldsymbol{\cdot} t \;=\; (s^{\mathbf{-}} \mathbf{+} t^{\mathbf{-}})^{\mathbf{-}}, \qquad s \boldsymbol{\oplus} t \;=\; (s^{\mathbf{-}} \boldsymbol{\odot} t^{\mathbf{-}})^{\mathbf{-}}, \qquad s \boldsymbol{\le} t \;=\; (s \mathbf{+} t \mathbf{=} t)$$

(so that $\boldsymbol{\le}$ is a binary operation mapping the set of ordered pairs of terms into the set of formulas), and finally we set

$$\mathbf{1} \;=\; \mathring{\mathbf{1}} \mathbf{+} \mathring{\mathbf{1}}^{\mathbf{-}}, \qquad \mathbf{0} \;=\; (\mathring{\mathbf{1}} \mathbf{+} \mathring{\mathbf{1}}^{\mathbf{-}})^{\mathbf{-}}, \qquad \mathring{\mathbf{0}} \;=\; \mathring{\mathbf{1}}^{\mathbf{-}}.$$

For various reasons the notion of a simple algebra, well known from the general theory of algebras, plays a prominent role in the theory of RA's. Simple RA's admit a very elementary characterization.

(iv) *An* RA $\mathfrak{A} = \langle A, +, {}^-, \odot, {}^\smile, \mathring{1} \rangle$ *is simple iff the conditions* $1 \odot x \odot 1 = 1$ *and* $x \neq 0$ *are equivalent for every* $x \in A$.

With the help of (iv) it is shown that:

(v) *The simple* RA*'s, the subdirectly indecomposable* RA*'s, and the directly indecomposable* RA*'s coincide.*

Hence, by a well-known result in Birkhoff [**1944**] we get

(vi) *Every* RA *is semisimple, i.e., is isomorphic to a subdirect product of simple* RA*'s.*

For the proofs of (iv)–(vi), see Jónsson–Tarski [**1952**], pp. 132–135.

Theorem (iv) has an important metamathematical implication:

(vii) *With every formula* X *in* $\mathcal{P}^{\mathsf{A}}$ *which is a quantifier-free combination of equations we can correlate an equation* Y *in* $\mathcal{P}^{\mathsf{A}}$ *such that the sentence* $[X \leftrightarrow Y]$ *holds in every simple* RA.

An immediate consequence of (vi) is:

(viii) *An equation is identically satisfied in every* RA *if (and only if) it is so satisfied in every simple* RA.

Theorem (viii) can be easily extended from equations to so-called conditional equations, i.e., formulas of the form $X_1 \wedge \cdots \wedge X_n \rightarrow X_{n+1}$ $(n = 1, 2, \ldots)$ where each X_i with $1 \leq i \leq n+1$ is an equation.

It was emphasized in §3.2 that the close relationship between the axioms of RA's and the logical axiom schemata of $\mathcal{L}^\times$ permits us to carry over various results from the mathematical theory of RA's to the metamathematical investigation of $\mathcal{L}^\times$, and conversely. Various results obtained in this way for $\mathcal{L}^\times$ have been used, implicitly or explicitly, throughout this book. Now, however, we are interested in the results which can be established for RA's using theorems about $\mathcal{L}^\times$. We discuss here a procedure for establishing such results.

We begin with a lemma upon which this procedure is essentially based. The lemma is of the type of results which originate with Birkhoff [**1935**], in fact, a result on a metamathematical construction of free algebras; specifically, in our context it will lead us to the construction of a free relation algebra with arbitrarily many generators. For the sake of generality we refer the construction, not to the formalism $\mathcal{L}^\times$, but to any of the formalisms $\mathcal{M}^\times$ correlated with formalisms $\mathcal{M}$ of predicate logic with a finite or infinite (even nondenumerable) set of nonlogical constants, all of which are binary atomic predicates. Formalisms of this type with finitely many nonlogical symbols were discussed in §7.1 and referred to as $\mathcal{M}^{(n)}$. It will be seen that the formalism $\mathcal{P}^{\mathsf{A}}$ is not used in this construction.

Let $\langle \boldsymbol{F}_i : i \in I \rangle$ be the indexed system of all distinct nonlogical atomic predicates of $\mathcal{M}^\times$. $\boldsymbol{\Pi}[\mathcal{M}^\times]$, or $\boldsymbol{\Pi}[\mathcal{M}^+]$, is the set of all predicates of $\mathcal{M}^\times$, or $\mathcal{M}^+$, and of course $\boldsymbol{\Pi}[\mathcal{M}^\times] = \boldsymbol{\Pi}[\mathcal{M}^+]$.

In the metamathematics of $\mathcal{M}^\times$ we construct the structure

$$\mathfrak{P} = \langle \boldsymbol{\Pi}[\mathcal{M}^\times], \boldsymbol{+}, \boldsymbol{^-}, \boldsymbol{\odot}, \boldsymbol{^\smile}, \mathring{\boldsymbol{1}} \rangle.$$

(It is important to observe that here $\boldsymbol{+}$, $\boldsymbol{^-}$, $\boldsymbol{\odot}$, and $\boldsymbol{^\smile}$ are operations on and to predicates of $\boldsymbol{\Pi}[\mathcal{M}^\times]$, and $\mathring{\boldsymbol{1}}$ is a distinguished predicate in $\boldsymbol{\Pi}[\mathcal{M}^\times]$. Thus, we are using them to refer to operations of the structure $\mathfrak{P}$, and *not* to the operation symbols of the formalism $\mathcal{P}^{\mathsf{A}}$.) The structure $\mathfrak{P}$ is clearly an algebra similar to RA's; it is, however, not an RA, but an absolutely free algebra of type $\langle 2,1,2,1,0 \rangle$ freely generated by the set $\{\boldsymbol{F}_i : i \in I\}$. (For a discussion of free algebras and related notions, compare Henkin–Monk–Tarski [**1971**], §0.4.) Thus, if $\mathfrak{A} = \langle A, +, ^-, \odot, ^\smile, \mathring{1} \rangle$ is any algebra of type $\langle 2,1,2,1,0 \rangle$ and $b = \langle b_i : i \in I \rangle$ is any system of elements of $\mathfrak{A}$ indexed by I, then there is a unique homomorphism H from $\mathfrak{P}$ into $\mathfrak{A}$ such that $H(\boldsymbol{F}_i) = b_i$ for each $i \in I$.

Given any set $\Psi \subseteq \Sigma[\mathcal{M}^\times]$, let $\simeq_\Psi^\times$ be the relation which holds between any $C, D \in \boldsymbol{\Pi}[\mathcal{M}^\times]$ iff

$$\Psi \vdash C \boldsymbol{=} D \ [\mathcal{M}^\times].$$

In case $\Psi = \varnothing$, we write "$C \simeq^\times D$" instead of "$C \simeq_\Psi^\times D$". (Technically, the symbols "$\simeq_\Psi^\times$" and "$\simeq^\times$" should exhibit some relativization to a given formalism $\mathcal{M}^\times$. However, for simplicity of notation, the relativization is not explicitly exhibited.)

The results in which we are interested can now be formulated as follows.

(ix) *For every $\Psi \subseteq \Sigma[\mathcal{M}^\times]$, $\simeq_\Psi^\times$ is a congruence relation on the algebra*

$$\mathfrak{P} = \langle \boldsymbol{\Pi}[\mathcal{M}^\times], \boldsymbol{+}, \boldsymbol{^-}, \boldsymbol{\odot}, \boldsymbol{^\smile}, \mathring{\boldsymbol{1}} \rangle,$$

and the quotient algebra $\mathfrak{P}/\simeq_\Psi^\times$ is an RA.

(x) *In particular, $\mathfrak{P}/\simeq^\times$ is an* RA *which is* RA*-freely generated by the set*

$$\{\boldsymbol{F}_i/\simeq^\times : i \in I\}.$$

The proof of (ix) is based primarily on the definition of derivability in $\mathcal{M}^\times$, and is routine. To prove (x) we fix an arbitrary RA $\mathfrak{A}$ and an arbitrary system $b = \langle b_i : i \in I \rangle$ of elements of $\mathfrak{A}$. Let H be the (unique) homomorphism from $\mathfrak{P}$ into $\mathfrak{A}$ such that $H(\boldsymbol{F}_i) = b_i$ for each $i \in I$. We then show by an easy induction on derivable sentences in $\mathcal{M}^\times$ that

(1) for any $C, D \in \boldsymbol{\Pi}[\mathcal{M}^\times]$ we have $H(C) = H(D)$ whenever $C \simeq^\times D$.

In view of (1), the mapping G from the quotient set $\boldsymbol{\Pi}[\mathcal{M}^\times]/\simeq^\times$ (i.e., the universe of the algebra $\mathfrak{P}/\simeq^\times$) to A (the universe of $\mathfrak{A}$), determined by the stipulation

$$G(C/\simeq^\times) = H(C) \qquad \text{for every} \quad C \in \boldsymbol{\Pi}[\mathcal{M}^\times],$$

is well defined; it is obviously a homomorphism from $\mathfrak{P}/\simeq^{\times}$ to $\mathfrak{A}$ which maps $\boldsymbol{F}_i/\simeq^{\times}$ to b_i for each $i \in I$. This shows that the set $\{\boldsymbol{F}_i/\simeq^{\times}: i \in I\}$ RA-freely generates $\mathfrak{P}/\simeq^{\times}$, as was to be proved.

It may be noticed that the theorem just proved could be given a stronger form by using the notion of a free algebra with defining relations. In this form the theorem would assert that, for every $\Psi \subseteq \Sigma[\mathcal{M}^{\times}]$, $\mathfrak{P}/\simeq^{\times}_{\Psi}$ is an RA that is RA-freely generated by the system $\langle \boldsymbol{F}_i/\simeq^{\times}_{\Psi}: i \in I\rangle$ under certain defining relations. Roughly speaking, these defining relations are those obtained from the equations in Ψ by replacing distinct binary predicates $\boldsymbol{F}_i$, $i \in I$, with distinct variables. A precise formulation of the improved Theorem (x) would be rather involved, and we leave it to the reader.

The notion of free algebras with defining relations will play some role in our subsequent discussion, namely in the proofs of 8.4(iii),(xiii) and 8.5(iv); the improved version of Theorem (x) would have the virtue that it could be directly applied in those proofs. In connection with the notion of a free algebra with defining relations, which is rather well known from contemporary literature, the reader may compare Henkin–Monk–Tarski [**1971**], pp. 146ff., although our use of this notion will not follow strictly the development in *op. cit.*

8.3. Representable relation algebras

An algebra $\mathfrak{A} = \langle A, +, ^-, \odot, ^\smile, \mathring{1}\rangle$ is called a *proper relation algebra* (or a *relation set algebra*) if its universe A is a nonempty family of binary relations between elements of a set U, and its fundamental notions $+$, $^-$, $\odot$, $^\smile$, $\mathring{1}$ respectively coincide with the appropriately restricted set-theoretical notions $\cup$, $\sim$, $|$, $^{-1}$, $U\!\upharpoonright\! Id$. Here $\sim R$, for any $R \in A$, is the complement of R relative to the relation $\bigcup A$. From this definition it is seen that a proper relation algebra $\mathfrak{A}$ is an RA; its universe A is a relation ring, if we agree to use this term in a somewhat wider sense than the one specified in §6.2 (namely, if we agree to replace in the definition of relation ring the expression "$U \times U \sim R$" by "$\bigcup F \sim R$"). Furthermore, we conclude that the relation $\bigcup A$ belongs to A and is, algebraically, the unit element of $\mathfrak{A}$; set-theoretically, $\bigcup A$ is an equivalence relation with domain U. In case $\bigcup A = U \times U$, A is a relation ring (on U) precisely in the sense of §6.2. In this case $\mathfrak{A}$ is called a (*proper*) RA *on the set* U; if A consists of all subrelations of $U \times U$, then $\mathfrak{A}$ is referred to as the *full* RA *on* U and is denoted by "$\mathfrak{Fr}(U)$".

An RA $\mathfrak{A}$ is said to be *representable* if it is isomorphic with a proper RA; RRA is the class of all representable RA's.

Simple RRA's can be characterized as follows.

(i) $\mathfrak{A}$ *is a simple* RRA *iff it is isomorphic with a proper* RA *on a nonempty set* U.

For a proof of (i), see Jónsson–Tarski [**1952**], pp. 141–143.

Other characterizations of RRA are provided by the following theorem, which is readily established directly, but can also be derived from (i) and 8.2(vi).

(ii) *The following three conditions are equivalent:*

(α) $\mathfrak{A} \in \mathsf{RRA}$;

(β) *$\mathfrak{A}$ is isomorphic with a subdirect product of some algebras $\mathfrak{B}_i$ (indexed by elements i of a set I), each of which is a proper* RA *on a nonempty set U_i;*

(γ) *$\mathfrak{A}$ is isomorphic with a subalgebra of the direct product of algebras $\mathfrak{B}_i$ ($i \in I$), each of which is the full* RA *on a nonempty set U_i.*

From (ii) we can easily derive the analogue for RRA's of 8.2(viii).

(iii) *An equation is identically satisfied in every* RRA *if (and only if) it is so satisfied in every full* RA *over some nonempty set U.*

Just as in the case of 8.2(viii), we can extend (iii) above to arbitrary conditional equations.

The main results concerning RRA's which are known from the literature are:

(iv) *There are algebras $\mathfrak{A}$ (both finite and infinite) such that $\mathfrak{A} \in \mathsf{RA}$ and $\mathfrak{A} \notin \mathsf{RRA}$.*

(v) RRA *is a variety.*

(vi) RRA *is not a finitely based variety.*

In connection with (iv), see Lyndon [**1950**] for the first examples of nonrepresentable RA's; compare also Jónsson [**1959**] and Lyndon [**1961**]. The simplest example of a finite nonrepresentable RA (an algebra with 16 elements and one generator) can be found in McKenzie [**1970**]. A proof of (v) is outlined in Tarski [**1955**], and (vi) is established in Monk [**1964**]. In connection with (v) and (vi) it may be mentioned that by using some arguments in Lyndon [**1956**] and Monk [**1969a**] we can obtain explicit examples of recursive infinite sets of equations characterizing the variety RRA; however, the structure of these sets is rather involved.

Using the notation of §8.2, define $\simeq^+_\Psi$ to be the relation which, for a given set $\Psi \subseteq \Sigma[\mathcal{M}^+]$, holds between any $C, D \in \mathbf{\Pi}[\mathcal{M}^\times]$ iff

$$\Psi \vdash C = D\ [\mathcal{M}^+],$$

and let $\simeq^+$ be the relation $\simeq^+_\Psi$ in case $\Psi = \varnothing$. We now establish for RRA analogues of 8.2(ix),(x).

(vii) *Let $\Psi \subseteq \Sigma[\mathcal{M}^+]$ and let $\mathfrak{P}$ be the algebra of* 8.2(ix).

(α) *$\simeq^+_\Psi$ is a congruence relation on $\mathfrak{P}$.*

(β) *For every model $\mathfrak{U} = \langle U, S_i\rangle_{i\in I}$ of Ψ there is a unique homomorphism G of $\mathfrak{P}/\simeq^+_\Psi$ into $\mathfrak{Fr}(U)$ such that*

$$G(\boldsymbol{F}_i/\simeq^+_\Psi) = S_i \quad \text{for each} \quad i \in I.$$

In fact, for each $C \in \mathbf{\Pi}[\mathcal{M}^+]$, $G(C/\simeq^+_\Psi)$ is just the denotation of C in $\mathfrak{U}$.

(γ) *$\mathfrak{P}/\simeq^+_\Psi$ is an* RRA.

The proof of (α) is routine. To prove (β) let H be the homomorphism of $\mathfrak{P}$ into $\mathfrak{Fr}(U)$ such that

$$H(\boldsymbol{F}_i) = S_i \quad \text{for each} \quad i \in I.$$

One easily shows by induction on predicates that, for every $C \in \boldsymbol{\Pi}[\mathcal{M}^+]$, $H(C)$ is just the denotation of C in $\mathfrak{U}$. Moreover, if $C \simeq^+_\Psi D$, i.e., if $\Psi \vdash C \boldsymbol{=} D$ $[\mathcal{M}^+]$, then C and D denote the same relation in $\mathfrak{U}$, and hence $H(C) = H(D)$. Therefore the mapping G from $\mathfrak{P}/\simeq^+_\Psi$ to $\mathfrak{Fr}(U)$ determined by

$$G(C/\simeq^+_\Psi) = H(C) \quad \text{for every} \quad C \in \boldsymbol{\Pi}[\mathcal{M}^+]$$

is well defined, and is easily seen to have the desired properties.

To prove (γ) we first construct, with the possible help of the axiom of choice, a system $\langle \mathfrak{U}_j : j \in J\rangle$ of models of Ψ such that each model $\mathfrak{M}$ of Ψ is elementarily equivalent to one of the structures $\mathfrak{U}_j$ (i.e., the sets of sentences of $\mathcal{M}^+$ that are true of $\mathfrak{M}$ and of $\mathfrak{U}_j$ coincide). Let $\mathfrak{S}$ be the direct product of the system $\langle \mathfrak{Fr}(U_j) : j \in J\rangle$, where U_j is the universe of $\mathfrak{U}_j$. In view of (ii), it suffices to construct an isomorphism L from $\mathfrak{P}/\simeq^+_\Psi$ into $\mathfrak{S}$. Let G_j be the homomorphism of $\mathfrak{P}/\simeq^+_\Psi$ into $\mathfrak{Fr}(U_j)$ obtained in (β) by taking $\mathfrak{U}_j$ for $\mathfrak{U}$. We define L by setting

$$L(C/\simeq^+_\Psi) = \langle G_j(C/\simeq^+_\Psi) : j \in J\rangle \quad \text{for every} \quad C \in \boldsymbol{\Pi}[\mathcal{M}^+].$$

Clearly L is a homomorphism from $\mathfrak{P}/\simeq^+_\Psi$ to $\mathfrak{S}$. Let $C, D \in \boldsymbol{\Pi}[\mathcal{M}^+]$ and suppose $C/\simeq^+_\Psi \neq D/\simeq^+_\Psi$. In view of the semantic completeness of $\mathcal{M}^+$ and the definition of the system $\langle \mathfrak{U}_j : j \in J\rangle$, there is a $j \in J$ such that C and D denote different relations in $\mathfrak{U}_j$. From (β) we now get that

$$G_j(C/\simeq^+_\Psi) \neq G_j(D/\simeq^+_\Psi),$$

and hence $L(C/\simeq^+_\Psi) \neq L(D/\simeq^+_\Psi)$. Thus L is one-one and therefore an isomorphism. This completes the proof of (vii).

(viii) $\mathfrak{P}/\simeq^+$ *is an* RRA *which is* RRA*-freely generated by the set* $\{\boldsymbol{F}_i/\simeq^+ : i \in I\}$.

That $\mathfrak{P}/\simeq^+ \in$ RRA follows at once from (vii)(γ). Let K be the class of all full RA's on sets. We shall show that

(1) $\mathfrak{P}/\simeq^+$ is K-freely generated by the set $\{\boldsymbol{F}_i/\simeq^+ : i \in I\}$.

To this end consider any nonempty set U and any system $S = \langle S_i : i \in I\rangle$ of binary relations on U, i.e., members of $\mathfrak{Fr}(U)$. Setting $\mathfrak{U} = \langle U, S_i\rangle_{i\in I}$, we apply (vii)($\beta$) (with $\Psi = \varnothing$) to obtain a homomorphism of $\mathfrak{P}/\simeq^+$ into $\mathfrak{Fr}(U)$ that maps $\boldsymbol{F}_i/\simeq^+$ to S_i for each $i \in I$. This proves (1). That $\mathfrak{P}/\simeq^+$ is RRA-freely generated by $\{\boldsymbol{F}_i/\simeq^+ : i \in I\}$ follows from (1) with the help of (iii) and some well-known facts about K-free generating sets (such as Theorem 0.4.26(ii) in Henkin–Monk–Tarski [**1971**]).

8.4. Q-relation algebras

The discussion in earlier parts of this work suggests in a natural way the introduction of the notion of conjugated quasiprojections in the theory of (abstract) relation algebras.

(i) *For any given* RA $\mathfrak{A} = \langle A, +, ^-, \odot, ^\smile, \mathring{1}\rangle$, *two elements* $a, b \in A$ *are called conjugated quasiprojections if*

(α) $[(a^\smile \odot a + b^\smile \odot b)^- + \mathring{1}] \cdot (a^\smile \odot b) = 1$,

or, equivalently, if

(β) $a^\smile \odot a \leq \mathring{1}$, $b^\smile \odot b \leq \mathring{1}$, *and* $a^\smile \odot b = 1$.

(ii) *An* RA $\mathfrak{A}$ *is called a* Q*-relation algebra if its universe contains some conjugated quasiprojections;* QRA *is the class of all* Q*-relation algebras.*

(Compare (i)(α),(β) with 4.1(i),(ii).)

The main contribution of this work to the theory of relation algebras is the following theorem.

(iii) *Every* QRA *is an* RRA.[1,2]

---

[1]*This statement is actually equivalent to the assertion that $\mathcal{L}^+$ and $\mathcal{L}^\times$ are equipollent in means of proof relative to sentences Q_{AB}; i.e., it is equivalent to Theorem 4.4(xxxvii) (or, alternately, it is equivalent to the semantical completeness of $\mathcal{L}^\times$ relative to sentences Q_{AB}, i.e., it is equivalent to Theorem 4.4(xl)). In fact, the proof of 8.4(iii) shows that 8.4(iii) is implied by 4.4(xxxvii). For a simple proof of the converse implication, we use the methods of the next section, and in particular 8.5(iv),(vii). Let $\Psi \subseteq \Sigma^\times$, $X \in \Sigma^\times$, and $A, B \in \Pi$, and suppose that

(1) $\Psi \cup \{Q_{AB}\} \vdash^+ X$,

with the goal of proving

(2) $\Psi \cup \{Q_{AB}\} \vdash^\times X$.

We begin by establishing

(3) $\Psi \cup \Xi \cup \{Q_{AB}\} \models X\ [\mathfrak{E}^{\mathsf{EA}}]$,

where Ξ is the set of axioms of RA. To this end, let $\mathfrak{A} = \langle A, +, ^-, \odot, ^\smile, \mathring{1}, E\rangle$ be an algebra of type $\langle 2, 1, 2, 1, 0, 0\rangle$, and suppose that

(4) $\mathfrak{A}$ is a model of $\Psi \cup \Xi \cup \{Q_{AB}\}$.

Consider first the special case of (4) when $\mathfrak{A}$ is subdirectly indecomposable. By 8.2(v), the algebra $\mathfrak{A}' = \langle A, +, ^-, \odot, ^\smile, \mathring{1}\rangle$ is simple. Since $\mathfrak{A}'$ is a QRA, by (4) and 8.4(ii), it is also representable, by 8.4(iii). Therefore, by 8.3(i) we may assume that $\mathfrak{A}'$ is a proper relation algebra on some nonempty set U. In other words, we may assume that A is a set of binary relations on U, and that

$$\mathfrak{A} = \langle A, \cup, \sim, |, ^{-1}, U 1 Id, E\rangle.$$

To get an idea of the proof, consider any QRA, $\mathfrak{C} = \langle C, +, ^-, \odot, ^\smile, \mathring{1}\rangle$, and any two fixed elements $a, b \in C$ satisfying (i)(α). We construct the formalisms $\mathcal{M}^+$ and $\mathcal{M}^\times$ in which the nonlogical atomic predicates are correlated in a one-one way with the elements of C. Let $\boldsymbol{P}_x$ be the predicate correlated with $x \in C$; in particular, let $A = \boldsymbol{P}_a$ and $B = \boldsymbol{P}_b$. $\boldsymbol{Q}_{AB}$ as defined in 4.1(i) is obviously a sentence of $\mathcal{M}^\times$; let $\Psi = \{\boldsymbol{Q}_{AB}\}$. Consider the algebra $\mathfrak{P}$ and the relation $\simeq_\Psi^\times$ introduced in §8.2; by 8.2(ix), $\simeq_\Psi^\times$ is a congruence relation on $\mathfrak{P}$, and $\mathfrak{P}/\simeq_\Psi^\times$ is an RA.

It is readily seen that $\mathfrak{P}/\simeq_\Psi^\times$ is what is called a free RA with one defining relation expressed by the condition (i)(α). Indeed, $\mathfrak{P}/\simeq_\Psi^\times$ is RA-freely generated, relative to (i)(α), by the indexed system $\langle \boldsymbol{P}_x/\simeq_\Psi^\times : x \in C\rangle$, and two terms of this system, $A/\simeq_\Psi^\times$ and $B/\simeq_\Psi^\times$, satisfy (i)(α). Hence, using the basic properties of freely generated algebras, we conclude that $\mathfrak{C}$ is a homomorphic image of $\mathfrak{P}/\simeq_\Psi^\times$. In connection with this argument, cf. the remarks following the proof of 8.2(x).

On the other hand, consider the relation $\simeq_\Psi^+$ defined in §8.3. By 4.4(xxxvii), the relations $\simeq_\Psi^\times$ and $\simeq_\Psi^+$ are identical and so are the algebras $\mathfrak{P}/\simeq_\Psi^\times$ and $\mathfrak{P}/\simeq_\Psi^+$. By 8.3(vii)($\gamma$) the algebra $\mathfrak{P}/\simeq_\Psi^+$ is an RRA. Thus, $\mathfrak{C}$ is a homomorphic image of an RRA and therefore, in view of 8.3(v), is itself an RRA, as was to be proved.

(In particular, E is a binary relation on U.) From this, (4), and 8.5(vii)(β) it follows that every sentence in $\Psi \cup \{Q_{AB}\}$ is true of the structure $\langle U, E\rangle$. Therefore, so is the sentence X, by (1) and the soundness of $\mathcal{L}^+$. Applying 8.5(vii)(β) again, we see that X is true of $\mathfrak{A}$, as was to be shown.

Consider now the general case of (4). By a well-known theorem of Birkhoff [**1944**], $\mathfrak{A}$ is a subdirect product of subdirectly indecomposable algebras. From (4) and some elementary properties of subdirect products, it follows that each of these subdirectly indecomposable algebras is a model of $\Psi \cup \Xi \cup \{Q_{AB}\}$, and hence is also a model of X, by the above argument. Therefore $\mathfrak{A}$ is a model of X. This proves (3).

Since every equational formalism is semantically complete (see §8.1), (3) at once gives us

(5) $\qquad \Psi \cup \Xi \cup \{Q_{AB}\} \vdash X \; [\mathcal{E}^{\mathsf{EA}}]$.

But (2) follows from (5) by 8.5(iv), since

$$(\Psi \cup \{Q_{AB}\})_0 = \Psi \cup \{Q_{AB}\} \subseteq \Sigma^\times = \Sigma^{\mathsf{EA}}.$$

[2]*A semi-associative relation algebra is an algebra of type $\langle 2,1,2,1,0\rangle$ which satisfies the conditions (Ra I)–(Ra III), (Ra V)–(Ra X); also the condition

$$(x \odot 1) \odot 1 = x \odot (1 \odot 1)$$

is satisfied for all elements x of the algebra. (The last equation replaces the associative law, (Ra IV).) The theory of semi-associative relation algebras bears the same relationship to the formalism $\mathcal{L}w^\times$ from §3.10 as the theory of relation algebras bears to $\mathcal{L}^\times$. Since we know that certain equipollence results established for $\mathfrak{Q}$-systems in $\mathcal{L}^\times$ also carry over to $\mathcal{L}w^\times$ (see the footnotes, pp. 90 and 143), it is an interesting question whether Theorem (iii) carries over to semi-associative relation algebras. Németi [**1985**] has shown that the answer to this question is negative: there is a semi-associative relation algebra that contains two conjugated quasiprojections and that is not an RRA, and not even an RA.

The reasoning just outlined uses essentially Theorem 4.4(xxxvii) and depends therefore on the heavy proof-theoretical argument by means of which that theorem has been established. On the other hand, in Maddux [**1978**] a substantial generalization of (iii) can be found which, moreover, is established by purely algebraic methods.[3]

That the converse of (iii) does not hold is seen from the following observation.

(iv) *The full* RA *on a set* U *is a* QRA *iff the set* U *is either infinite or else consists of at most one element.*

This is an easy consequence of 4.1(v).

In view of (iii), every subalgebra of a QRA is also representable. However, such a subalgebra does not have to be a QRA itself. For example, if $\mathfrak{A}$ is any full RA on an infinite set, then $\mathfrak{A}$ is a QRA, but the subalgebra of $\mathfrak{A}$ with universe $\{0, 1, \mathring{0}, \mathring{1}\}$ is not a QRA. Let **SQRA** be the class of all subalgebras of QRA's. The following theorem characterizes **SQRA**'s in terms of their representations as subdirect products of proper RA's on certain sets.

(v) *For* $\mathfrak{A}$ *to be an* **SQRA** *it is necessary and sufficient that* $\mathfrak{A}$ *be isomorphic with a subdirect product of proper* RA*'s* $\mathfrak{B}_i$ *on sets* U_i $(i \in I)$, *where each set* U_i *is either infinite or consists of a single element.*

The proof uses (iii), (iv), and 8.3(ii), and is straightforward.

The necessary and sufficient condition in (v) for $\mathfrak{A}$ to be a **SQRA** can be reformulated as follows: $\mathfrak{A}$ is isomorphic to a proper RA with a unit (equivalence) relation V such that each of the equivalence sets into which V partitions its domain is either infinite or a singleton.

The class QRA is clearly closed under the formation of direct products and homomorphic images. However, as mentioned above, it is not closed under the formation of subalgebras, and hence is not a variety. On the other hand, we have the following results, (vi), (vii), (ix), (x), obtained by Givant around 1973, and briefly referred to in Maddux [**1978a**], p. 100.

(vi) SQRA *is a variety.*

The original proof of (vi) was closely related to that of 8.3(v). However, using a result of Maddux [**1978**], Lemma 10, one can give a much simpler proof of (vi) (cf. Maddux [**1978a**], pp. 98–99). Maddux showed that the homomorphic image of a subalgebra of an RA $\mathfrak{A}$ is a subalgebra of a homomorphic image of $\mathfrak{A}$. From this it easily follows that **SQRA** is closed under the formation of homomorphic

[3]*In view of the observation made in footnote 1* on p. 242, Maddux's algebraic proof of 8.4(iii) gives a semantical proof of the relative equipollence of $\mathcal{L}^{\times}$ and $\mathcal{L}^{+}$ in means of proof, i.e., of Theorem 4.4(xxxvii). Their relative equipollence in means of expression, Theorem 4.4(xxxvi), was already established by semantical methods in 4.4(xiv). Of course, Maddux's proof also gives us a semantical proof of the various properties of the translation mappings $\boldsymbol{K}_{AB}$ (cf. 2.4(vi)).

images, subalgebras, and direct products, and is therefore a variety by a well-known theorem of Birkhoff [**1935**].

The above argument can actually be used to show that the class **SK** of all subalgebras of algebras in K is a variety whenever K is a class of RA's closed under the formation of homomorphic images and direct products (cf. Maddux [**1978a**], p. 99).

(vii) SQRA *is not a finitely based variety; in fact, it is not even finitely based relative to* RRA, *i.e., there is no finite set* Γ *of equations such that* SQRA *is just the class of* RRA*'s in which all equations of* Γ *are identically satisfied.*

The proof of (vii) involves the construction of an ultraproduct. We assume that the reader is familiar with this construction (see, e.g., Henkin–Monk–Tarski [**1971**], pp. 106–111). Let F be a nonprincipal ultrafilter on the set $\{3, 4, 5, \ldots\}$, and let $\mathfrak{B}$ be the ultraproduct over F of the algebras $\mathfrak{Fr}(n)$, $n = 3, 4, 5, \ldots$, where $\mathfrak{Fr}(n)$ is the full RA on the set $n = \{0, \ldots, n-1\}$. Since each $\mathfrak{Fr}(n)$ is simple and representable, it follows from 8.2(iv) and 8.3(v) that the same is true of $\mathfrak{B}$. Thus by 8.3(i), $\mathfrak{B}$ is isomorphic with a proper RA on a nonempty set U. But $\mathfrak{B}$ is easily seen to be infinite, so U must be infinite. We conclude from (v) that $\mathfrak{B} \in$ **SQRA**. On the other hand, one readily shows with the help of (v) and 8.3(i) that $\mathfrak{Fr}(n) \notin$ **SQRA** for $n = 3, 4, 5, \ldots$.

Suppose now that there is a finite set Γ of equations such that $\mathfrak{A} \in$ **SQRA** iff $\mathfrak{A} \in$ RRA and each equation of Γ is identically satisfied in $\mathfrak{A}$. Let X be the universal closure of the conjunction of the equations in Γ. Then X is true of $\mathfrak{B}$, since $\mathfrak{B} \in$ **SQRA**. On the other hand, $\neg X$ is true of $\mathfrak{Fr}(n)$ for $n = 3, 4, 5, \ldots$, since $\mathfrak{Fr}(n) \notin$ **SQRA**, so $\neg X$ is true of $\mathfrak{B}$. Thus we have a contradiction. This completes the proof of (vii).

In Theorem (x) below we extend (vi) and (vii) to a naturally defined subclass, IRRA, of RRA. The definition of IRRA follows.

(viii) IRRA *is the class of algebras* $\mathfrak{A}$ *that are isomorphic with a subdirect product of proper* RA*'s* $\mathfrak{B}_i$ *on sets* U_i $(i \in I)$, *where each set* U_i *is infinite.*

It is easily seen that $\mathfrak{A} \in$ IRRA iff it is isomorphic to a proper RA with a unit (equivalence) relation V such that each of the equivalence sets into which V partitions its domain is infinite. An immediate corollary of (v) and (viii) is:

(ix) IRRA *is the class of all* $\mathfrak{A} \in$ **SQRA** *such that the equation* $\mathbf{\mathring{0}} \odot \mathbf{\mathring{0}} = \mathbf{1}$ *is true of* $\mathfrak{A}$.

(x) *Statements* (vi) *and* (vii) *continue to hold if* **SQRA** *is replaced everywhere by* IRRA.

That IRRA is a variety follows at once from (vi) and (ix). The proof of (vii), with **SQRA** replaced everywhere by IRRA, actually shows that IRRA is not finitely based relative to RRA. This completes the proof of (x).

For each $n = 2, 3, 4, \ldots$ one can construct an explicit equation Y_n which is identically satisfied in all **SQRA**, but which fails in $\mathfrak{Fr}(n)$, and hence in some RRA. Consider, for example, the case when $n = 2$. Let Z_2 be the formula determined by

$$Z_2 = [(x \odot 1 \odot x \leq \mathring{1} \wedge y \odot 1 \odot y \leq \mathring{1} \wedge x \cdot y = 0 \wedge x + y = \mathring{1}) \rightarrow x = 0 \vee y = 0].$$

In application to proper RA's the sentence $\forall_{xy} Z_2$ expresses the fact that the identity relation is not the disjoint union of two binary relations, each consisting of just one ordered pair. Hence $\forall_{xy} Z_2$ fails in $\mathfrak{Fr}(2)$, but is true of every simple **SQRA**. Now by 8.2(vii), the formula Z_2 is equivalent in all simple RA's to an equation Y_2. Thus Y_2 is identically satisfied in all **SQRA**'s, by (v), but fails in $\mathfrak{Fr}(2)$ by 8.3(i).[4]

In connection with the equations Y_n referred to above, it may be interesting to mention that every equation which is identically satisfied in all **SQRA**'s, or just in all IRRA's, but which fails to be so satisfied in all RRA's, necessarily contains the constant $\mathring{1}$. This is an immediate consequence of our next theorem.

(xi) *For any equation X in which the constant $\mathring{1}$ does not occur, the following conditions are equivalent:*

(α) *X is identically satisfied in every* RRA;
(β) *X is identically satisfied in every* QRA;
(γ) *X is identically satisfied in every* IRRA;
(δ) *X is identically satisfied in $\mathfrak{Fr}(U)$ for some infinite set U.*

The implications from (α) to (β), from (β) to (γ), and from (γ) to (δ) are immediate consequences of (iii), (v), and (viii).

In establishing the implication from (δ) to (α), we shall restrict ourselves to the case of $U = \omega$. (The proof of the general case requires some rather easy modifications of our argument.) Suppose that X is identically satisfied in $\mathfrak{Fr}(\omega)$. By 8.3(iii) it suffices to show that X is so satisfied in $\mathfrak{Fr}(V)$ for every set V.

Suppose first that V is infinite. Then

---

[4] *This construction can also be used to give a simple proof of the known result (see Jónsson [**1982**]) that there are continuum many subvarieties of RRA. For each $n \in \omega \sim \{0\}$, let Z_n be the quantifier-free formula with n variables, defined analogously to Z_2, such that, in application to proper relation algebras, $[Z_n]$ expresses the fact that the identity relation is not the disjoint union of n binary relations, each consisting of just one ordered pair. Let Y_n be the equation that is equivalent to Z_n in all simple RA's. Then for any (finite or infinite) cardinal number κ,

(1) $\quad Y_n$ is identically satisfied in $\mathfrak{Fr}(\kappa)$ iff $\kappa \neq n$.

Given a set $S \subseteq \omega \sim \{0\}$, define K_S to be the class of RRA's that are models of the set of equations $\{Y_n : n \in S\}$. In view of 8.3(v), it is obvious that K_S is a variety. Moreover, it follows immediately from (1) that

$$\mathfrak{Fr}(\kappa) \in K_S \quad \text{iff} \quad \kappa \notin S.$$

Hence, for distinct subsets S and T of $\omega \sim \{0\}$, the varieties K_S and K_T are distinct. (Actually, Jónsson [**1982**] proves that there are continuum many varieties of *symmetric* RRA's, i.e., RRA's in which the equation $x^{\smile} = x$ is identically satisfied.)

(1) every countable subalgebra of $\mathfrak{Fr}(V)$ is isomorphic to a subalgebra of $\mathfrak{Fr}(\omega)$.

Indeed, consider any countable subalgebra $\mathfrak{A}$ of $\mathfrak{Fr}(V)$ with universe A. To prove (1) we assume that the set V is well ordered; let P be any denumerable subset of V (e.g., the set of the first ω elements in the well ordering of V). Similarly, we assume that the set $V \times V$ is well ordered (for instance, lexicographically, using the well ordering of V); let Q be the set of elements q in $V \times V$ such that, for some distinct $S, T \in A$, q is the first element of $(S \sim T) \cup (T \sim S)$ in the well ordering of $V \times V$. We now define recursively an infinite sequence of denumerable subsets $Z_0, Z_1, \ldots, Z_n, \ldots$ of V by stipulating:

$$z \in Z_0 \quad \text{iff} \quad z \in P \quad \text{or} \quad z \quad \text{is in the field of} \quad Q \text{ (i.e., } z \in \bigcup\bigcup Q\text{)},$$

and for every $n \in \omega$,

> $z \in Z_{n+1}$ iff either $z \in Z_n$ or else z is the first element of V such that, for some $S, T \in A$ and $\langle x, y\rangle \in (S|T) \cap (Z_n \times Z_n)$, we have xSz and zTy.

Using the fact that A is countable, one easily shows by induction that the sets $Z_0, Z_1, \ldots$ are denumerable.

Set $K = \bigcup\{Z_n : n \in \omega\}$, and define a mapping G from A into $\mathit{Sb}(K \times K)$ by stipulating

$$GS = S \cap (K \times K) \quad \text{for every} \quad S \in A.$$

G is one-one, since $Z_0 \subseteq K$, and G is easily seen to preserve the operations of Boolean addition, complementation, and conversion, and the identity element. That G preserves Peircean multiplication follows readily from our construction of K. Thus, G maps $\mathfrak{A}$ isomorphically onto a subalgebra of $\mathfrak{Fr}(K)$. Since K is denumerable, $\mathfrak{Fr}(K)$ is obviously isomorphic to $\mathfrak{Fr}(\omega)$. This establishes (1).

Since X is identically satisfied in $\mathfrak{Fr}(\omega)$, it follows from (1) that X is identically satisfied in every finitely generated subalgebra of $\mathfrak{Fr}(V)$, and hence in $\mathfrak{Fr}(V)$ itself.

We now turn to the case when V is finite. Without loss of generality, we may assume that $V = n$ for some $n \in \omega$. With every relation $S \subseteq n \times n$ we correlate a relation $FS \subseteq \omega \times \omega$ defined as follows:

(2) $pFSq$ iff there are numbers $k, l < n$ such that p and q are respectively congruent (in the sense of number theory) to k and l modulo n, and kSl.

It is not difficult to prove that F is a function that maps the set $\mathit{Sb}(n \times n)$ into the set $\mathit{Sb}(\omega \times \omega)$ in a one-one way, and preserves all the fundamental operations of the corresponding full RA's $\mathfrak{Fr}(n)$ and $\mathfrak{Fr}(\omega)$ except the nullary operation $\mathring{1}$. Since $\mathring{1}$ does not occur in X, and X is identically satisfied in $\mathfrak{Fr}(\omega)$, we conclude that X is so satisfied in $\mathfrak{Fr}(n)$. The proof of the theorem is thus complete.

The proof that (δ) implies (α) (at least in the case where U is denumerable) can also be obtained from a familiar metalogical result by which a sentence in the first-order predicate calculus without identity is universally valid (in the sense of Hilbert–Ackerman [**1950**], p. 68) in every domain of individuals iff it is universally valid in some denumerably infinite domain; cf. *op. cit.*, p. 115. It seems that such a proof would involve formalisms different in nature from those discussed in this work (for instance, formalisms obtained from $\mathcal{L}$, $\mathcal{L}^+$, and $\mathcal{L}^\times$ by providing them with arbitrarily many variables ranging over binary relations).

In the next two theorems we state some results concerning simple QRA's.

(xii) *Let $\mathfrak{A}$ be a simple* QRA *with universe A, and suppose $a, b \in A$ are conjugated quasiprojections. For any $x, y \in A$ set*

(α) $x \mathbin{\square} y = (a \odot x \odot a^\smile) \cdot (b \odot y \odot b^\smile)$.

We then have

(β) $a^\smile \odot (x \mathbin{\square} y) \odot a = x$ *in case* $y \neq 0$;

(γ) $b^\smile \odot (x \mathbin{\square} y) \odot b = y$ *in case* $x \neq 0$.

Hence, for any $x, y, x', y' \in A \sim \{0\}$, the formula $x \mathbin{\square} y = x' \mathbin{\square} y'$ implies $x = x'$ and $y = y'$.

Indeed by (α), Definition (i)(β), and a familiar law from the theory of relation algebras whose interpretation in $\mathcal{L}^\times$ we have given as 3.2(xxviii), we easily obtain

$$\begin{aligned} a^\smile \odot (x \mathbin{\square} y) \odot a &= [a^\smile \odot (b \odot y \odot b^\smile) \cdot (x \odot a^\smile)] \odot a \\ &= a^\smile \odot (b \odot y \odot b^\smile) \odot a \cdot x \\ &= 1 \odot y \odot 1 \cdot x. \end{aligned}$$

Hence, $\mathfrak{A}$ being simple, we obtain (β) by 8.2(iv). In a completely symmetrical way we derive (γ).

Thus, in a simple QRA we can introduce a binary operation $\square$ which functions like the ordered couple operation with respect to all nonzero elements. By induction this result can be extended from ordered couples to ordered m-tuples, cf. 4.2(viii),(x).

(xiii) *If $\mathfrak{A}$ is a finitely generated simple* QRA, *then there is a single element c that generates $\mathfrak{A}$.*

With the essential help of (xii) we could give a direct algebraic proof of this theorem by imitating the proof of 7.1(ii), and actually that part of the proof comprehending steps (1) through (23).

However, we can also derive (xiii) from the metamathematical result 7.1(ii) by using the method that is based upon 8.2(ix),(x) and that was applied in proving Theorem (iii). Since some changes in the statement of 7.1(ii) are needed to adapt it to our purposes, we first explicitly restate in (1) below what is achieved in that portion of the proof of 7.1(ii) which is relevant to us, namely, steps (1)–(23).

(1) *Let $\mathcal{M}$ be a formalism of predicate logic with $n+1$ nonlogical binary atomic predicates $\boldsymbol{F}_0, \ldots, \boldsymbol{F}_n$, and let $\mathcal{T}$ be a $\mathcal{Q}$-system in $\mathcal{M}$ such that*

$$\mathbf{A}\boldsymbol{\xi}[\mathcal{T}] \vdash^{+} \{\mathbf{1} \odot \boldsymbol{F}_k \odot \mathbf{1} = \mathbf{1} \colon k = 0, \ldots, n\}.$$

Then there is a predicate E in $\mathbf{\Pi}[\mathcal{M}^{+}]$ from which we can construct some further predicates $\boldsymbol{F}_0^{\nabla}, \ldots, \boldsymbol{F}_n^{\nabla}$, using exclusively the operators $\boldsymbol{+}$, $^{-}$, $\odot$, $^{\smile}$, $\mathring{\mathbf{1}}$ (thus, without the additional help of $\boldsymbol{F}_0, \ldots, \boldsymbol{F}_n$), such that $\boldsymbol{F}_0^{\nabla}, \ldots, \boldsymbol{F}_n^{\nabla}$ satisfy the condition

$$\mathbf{A}\boldsymbol{\xi}[\mathcal{T}] \vdash^{+} \{\boldsymbol{F}_k = \boldsymbol{F}_k^{\nabla} \colon k = 0, \ldots, n\}.$$

Taking up now the proof of (xiii), let $\mathfrak{A}$ be a finitely generated simple QRA, say with generators $f_0, \ldots, f_n$ and quasiprojections a', b'. Without loss of generality we may assume that each $f_k \neq 0$, so that by 8.2(iv), $1 \odot f_k \odot 1 = 1$. Since $f_0, \ldots, f_n$ generate $\mathfrak{A}$, we can obtain a' and b' from these generators by repeated applications of the fundamental operations of $\mathfrak{A}$.

Now let $\mathcal{M}$ be the formalism introduced in the hypothesis of (1). We construct in $\mathcal{M}^{\times}$ predicates A' and B' from the atomic predicates $\boldsymbol{F}_0, \ldots, \boldsymbol{F}_n$ in exactly the same way as a' and b' are constructed from $f_0, \ldots, f_n$ in $\mathfrak{A}$. Let Ψ be the set consisting of the following $n+2$ sentences of $\mathcal{M}^{\times}$:

(2) $\mathbf{1} \odot \boldsymbol{F}_0 \odot \mathbf{1} = \mathbf{1}, \quad \ldots, \quad \mathbf{1} \odot \boldsymbol{F}_n \odot \mathbf{1} = \mathbf{1}, \quad Q_{A'B'}.$

Consider the algebra $\mathfrak{P}$ and the relation $\simeq_{\Psi}^{\times}$ introduced in §8.2. By 8.2(ix), $\mathfrak{P}/\simeq_{\Psi}^{\times}$ is an RA. It is readily seen that $\mathfrak{P}/\simeq_{\Psi}^{\times}$ is a free RA with the $n+2$ defining relations obtained from (2) in the manner described in the remarks after 8.2(x); indeed, $\mathfrak{P}/\simeq_{\Psi}^{\times}$ is RA-freely generated under these relations by the indexed system $\langle \boldsymbol{F}_k/\simeq_{\Psi}^{\times} \colon k = 0, \ldots, n\rangle$. Using the basic properties of freely generated algebras (and the fact that $f_0, \ldots, f_n, a', b'$ satisfy the same defining conditions in $\mathfrak{A}$) we conclude that there is a homomorphism H from $\mathfrak{P}/\simeq_{\Psi}^{\times}$ onto $\mathfrak{A}$ such that $H(\boldsymbol{F}_k/\simeq_{\Psi}^{\times}) = f_k$ for $k = 0, \ldots, n$ and $H(A'/\simeq_{\Psi}^{\times}) = a'$, $H(B'/\simeq_{\Psi}^{\times}) = b'$.

Let $\mathcal{T}$ be the system in $\mathcal{M}$ determined by the formula

$$\mathbf{A}\boldsymbol{\xi}[\mathcal{T}] = \boldsymbol{G}^{*}\Psi;$$

here $\boldsymbol{G}$ is the translation mapping originally defined in §2.3 for $\mathcal{L}^{+}$ and appropriately extended to formalisms of the type of $\mathcal{M}^{+}$. Clearly

(3) $\mathbf{A}\boldsymbol{\xi}[\mathcal{T}] \equiv^{+} \Psi,$

so that $\mathcal{T}$ satisfies the hypotheses of (1). Thus, there is a predicate E in $\mathbf{\Pi}[\mathcal{M}^{\times}]$ from which we can construct further predicates $\boldsymbol{F}_0^{\nabla}, \ldots, \boldsymbol{F}_n^{\nabla}$, as indicated in (1), such that

(4) $\mathbf{A}\boldsymbol{\xi}[\mathcal{T}] \vdash^{+} \{\boldsymbol{F}_k = \boldsymbol{F}_k^{\nabla} \colon k = 0, \ldots, n\}.$

By (3) we may replace $\mathbf{A}\mathfrak{E}[\mathfrak{T}]$ with Ψ in (4). Since $\boldsymbol{Q}_{A'B'} \in \Psi$, we conclude with the help of 4.4(xxxvii) that

$$\Psi \vdash^{\times} \{\boldsymbol{F}_k = \boldsymbol{F}_k^{\nabla} : k = 0, \ldots, n\},$$

i.e.,

$$\boldsymbol{F}_k/\simeq_{\Psi}^{\times} \; = \; \boldsymbol{F}_k^{\nabla}/\simeq_{\Psi}^{\times} \quad \text{for} \quad k = 0, \ldots, n.$$

Thus $E/\simeq_{\Psi}^{\times}$ generates the algebra $\mathfrak{P}/\simeq_{\Psi}^{\times}$, and hence $H(E/\simeq_{\Psi}^{\times})$ generates $\mathfrak{A}$. This completes the proof of (xiii).

We shall draw from (xiii) an interesting consequence, (xv), an unpublished result of Givant (obtained in 1972). We begin with a simple observation.

(xiv) *For every finite set U, the algebra $\mathfrak{Fr}(U)$ is generated by one element.*

In fact, since U is finite, there is a binary relation R on U which is one-one, whose domain contains all elements of U with one exception, and such that no nonempty subrelation of R is a permutation (of some subset of U). As is easily seen, every such relation R generates $\mathfrak{Fr}(U)$.

(xv) *If* K *is any one of the three classes of algebras* RRA, SQRA *or* IRRA, *then* K *is the least variety that contains all simple* RA*'s in* K *with one generator.*

To prove (xv) we take L to be the least variety containing all simple algebras in K that are generated by one element. Clearly it suffices to show that $\mathsf{K} \subseteq \mathsf{L}$, and this will follow if we prove that every finitely generated member of K is in L. Suppose first that $\mathfrak{C}$ is a finitely generated simple algebra in K. Since $\mathsf{K} \subseteq \mathsf{RRA}$, we may assume without loss of generality that $\mathfrak{C}$ is a proper RA on a set U, by 8.3(i), and, in fact, we can choose U so that $\mathfrak{Fr}(U)$ is also in K; this is obvious in the case $\mathsf{K} = \mathsf{RRA}$, and it is also easy to prove in the other two cases. If U is finite, then $\mathfrak{Fr}(U)$, and hence its subalgebra $\mathfrak{C}$, are in L by (xiv). Suppose now that U is infinite. Then there is a pair of conjugated quasiprojections A, B on U. Let $\{R_0, \ldots, R_n\}$ be a set of generators for $\mathfrak{C}$, and let $\mathfrak{C}'$ be the subalgebra of $\mathfrak{Fr}(U)$ generated by $R_0, \ldots, R_n, A, B$. Then $\mathfrak{C}'$ is also in K, and hence it is in L by (xiii). Thus $\mathfrak{C}$ is in L. This shows that every finitely generated simple algebra in K is also in L. But with the help of 8.2(vi) we see that every finitely generated algebra in K is a subdirect product of such simple algebras, and hence is also in L. This completes the proof of (xv).

This theorem can be equivalently formulated in metamathematical terms as follows.

(xvi) *Let* K *be as in* (xv). *If an equation X is true of every simple algebra in* K *with one generator, then X is true of every algebra in* K.

The problem is open whether (xv) and (xvi) can be extended to the class K of all RA's.

8.5. Decision problems for varieties of relation algebras

While in the preceding section we have been primarily interested in the application of our main results to the mathematical theory of RA's, in this section we shall discuss applications to some metamathematical problems concerning RA's. Specifically, we shall be concerned here with the equational theories of the class RA and some related classes. Our main purpose will be to carry over to these theories the undecidability results established for $\mathcal{L}^\times$ in 4.7(v)–(vii). The history of the results established in this and the following sections is somewhat involved. Various observations regarding this matter will be given in §8.7.

The equational formalism appropriate for the discussion of RA's is the formalism $\mathcal{E}^{\mathsf{A}}$ with the same nonlogical constants, $\mathbf{+}$, $\bar{\ }$, $\boldsymbol{\odot}$, $\smile$, $\mathring{\mathbf{1}}$, as in $\mathcal{P}^{\mathsf{A}}$. Recall that $\boldsymbol{\Phi}^{\mathsf{A}}$ is the set of equations, and Σ^{A} the set of variable-free equations, in $\mathcal{E}^{\mathsf{A}}$. For any set $\Psi \subseteq \boldsymbol{\Phi}^{\mathsf{A}}$, we write "$\boldsymbol{\Theta\eta}^{\mathsf{A}}\Psi$" as an abbreviation for "$\boldsymbol{\Theta\eta}\Psi[\mathcal{E}^{\mathsf{A}}]$", and similarly in analogous situations.

Let Ξ be the set of postulates (axioms) for RA, i.e., the set of equations (Ra I)–(Ra X) used in 8.2(i) to characterize the class RA. The equational theory of RA, in symbols $\boldsymbol{\Theta\rho}\mathsf{RA}$, is thus the set of all equations in $\boldsymbol{\Phi}^{\mathsf{A}}$ that are semantical consequences of Ξ. In view of the semantical completeness of $\mathcal{E}^{\mathsf{A}}$ (cf. the remarks near the end of §8.1), this theory can also be characterized as the set of equations derivable in $\mathcal{E}^{\mathsf{A}}$ from Ξ; in other words,

(i) $\boldsymbol{\Theta\rho}\mathsf{RA} = \boldsymbol{\Theta\eta}^{\mathsf{A}}\Xi$.

To carry over the results of 4.7(v)–(vii), we need to consider certain algebras that are expansions of RA's, in fact algebras

(ii) $\mathfrak{A} = \langle A, +, ^-, \odot, ^\smile, \mathring{1}, E\rangle$,

where $\langle A, +, ^-, \odot, ^\smile, \mathring{1}\rangle$ is an RA and E an additional distinguished element. An equational formalism appropriate for such algebras will be, of course, the formalism $\mathcal{E}^{\mathsf{EA}}$ which differs from $\mathcal{E}^{\mathsf{A}}$ only in that one additional individual nonlogical constant $\boldsymbol{E}$ has been included in its vocabulary. We are purposely choosing this constant to coincide with the nonlogical atomic predicate of $\mathcal{L}^\times$; cf. the remarks after 8.2(iii). (It may be mentioned that, with small changes, the whole subsequent discussion in this section could be referred to equational formalisms obtained from $\mathcal{E}^{\mathsf{A}}$ by adjoining not just one, but any finite number of individual nonlogical constants.)

We want now to establish a precise relationship between $\mathcal{L}^\times$ and $\mathcal{E}^{\mathsf{EA}}$ (treating them as proper, uninterpreted formalisms; cf. §1.6). Because of its general structure $\mathcal{L}^\times$ can undoubtedly be treated as a kind of equational formalism. It differs, however, from equational formalisms $\mathcal{E}$ as we have described them in §8.1, and indeed even from the formalism $\mathcal{E}^{\mathsf{EA}}$ which has the same constants as $\mathcal{L}^\times$. One important difference consists in the fact that, in opposition to $\mathcal{E}^{\mathsf{EA}}$,

$\mathcal{L}^\times$ is not provided with any variables. To remove this difference we first concentrate on the variable-free expressions of $\mathcal{E}^{\mathsf{EA}}$. Observe that the predicates of $\mathcal{L}^\times$ coincide with the variable-free terms of $\mathcal{E}^{\mathsf{EA}}$, and, as a consequence, the equations of $\mathcal{L}^\times$ coincide with the sentences of $\mathcal{E}^{\mathsf{EA}}$, i.e.,

$$\mathbf{\Phi}^\times = \mathbf{\Sigma}^\times = \mathbf{\Sigma}^{\mathsf{EA}}.$$

Apart from the rule of substitution given in 8.1(iii)(γ), which is inapplicable to $\mathcal{L}^\times$, there are still some differences between the definitions of derivability in $\mathcal{L}^\times$ and in $\mathcal{E}^{\mathsf{EA}}$. One difference is superficial, and vanishes if we assume that in defining derivability for $\mathcal{L}^\times$ we have replaced conditions 3.1(iii)(δ),(ε) by the equivalent condition 3.1(iv). Another, more substantial difference is that condition 3.1(iii)(β) does not occur in the definition of derivability for $\mathcal{E}^{\mathsf{EA}}$, and what is more, no sentence of $\mathbf{\Lambda}^\times$ which is not a tautology is even logically derivable in $\mathcal{E}^{\mathsf{EA}}$. (The treatment of sentences of $\mathbf{\Lambda}^\times$ as logical axioms of $\mathcal{L}^\times$ is proper only in view of our specific, rather narrow, semantical interpretation of the formalism $\mathcal{L}^\times$.) However, we do have the following theorem, which is proved by induction on derivable sentences in $\mathcal{L}^\times$.

(iii) *Given any $\Delta \subseteq \mathbf{\Sigma}^\times$ we have, for every $X \in \mathbf{\Sigma}^\times$, that $\Delta \cup \mathbf{\Lambda}^\times \vdash X$ $[\mathcal{E}^{\mathsf{EA}}]$ whenever $\Delta \vdash^\times X$; or, equivalently,*

$$\mathbf{\Theta\eta}^\times\Delta \subseteq \mathbf{\Theta\eta}^{\mathsf{EA}}(\Delta \cup \mathbf{\Lambda}^\times) \cap \mathbf{\Sigma}^{\mathsf{EA}}.$$

It turns out that the converse of (iii) is also true, and in fact, it is possible to state both (iii) and its converse in a stronger form that involves arbitrary subsets of $\mathbf{\Phi}^{\mathsf{EA}}$ and not just subsets of $\mathbf{\Sigma}^{\mathsf{EA}}$ ($= \mathbf{\Sigma}^\times$). To state this theorem we introduce a temporary notation. For any set $\Psi \subseteq \mathbf{\Phi}^{\mathsf{EA}}$ we let Ψ_0 be the set of all sentences in $\mathbf{\Sigma}^{\mathsf{EA}}$ that are substitution instances of equations in Ψ. Thus, for instance, $\Xi_0 = \mathbf{\Lambda}^\times$.

(iv) *Given any $\Psi \subseteq \mathbf{\Phi}^{\mathsf{EA}}$ we have*

$$\mathbf{\Theta\eta}^\times\Psi_0 \;=\; \mathbf{\Theta\eta}^{\mathsf{EA}}(\Psi \cup \Xi) \cap \mathbf{\Sigma}^{\mathsf{EA}}.$$

That $\mathbf{\Theta\eta}^\times\Psi_0$ is included in $\mathbf{\Theta\eta}^{\mathsf{EA}}(\Psi \cup \Xi) \cap \mathbf{\Sigma}^{\mathsf{EA}}$ is an almost immediate consequence of (iii). To obtain the reverse inclusion, let $\mathfrak{P}$ be the absolutely free algebra constructed in §8.2 for the formalism $\mathcal{L}^\times$, and take $\mathfrak{P}'$ to be the expansion of $\mathfrak{P}$ obtained by adjoining $\boldsymbol{E}$ as a new individual constant to the fundamental operations of $\mathfrak{P}$. Thus $\mathfrak{P}'$ is a realization of the formalism $\mathcal{E}^{\mathsf{EA}}$. Ψ_0 being a subset of $\mathbf{\Sigma}^\times$, we recall from 8.2(ix) that $\simeq^\times_{\Psi_0}$ is a congruence relation on $\mathfrak{P}$, and hence also on $\mathfrak{P}'$; furthermore, we recall that $\mathfrak{P}/{\simeq^\times_{\Psi_0}}$ is an RA, and therefore

(1) $\mathfrak{P}'/{\simeq^\times_{\Psi_0}}$ is a model of Ξ.

Finally, by an easy induction on predicates, we have

(2) any predicate $A \in \mathbf{\Pi}$, treated as a term of $\mathcal{E}^{\mathsf{EA}}$, denotes the element $A/{\simeq^\times_{\Psi_0}}$ in $\mathfrak{P}'/{\simeq^\times_{\Psi_0}}$.

Suppose now that $X \in \Psi$ and that $\langle A_0, \ldots, A_n, \ldots \rangle$ is an infinite sequence of predicates in $\mathbf{\Pi}$. Treating the A_n's as (variable-free) terms of $\mathcal{E}^{\mathsf{EA}}$, we form the equation X' from X by simultaneously substituting $A_0, \ldots, A_n, \ldots$ for the variables $\boldsymbol{v}_0, \ldots, \boldsymbol{v}_n, \ldots$. Clearly, the sentence X' is in Ψ_0, and is therefore valid in $\mathfrak{P}'/{\simeq^{\times}_{\Psi_0}}$, by (2) and the definition of $\simeq^{\times}_{\Psi_0}$ in §8.2. Put another way, the sequence $\langle A_0/{\simeq^{\times}_{\Psi_0}}, \ldots, A_n/{\simeq^{\times}_{\Psi_0}}, \ldots \rangle$ satisfies the equation X in $\mathfrak{P}'/{\simeq^{\times}_{\Psi_0}}$. Since the sequence $\langle A_0, \ldots, A_n, \ldots \rangle$ was arbitrary, we conclude that X is identically satisfied in $\mathfrak{P}'/{\simeq^{\times}_{\Psi_0}}$. Combining this result with (1) we obtain

(3) $\quad \mathfrak{P}'/{\simeq^{\times}_{\Psi_0}}$ is a model of $\mathbf{\Theta\eta}^{\mathsf{EA}}(\Psi \cup \Xi)$.

To complete the proof, let $A \boldsymbol{=} B$ be an equation in $\mathbf{\Theta\eta}^{\mathsf{EA}}(\Psi \cup \Xi) \cap \Sigma^{\mathsf{EA}}$. By (3), $A \boldsymbol{=} B$ is true of $\mathfrak{P}'/{\simeq^{\times}_{\Psi_0}}$, so that by (2) we have $A/{\simeq^{\times}_{\Psi_0}} = B/{\simeq^{\times}_{\Psi_0}}$. From the definition of $\simeq^{\times}_{\Psi_0}$ we now get $A \boldsymbol{=} B \in \mathbf{\Theta\eta}^{\times}\Psi_0$, and this is just what was to be shown.

Theorem (iv) has several interesting corollaries. For the purposes of our further discussion we shall state here two of them.

(v) *If Ψ is a theory in $\mathcal{E}^{\mathsf{EA}}$ which includes Ξ, then $\Psi \cap \Sigma^{\mathsf{EA}}$ is a theory in $\mathcal{L}^{\times}$, and in fact $\Psi \cap \Sigma^{\mathsf{EA}} = \Psi_0$.*

This is an immediate consequence of (iv) and the definition of Ψ_0.

(vi) *Suppose $\Psi, \Delta \subseteq \mathbf{\Phi}^{\mathsf{EA}}$ and $\Xi \subseteq \mathbf{\Theta\eta}^{\mathsf{EA}}\Psi$. Then Ψ is compatible with Δ (in $\mathcal{E}^{\mathsf{EA}}$) iff Ψ_0 is compatible with Δ_0 (in $\mathcal{L}^{\times}$).*

In fact, by (iv) and our hypotheses we have

$$\mathbf{\Theta\eta}^{\mathsf{EA}}(\Psi \cup \Delta) \cap \Sigma^{\mathsf{EA}} = \mathbf{\Theta\eta}^{\times}(\Psi_0 \cup \Delta_0).$$

Because the equation $\mathbf{0} \boldsymbol{=} \mathbf{1}$ is in Σ^{EA}, it will therefore be in $\mathbf{\Theta\eta}^{\mathsf{EA}}(\Psi \cup \Delta)$ iff it is in $\mathbf{\Theta\eta}^{\times}(\Psi_0 \cup \Delta_0)$, which proves (vi).

We turn now to the relationship between $\mathcal{L}^{\times}$ and $\mathcal{E}^{\mathsf{EA}}$, treated as interpreted formalisms. Recall that realizations of $\mathcal{L}^{\times}$ are structures $\langle U, E \rangle$, where E is a binary relation on U. On the other hand, realizations of $\mathcal{E}^{\mathsf{EA}}$ are algebras of type (ii), i.e., algebras of type $\langle 2, 1, 2, 1, 0, 0 \rangle$. The basic semantical notion in $\mathcal{L}^{\times}$ is that of denotation, which was defined precisely in 6.1(i). In an analogous way we can define what it means for a term to denote an element in a given realization of $\mathcal{E}^{\mathsf{EA}}$. An equation, i.e., a sentence, in $\mathcal{L}^{\times}$ is true of a realization $\langle U, E \rangle$ iff both sides of this equation denote the same relation on U. Similarly, this equation is true (i.e., identically satisfied) in an algebra of type (ii) iff both sides of the equation denote the same element. (In $\mathcal{E}^{\mathsf{EA}}$ the definition of the notion of an equation being identically satisfied in a structure is more involved.) However, close semantical connections between $\mathcal{L}^{\times}$ and $\mathcal{E}^{\mathsf{EA}}$ can be established only if we restrict ourselves to special realizations of $\mathcal{E}^{\mathsf{EA}}$, namely to proper relation algebras.

(vii) *Let U be a nonempty set and E a binary relation on U; set $\mathfrak{U} = \langle U, E\rangle$ and let $\mathfrak{A} = \langle A, \cup, \sim, |, {}^{-1}, U 1 Id, E\rangle$ be any algebra of type $\langle 2,1,2,1,0,0\rangle$ such that $\langle A, \cup, \sim, |, {}^{-1}, U 1 Id\rangle$ is a proper relation algebra on the set U and $E \in A$.*

(α) *For every predicate $B \in \mathbf{\Pi}$, the relation in $\mathfrak{U}$ denoted by B (in $\mathcal{L}^{\times}$) coincides with the relation in $\mathfrak{A}$ denoted by B (in $\mathcal{E}^{\mathsf{EA}}$).*

(β) *An equation X in $\Sigma^{\times}$ is true of $\mathfrak{U}$ iff it is true of $\mathfrak{A}$.*

Indeed, (α) is an obvious consequence of Definition 6.1(i), while (β) follows directly from (α).

We shall use the relationships between the formalisms $\mathcal{L}^{\times}$ and $\mathcal{E}^{\mathsf{EA}}$ discussed in (v)–(vii) to establish the undecidability results mentioned at the beginning of this section. The reader should now recall the notion of dual decidability introduced in 8.1(iv) and the definition of the set $\mathbf{\Omega}$ of three equations of $\mathcal{L}^{\times}$ given in 4.7(vii).

(viii) (α) *Let $\mathcal{F}$ be any equational formalism whose vocabulary includes that of $\mathcal{E}^{\mathsf{EA}}$, and let Ψ be a theory in $\mathcal{F}$. If $\Psi \supseteq \Xi$ and Ψ is compatible with $\mathbf{\Omega}$, then the set $\Psi \cap \Sigma^{\mathsf{EA}}$ is an undecidable theory in $\mathcal{L}^{\times}$, and hence the theory Ψ is undecidable.*

(β) *If, moreover, $\mathcal{F}$ coincides with $\mathcal{E}^{\mathsf{EA}}$, then Ψ is also dually undecidable.*

Indeed, assume that Ψ satisfies the hypotheses of (α). Then $\Psi \cap \mathbf{\Phi}^{\mathsf{EA}}$ is readily seen to be a theory of $\mathcal{E}^{\mathsf{EA}}$ that includes Ξ and is compatible with $\mathbf{\Omega}$. By (v) and (vi), with $\Psi \cap \mathbf{\Phi}^{\mathsf{EA}}$ and $\mathbf{\Omega}$ in place of Ψ and Δ, we get that $\Psi \cap \Sigma^{\mathsf{EA}}$ is a theory of $\mathcal{L}^{\times}$ which is compatible with $\mathbf{\Omega}$. Therefore we can apply 4.7(vii) to obtain the undecidability of $\Psi \cap \Sigma^{\mathsf{EA}}$, and this in turn implies the undecidability of Ψ.

To prove (β) observe that any given sentence X in Σ^{EA} is incompatible with Ψ iff it is incompatible with Ψ_0; this is a direct consequence of (vi) with Δ replaced by $\{X\}$. Therefore the dual decidability of Ψ would carry with it the dual decidability of $\Psi \cap \Sigma^{\mathsf{EA}}$ (in $\mathcal{L}^{\times}$). But $\Psi \cap \Sigma^{\mathsf{EA}}$ is undecidable by (α), and hence dually undecidable, since the notions of undecidability and dual undecidability coincide in $\mathcal{L}^{\times}$ by 3.3(vi). The conclusion of (β) follows at once.

(ix) *By an omega-*RA *we shall understand any algebra of type $\langle 2,1,2,1,0,0\rangle$ in which all the equations of $\Xi \cup \mathbf{\Omega}$ are identically satisfied. The class of all such algebras will be denoted by* ORA.

Thus, $\Xi \cup \mathbf{\Omega}$ may be called the postulate set for ORA, and in analogy to (i) we have

(x) $\mathbf{\Theta}\rho\mathsf{ORA} = \mathbf{\Theta}\eta^{\mathsf{EA}}(\Xi \cup \mathbf{\Omega})$.

As an immediate consequence of (viii)–(x) we get

(xi) $\mathbf{\Theta}\rho\mathsf{ORA}$ *is a finitely based, essentially undecidable, and essentially dually undecidable equational theory.*

The main result of this section is

(xii) (α) *Every theory* Θ *in* $\mathcal{E}^{\mathsf{A}}$ *such that* $\boldsymbol{\Xi} \subseteq \Theta \subseteq \boldsymbol{\Theta\eta}^{\mathsf{EA}}(\boldsymbol{\Xi} \cup \boldsymbol{\Omega})$ *is undecidable.*

(β) *Every theory* Θ *in* $\mathcal{E}^{\mathsf{A}}$ *such that* $\boldsymbol{\Theta\rho}\mathsf{RA} \subseteq \Theta \subseteq \boldsymbol{\Theta\rho}\mathsf{IRRA}$ *is undecidable. In particular,* $\boldsymbol{\Theta\rho}\mathsf{RA}$, $\boldsymbol{\Theta\rho}\mathsf{RRA}$, $\boldsymbol{\Theta\rho}\mathsf{QRA}$, *and* $\boldsymbol{\Theta\rho}\mathsf{IRRA}$ *are all undecidable.*

Indeed, given any theory Θ in $\mathcal{E}^{\mathsf{A}}$ satisfying the hypotheses of (α), we set

(1) $$\Psi = \boldsymbol{\Theta\eta}^{\mathsf{EA}}\Theta.$$

The theory Ψ so defined is what is usually called an inessential extension of Θ. By using a well-known theorem on inessential extensions given, e.g., in Tarski–Mostowski–Robinson [**1953**], Theorem 4, pp. 16f. (which can be referred to equational formalisms with practically no changes in its proof), we conclude that Θ is undecidable iff Ψ is undecidable. From (1) and the hypotheses of (α) we easily see that

$$\boldsymbol{\Xi} \subseteq \Psi \subseteq \boldsymbol{\Theta\eta}^{\mathsf{EA}}(\boldsymbol{\Xi} \cup \boldsymbol{\Omega}).$$

Hence, by (viii)(α), Ψ is undecidable. Therefore Θ is undecidable, which completes the proof of (α).

To prove the first assertion of (β) we set $\Theta = \boldsymbol{\Theta\rho}\mathsf{IRRA}$ and proceed to verify that Θ satisfies the double inclusion stated in the hypothesis of (α). The first half of this double inclusion being obvious, we shall concern ourselves only with the second half. To this end consider any subdirectly indecomposable algebra

$$\mathfrak{A} = \langle A, +, ^{-}, \odot, ^{\smile}, \mathring{1}, E\rangle$$

that is a model of $\boldsymbol{\Theta\eta}^{\mathsf{EA}}(\boldsymbol{\Xi} \cup \boldsymbol{\Omega})$. Then

$$\mathfrak{A}' = \langle A, +, ^{-}, \odot, ^{\smile}, \mathring{1}\rangle$$

is a subdirectly indecomposable RA, and hence is simple by 8.2(v). Also, $\mathfrak{A}$ is a model of the sentence $\boldsymbol{Q}_{AB}$ in $\boldsymbol{\Omega}$, so that $\mathfrak{A}'$ is a QRA, by 8.4(ii), and therefore an RRA, by 8.4(iii). Now every simple RRA is isomorphic to a proper RA on a nonempty set, by 8.3(i). Thus, without loss of generality we may assume $\mathfrak{A}'$ is in fact a proper RA on some nonempty set U, so that A is a set of binary relations on U and

$$\mathfrak{A} = \langle A, \cup, \sim, |, ^{-1}, U\upharpoonleft Id, E\rangle,$$

where $\sim$ is complementation relative to $U \times U$, and E is a binary relation on U. Because $\mathfrak{A}$ is a model of $\boldsymbol{\Omega}$ we see from (vii)(β) that $\langle U, E\rangle$ will be a model of $\boldsymbol{\Omega}$. With the help of 4.1(v) we conclude that U is infinite. This proves that $\mathfrak{A}' \in \mathsf{IRRA}$ and hence that $\mathfrak{A}$ is a model of Θ. We have thus proved that every subdirectly indecomposable model of $\boldsymbol{\Theta\eta}^{\mathsf{EA}}(\boldsymbol{\Xi} \cup \boldsymbol{\Omega})$ is a model of Θ. Using the result of Birkhoff [**1944**] by which every algebra is isomorphic to a subdirect product of subdirectly indecomposable algebras, we conclude that every model of $\boldsymbol{\Theta\eta}^{\mathsf{EA}}(\boldsymbol{\Xi} \cup \boldsymbol{\Omega})$ is also a model of Θ, as was to be shown.

The second assertion in (β) follows from the first, in view of 8.4(iii),(ix).

In connection with this theorem, several questions arise. (1) Is $\mathbf{\Theta}\rho\mathsf{RA}$ essentially undecidable? (2) If not, does $\mathbf{\Theta}\rho\mathsf{RA}$ have an essentially undecidable extension in the formalism of $\mathcal{E}^{\mathsf{A}}$? (3) Is $\mathbf{\Theta}\rho\mathsf{RA}$ dually undecidable? (4) If not, does $\mathbf{\Theta}\rho\mathsf{RA}$ have a dually undecidable extension in $\mathcal{E}^{\mathsf{A}}$? It was pointed out by Givant that, somewhat surprisingly, all these questions have negative solutions. To show this, we first state the following known result (see Tarski [**1956a**], where a proof of the result can also be found). Recall that $\mathfrak{Fr}(U)$ is the full RA on U, that an equational theory Θ is *complete* if it is consistent and if every equation compatible with Θ is in Θ (cf. §1.3), and that every natural number n coincides with the set $\{0,\ldots,n-1\}$, so that $1=\{0\}$, $2=\{0,1\}$, and $3=\{0,1,2\}$.

(xiii) (α) *The theory of* RA*'s,* $\mathbf{\Theta}\eta^{\mathsf{A}}\Xi$*, has just three complete extensions,* Θ_1, Θ_2, Θ_3, *in* $\mathcal{E}^{\mathsf{A}}$*, determined by*

$$\Theta_k = \mathbf{\Theta}\eta^{\mathsf{A}}(\Xi\cup\{X,Y_k\}) \quad \textit{with} \quad k=1,2,3,$$

where

$$X = (\boldsymbol{x}\odot\mathbf{1}\odot\boldsymbol{x}^{\smile}\odot\mathbf{1}\odot(\boldsymbol{x}\cdot\mathring{\mathbf{1}}+\boldsymbol{x}^{\smile}\cdot\mathring{\mathbf{0}})\odot\mathbf{1}\odot(\boldsymbol{x}\cdot\mathring{\mathbf{0}}+\boldsymbol{x}^{\smile}\cdot\mathring{\mathbf{1}})=\mathbf{0}),$$

$$Y_1 = (\mathring{\mathbf{0}}\odot\mathring{\mathbf{0}}=\mathbf{0}), \qquad Y_2 = (\mathring{\mathbf{0}}\odot\mathring{\mathbf{0}}=\mathring{\mathbf{1}}), \qquad Y_3 = (\mathring{\mathbf{0}}\odot\mathring{\mathbf{0}}=\mathbf{1}).$$

(β) *Each of the theories* $\Theta_1,\Theta_2,\Theta_3$ *(in* (α)*) is the theory of a finite* RA*; in fact, for* $k=1,2,3$*, we have* $\Theta_k=\mathbf{\Theta}\rho\mathfrak{A}_k$*, where* $\mathfrak{A}_k$ *is the subalgebra of* $\mathfrak{Fr}(k)$ *with universe* $\{0,1,\mathring{1},\mathring{0}\}$.

(γ) *All three theories* $\Theta_1,\Theta_2,\Theta_3$ *are decidable.*

The negative solutions of (1)–(4) are easy consequences of (xiii). We state these explicitly in the next two corollaries.

(xiv) *No theory in* $\mathcal{E}^{\mathsf{A}}$ *which includes* Ξ *is essentially undecidable.*

In fact, every consistent theory that includes Ξ can be extended to a complete theory, which, by (xiii), must coincide with one of the three theories $\Theta_1,\Theta_2,\Theta_3$, and hence must be decidable.

(xv) (α) *For every consistent theory* Ψ *in* $\mathcal{E}^{\mathsf{A}}$ *which includes* Ξ*, the set of equations compatible with* Ψ *coincides with one of the seven sets* Θ_1, Θ_2, Θ_3, $\Theta_1\cup\Theta_2$, $\Theta_1\cup\Theta_3$, $\Theta_2\cup\Theta_3$, *or* $\Theta_1\cup\Theta_2\cup\Theta_3$.

(β) *Every theory in* $\mathcal{E}^{\mathsf{A}}$ *that includes* Ξ *is dually decidable.*

To prove (α) let

$$J=\{j\in 3\colon \Psi\subseteq\Theta_j\}.$$

Using (xiii)(α) and the well-known result of Lindenbaum, by which every consistent theory has a complete extension, we obtain $X\in\bigcup\{\Theta_j\colon j\in J\}$ whenever X is compatible with Ψ. The proof of the converse of this implication is even easier, and the two implications together yield the conclusion of (α). Since the theories Θ_1,Θ_2, and Θ_3 are recursive, by (xiii)(γ), so are their complements (in $\mathbf{\Phi}^{\mathsf{A}}$); hence ($\beta$) follows immediately from (α).

We see thus that the equational theory of RA and all of its extensions mentioned in (xii)(β) are examples of equational theories that are undecidable and dually decidable; they could be referred to as theories which are *essentially dually decidable*, since all their extensions are dually decidable. It may be noticed that no equational theory can simultaneously be essentially undecidable and essentially dually decidable, nor essentially decidable and essentially dually undecidable.

Theorem (viii) also has other consequences of metalogical interest. These are (negative) solutions of some decision problems of the following type: let S be the set of all finite bases of theories in a given formalism which have a prescribed property; is S recursive? Such problems are sometimes referred to as decision problems of the second degree. For examples of results of this kind concerning theories formalized in predicate logic, see Tarski–Mostowski–Robinson [**1953**], pp. 34f. We state here several analogous results for theories formalized in $\mathcal{E}^{\mathsf{EA}}$.

(xvi) (α) *Let Σ be the set of all equations X in $\mathcal{E}^{\mathsf{EA}}$ such that $\mathbf{\Theta\eta}^{\mathsf{EA}}X$ is consistent. Then Σ is not recursive.*

(β) *Statement (α) continues to hold if, in its formulation, "consistent" is replaced by "complete", "decidable", "dually decidable", "essentially undecidable", or "essentially dually undecidable".*

(γ) *Statements (α) and (β) continue to hold if, instead of the set Σ of all equations X, we consider the set S of all finite sets Φ of equations in $\mathcal{E}^{\mathsf{EA}}$, and we replace "$\mathbf{\Theta\eta}^{\mathsf{EA}}X$" by "$\mathbf{\Theta\eta}^{\mathsf{EA}}\Phi$".*

To prove (α) we first observe that, for every equation $Y \in \mathbf{\Phi}^{\mathsf{EA}}$, the theory $\mathbf{\Theta\eta}^{\mathsf{EA}}(\Xi \cup \mathbf{\Omega} \cup \{Y\})$ is an extension of the equational theory of Boolean algebras. Therefore, by Theorems 3, 4, and the subsequent remarks in Tarski [**1968**], we can recursively correlate with Y an equation $FY \in \mathbf{\Phi}^{\mathsf{EA}}$ such that

$$\mathbf{\Theta\eta}^{\mathsf{EA}}FY = \mathbf{\Theta\eta}^{\mathsf{EA}}(\Xi \cup \mathbf{\Omega} \cup \{Y\}).$$

We conclude that $\mathbf{\Theta\eta}^{\mathsf{EA}}FY$ is consistent iff Y is compatible with $\mathbf{\Theta\eta}^{\mathsf{EA}}(\Xi \cup \mathbf{\Omega})$. Hence, if the set Σ were recursive, so would be the set of all equations compatible with $\mathbf{\Theta\eta}^{\mathsf{EA}}(\Xi \cup \mathbf{\Omega})$. This, however, is impossible, since this latter theory is dually undecidable by (x), (xi). Thus (α) has been established.

To obtain (β) it now suffices to show that, for every equation Y, the following six conditions on the set $\Theta = \mathbf{\Theta\eta}^{\mathsf{EA}}(\Xi \cup \mathbf{\Omega} \cup \{Y\})$ are equivalent:

(1) Θ is consistent,
(2) Θ is consistent but not complete,
(3) Θ is undecidable,
(4) Θ is dually undecidable,
(5) Θ is essentially undecidable,
(6) Θ is essentially dually undecidable.

Since Θ is an extension of $\mathbf{\Theta\eta}^{\mathsf{EA}}(\Xi \cup \mathbf{\Omega})$, most of the implications needed for establishing the equivalence of (1)–(6) follow immediately from (x), (xi). The implication from (3) to (2) is a consequence of a well-known result to the effect that every finitely based and complete equational theory is decidable.

Statements (α) and (β) obviously continue to hold if, instead of the set Σ of all equations X, we consider the set S' of all sets $\{X\}$. Since the set S' is the intersection of S with the (recursive) set of all singletons $\{X\} \subseteq \mathbf{\Phi}^{\mathsf{EA}}$, we at once arrive at ($\gamma$).

8.6. Decision problems for varieties of groupoids

In the preceding section we have established the existence of a finitely based undecidable equational theory, namely $\mathbf{\Theta}\rho\mathsf{RA}$, and even more, the existence of a finitely based essentially undecidable, and essentially dually undecidable equational theory, namely $\mathbf{\Theta}\rho\mathsf{ORA}$. Independent of the specific mathematical content of the theories just mentioned, the mere existence of equational theories with these properties presents some interest for the general metatheory of equational formalisms. A defect of our results from this point of view is the complicated similarity type of the algebras and formalisms involved, namely $\langle 2,1,2,1,0\rangle$ for RA's and $\langle 2,1,2,1,0,0\rangle$ for ORA's.

The situation can be improved by construing RA's in a different, but definitionally equivalent way (in the sense of equational logic, i.e., polynomially or rationally equivalent; cf. Henkin–Monk–Tarski [**1971**], pp. 125–127). For instance, it is seen from the discussion in §5.2 that RA can be construed as a variety of type $\langle 2,2,0\rangle$, and in fact as a class of algebras $\langle A,\dagger,\odot,\mathring{1}\rangle$, where $\dagger$ is the binary operation definable (in the original formalism $\mathcal{E}^{\mathsf{A}}$) by means of the formula

$$x \dagger y \;=\; x^{-} + y^{-\smile}.$$

ORA's become then algebras $\langle A,\dagger,\odot,\mathring{1},E\rangle$ of type $\langle 2,2,0,0\rangle$. On the other hand, from the postulates characterizing ORA's we can easily derive the formula

$$\mathring{1} = (E \odot E^{-\smile})^{-},$$

from which we conclude that $\mathring{1}$ is definable in terms of $\dagger$, $\odot$, and E. Thus, eventually, ORA's can be construed as algebras $\langle A,\dagger,\odot,E\rangle$ of type $\langle 2,2,0\rangle$ (cf. §5.3). Actually, for our purposes, it proves more convenient to construe ORA's as algebras $\langle A,\dagger,\odot,{}^{\vartriangle}\rangle$ of type $\langle 2,2,1\rangle$, where ${}^{\vartriangle}$ is the unary operation on A such that $x^{\vartriangle} = E$ for every $x \in A$. Rather as a curiosity we may mention that ORA's can also be construed as algebras of type $\langle 2,1,1,0,0\rangle$ (this is related to the results established in 4.1(x)(α) and in §5.2).

It does not seem to be plausible that any further essential progress can be achieved with the exclusive help of similar devices. In particular, it is not known whether there exists a variety of groupoids, i.e., algebras of type $\langle 2\rangle$, which is definitionally (more precisely, polynomially) equivalent with RA or ORA. An affirmative solution of this problem seems however unlikely; compare the last paragraph of §5.2.[5]

[5] *Börner [**1986**] has shown that RA is polynomially equivalent to a variety of type $\langle 2\rangle$. (See the footnote, p. 153.)

Nevertheless, we shall be able to extend our results to equational theories of algebras of type $\langle 2\rangle$ by using a different method. Generally speaking, this method consists in constructing, for any given theory Θ in an arbitrary equational formalism $\mathcal{E}$, a new theory Θ' in the equational formalism $\mathcal{E}'$ of type $\langle 2\rangle$ in such a way that Θ' preserves various properties of Θ. We shall take for $\mathcal{E}$ a formalism adequate for the discussion of ORA's, and among the properties of Θ which are to be preserved we shall be primarily concerned with undecidability, essential undecidability, dual undecidability, essential dual undecidability, and finite axiomatizability.

To simplify our exposition we assume, in accordance with the earlier remarks of this section, that the theory of ORA's has been equipollently reconstructed in a formalism of type $\langle 2,2,1\rangle$, and we take for $\mathcal{E}$ just such a formalism. Thus $\mathcal{E}$ has three nonlogical constants, two binary operation symbols, $\dagger$ and $\odot$, and one unary operation symbol $^{\vartriangle}$. In addition, we pick a binary operation symbol, say $\square$, not occurring in $\mathcal{E}$. Let $\mathcal{E}'$ be the equational formalism with $\square$ as the only operation symbol. To improve the readability of various formulas below we shall adopt some abbreviations. For example, we shall write "$\boldsymbol{\Phi}$", "$\boldsymbol{\Theta}\eta$", and "$\vdash$" instead of "$\boldsymbol{\Phi}[\mathcal{E}]$", "$\boldsymbol{\Theta}\eta[\mathcal{E}]$", and "$\vdash [\mathcal{E}]$". Similarly, we write "$\boldsymbol{\Phi}'$",... instead of "$\boldsymbol{\Phi}[\mathcal{E}']$",.... Analogous abbreviations will also be used for the formalism $\mathcal{E}''$ to be subsequently introduced.

Next we shall choose three equations, H_1, H_2, and H_3 (in some equational formalism that is a common extension of $\mathcal{E}$ and $\mathcal{E}'$) which have the form of possible definitions of $\dagger$, $\odot$, and $^{\vartriangle}$ in $\mathcal{E}'$. The choice of these definitional equations is of fundamental significance for the whole procedure. The reader should not ascribe some intuitive meaning to them; what matters is their formal structure, and not their mathematical content. The construction of these equations aims at obtaining possible definitions of $\dagger$, $\odot$, and $^{\vartriangle}$ in $\mathcal{E}'$ that have the least deductive power. In fact, they are chosen in such a way that no equation of $\mathcal{E}$ which is not derivable in a given theory Θ in $\mathcal{E}$ can be derived from $\Theta \cup \{H_1, H_2, H_3\}$. This will be achieved here by choosing for the *definientes*, H_i^r, terms satisfying the condition: H_1^r, H_2^r, H_3^r are not variables; moreover, for any $i, j = 1,2,3$, if t is a subterm of H_i^r that is not a variable, and some substitution instance of t coincides with a substitution instance of H_j^r, then $i = j$ and $t = H_i^r$. This condition may be called a *subterm condition* following a suggestion of McKenzie. (That this condition, or a variant of it, is necessary is seen from the following rather trivial example. If we set $H_1 = [\boldsymbol{x} \dagger \boldsymbol{y} = (\boldsymbol{x} \square \boldsymbol{y}) \square (\boldsymbol{x} \square \boldsymbol{y})]$, $H_2 = [\boldsymbol{x} \odot \boldsymbol{y} = (\boldsymbol{x} \square \boldsymbol{x}) \square (\boldsymbol{y} \square \boldsymbol{y})]$, and let H_3 be arbitrary, we can easily derive $\boldsymbol{x} \dagger \boldsymbol{x} = \boldsymbol{x} \odot \boldsymbol{x}$, independent of whether this equation belongs to the original theory Θ.) For H_1, H_2, and H_3, we shall actually take the following equations:

$$\begin{aligned} H_1 &= [\boldsymbol{x} \dagger \boldsymbol{y} = ((\boldsymbol{x} \square \boldsymbol{x}) \square \boldsymbol{x}) \square (\boldsymbol{y} \square \boldsymbol{y})], \\ H_2 &= \big[\boldsymbol{x} \odot \boldsymbol{y} = \big(\boldsymbol{x} \square [(\boldsymbol{x} \square \boldsymbol{x}) \square \boldsymbol{x}]\big) \square (\boldsymbol{y} \square \boldsymbol{y})\big], \\ H_3 &= \big[\boldsymbol{x}^{\vartriangle} = \boldsymbol{x} \square \big(\boldsymbol{x} \square [(\boldsymbol{x} \square \boldsymbol{x}) \square \boldsymbol{x}]\big)\big]. \end{aligned}$$

To proceed formally, we now introduce the equational formalism $\mathcal{E}''$ with the nonlogical constants $\dagger$, $\odot$, ${}^\vartriangle$, and $\square$. Let T be the mapping which, loosely speaking, eliminates from all meaningful expressions of $\mathcal{E}''$ (i.e., from all terms and equations) the operation symbols $\dagger$, $\odot$, and ${}^\vartriangle$ on the basis of H_1, H_2, and H_3. Precisely, T is defined by recursion on the set $\mathbf{T}\boldsymbol{\mu}'' = \mathbf{T}\boldsymbol{\mu}[\mathcal{E}'']$ (cf. §8.1) as the unique operation satisfying, for any variable $x \in \Upsilon$ and any terms $t, s \in \mathbf{T}\boldsymbol{\mu}''$, the conditions:

$$\begin{aligned} Tx &= x, \\ T(t\,\square\, s) &= Tt\,\square\, Ts, \\ T(t\dagger s) &= [(Tt\,\square\, Tt)\,\square\, Tt]\,\square\,(Ts\,\square\, Ts), \\ T(t\odot s) &= (Tt\,\square\,[(Tt\,\square\, Tt)\,\square\, Tt])\,\square\,(Ts\,\square\, Ts), \\ T(t^\vartriangle) &= Tt\,\square\,(Tt\,\square\,[(Tt\,\square\, Tt)\,\square\, Tt]), \end{aligned}$$

and finally

$$T(t = s) = (Tt = Ts).$$

From the definition of T we can derive by an easy induction on terms various elementary lemmas, of which we state the following.

(i) *For any finite sequence $\langle x_0, \ldots, x_{n-1}\rangle$ of distinct variables, any term t, and any finite sequence $\langle s_0, \ldots, s_{n-1}\rangle$ of terms (of $\mathcal{E}''$),*

$$(Tt)[x_0/Ts_0, \ldots, x_{n-1}/Ts_{n-1}] = T(t[x_0/s_0, \ldots, x_{n-1}/s_{n-1}]).$$

(ii) *For $i = 1, 2, 3$, TH_i is a tautology, and in fact $TH_i = (H_i^r = H_i^r)$.*

(iii) (α) *$TX = X$ for every $X \in \boldsymbol{\Phi}'$,*
(β) *$TX \equiv''_{\{H_1,H_2,H_3\}} X$ for every $X \in \boldsymbol{\Phi}''$.*

From (i)–(iii), using induction on derivable sentences, we obtain the following mapping theorem from $\mathcal{E}''$ to $\mathcal{E}'$.

(iv) *For any $\Delta \subseteq \boldsymbol{\Phi}''$ and $X \in \boldsymbol{\Phi}''$, we have*

$$\Delta \cup \{H_1, H_2, H_3\} \vdash'' X \text{ iff } T^*\Delta \vdash' TX.$$

With any given theory Θ in $\mathcal{E}$ we correlate a theory Θ' in $\mathcal{E}'$ and a theory Θ'' in $\mathcal{E}''$ by means of the following stipulations:

(v) (α) $\Theta'' = \boldsymbol{\Theta\eta}''(\Theta \cup \{H_1, H_2, H_3\})$,
(β) $\Theta' = \Theta'' \cap \boldsymbol{\Phi}'$.

It may be mentioned that the theory Θ'' (and the formalism $\mathcal{E}''$) play no essential part in our method of constructing Θ'. It seems, however, that the introduction of Θ'' simplifies the description of the method and makes it more elegant. The role of Θ'' is somewhat analogous to that of the formalism $\mathcal{L}^+$ in our earlier discussion.

Definition (v) has numerous consequences. Most important for our purposes is:

(vi) (α) $\Theta'' = \mathbf{\Theta}\boldsymbol{\eta}''(\Theta' \cup \{H_1, H_2, H_3\})$,
(β) $\Theta = \Theta'' \cap \mathbf{\Phi}$.

(Notice the structural similarity between pairs (α) and (β) of this theorem and the corresponding parts (α) and (β) of Definition (v).)

The proof of (α) is quite simple. Indeed we get directly from (v)(α),(β) that

$$\mathbf{\Theta}\boldsymbol{\eta}''(\Theta' \cup \{H_1, H_2, H_3\}) \subseteq \Theta''.$$

To obtain the inclusion in the opposite direction, notice that for $X \in \Theta''$ we have $TX \in \Theta''$ by (iii)(β) and (v)(α), so that $TX \in \Theta'$ by (v)(β). We therefore conclude by (iii)(β) that $X \in \mathbf{\Theta}\boldsymbol{\eta}''(\Theta' \cup \{H_1, H_2, H_3\})$, which completes the proof of (α).

The proof of (β) is more involved. Let K be the class of all models of Θ. Thus K consists of algebras $\mathfrak{A}$ of the type $\mathfrak{A} = \langle A, \dagger, \odot, {}^{\vartriangle} \rangle$. We choose a particular algebra $\mathfrak{A}$ in K, namely a K-free algebra with ω generators. The following is a basic and well-known property of $\mathfrak{A}$.

(1) An equation X in $\mathcal{E}$ is identically satisfied in $\mathfrak{A}$ iff $X \in \Theta$. In particular, $\mathfrak{A}$ is a model of Θ.

The crux of the proof of (vi)(β) is the construction of a binary operation $\square$ on A such that $\langle A, \dagger, \odot, {}^{\vartriangle}, \square \rangle$ is a model of H_1, H_2, H_3, and hence of Θ''.

The universe A of $\mathfrak{A}$ is of course denumerable, and by passing from one K-free algebra on ω generators to an isomorphic image, we can assume that A is any denumerable set given in advance. It proves convenient for our purposes to take for A the set of all terms of $\mathcal{E}'$. Since $\boldsymbol{\square}$ is an operation symbol (and in fact the only operation symbol) of the formalism $\mathcal{E}'$, we can treat $\boldsymbol{\square}$ as a binary operation on A which, when applied to two terms s, t in A, yields the term $s \boldsymbol{\square} t$ in A. On the other hand, the operations $\dagger$, $\odot$ of $\mathfrak{A}$ are also binary operations on A, yielding, for any two terms s, t in A, new terms $s \dagger t$ and $s \odot t$ in A. However, $\dagger$ and $\odot$ have nothing to do with the metamathematical operations $\boldsymbol{\dagger}$ and $\boldsymbol{\odot}$; in particular, the terms $s \dagger t$ and $s \odot t$ do not coincide with the terms $s \boldsymbol{\dagger} t$ and $s \boldsymbol{\odot} t$, and in fact, these latter terms are not elements of A, and not even expressions in the formalism $\mathcal{E}'$ (although they are terms in $\mathcal{E}''$).

The following two properties of terms in $\mathcal{E}'$ are obvious, and it is superfluous to prove them formally.

(2) No term is a proper segment of itself.

(3) For any terms s, s', t, t' of $\mathcal{E}'$, if $s \boldsymbol{\square} t = s' \boldsymbol{\square} t'$, then $s = s'$ and $t = t'$.

We list several easy consequences of (2), (3) with a more specialized character. For any terms s, t of $\mathcal{E}'$ we have:

(4) if $(s \boldsymbol{\square} s) \boldsymbol{\square} s = (t \boldsymbol{\square} t) \boldsymbol{\square} t$, then $s = t$;

(5) if $s \boldsymbol{\square} [(s \boldsymbol{\square} s) \boldsymbol{\square} s] = t \boldsymbol{\square} [(t \boldsymbol{\square} t) \boldsymbol{\square} t]$, then $s = t$;

(6) $(s \boldsymbol{\square} s) \boldsymbol{\square} s \neq t \boldsymbol{\square} t$;

(7) $s \boldsymbol{\square} [(s \boldsymbol{\square} s) \boldsymbol{\square} s] \neq t \boldsymbol{\square} t$;

(8) $s \boldsymbol{\square} [(s \boldsymbol{\square} s) \boldsymbol{\square} s] \neq (t \boldsymbol{\square} t) \boldsymbol{\square} t$.

From (2)–(8) we obtain the following conclusion.

(9) For any terms $p, q \in A$ there is just one term r satisfying, for all terms $s, t \in A$, the four conditions:

(a) $r = s \dagger t$, assuming that $p = (s \boldsymbol{\square} s) \boldsymbol{\square} s$ and $q = t \boldsymbol{\square} t$;
(b) $r = s \odot t$, assuming that $p = s \boldsymbol{\square} [(s \boldsymbol{\square} s) \boldsymbol{\square} s]$ and $q = t \boldsymbol{\square} t$;
(c) $r = s^{\triangle}$, assuming that $p = s$ and $q = s \boldsymbol{\square} [(s \boldsymbol{\square} s) \boldsymbol{\square} s]$;
(d) $r = p \boldsymbol{\square} q$, in case none of the above assumptions on p, q, s, and t hold.

Statement (9) justifies the definition of a new binary operation, $\square$, on A.

(10) For any $p, q \in A$ we set $p \square q = r$ in case r is the unique term satisfying, for all $s, t \in A$, the conditions (a)–(d) in (9).

The operations $\square$ and $\boldsymbol{\square}$ on A do not in general coincide. However, they do yield the same results when applied to certain pairs of elements from A. This is seen from the next three statements.

For every $s \in A$ we have

(11) $s \square s = s \boldsymbol{\square} s$,

(12) $(s \square s) \square s = (s \boldsymbol{\square} s) \boldsymbol{\square} s$,

(13) $s \square [(s \square s) \square s] = s \boldsymbol{\square} [(s \boldsymbol{\square} s) \boldsymbol{\square} s]$.

Indeed, taking p and q in (10) to be s, we see from (6)–(8) that none of the assumptions (a)–(c) on p and q are satisfied. Thus condition (d) applies and we at once obtain (11). The proofs of (12) and (13) are completely analogous; in (12) we make use of (11), and in (13) of (12).

(14) $\mathfrak{A}' = \langle A, \dagger, \odot, ^{\triangle}, \square \rangle$ is a model of Θ''.

By (1), $\mathfrak{A}'$ is a model of Θ. Using (10)–(13) we readily show that H_1, H_2, H_3 are identically satisfied in $\mathfrak{A}'$. For example, to verify H_2 let s and t be any elements in A. By (11), (13), and condition (b) respectively, we have

$$t \square t = t \boldsymbol{\square} t,$$

$$s \square [(s \square s) \square s] = s \boldsymbol{\square} [(s \boldsymbol{\square} s) \boldsymbol{\square} s],$$

$$(s \boldsymbol{\square} [(s \boldsymbol{\square} s) \boldsymbol{\square} s]) \square (t \boldsymbol{\square} t) = s \odot t.$$

From these equations it follows at once that s, t satisfy H_2 in $\mathfrak{A}'$, as was to be shown.

We are now in a position to directly establish (vi)(β). The inclusion $\Theta \subseteq \Theta'' \cap \mathbf{\Phi}$ is an immediate consequence of (v)(α). To establish the reverse inclusion suppose X is an equation of $\mathcal{E}$ that is in Θ''. By (14), X is identically satisfied in $\mathfrak{A}'$, and hence also in $\mathfrak{A}$. By (1) we have $X \in \Theta$. This completes the proof of (vi)(β), and therefore of (vi).

Using (iv)–(vi) we now obtain:

(vii) *For every* $\Gamma \subseteq \mathbf{\Phi}$ *the following conditions are equivalent:*

(α) Γ *is a base of* Θ;

(β) $T^*\Gamma$ *is a base of* Θ';

(γ) $T^*\Gamma \cup \{H_1, H_2, H_3\}$ *is a base of* Θ''.

Indeed, (α) obviously implies (γ) by (iii)(β), (v)(α). Using (iii)(α) we conclude from (iv) (with Δ replaced by $T^*\Gamma$) that

$$T^*\Gamma \cup \{H_1, H_2, H_3\} \vdash'' X \ \textit{ iff } \ T^*\Gamma \vdash' X,$$

for every $X \in \mathbf{\Phi}'$, i.e.,

$$\mathbf{\Theta}\boldsymbol{\eta}''(T^*\Gamma \cup \{H_1, H_2, H_3\}) \cap \mathbf{\Phi}' = \mathbf{\Theta}\boldsymbol{\eta}'(T^*\Gamma).$$

Hence, by (v)(β) we see that (γ) implies (β).

It remains to show that (β) implies (α). Set

$$(1) \qquad \Delta = \mathbf{\Theta}\boldsymbol{\eta}\Gamma.$$

Then

$$\begin{aligned} \Theta'' &= \mathbf{\Theta}\boldsymbol{\eta}''(\Theta' \cup \{H_1, H_2, H_3\}) && \text{by (vi)}(\alpha), \\ &= \mathbf{\Theta}\boldsymbol{\eta}''(T^*\Gamma \cup \{H_1, H_2, H_3\}) && \text{by } (\beta), \\ &= \mathbf{\Theta}\boldsymbol{\eta}''(\Gamma \cup \{H_1, H_2, H_3\}) && \text{by (iii)}(\beta), \\ &= \mathbf{\Theta}\boldsymbol{\eta}''(\Delta \cup \{H_1, H_2, H_3\}) && \text{by (1)}, \\ &= \Delta'' && \text{by (v)}(\alpha). \end{aligned}$$

Hence, by applying (vi)(β) twice we conclude that

$$\Theta = \Theta'' \cap \mathbf{\Phi} = \Delta'' \cap \mathbf{\Phi} = \Delta.$$

In view of (1) this leads to (α), and the proof of our theorem is completed.

From (iii)–(vii) we easily conclude that theories Θ' and Θ'' preserve various properties of the original theory Θ. We shall deal with such properties in the next few theorems.

(viii) *If one of the theories* Θ, Θ', *and* Θ'' *has any of the properties of consistency, finite axiomatizability, decidability, or essential undecidability, then all three of the theories possess this property.*

To show this we first notice that the conclusion of our theorem with respect to Θ' and Θ'' follows easily from the fact that Θ'' is a definitional extension of

Θ', and from various related properties; cf. (iii), (iv), (v)(β), and (vi)(α). Thus, we restrict ourselves to the relationship between Θ and Θ''. By (vi)(β), one of the theories Θ and Θ'' is consistent just in case the other is.

By (vii), if Θ is finitely based, then so is Θ''. Suppose, on the other hand, that Θ'' is finitely based. Then by (v)(α) there is a finite set $\Delta \subseteq \Theta$ such that

$$(1) \qquad \Theta'' = \boldsymbol{\Theta\eta}''(\Delta \cup \{H_1, H_2, H_3\}).$$

Using (v)(α) and (vi)(β), with Θ replaced by $\boldsymbol{\Theta\eta}\Delta$, we see that

$$\boldsymbol{\Theta\eta}\Delta = \boldsymbol{\Theta\eta}''(\Delta \cup \{H_1, H_2, H_3\}) \cap \boldsymbol{\Phi},$$

whence, by (1), $\boldsymbol{\Theta\eta}\Delta = \Theta'' \cap \boldsymbol{\Phi}$. Therefore we can conclude from (vi)(β) that $\Theta = \boldsymbol{\Theta\eta}\Delta$, and thus Θ is also finitely based.

If Θ'' is recursive, then by (vi)(β), Θ is the intersection of two recursive sets, and hence is itself recursive. The implication in the opposite direction is, in this case, less obvious. It is an immediate consequence of a more general result of Pigozzi [**1979**], Theorem 4.2 and the remark following it, by which Θ and Θ'' have the same degree of unsolvability.

Finally, from the implications involving decidability, we derive in a straightforward manner those involving essential undecidability.

(ix) *If Θ is dually undecidable, then so are Θ' and Θ''.*

This is a simple consequence of (vi)(β) and the fact that Θ'' is a definitional extension of Θ'.

We do not know whether the converse of (ix) is true. As regards essential dual undecidability, we do not know whether (ix) or its converse extends to this notion.

Our discussion so far has concerned an arbitrary theory Θ in an appropriate formalism $\mathcal{E}$. We now apply the results obtained to a specific theory $\Theta = \boldsymbol{\Gamma}$ in $\mathcal{E}$. In fact, we take for $\boldsymbol{\Gamma}$ the equational theory of ORA's. Following the assumption made in the first part of this section, we continue to treat ORA's as algebras of type $\langle 2, 2, 1\rangle$ whose equational theory $\boldsymbol{\Gamma}$ has been developed in the formalism $\mathcal{E}$. From (viii), (ix), and 8.5(xi) we conclude

(x) *$\boldsymbol{\Gamma}'$ is a finitely based, essentially undecidable, and dually undecidable equational theory of groupoids.*

We have not investigated the problem whether $\boldsymbol{\Gamma}'$ is essentially dually undecidable. However, from (ix) and 8.5(xi) we at once get the following weaker result.

(xi) *For every consistent theory Δ in $\mathcal{E}$ extending $\boldsymbol{\Gamma}$, the corresponding Δ' in $\mathcal{E}'$ is a dually undecidable extension of $\boldsymbol{\Gamma}'$.*

There are some further properties of the theory $\boldsymbol{\Gamma}$ that can be extended to $\boldsymbol{\Gamma}'$ and $\boldsymbol{\Gamma}''$. Two such properties will be presented in the next two theorems.

(xii) *The set of all equations with just one variable in the theory* $\mathbf{\Gamma}'$ *is not recursive. The same applies to every consistent theory in* $\mathcal{E}'$ *extending* $\mathbf{\Gamma}'$.

In fact, it is easily seen from 8.5(viii)(α) that, in every consistent theory Δ in $\mathcal{E}''$ extending $\mathbf{\Gamma}''$, the set of all equations in Δ with just one variable is not recursive, i.e., our theorem proves to hold, not for the theories in $\mathcal{E}'$ extending $\mathbf{\Gamma}'$, but for those in $\mathcal{E}''$ extending $\mathbf{\Gamma}''$. However, using various elementary properties of the translation mapping T, we can readily carry over this result to theories in $\mathcal{E}'$ extending $\mathbf{\Gamma}'$.

(xiii) *The theory* $\mathbf{\Gamma}'$ *has a base consisting of a single equation, and the same applies to all finitely based theories that are extensions of* $\mathbf{\Gamma}'$ *in* $\mathcal{E}'$, *as well as to all theories (developed in an equational formalism with only finitely many operation symbols) that are definitionally equivalent with such extensions of* $\mathbf{\Gamma}'$.

To show this we first observe that, by results announced in Tarski [**1968**], p. 281, Theorems 3 and 4, our Theorem (xiii) above holds if we respectively replace $\mathbf{\Gamma}'$ and $\mathcal{E}'$ by either $\mathbf{\Gamma}$ and $\mathcal{E}$ or else by $\mathbf{\Gamma}''$ and $\mathcal{E}''$. In view of Theorem 4 in *op. cit.* and the definitional equivalence of $\mathbf{\Gamma}'$ with $\mathbf{\Gamma}''$, we get the desired conclusion for $\mathbf{\Gamma}'$ and $\mathcal{E}'$.

In connection with this argument it may be of some interest to recall the following fact: the property that an equational theory has a base consisting of one equation is not, in general, preserved under definitional equivalence; see *op. cit.*, p. 280.

We now take up decision problems of the kind which were discussed in 8.5(xvi) for the formalism $\mathcal{E}^{\mathsf{EA}}$ and which we wish to transfer to the formalism $\mathcal{E}'$ of type $\langle 2\rangle$. It turns out that this can be done for almost all of the notions involved.

(xiv) (α) *Let* Σ *be the set of all equations* X *in* $\mathcal{E}'$ *such that* $\mathbf{\Theta}\boldsymbol{\eta}'X$ *is consistent. Then* Σ *is not recursive.*

(β) *Statement* (α) *continues to hold if, in its formulation, "consistent" is replaced by "complete", "decidable", "dually decidable", or "essentially undecidable".*

(γ) *Statements* (α) *and* (β) *continue to hold if, instead of the set* Σ *of all equations* X, *we consider the set* S *of all finite sets* Φ *of equations in* $\mathcal{E}'$, *and we replace "*$\mathbf{\Theta}\boldsymbol{\eta}'X$*" by "*$\mathbf{\Theta}\boldsymbol{\eta}'\Phi$*".*

To prove (α), we apply (xiii) to correlate recursively with each $Y \in \mathbf{\Phi}'$ an equation FY in $\mathbf{\Phi}'$ such that

$$(1) \qquad \mathbf{\Theta}\boldsymbol{\eta}'FY = \mathbf{\Theta}\boldsymbol{\eta}'(\mathbf{\Gamma}' \cup \{Y\}).$$

We conclude that $\mathbf{\Theta}\boldsymbol{\eta}'FY$ is consistent iff Y is compatible with $\mathbf{\Gamma}'$ (in $\mathcal{E}'$). Hence, if the set Σ were recursive, so would be the set of all equations compatible with $\mathbf{\Gamma}'$, in contradiction to the fact that $\mathbf{\Gamma}'$ is dually undecidable by (x). This proves (α).

We wish now to establish (α) with "consistent" replaced by "dually decidable". Consider any fixed equation Z in $\mathcal{E}$. By arguing as in the proof of the first part of (viii), we derive, with the help of (v) and (vi),

(2) Z is incompatible with $\mathbf{\Gamma}$ iff TZ is incompatible with $\mathbf{\Gamma}'$.

We set

(3) $\Delta = \mathbf{\Theta}\boldsymbol{\eta}(\mathbf{\Gamma} \cup \{Z\})$,

and observe, with the help of (iv) and (v), that the associated theory Δ' satisfies

(4) $\Delta' = \mathbf{\Theta}\boldsymbol{\eta}'(\mathbf{\Gamma}' \cup \{TZ\})$.

By (3), (4), and Theorem (xi), if Δ' is dually decidable, then Δ is inconsistent, and hence, in view of (2), Δ' is inconsistent. The implication in the opposite direction is obvious, and we arrive at

(5) Δ' is dually decidable iff it is inconsistent.

By (1), (4) we have

$$\mathbf{\Theta}\boldsymbol{\eta}'FTZ \text{ is dually decidable iff } \Delta' \text{ is dually decidable }.$$

Hence, by (5),

$$\mathbf{\Theta}\boldsymbol{\eta}'FTZ \text{ is dually decidable iff } \Delta' \text{ is inconsistent }.$$

Therefore, by (2)–(4),

$$\mathbf{\Theta}\boldsymbol{\eta}'FTZ \text{ is dually decidable iff } Z \text{ is incompatible with } \mathbf{\Gamma}.$$

From this we infer that if the set Σ determined by

$$\Sigma = \{X \in \mathbf{\Phi}' : \mathbf{\Theta}\boldsymbol{\eta}'X \text{ is dually decidable}\}$$

were recursive, so would be the set of all equations Z in $\mathcal{E}$ that are incompatible with $\mathbf{\Gamma}$, which contradicts the dual undecidability of $\mathbf{\Gamma}$ established in 8.5(xi). Consequently Σ is not recursive, which is just what we wanted to show.

To complete the proof of (β), we first notice that, by imitating the proof of 8.5(xvi)(β), we obtain the following statement.

(6) For a given $Y \in \mathbf{\Phi}'$, suppose that the theory $\mathbf{\Theta}\boldsymbol{\eta}'(\mathbf{\Gamma}' \cup \{Y\})$ satisfies one of the four conditions: is consistent; is consistent, but not complete; is undecidable; is essentially undecidable. Then it satisfies all of these conditions.

In view of (1) we may replace in (6) the expression "$\mathbf{\Theta}\boldsymbol{\eta}'(\mathbf{\Gamma}' \cup \{Y\})$" by "$\mathbf{\Theta}\boldsymbol{\eta}'FY$". As a consequence, if for instance the set Σ determined by

$$\Sigma = \{X \in \mathbf{\Phi}' : \mathbf{\Theta}\boldsymbol{\eta}'X \text{ is undecidable}\}$$

were recursive, then we could decide, for any given $Y \in \mathbf{\Phi}'$, whether $\mathbf{\Theta}\eta' FY$ is or is not undecidable and hence consistent. As in the proof of (α), this would contradict the dual undecidability of $\mathbf{\Gamma}'$. This shows that (α) holds with "consistent" replaced by "undecidable". The remaining parts of (β) are established in an entirely analogous way.

Finally, (γ) can be derived from (α) and (β) just as in the proof of 8.5(xvi).

The proof of that part of (xiv)(β) which concerns dual decidability is due to Givant. (It uses some ideas, originating with Tarski, which were applied in the proof of 8.5(xvi)(α).)

In the present section the undecidability results stated in 8.5(xi),(xvi) have been extended in (x) and (xiv) to equational formalisms of type $\langle 2 \rangle$. The only exceptions are the results involving the notion of essential dual undecidability (but not those involving only dual undecidability); the problem of an analogous extension of the latter results remains open. Thus, for instance, we do not know an example of a finitely based equational theory of groupoids that is essentially dually undecidable.

The problem naturally arises whether the results in (x) and (xiv) can be carried over to equational formalisms of arbitrary finite similarity types. An affirmative solution to this problem is easily obtained for formalisms of richer type, i.e., for formalisms with at least one operation symbol of rank ≥ 2, since in such formalisms we can readily construct a theory definitionally equivalent with the theory $\mathbf{\Gamma}'$ involved in (x) and in the proof of (xiv).

For the remaining equational formalisms of finite type the situation is at present less clear. Here the main result is the theorem in Mal′cev [**1966**] to the effect that in the formalism with just two operation symbols, both of which are unary, there is a finitely based undecidable theory. Again, by constructing definitionally equivalent theories, this result can be extended to all formalisms with at least two unary operation symbols. McNulty has informed us that in every such formalism there is also a finitely based dually undecidable theory. The problem is still open whether, in an equational theory with at least two operation symbols of rank one and without operation symbols of higher rank, there is a finitely based theory that is essentially undecidable or essentially dually undecidable. In McNulty [**1972**], Theorem 3.0, it is shown that the parts of (xiv)(γ) which refer to consistency, completeness, and decidability hold for formalisms with at least two unary operation symbols. It is not known whether the other parts of (xiv) hold for these formalisms as well.

It is readily seen that the results of 8.5(xi),(xvi) cannot be extended to equational formalisms of still poorer type, i.e., with only individual constants and possibly one unary operation symbol. Indeed, every finitely based theory in such a formalism is both decidable and dually decidable (see Ehrenfeucht [**1959**]). Hence, for such formalisms all of the sets defined in 8.5(xvi) turn out to be recursive.

8.7. Historical remarks regarding the decision problems

The results of §§8.5 and 8.6 call for some historical remarks, especially in view of the fact that not all of these results are due to Tarski.

Theorems 8.5(xi),(xii) imply the existence of many different finitely based undecidable equational theories; $\mathbf{\Theta}\rho\mathsf{RA}$ appears to be the simplest and most natural example. Tarski's proof of the undecidability of $\mathbf{\Theta}\rho\mathsf{RA}$ appears here in print for the first time.[6] However, the result was announced a long time ago, in fact, in a doctoral dissertation by Chin [**1948**], pp. 2–3, and in Chin–Tarski [**1951**], pp. 341–342, in both cases credited to Tarski. The abstract Tarski [**1953**] contains statements of this result and of several lemmas upon which the proof is based. From Chin–Tarski [**1951**], p. 343, footnote, it appears that the result was already known in 1945 and reported in Tarski's seminar on relation algebras held at the University of California, Berkeley, during that year; according to Tarski's recollection it was obtained around 1943–1944.[7] (An earlier result in Tarski [**1941**], p. 88, $\ell\ell$. 15–25, states in an equivalent form the undecidability of the equational theory $\mathbf{\Theta}\rho\mathsf{RRA}$, which by Monk [**1964**], turns out not to be finitely based.)

However, some results and discussions are known in the literature which appeared earlier than the announcements of Tarski's results cited above, and from which the existence of an undecidable finitely based equational theory can readily be derived. We have in mind the articles Markov [**1947**], [**1947a**] and Post [**1947**]; compare also Markov [**1961**], Chapter VI, especially §6.

In current terminology we could say that the main results of these papers give a negative solution to the word problem for a particular variety of algebras, namely the variety of semigroups. By analyzing the results and proofs of Markov and Post (cf., in particular, Chapter VI of Markov [**1961**]) we readily derive the existence of undecidable finitely based equational theories in a formalism with one binary operation symbol and finitely many individual constants. In fact, however, this connection between word problems and decision problems for equational theories was not recognized for a long time. (We shall return to this matter below.) Post [**1947**] refers to earlier papers going in this same direction, namely Post [**1943**], [**1946**].

From the remarks made above it seems to be clear that the results concerning the negative solution of the word problem, as well as the one establishing the undecidability of $\mathbf{\Theta}\rho\mathsf{RA}$, originate roughly from the same period. Moreover, in opposition to the result of Tarski, those of Markov and Post were published with proofs soon after their discovery. Nevertheless, $\mathbf{\Theta}\rho\mathsf{RA}$ seems to be the first example of a finitely based undecidable equational theory that was recognized as such and explicitly announced in the literature.

---

[6] *A different proof of the undecidability of $\mathbf{\Theta}\rho\mathsf{RA}$, and actually of 8.5(xii)(β), has been published in Maddux [**1978a**], pp. 220–222.

[7] *See footnote 3*, p. 168.

It should also be pointed out that $\boldsymbol{\Theta}\rho$ORA appears to be the first example of a finitely based essentially undecidable equational theory. So far as Tarski remembers, this result (in a slightly different form) was found around 1954, but was never announced by him. It is referred to in McNulty [**1972**], p. 53, but the formulation there is erroneous.

Subsequently, examples of finitely based undecidable equational theories have been constructed in formalisms with simpler similarity types. Indeed, in Mal′cev [**1966**], submitted for publication in 1965, examples of such theories are constructed in formalisms of types $\langle 1,1\rangle$ and $\langle 2\rangle$, thus in the simplest possible types.

It is interesting that Mal′cev bases his construction essentially on the results of Markov and Post, as well as Hall [**1949**]. However, various remarks in his paper (see, in particular, those from p. 285, ℓ. 31 through p. 286, ℓ. 14 in the English translation Mal′cev [**1971**]) seem to show that he did not see any possibility of deriving directly the existence of finitely based undecidable equational theories (in somewhat more complicated similarity types) from the known results concerning the negative solution of the word problem for the variety of semigroups. Thus, he expresses the belief that the varieties constructed by him are the first examples of finitely based varieties with undecidable equational theories that have appeared in the literature. This can serve as an illustration of the observation made above to the effect that the connection between word problems and decision problems for equational theories was not recognized for a long time. (We may mention that, according to an oral communication of Mal′cev, when publishing *op. cit.* he was unaware of the results concerning the undecidability of the equational theory of RA.)

Independently of Mal′cev, and about the same time, Perkins also constructed a finitely based undecidable theory in an equational formalism of type $\langle 2\rangle$; see the doctoral dissertation Perkins [**1966**], Theorem 35, as well as Perkins [**1967**], Theorem 12. Upon learning from Perkins about his result (sometime in 1965), Tarski realized very quickly that it could also be obtained from the undecidability of $\boldsymbol{\Theta}\rho$RA by applying the method outlined in §8.6. By means of the same method, Tarski simultaneously extended his results concerning the essential undecidability of $\boldsymbol{\Theta}\rho$ORA to formalisms of type $\langle 2\rangle$ (cf. 8.6(x)); actually, the equations $H_1 - H_3$ given in §8.6 are just the ones that were used by Tarski in his original argument. In contrast to Mal′cev, neither Perkins nor Tarski concerned themselves with the formalism of type $\langle 1,1\rangle$.

As regards the undecidability results of the second degree stated in 8.5(xvi) and 8.6(xiv), several of them are known from the literature. Thus, in Perkins [**1966**] (see also Perkins [**1967**]) we find a theorem to the effect that the set of all finite sets Φ of equations in the formalism of type $\langle 2\rangle$ such that $\boldsymbol{\Theta}\boldsymbol{\eta}\Phi$ is consistent is not recursive (cf. 8.5(xvi)(γ) and 8.6(xiv)(γ)). The analogous results concerning the properties of being complete or decidable were originally established in *op. cit.* for the similarity type $\langle 2,2,0,0\rangle$, and were extended in McNulty [**1972**], [**1976a**] to the similarity type $\langle 2\rangle$. Theorem 8.6(xiv)(α) and that part of 8.6(xiv)(β) which refers to decidability were also first established in the latter papers. Except for the part of 8.6(xiv)(β) which concerns dual decidability (cf. the remarks following the proof of that theorem), the remaining results in 8.5(xvi) and 8.6(xiv) are due to Tarski and appear here in print for the first time. According to Tarski's recollection, it seems that these results were obtained sometime during the period 1972–1973.

The method applied in §8.6 in the construction and discussion of the theory Θ', together with its many different variants, will be referred to here for brevity as the reduction method. It is used primarily to construct sentences and sets of sentences with a restricted deductive power. It seems that the underlying ideas have occurred independently to many people, and it would be difficult to ascribe priority to any particular person.

Arguments using these ideas can be traced as far back as the years 1920–1930; they originate not with equational, but with sentential metalogic. As examples of results whose proofs use these ideas we may mention Theorems 11–13, 27, 28 in Łukasiewicz–Tarski [**1930**]. (An English translation of this paper occurs as Article IV in Tarski [**1956**].) Since the rules for constructing and deriving formulas in sentential and in equational logic are very different, the proofs of these theorems are based, not on a subterm condition, but on an entirely analogous *antecedent condition.* The proof of Theorem 28, given in Tarski [**1956**], p. 51, footnote, may be illuminating for the reader in this context.

Only at a later date did the reduction method find applications in the study of equational logic. Some results obtained with its help were strict analogues of theorems about sentential logic mentioned above. For instance, strict analogues of Theorems 27 and 28 in *op. cit.* were established for the equational formalism of type $\langle 2\rangle$ in Kalicki [**1955**]. Moreover, in equational metalogic there emerges a new (and perhaps more important) direction for applications of the reduction method, namely the study of various decision and recursiveness problems. Thus, some elements of the reduction method are involved in Post [**1947**] and clearly appear in Hall [**1949**]. The procedure used in Mal′cev [**1971**], pp. 392ff., to construct a finitely based and undecidable theory of type $\langle 2\rangle$ from a theory with the same property of type $\langle 1, 1\rangle$ can be regarded as a specialized and modified form of the reduction method.

In Perkins [**1966**], pp. 49–51, the reduction method is described for the first time in general terms as a method applicable to arbitrary equational theories of all possible (finite) similarity types. The general notion of the subterm condition is not introduced; instead, it is indicated how a suitable set of terms (which actually satisfies the subterm condition) can be constructed for any given finite similarity type.

More recently, the reduction method has been thoroughly and broadly analyzed and applied in McNulty [**1972**], [**1976**]; a credit is given there to McKenzie for influencing this analysis. McNulty extensively applies the reduction method to study a variety of decision problems; the applications are not restricted to the cases when a result originally established for some complicated equational formalism without the help of this method is subsequently extended to the formalism of type $\langle 2 \rangle$. We may mention that McNulty's work does not follow strictly the same lines as our development in §8.6; for instance, the auxiliary theory Θ'' is not used in his arguments.

To conclude, we wish to recall that at the end of the preceding section we gave a survey of problems involved in the discussions of §§8.5 and 8.6 which still remain open. The most interesting of these problems seem to us to be the following ones:

(1) Does there exist a finitely based essentially undecidable theory in the equational formalism of type $\langle 1, 1 \rangle$? If the answer is negative, does it extend to equational formalisms with any finite number of operation symbols, all of which are unary?

(2) Does there exist a finitely based essentially dually undecidable theory in the equational formalism of type $\langle 2 \rangle$?

Bibliography

Ackermann, W.

[1937] *Die Widerspruchsfreiheit der allgemeinen Mengenlehre*, ***Math. Ann.***, vol. 114 (1937), pp. 305–315.

[1952] Review of G. H. von Wright's *On double quantification*, ***J. Symbolic Logic***, vol. 17 (1952), pp. 201–203.

[1954] ***Solvable cases of the decision problem***, North-Holland Publishing Company, Amsterdam, 1954, viii+114 pp.

[1956] *Zur Axiomatik der Mengenlehre*, ***Math. Ann.***, vol. 131 (1956), pp. 336–345.

Andréka, H., Comer, S. D., and Németi, I.

[1985] *Clones of operations on relations*, ***Universal algebra and lattice theory: Proceedings, Charleston, 1984***, ed. S. D. Comer, **Lecture Notes in Math.**, vol. 1149, Springer-Verlag, Berlin, 1985, pp. 17–21.

Bernays, P.

[1937] *A system of axiomatic set theory – Part I*, ***J. Symbolic Logic***, vol. 2 (1937), pp. 65–77.

[1941] *A system of axiomatic set theory – Part II*, ***J. Symbolic Logic***, vol. 6 (1941), pp. 1–17.

[1942] *A system of axiomatic set theory – Part III*, ***J. Symbolic Logic***, vol. 7 (1942), pp. 65–89.

[1942a] *A system of axiomatic set theory – Part IV*, ***J. Symbolic Logic***, vol. 7 (1942), pp. 133–145.

[1943] *A system of axiomatic set theory – Part V*, ***J. Symbolic Logic***, vol. 8 (1943), pp. 89–106.

[1948] *A system of axiomatic set theory – Part VI*, ***J. Symbolic Logic***, vol. 13 (1948), pp. 65–79.

[1954] *A system of axiomatic set theory – Part VII*, ***J. Symbolic Logic***, vol. 19 (1954), pp. 81–96.

[1959] *Über eine natürliche Erweiterung des Relationenkalküls*, ***Constructivity in mathematics*: *Proceedings of the colloquium held at Amsterdam, 1957***, ed. A. Heyting, North-Holland Publishing Co., Amsterdam, 1959, pp. 1–14.

Birkhoff, G.

[1935] *On the structure of abstract algebras*, ***Proc. Cambridge Philos. Soc.***, vol. 31 (1935), pp. 433–454.

[1944] *Subdirect unions in universal algebras*, ***Bull. Amer. Math. Soc.***, vol. 50 (1944), pp. 764–768.

Börner, F.

[1986] *One-generated clones of operations on binary relations*, ***Beiträge Algebra Geom.***, vol. 23 (1986), pp. 73–84.

Chin, L. H.

[1948] ***Distributive and modular laws in relation algebras***, Doctoral dissertation, University of California, Berkeley, 1948, 62 pp.

Chin, L. H. and Tarski, A.

[1951] *Distributive and modular laws in the arithmetic of relation algebras*, ***University of California Publications in Mathematics***, new series, vol. 1, no. 9 (1951), pp. 341–384.

Chuaqui, R.

[1980] *Internal and forcing models for the impredicative theory of classes*, ***Dissertationes Mat.*** (=***Rozprawy Mat.***), vol. 176 (1980), 65 pp.

Church, A.

[1936] *A note on the Entscheidungsproblem*, ***J. Symbolic Logic***, vol. 1 (1936), pp. 40–41. Correction, *ibid.*, pp. 101–102.

[1956] ***Introduction to mathematical logic***, vol. 1, Princeton University Press, Princeton, 1956, x+378 pp.

Couturat, L.

[1914] ***The algebra of logic***, translated by L. G. Robinson, Open Court Publishing Co., London, 1914, xiv+98 pp. [English translation of ***L'algèbre de la logique***, Gauthier-Villars, Paris, 1905, 100 pp.]

Craig, W.

[1953] *On axiomatizability within a system*, ***J. Symbolic Logic***, vol. 18 (1953), pp. 30–32.

[1974] ***Logic in an algebraic form. Three languages and theories***, North-Holland Publishing Co., Amsterdam, 1974, viii+204 pp.

Curry, H. B. and Feys, R.

[1958] ***Combinatory logic***, vol. I, North-Holland Publishing Co., Amsterdam, 1958, xvi+417 pp.

Curry, H. B., Hindley, J. R., and Seldin, J. P.

[1972] ***Combinatory logic***, vol. II, North-Holland Publishing Co., Amsterdam, 1972, xiv+520 pp.

Ehrenfeucht, A.

[1959] *Decidability of the theory of one function*, ***Notices Amer. Math. Soc.***, vol. 6 (1959), p. 268.

Everett, C. J. and Ulam, S.

[1946] *Projective Algebra I*, ***Amer. J. Math.***, vol. 68 (1946), pp. 77–88.

Frege, G.

[1879] ***Begriffschrift, eine der arithmetischen nachgebildete Formelsprache des reinen Denkens***, L. Niebert, Halle, 1879, 88 pp.

Givant, S. and Tarski, A.

[1977] *Peano arithmetic and the Zermelo-like theory of sets with finite ranks*, ***Notices Amer. Math. Soc.***, vol. 24 (1977), p. A–437.

Gödel, K.

[1930] *Die Völlstandigkeit der Axiome des logischen Funktionenkalküls*, ***Monatshefte für Mathematik und Physik***, vol. 37 (1930), pp. 349–360.

[1933] *Zum Entscheidungsproblem des logischen Funktionenkalküls*, ***Monatshefte für Mathematik und Physik***, vol. 40 (1933), pp. 433–443.

[1940] ***The consistency of the axiom of choice and of the generalized continuum hypothesis with the axioms of set theory***, **Annals of Math. Studies**, no. 3, Princeton University Press, Princeton, 1940, 66 pp.

Goldfarb, W. D.

[1984] *The unsolvability of the Gödel class with identity*, ***J. Symbolic Logic***, vol. 49 (1984), pp. 1237–1252.

Hall, M.

[1949] *The word problem for semigroups with two generators*, ***J. Symbolic Logic***, vol. 14 (1949), pp. 115–118.

Halmos, P. R.

[1962] ***Algebraic logic***, Chelsea Publishing Co., New York, 1962, 271 pp.

Henkin, L.

[1973] *Internal semantics and algebraic logic*, ***Truth, syntax and modality***, ed. H. LeBlanc, North-Holland Publishing Co., Amsterdam, 1973, pp. 111–127.

Henkin, L., Monk, J. D., and Tarski, A.

[1971] ***Cylindric algebras. Part I***, North-Holland Publishing Co., Amsterdam, 1971, vi+508 pp.

[1985] ***Cylindric algebras. Part II***, North-Holland Publishing Co., Amsterdam, 1985, vii+302 pp.

Henkin, L. and Tarski, A.

[1961] *Cylindric algebras*, ***Lattice theory***, **Proc. Sympos. Pure Math.**, vol. 2, ed. R. P. Dilworth, American Mathematical Society, Providence, 1961, pp. 83–113.

Hilbert, D. and Ackermann, W.

[1950] ***Mathematical logic***, Chelsea Publishing Co., New York, 1950, xii+172 pp.

Hilbert, D. and Bernays, P.

[1968] ***Grundlagen der Mathematik***, vol. 1, Springer-Verlag, Berlin, 1968, xv+473 pp.

Huntington, E. V.

[1933] *New sets of independent postulates for the algebra of logic, with special reference to Whitehead and Russell's Principia Mathematica*, ***Trans. Amer. Math. Soc.***, vol. 35 (1933), pp. 274–304.

[1933a] *Boolean algebra. A correction*, ***Trans. Amer. Math. Soc.***, vol. 35 (1933), pp. 557–558.

Jónsson, B.

[1959] *Representation of modular lattices and of relation algebras*, ***Trans. Amer. Math. Soc.***, vol. 92 (1959), pp. 449–464.

[1982] *Varieties of relation algebras*, ***Algebra Universalis***, vol. 15 (1982), pp. 273–298.

Jónsson, B. and Tarski, A.

[1952] *Boolean algebras with operators. Part II*, ***Amer. J. Math.***, vol. 74 (1952), pp. 127–162.

Kalicki, J.

[1955] *The number of equationally complete classes of equations*, ***Nederl. Akad. Wetensch. Proc. Ser. A.***, vol. 58 (=***Indag. Math.***, vol. 17) (1955), pp. 660–662.

Kalish, D. and Montague, R.

[1965] *On Tarski's formalization of predicate logic with identity*, ***Arch. Math. Logik Grundlag.***, vol. 7 (1965), pp. 81–101.

Kalmár, L.

[1936] *Zurückführung des Entscheidungsproblems auf den Fall von Formeln mit einer einzigen, binären, Funktionsvariablen*, ***Compositio Math.***, vol. 4 (1936), pp. 137–144.

Kelley, J. L.

[1955] ***General topology***, D. Van Nostrand Co., Inc., Princeton, 1955, xiv+298 pp.

Kwatinetz, M. K.

[1981] ***Problems of expressibility in finite languages***, Doctoral dissertation, University of California, Berkeley, 1981, xi+106 pp.

Levy, A.

[1965] *A hierarchy of formulas in set theory*, ***Memoirs Amer. Math. Soc.***, no. 57 (1965), 76 pp.

Lewis, C. I.

[1918] ***A survey of symbolic logic***, University of California Press, Berkeley, 1918, vi+406 pp. [Second edition published by Dover, New York, 1960]

Lindenbaum, A. and Tarski, A.

[1936] *Über die Beschränktheit der Ausdrucksmittel deduktiver Theorien*, ***Ergeb. Math. Kolloq.***, vol. 7 (1936), pp. 15–22.

Linial, S. and Post, E. L.

[1949] *Recursive unsolvability of the deducibility, Tarski's completeness, and independence of axioms problems of the propositional calculus*, ***Bull. Amer. Math. Soc.***, vol. 55 (1949), p. 50.

Löwenheim, L.

[1915] *Über Möglichkeiten im Relativkalkül*, ***Math. Ann.***, vol. 76 (1915), pp. 447–470.

Łukasiewicz, J. and Tarski, A.

[1930] *Untersuchungen über den Aussagenkalkül*, ***Comptes Rendus des seances de la Société des Sciences et des Lettres de Varsovie, Class III***, vol. 23 (1930), pp. 30–50.

Lyndon, R. C.

[1950] *The representation of relational algebras*, ***Ann. of Math.***, ser. 2, vol. 51 (1950), pp. 707–729.

[1956] *The representation of relation algebras, II*, ***Ann. of Math.***, ser. 2, vol. 63 (1956), pp. 294–307.

[1961] *Relation algebras and projective geometries*, ***Michigan Math. J.***, vol. 8 (1961), pp. 21–28.

Maddux, R.

[1978] *Some sufficient conditions for the representability of relation algebras*, ***Algebra Universalis***, vol. 8 (1978), pp. 162–172.

[1978a] ***Topics in relation algebras***, Doctoral dissertation, University of California, Berkeley, 1978, iii+241 pp.

[1987] *Nonfinite axiomatizability results for cylindric and relation algebras*, submitted for publication.

Mal′cev, A. I.

[1966] *Toždestvennye sootnošenija na mnogoobrazijah kvazigrupp*, ***Mat. Sbornik***, new series, vol. 69 (1966), pp. 3–12. English translation:

[1971] *Identical relations in varieties of quasigroups*, ***The metamathematics of algebraic systems—collected papers: 1936–1967***, translated by B. Wells, III, North-Holland Publishing Co., Amsterdam, 1971, pp. 384–395.

Markov, A. A.

[1947] *The impossibility of some algorithms in the theory of associative systems* (Russian), ***Doklady AN SSSR***, vol. 55 (1947) pp. 587–590.

[1947a] *The impossibility of certain algorithms in the theory of associative systems* (Russian), ***Doklady AN SSSR***, vol. 58 (1947), pp. 353–356.

[1961] ***Theory of algorithms***, The Israel Program for Scientific Translations, Jerusalem, 1961, vii+444 pp. [English translation of *Teorija algorifmov*, ***Trudy Mat. Inst. im. Steklova***, vol. 42 (1954), 375 pp.]

McKenzie, R.

[1970] *Representations of integral relation algebras*, ***Michigan Math. J.***, vol. 17 (1970), pp. 279–287.

McNulty, G. F.

[1972] ***The decision problem for equational bases of algebras***, Doctoral dissertation, University of California, Berkeley, 1972, ix+125 pp.

[1976] *The decision problem for equational bases of algebras*, ***Ann. Math. Logic***, vol. 11 (1976), pp. 193–259.

[1976a] *Undecidable properties of finite sets of equations*, ***J. Symbolic Logic***, vol. 41 (1976), pp. 589–604.

Mendelson, E.

[1964] ***Introduction to mathematical logic***, D. Van Nostrand Co., Inc., Princeton, 1964, x+300 pp.

Monk, J. D.

[1964] *On representable relation algebras*, ***Michigan Math. J.***, vol. 11 (1964), pp. 207–210.

[1969] ***Introduction to set theory***, McGraw-Hill Book Co., New York, 1969, xii+193 pp.

[1969a] *Nonfinitizability of classes of representable cylindric algebras*, ***J. Symbolic Logic***, vol. 34 (1969), pp. 331–343.

[1971] *Provability with finitely many variables*, ***Proc. Amer. Math. Soc.***, vol. 27 (1971), pp. 353–358.

[1976] ***Mathematical logic***, Springer-Verlag, New York, 1976, x+231 pp.

Montague, R. M.

[1961] *Semantical closure and non-finite axiomatizability I*, ***Infinitistic methods: Proceedings of the symposium on foundations of mathematics, Warsaw, 2–9 September 1959***, Pergamon Press, Oxford, 1961, pp. 45–69.

[1962] *Fraenkel's addition to the axioms of Zermelo*, ***Essays on the foundations of mathematics***, eds. Y. Bar-Hillel, E. I. J. Poznanski, M. O. Rabin, and A. Robinson, North-Holland Publishing Co., Amsterdam, 1962, pp. 91–114.

Montague, R. M. and Vaught, R. L.

[1959] *Natural models of set theory*, ***Fund. Math.***, vol. 47 (1959), pp. 219–242.

Morse, A.

[1965] ***A theory of sets***, Academic Press, New York, 1965, xxxi+130 pp.

Mortimer, M.

[1975] *On languages with two variables*, ***Z. Math. Logik Grundlag. Math.***, vol. 21 (1975), pp. 135–140.

Mostowski, A.

[1939] *Über die Unabhängigkeit des Wohlordnungssatzes vom Ordnungsprinzip*, ***Fund. Math.***, vol. 32 (1939), pp. 201–252.

[1954] *Der gegenwärtige Stand der Grundlagenforschung in der Mathematik*, ***Die Hauptreferate des 8. Polnischen Mathematikerkongresses vom 6. bis 12. September 1953 in Warschau***, Deutscher Verlag der Wissenschaften, Berlin, 1954, pp. 11–44. [An English version of this report appeared as: *The present state of investigations on the foundations of mathematics*, ***Dissertationes Mat.*** (=***Rozprawy Mat.***), vol. 9 (1955), 48 pp.]

Myhill, J. R.

[1950] *A reduction in the number of primitive ideas of arithmetic*, ***J. Symbolic Logic***, vol. 15 (1950), p. 130.

Németi, I.

[1985] *Logic with three variables has Gödel's incompleteness property—thus free cylindric algebras are not atomic*, Preprint, Mathematical Institute of the Hungarian Academy of Sciences, Budapest, 1985, 92 pp.

[1987] *Decidability of relation algebras with weakened associativity*, ***Proc. Amer. Math. Soc.***, vol. 100 (1987), pp. 340–344.

Peirce, C. S.

[1882] *The logic of relatives*, ***Studies in logic***, Johns Hopkins University, pp. 187–203.

Perkins, P.

[1966] ***Decision problems for equational theories of semigroups and general algebras***, Doctoral dissertation, University of California, Berkeley, 1966, 54 pp.

[1967] *Unsolvable problems for equational theories*, ***Notre Dame J. Formal Logic***, vol. 8 (1967), pp. 175–185.

Pigozzi, D.

[1979] *Universal equational theories and varieties of algebras*, ***Ann. Math. Logic***, vol. 17 (1979), pp. 117–150.

Post, E. L.

[1943] *Formal reductions of the general combinatorial decision problem*, ***Amer. J. Math.***, vol. 65 (1943), pp. 197–215.

[1946] *A variant of a recursively unsolvable problem*, ***Bull. Amer. Math. Soc.***, vol. 52 (1946), pp. 264–268.

[1947] *Recursive unsolvability of a problem of Thue*, ***J. Symbolic Logic***, vol. 12 (1947), pp. 1–11.

Quine, W. V. O.

[1937] *New foundations for mathematical logic*, ***Amer. Math. Monthly***, vol. 44 (1937), pp. 70–80.

[1951] ***Mathematical logic***, revised edition, Harvard University Press, Cambridge, 1951, xii+346 pp.

[1960] *Variables explained away*, ***Proc. Amer. Philos. Soc.***, vol. 140 (1960), pp. 343–347.

[1969] ***Set theory and its logic***, revised edition, Harvard University Press, Cambridge, 1969, xviii+361 pp.

[1971] ***Algebraic logic and predicate functors***, Bobbs-Merrill Publishing Co., 1971, 25 pp.

Robinson, J. B.

[1949] *Definability and decision problems in arithmetic*, ***J. Symbolic Logic***, vol. 14 (1949), pp. 98–114.

Schönfinkel, M.

[1924] *Über die Bausteine der mathematischen Logik*, ***Math. Ann.***, vol. 92 (1924), pp. 305–316.

Schröder, E.

[1891] ***Vorlesungen über die Algebra der Logik*** (***exakte Logik***), vol. 2, part 1, B. G. Teubner, Leipzig, 1891, xiii+400 pp. [Reprinted by Chelsea Publishing Co., New York, 1966.]

[1895] ***Vorlesungen über die Algebra der Logik*** (***exakte Logik***), vol. 3, ***Algebra und Logik der Relative***, part 1, B. G. Teubner, Leipzig, 1895, 649 pp. [Reprinted by Chelsea Publishing Co., New York, 1966.]

Scott, D. S.

[1961] *Measurable cardinals and constructive sets*, ***Bull. Acad. Polon. Sci. Sér. Sci. Math. Astronom. Phys.***, vol. 9 (1961), pp. 521–524.

[1962] *A decision method for validity of sentences in two variables*, ***J. Symbolic Logic***, vol. 27 (1962), p. 477.

Sierpiński, W.

[1958] ***Cardinal and ordinal numbers***, Państwowe Wydawnictwo Naukowe, Warsaw, 1958, 487 pp.

Singletary, W. E.

[1964] *A complex of problems proposed by Post*, ***Bull. Amer. Math. Soc.***, vol. 70 (1964), pp. 105–109. Correction, *ibid.*, p. 826.

Skolem, T.

[1923] *Einige Bemerkungen zur axiomatischen Begründung der Mengenlehre*, ***Wiss. Vorträge 5. Kongress der skandinav. Mathematiker in Helsingfors 1922***, Akadem. Buchh., Helsingfors, 1923, pp. 217–232.

Suppes, P.

[1960] ***Axiomatic set theory***, D. Van Nostrand Co., Inc., Princeton, 1960, xii+265 pp.

Szmielew, W. and Tarski, A.

[1952] *Mutual interpretability of some essentially undecidable theories*, ***Proceedings of the International Congress of Mathematicians, Cambridge, Massachusetts, USA, August 30-September 6, 1950***, vol. 1, American Mathematical Society, Providence, 1952, p. 734.

Tarski, A.

[1924] *Sur les ensembles finis*, ***Fund. Math.***, vol. 6 (1924), pp. 45–95.

[1941] *On the calculus of relations*, ***J. Symbolic Logic***, vol. 6 (1941), pp. 73–89.

[1951] *Remarks on the formalization of the predicate calculus*, ***Bull. Amer. Math. Soc.***, vol. 57 (1951), pp. 81–82.

[1953] *Some metalogical results concerning the calculus of relations*, ***J. Symbolic Logic***, vol. 18 (1953), pp. 188–189.

[1953a] *A formalization of set theory without variables*, ***J. Symbolic Logic***, vol. 18 (1953), p. 189.

[1953b] *An undecidable system of sentential calculus*, ***J. Symbolic Logic***, vol. 18 (1953), p. 189.

[1954] *A general theorem concerning the reduction of primitive notions*, ***J. Symbolic Logic***, vol. 19 (1954), p. 158.

[1954a] *On the reduction of the number of generators in relation rings*, ***J. Symbolic Logic***, vol. 19 (1954), pp. 158–159.

[1955] *Contributions to the theory of models. III*, ***Nederl. Akad. Wetensch. Proc. Ser. A.***, vol. 58 (=***Indag. Math.***, vol. 17) (1955), pp. 56–64.

[1956] ***Logic, semantics, metamathematics: Papers from 1923 to 1938***, translated by J. H. Woodger, Clarendon Press, Oxford, 1956, xiv+471 pp. [Second, revised edition, ed. J. Corcoran, Hackett Publishing Co., Indianapolis, 1983, xxx+506 pp.]

[1956a] *Equationally complete rings and relation algebras*, ***Nederl. Akad. Wetensch. Proc. Ser. A***, vol. 59 (=***Indag. Math.***, vol. 18) (1956), pp. 39–46.

[1956b] *Notions of proper models for set theories*, ***Bull. Amer. Math. Soc.***, vol. 62 (1956), p. 601.

[1965] *A simplified formulation of predicate logic with identity*, ***Arch. Math. Logik Grundlag.***, vol. 7 (1965), pp. 61–79.

[1965a] ***Introduction to logic and to the methodology of deductive sciences***, Oxford University Press, Inc., New York, 1965, xviii+252 pp.

[1968] *Equational logic and equational theories of algebras*, ***Contributions to Mathematical Logic***, ed. K. Schütte, North-Holland Publishing Co., Amsterdam, 1968, pp. 275–288.

[1986] *What are logical notions?*, ed. J. Corcoran, ***History and Philosophy of Logic***, vol. 7 (1986), pp. 143–154.

Tarski, A., Mostowski, A., and Robinson, R. M.

[1953] ***Undecidable theories***, North-Holland Publishing Co., Amsterdam, 1953, xii+98 pp.

Thiele, E. J.

[1968] *Über endlich-axiomatisierbare Teilsysteme der Zermelo-Fraenkelschen Mengenlehre*, ***Z. Math. Logik Grundlag. Math.***, vol. 14 (1968), pp. 39–58.

Vaught, R. L.

[1962] *On a theorem of Cobham concerning undecidable theories*, ***Logic, methodology, and philosophy of science*: *Proceedings of the 1960 International Congress***, eds. E. Nagel, P. Suppes, and A. Tarski, Stanford University Press, Stanford, 1962, pp. 14–25.

von Wright, G. H.

[1950] *On the idea of logical truth (II)*, ***Soc. Sci. Fenn. Comment. Phys.-Math.***, vol. 15, no. 10 (1950), 45 pp.

[1952] *On double quantification*, ***Soc. Sci. Fenn. Comment. Phys.-Math.***, vol. 16, no. 3 (1952), 14 pp. [Reprinted in G. H. von Wright, ***Logical studies***, The Humanities Press, Inc., New York, and Routledge and Kegan Paul Ltd., London, 1957, pp. 44–57.]

Vopěnka, P. and Hájek, P.

[1972] ***The theory of semisets***, Academia, Prague, 1972, 332 pp.

Yntema, M. K.

[1964] *A detailed argument for the Post-Linial theorems*, ***Notre Dame J. Formal Logic***, vol. 5 (1964), pp. 37–50.

Wostner, U.

[1976] *On equationally definable classes of partial ordering relations*, ***Notices Amer. Math. Soc.***, vol. 23 (1976), p. A-643.

Zermelo, E.

[1908] *Untersuchungen über die Grundlagen der Mengenlehre. I*, ***Math. Ann.***, vol. 65 (1908), pp. 261–281.

Index of Symbols

Set-theoretical notions

Formalisms

Systems

Translation mappings

Primitive and defined symbols of formalisms

General syntactical and semantical notions

Special compound expressions

Special sets of sentences

Numeration of schemata

Index of Names

Index of Subjects

The main reference to a subject is given in boldface type.

Index of Numbered Items

BCDEFGHIJ-898